D0894324

Applied Probability and Statistics (Continued)

CHERNOFF and MOSES · Elementary Decisi

CHIANG · Introduction to Stochastic Processes

CLELLAND, deCANI, BROWN, BURSK, and ....
Statistics with Business Applications

COCHRAN · Sampling Techniques, *Second Edition*

COCHRAN and COX · Experimental Designs, *Second Edition*

COX · Planning of Experiments

COX and MILLER · The Theory of Stochastic Processes

DAVID · Order Statistics

DEMING · Sample Design in Business Research

DODGE and ROMIG · Sampling Inspection Tables, *Second Edition*

DRAPER and SMITH · Applied Regression Analysis

ELANDT-JOHNSON · Probability Models and Statistical Methods in
Genetics

GOLDBERGER · Econometric Theory

GUTTMAN, WILKS and HUNTER · Introductory Engineering Statistics, *Second Edition*

HAHN and SHAPIRO · Statistical Models in Engineering

HALD · Statistical Tables and Formulas

HALD · Statistical Theory with Engineering Applications

HOEL · Elementary Statistics, *Third Edition*

HUANG · Regression and Econometric Methods

JOHNSON and LEONE · Statistics and Experimental Design: In Engineering and the Physical Sciences, Volumes I and II

LANCASTER · The Chi Squared Distribution

MILTON · Rank Order Probabilities: Two-Sample Normal Shift
Alternatives

PRABHU · Queues and Inventories: A Study of Their Basic Stochastic
Processes

RAO and MITRA · Generalized Inverse of Matrices and Its Applications

SARD and WEINTRAUB · A Book of Splines

SARHAN and GREENBERG · Contributions to Order Statistics

SEAL · Stochastic Theory of a Risk Business

SEARLE · Linear Models

THOMAS · An Introduction to Applied Probability and Random
Processes

WHITTLE · Optimization under Constraints

WILLIAMS · Regression Analysis

WOLD and JUREEN · Demand Analysis

WONNACOTT and WONNACOTT · Introduction to Econometric
Methods

YOUDEN · Statistical Methods for Chemists

ZELLNER · An Introduction to Bayesian Inference in Econometrics

Tracts on Probability and Statistics

BILLINGSLEY · Ergodic Theory and Information

BILLINGSLEY · Convergence of Probability Measures

CRAMÉR and LEADBETTER · Stationary and Related Stochastic
Processes

JARDINE and SIBSON · Mathematical Taxonomy

RIORDAN · Combinatorial Identities

TAKÁCS · Combinatorial Methods in the Theory of Stochastic
Processes

Sample Design
in Business Research

Sample Design in Business Research

W. EDWARDS DEMING

Consultant in Statistical Surveys

and

Professor of Statistics
Graduate School of Business Administration
New York University

John Wiley & Sons, Inc.

New York · London

ISBN 0 471 20724 1

Library of Congress Catalog Card Number: 60–6451

PRINTED IN THE UNITED STATES OF AMERICA

Preface

This book is a textbook, written from the point of view of the theoretical statistician in industry, which includes surveys of consumers, surveys in psychological problems, surveys of attitudes and opinions, surveys of business establishments, statistical problems in accounting and in auditing, evaluation of inventory, appraisal of property, and the inspection and testing of physical materials. A theoretical statistician is one who guides his practice with theory. The theoretical statistician is the practical man, as he has a better guide for practice than the errors of his forefathers (to borrow from Huxley).

Part I establishes standards of professional statistical practice, the aim being to enhance the usefulness of statistical methods. This part of the book is directed especially to the needs of the executive in business, and to the expert in a substantive field (consumer research, engineering, production, accounting, psychology, law, etc.), although statisticians themselves have perhaps even more need for this kind of advice. Statistical theory tells us what it is that we need to know about a problem in order to carry out a proposed study effectively. Statistical theory shows how mathematics, judgment, and substantive knowledge work together to the best advantage. The statistician is thus the logician and the architect of a statistical study.

Parts I and II together provide principles and examples in the planning of surveys and in the statistical interpretation of results. The executive, or the expert in a substantive field, even though he lacks formal statistical education, may learn to appreciate sampling as a trustworthy instrument, and as an indispensable tool of scientific inference in management and in research.

My recommendation, for use of the book as a textbook, is to begin with Part II and to study it concurrently with Part I, with daily lessons in Part III. The order of the chapters is purposive, not random, to enable nonmathematical readers to begin with Chapter 1 and to

continue far enough to appreciate sampling as the scientific method of
inference before the mathematics looks too heavy, the premise being
that one must learn to hum or strum a tune and must like music before
he can derive much good from study of harmony and counterpoint.

It may be obvious from Parts I and II that the boundary of success
in the statistician's application of statistical techniques is his knowledge
of theory. There has been too much noise in the past about the
statistician's need for knowledge of the subject-matter and of the
material that he works with. He will pick up what he needs of this
knowledge readily enough. He should depend on the experts that he
works with to supply knowledge of the subject-matter, and to assume
the responsibilities that are rightfully theirs. The statistician's first
duty is to provide the requisite statistical theory.

Sampling is not a procedure for selecting a part from the whole, nor
a substitute for a complete investigation. Rather, sampling is the
scientific method of investigation and inference, in which demonstrable
reliability is requisite. Indeed, the major portion of the content of
new techniques in the management sciences, called by various names
such as operational research, queuing theory, and methods of simu-
lation, consists mostly of applications of the theory of sampling.

The statistical consultant in industry and in government today faces
requirements and responsibilities that have not confronted him in pure
science or in teaching. A sampling plan, to be satisfactory today, must
not only produce a figure for an estimate of the value of an inventory,
or of a certain class of account, or of the number of consumers that have
some specific characteristic, or a measure of the performance of a
product in service, or of a medical treatment, to name a few examples;
it must also provide demonstrable measures of the reliability of this
estimate. The same thing is true of figures issued by a government
agency on employment, unemployment, expenditures, production of
lumber, retail sales for the month, housing conditions, etc. This book
teaches, I hope, by precept and example, necessary standards of
statistical practice for today's needs.

The main reason for writing this book was for the sheer pleasure of
doing it. Secondary reasons were to record new methods of sampling,
new concepts and operational definitions, and standards of professional
statistical practice.

In respect to new methods, the chief contribution is replication by
procedures that maintain efficiency, yet facilitate estimation of the
standard error and of any mathematical bias that exists in the estimator used. The method is essentially Mahalanobis's interpenetrating
samples, which he introduced in 1936 in his surveys of Bengal. Acting

on a suggestion from my friend Professor John W. Tukey in 1949, I have simplified the theory and the practice, especially for the estimates of the standard error, through replicated selection of sampling units by serial number in equal paper zones. The simple invention of selecting the final sampling units by serial number draws primary units with probabilities in proportion to their sizes, and erases the complex theory for the variance of multistage sampling.

A further contribution is installation of the audit or statistical control, as a regular part of a sampling plan, for evaluation of the nonsampling errors of operation.

An important new concept is the equal complete coverage, which leads to an operational definition of the standard error of a sampling procedure, as given by Hansen and me in 1950. It leads also to operational definitions of accuracy and bias (Chapter 4). In place of the nonexistent and undefinable "true value," we now have the result of the equal complete coverage. In place of a "representative sample," which also defies definition, we have now, from the sample itself, the objective and operationally verifiable margin of difference, for a stated probability, between the result of the sample and the result of the equal complete coverage.

The implication for management, legal, and scientific purposes is immediate—a sample is acceptable in place of the equal complete coverage if the margin of sampling error is not too big, and if the field-work and the processing of the data depart not too far from the procedures specified. Defense of the equal complete coverage is the responsibility of the expert in the subject-matter. The equal complete coverage will stand or fall on the choice of frame, on the content of the questionnaire, on the method of test or interview, and on the statistical control or audit, which tells us whether the project was carried out with sufficient care. This judgment would be the same whether sampling were used or not.

Every exercise and every example in the book comes from experience. The examples are not all glistening examples of efficiency; they are, instead, examples of valid methods used under various restrictions and conditions of difficulty.

The purpose of the examples is to assist understanding of theory. Examples by themselves, without theory, will not teach anything. It is for this reason that I add a word here about the hazards of copying sample-designs and field-instructions. There are no simple rule-books nor ready-made sample-designs, and there never will be. The layman dare not copy sampling plans and instructions any more than he would dare look through a druggist's prescriptions, hoping to find one that suits his ailment.

The field-procedures in this book for the sampling of human populations make use of segments of area that the interviewers create in the field, by use of the half-open interval. Interviews take place on the spot in segments designated by random numbers. For the benefit of anyone who wishes to criticize these instructions, or who plans to copy them, I could add that I myself have revised them drastically every time that I have used them since the book was in proof.

The fiducial argument seen in Chapters 17 and 18 will, I believe, provide the scientist, the executive, and the lawyer with the kind of margin of error that they require in problems of estimation. What the lawyer or executive needs to know is the wildest shot that the sample may lead him into, and he relies on the statistician to tell him.

The scope of the book does not include some special but important problems, such as regression estimates, and continuing surveys in which some areas or establishments drop out of the sample every month as new units come in. The book emphasizes and uses cost functions and ratio-estimates, but stops short of complex treatments. The specialist has already at his disposal masterful treatments in the books by Yates, by Hansen, Hurwitz, and Madow, by Cochran, and by Sukhatme, plus the works of Mahalanobis, none of which this book will ever displace.

And now comes the part of the preface that is a special pleasure to write, namely, expression of my indebtedness to people whose help has made the book possible. Miss Josephine D. Cunningham assumed with untiring patience the burden of many experiments, computations and all, through the earlier years in which the replicated method was in formation. Her calculations also made possible the probability model mentioned in Chapter 5 for the best allocation of effort to combat the evil of nonresponse. Miss Margaret A. Ross worked with me regularly and faithfully for 3 years, rain or shine, in the preparation of manuscript. Her sound judgment in respect to order, clarity, and mode of presentation was invariably dependable. Dr. Jean Namias read and criticized early drafts of Part I. Miss Natalie Calabro worked especially on Chapter 9. My friend Mr. F. Newell Campbell of the Bureau of Internal Revenue was kind enough to read and to criticize two drafts of Part I, from the standpoint of an executive who uses and appreciates statistical methods.

My colleague Mr. Harold Nisselson of the Census read the whole book as it stood in 1958, and discovered a number of technical flaws, some of which, had they gone uncorrected, would have been troublesome to the reader and perennial embarrassment to the author. Dr. Tore Dalenius of Stockholm contributed the essential theory of Chapter 20, on the optimum dividing lines between strata. Mr. Leon Kilbert contributed the labor and most of the theory in Chapter 21 on the numerical comparison of zoning intervals. My wife, Lola S. Deming, with her special skills, absorbed overloads of calculation, verification, and proofreading. The production of the manuscript, with continual revisions that I scribbled patch on patch these 9 years on trains and on my lap in stations, could have been accomplished only by my able secretary Cecelia S. Kilian.

Various footnotes and references within the book record my appreciation and indebtedness for special assistance from other kind friends, and for permission from clients to use examples. Finally comes a tribute to other clients unnamed, who will smile as they recognize their own problems here and there disguised to form exercises and examples.

W. EDWARDS DEMING

Washington
July 1960

Contents

PART I   STANDARDS OF STATISTICAL PRACTICE

PART III SOME THEORY USEFUL IN SAMPLING

PART I

Standards of Statistical Practice

CHAPTER 1

Responsibilities in Planning a Survey

The statistician who supposes that his main contribution to the planning of an experiment will involve statistical theory, finds repeatedly that he makes his most valuable contribution simply by persuading the investigator to explain why he wishes to do the experiment, by persuading him to justify the experimental treatments, and to explain why it is that the experiment, when completed, will assist him in his research.—Gertrude M. Cox, Lecture in Washington, Department of Agriculture, 11 January 1951.

For that theory (mathematical theory of statistics) is solely concerned with working out the properties of the theoretical models, whereas what matters—and what in one sense is most difficult—is to decide what theoretical model best corresponds to the real world-situation to which statistical methods must be applied. There is great danger that mathematical pupils will imagine that a knowledge of mathematical statistics alone makes a statistician.—D. G. Champernowne, in a discussion on the teaching of mathematical statistics, *Journal of the Royal Statistical Society*, volume 118, 1955: page 203.

Need for understanding of responsibilities. Good statistical work is the product of several kinds of knowledge working skilfully together. The statistician needs to know what aspects of a statistical problem require statistical theory for solution; and equally, what contributions to the problem must come from knowledge of the subject-matter. Statistical theory is not a substitute for knowledge of subject-matter. Conversely, skill in subject-matter will not solve statistical problems. We begin in this chapter by learning how the various skills work together.

Origin of a problem. Before there is any thought of a survey or experiment, someone has a problem, and makes some effort to state what it is. The problem may originate in some substantive field, such as chemistry, engineering, production, consumer research, sociology, medicine. In fact,

3

one might say that good science and good management start with the recognition of problems, and clear statements thereof; followed by some solid thought concerning possible information that might throw light on these problems, and (in business and production) a statement of possible alternative decisions that one might make, and of the losses and gains that these decisions might lead to.

A problem exists only if there is more than 1 answer.   If there is only 1 decision to make, as when an automobile bears down upon you, there is no question about what to do—jump.   But if there are 2 or more possible decisions, with different losses or gains, and with different probabilities, we then face the problem of making a rational decision under uncertainty. A rational solution is one that we can trace by theory from postulates and data to a conclusion, generalization, or decision.   A rational solution requires a statement of aims, and a criterion for evaluating the effectiveness of the system of solution.   An irrational solution is one that no one can explain, neither retrace, nor understand.

It may be, in the judgment of management, or of experts in the subject-matter, that the chance of a wrong solution to a problem will be lessened by the collection and use of statistical information not now at hand.   It is at this point that consideration of a possible survey or experiment commences. Sooner or later the question will arise: how much is new information worth and how much will it cost?

The test of talent in management is ability to perceive and to formulate questions that if answered would help to find out what is wrong, or would improve the business, or would discover what we need to know; and in addition, wisdom enough to use the results of a statistical survey or experiment that is designed to throw light on the problem.   This view is contrary to the evaluation that some people make of themselves when they claim ability to make the right decision on the basis of experience.   Actually, the only safe use of experience is to perceive and to formulate questions, and to predict the losses and gains from the various decisions that are possible.

The information that a statistical survey or experiment provides is not right because someone agrees with it, nor wrong because he does not: the information is useful or not useful depending on the information and the questions that were built in initially in the statement of the problem, and on the skill that went into the preparation, execution, and summary of the survey or experiment.   The burden rests with the expert in the subject-matter, and with management, in the preparation of the questionnaire or of the method of test, to get information that they can apply to the problem when the results of the survey or experiment come in.

Fusing statistical knowledge to the substantive requirements.   A survey or experiment is neither right nor wrong except in the light of the purpose intended. What is the purpose? How may the results of the survey answer the questions that have come up? The statistician, to understand the problem, should try to place himself in the position of the administrator or expert in the subject-matter who needs the information and runs the risk of using it.

The statistician begins by reformulating the problem in statistical terms, so that it will be possible to decide whether and how any statistical information might help to solve the problem. Statistical information will be in terms of an estimate or test of hypothesis, and (for business use) a rule of decision based on the theory of probability by which to use the results. Statisticians speak of this formulation as finding a suitable statistical model for the problem.

The statistician will translate the needs of information into a plan by which to acquire estimates of totals, or of proportions (your share of the market), month-to-month changes (are you gaining or losing?), year-to-year changes, or comparisons of one kind or another, or into tests of decision-parameters. The survey should perhaps be a single survey or perhaps it should be a monthly or quarterly series. In the case of information from current business transactions, it may be continuous. Knowledge that the statistician may have accumulated concerning efficiencies and costs, and of the various types of estimate that are possible, will often simplify the content of the proposed questionnaire or method of test, as well as the administration of the study.

The translation of the substantive expert's problem into statistical terms to see whether any new statistical information will throw light on the problem is frequently the most difficult part of the statistician's work. It is at this point, as the quotation from Dr. Cox at the heading of this chapter states, that the statistician often makes his greatest contribution.*

The hardest job is to find out just what the problem is, or if there really is one. One procedure that the statistician may find helpful in the formative stages of a problem is to decide, with the aid of the experts in the subject-matter, on the tabulations that will be useful, and how one would use them. This is not as simple as it sounds. Even when a decision is reached on the kind of information that would be useful, and how to use it, problems of definition, simple as they may seem, rise up to cause trouble and delay. Unless one had been exposed to the statistical problems of the labor force, for example, he might suppose that it would be simple to count the number of people employed, or the number of people unemployed, or the number seeking work. It would seem equally simple to find out how many

* An excellent account of the working relationship between the statistician and the scientist is an article by William Lurie, "The impertinent questioner: the scientist's guide to the statistician's mind," *Amer. Scient.*, vol. 46, 1958: pp. 57–61.

people will purchase certain household durable goods next year.   All you have to do is to ask them, someone said.   One might suppose that it would be simple to define the density of a material, but how would you define the density of steel wool?   How would you define a quart of oysters?   How much liquid is permissible and how would you measure it?

It is humiliating, but a fact, that in spite of all the wonderful scientific apparatus in the world today, and in spite of improved knowledge of psychology and methods of interviewing, no known physical or chemical measure will predict with certainty that one design or another, even of some of the simplest household or personal items, will sell better than another.   People might say that the General Electric refrigerator is the best, yet go right on buying Frigidaires.   There are instruments that will measure color, flocculence, torsion, hardness, tensile strength, and will make any kind of chemical analysis; yet the fact is that none of these measures will tell a manufacturer whether one rug will be more acceptable than another to the people that are about to buy a rug, or a suit of clothes. No chemical analysis can predict whether one way of flavoring certain foods will attract more customers than another way.

No one knows today what questions to ask to find out from Mr. A. on Wednesday how he will vote next Tuesday, nor whether he will buy a certain item next year; nor even with exactness what proportion of people of his age, sex, and income will vote this way or that way.

These examples, and countless others that one could mention, illustrate the difficulties that confront management and the expert in the subject-matter.   It is important that the statistician and the people that he works with should not depend on statistical theory to supply the imagination that will solve these difficulties.   Statistical theory has power, but also limitations.   It is important not to misuse either substantive knowledge or statistical theory.

In respect to the power and the limitations of statistical theory, I may point out that statistical theory will not of itself originate recognition of a problem; yet only with statistical theory can one arrive at a statement of a problem in the form of a probability model by which tests and answers to questions may be feasible and meaningful.   Statistical theory can not generate a questionnaire nor a satisfactory test of hardness, or of acceptability of color in dishes or in a carpet; nor does statistical theory originate a way to teach field-workers or inspectors how to do their work properly, nor what characteristics of the work or of the workers to measure in the supervision of such work.   Yet it is only with the aid of statistical theory that one may learn reliably and economically the magnitude of the difference between the results of the 2 proposed questionnaires, or between 2 proposed tests, or between 2 proposed ways to teach workers how to do their work. Only statistical theory can provide a safe supervisory tool to help to achieve uniform performance of the workers (Ch. 13).

The statistician's insistence on the evaluation of all kinds of operational uncertainties (Ch. 5), and his specialized knowledge of variability, will frequently make an important contribution to the substance of the investigation. It was, for example, the statistician engaged on a large-scale test of fertilizers who insisted that the laboratories not only should carry out duplicate tests of the samples of fertilizer on any one day, but that they should repeat the tests of each sample a week later. It was he who insisted that the laboratories report all tests, not merely the usual best 2 out of 3.* It turned out that the tests a week apart of material out of the same bottle led to important discoveries about packaging the product, as the tests a week apart showed greater variability than duplicate tests run on the same day. The statistician's experience with variability led to better chemistry than would have been possible by knowledge of chemistry alone.†

A statistician may deal with a wide variety of applications during the course of a week. These might well range from medical experimentation, psychological surveys, problems in accounting, inventory, studies of records, sampling of physical materials, rate-structure of traffic by rail or by truck, flow of this traffic by mileage and commodity, consumer research, spending, saving, sickness, and other sociological and demographic studies. Naturally, he can not be an expert in such a wide variety of subject-matter. His knowledge of the subject-matter and of the material to be tested or interviewed will become intricate, but not necessarily profound (as examples of instructions here and there in the book may illustrate).

The fact is that he need not be an expert in the subject-matter in order to make his best contribution to a survey or experiment (Remark 1). His knowledge lies in statistical theory. A good part of his contribution is to point out what is substantive and what is statistical, so that everyone concerned may see where and how his own knowledge and efforts can be effective.

The statistician must of course learn enough of the subject-matter to be able to fuse his statistical knowledge into the problem. He must learn about the variability of the material, and he must protect the design against possible freakish variability and concentrations of special characteristics (see Chapter 14 on the sampling of new material). He must

---

* W. J. Youden, "The fallacy of the best two out of three," *News Bulletin, National Bureau of Standards*, vol. 33, 1949: p. 77. "Sets of three measurements," *The Scientific Monthly*, vol. lxxvii, 1953: pp. 143–147.

† Student, "Errors of routine analysis," *Biometrika*, vol. xix, 1927: pp. 151–164.

learn about any special difficulties of carrying out the investigation, and avoid them.   The statistician is responsible for the sample-design, and he must not agree to procedures that he believes will encounter difficulties or exorbitant costs.   At the same time, he must be careful not to let himself be argued out of a good survey, just because someone has never done it the statistician's way before and dreams up a host of nonexistent operational and administrative difficulties.

> **Remark 1.**   In support of the statement that the statistician need not be an expert in the subject-matter in order to make his contribution, I may point out this sobering fact: that it is statisticians and not experts in the subject-matter who have, during the past 2 decades, revolutionized the following pursuits through application of sampling and design of experiment:
>
> > Industrial production
> > Agricultural experimentation
> > Medical experimentation
> > Industrial experimentation
> > Design of product
> > Specifications for product
> > Testing and acceptance of product
> > Administration
> > Population censuses
> > Agricultural censuses
> > Current surveys of population for the labor force,
> >   housing, prices, savings, sickness, etc.
> > Marketing and consumer research
> > Psychological studies
>
> **Remark 2.**   In a legal case, or in a dispute within a company, the decision on the statistical information that is needed to answer the questions, and to reach a rational decision, should be settled preferably in advance of the survey, by agreement of both sides; otherwise, the results may only be a football to play with and to argue over.   One does not match a coin with a friend and decide after he looks at the coin whether to take heads or tails, but this is precisely what people do when they conduct a survey or experiment to settle a question, and decide afterward how to use the results.

The universe and the frame.   If it appears that a survey or experiment may be worth discussion, the next step is to define carefully the universe. The universe is all the people or firms or material, conditions, concentrations, models, levels, etc., that one wishes to learn about, whether accessible or not.   The universe will be clear from a careful statement of the problem.   An example is all the firms that make a certain product, or that may buy it; or all housewives, all school children, or all the material or piece-parts covered by a certain contract or specification.   The universe may be certain records within a company, where the aim of the study is to estimate the revenue from certain types of transactions.

Then comes the decision on the frame, which we pursue in further detail in Chapter 3. The frame is a set of physical materials (census statistics, maps, lists, directories, records) that enables us to take hold of the universe piece by piece.* The question that is uppermost in the early stages of planning is whether a suitable frame will cost too much, or whether one is obtainable at any price. The frame must cover enough of the universe to make the study worthwhile. What areas, classes, or conditions does a proposed frame fail to cover completely, and what proportions does it omit? A frame must also provide suitable sampling units, concerning which this book will have more to say later. In studies of records, the statistician must find a spot in the flow of the records at which one may conveniently conduct the study, preferably after final corrections have been made in the natural course of business. The final decision on whether a proposed frame covers enough of the universe belongs to the client (p. 16).

The methods of statistical inference, once the results of the survey are tabulated, will apply only to the frame sampled. Generalizations to other material, areas, methods, levels, and conditions can be only substantive. It is therefore necessary to take a good look first, before it is too late, to decide whether a proposed frame will be satisfactory.

The decision on whether a proposed frame is satisfactory depends on a balance between the hazard of getting wrong information about the universe and the cost of improving the frame. No question of sampling arises at this point. Rather, the question is whether a proposed frame, if covered completely (that is by a 100% sample), would supply useful information, and whether it would pay to improve the frame.

Consideration of costs. The statistician has an obligation to persuade the client to compute the probable total cost before talk of a survey goes too far. Statistical theory enables the statistician to trim the size of the sample, and hence its cost, to the minimum consistent with the precision required in the results. The final decision on whether to proceed rests with the man who is going to pay the bill.

There will be fixed costs and variable costs. The variable cost naturally depends to some extent on the precision aimed at, and the precision in turn depends on the size of sample and on the procedure of estimation. The tabulations are to some extent elastic. The cost of the frame, once decided, will often be the same whether you take a complete coverage of the frame or a small sample. Many other items of overhead, such as the time that the statistician and other executives spend on the job, will be about the same regardless of the size of the sample.

---

* The concept of the frame was first stated by Frederick F. Stephan in "Practical problems of sampling procedure," *Amer. Soc. Rev.*, vol. 1, 1936: pp. 569–580.

Cost is often the determining factor in the size of the sample. The question then arises whether the accuracy obtainable at this cost will be satisfactory. Statistical theory and knowledge of costs make possible prediction of the precision obtainable for a given total cost.

Responsibilities of the statistician.* The statistician accepts in any engagement certain responsibilities and obligations to his client† and to the people that he works with. In the first place, he is the architect of a survey or experiment. It is his business to fit the various skills together to make them effective. It is important that he clarify the various responsibilities at the outset of a study. In doing so, he will (*a*) enhance the usefulness of statistical theory and practice; and will (*b*) forestall disappointment on the part of the people that he serves. If they fail to exercise their own responsibilities in the planning of a survey or experiment, they may not realize in the end its fullest possibilities, or may discover too late that certain uses that they intended to make of the results are impossible.

The salaried statistician should have the same responsibilities and standards of workmanship as the consulting statistician, I believe; and the company has the same obligations to the statistician either way, whether he is on a salary or is engaged as a consultant.

The statistician's responsibility in a survey or experiment is for its statistical aspects; specifically:

1. To assist the client to formulate his problem in terms that will indicate whether and how statistical information might be useful on the problem. To formulate a choice of possible statistical surveys or experiments, if it appears that new information might be worth more than its cost. To explain to him the statistical procedure, cost, and utility of these plans.

2. To explain to him the use and importance of the frame; to explore with him the various frames that might be suitable; and to explain to him that the results of any survey or experiment may be limited if a proposed frame and experimental conditions fail to include all the materials, areas, methods, levels, types, and conditions concerning which he desires information. To explain that the objective inferences that one may draw by statistical theory from the results can only cover the frame and the material, areas, methods, levels, types, and conditions presented for

---

* Extracted from the author's code of professional conduct, and from a paper entitled, "On a formal structure of professional practice in sampling," meeting of the Institute of Mathematical Statistics, Boston, 29 August 1958.

† The word client is here a symbol which means a company, agency, or person that seeks advice from the statistician.

sampling;* that generalizations to other materials, areas, levels, conditions, questions, or methods of test can be made only through knowledge of the subject-matter, not by statistical theory; that the statistician can take no responsibility for these generalizations.

3. To design a sample of specified material, covered by the frame (Ch. 3) that the client certifies as necessary and sufficient for study, to reach precision agreeable to the client; or to design an experiment to reach significance in the tests of materials, methods, levels, types, and conditions that the client certifies are necessary and sufficient.

4. The statistician's instructions will include detailed procedures for use in the office and in the field for the delineation, definition, classification, and serialization of the sampling units. They will describe the selection of the sample and use of the sampling table that the statistician will furnish. They will include provision for a system of search for items not in place, and for a system of recalls or equivalent adjustment for persons not at home at 1st call, procedures for the calculation of the estimates desired, and for any standard errors or other measures of statistical significance that may be needed. The statistician will furnish the sampling table by which to draw the sample after the client certifies that the preparation of the frame, including the scheme for the serial numbers, is complete, or will be independent of the sampling table. The statistician will in some cases, instead of furnishing the sampling table, specify the exact procedure by which to construct it from a specified table of random numbers.

5. The statistician will furnish statistical plans in writing when they are finally fixed, and the client will thereafter make no changes in procedure without further instruction from the statistician so long as the statistician's responsibility remains in force.

6. The statistician will explain to the client the effect of departures from instructions. He may require probes of the sampling procedure, and of the coverage, and of the consistency of the tests or interviews, in order to detect and to evaluate any difficulties or departures from the procedure (Ch. 5). He may also require tests of the computations. The client will carry out these probes according to instructions, and he will furnish to the statistician the results thereof.

7. The statistician will furnish at the client's request statistical tests to use as an aid in supervision of the work, to attain more uniform performance in the testing or in the interviewing and in coverage than would be possible otherwise (Ch. 13).

8. The client will arrange for the statistician to have direct access at any

---

* An exception occurs in a state of statistical control, where prediction can safely be made so long as no new special or common causes of variation come into play.

time to the people that carry out the preparation of the sample, the testing or the interviewing, the supervision, the coding, and the computations.

9. The statistician will explain to the client, when the tabulations are finished, the meaning of the results of the survey in terms of their statistical significance. However, the statistician will not recommend that the client adopt any specific administrative action or policy. The uses of the data obtained by a survey or experiment are in the end entirely up to the client.

The above paragraphs give a bare outline of the statistician's responsibilities. They do not say how he should do his work. They do not deny the statistician the privilege of taking an interest in the subject-matter, nor from becoming expert in it. He will of course be concerned about the quality of the questionnaire and of the field-work, and he will help in any way within his power to achieve quality. He can do this without usurping responsibility that belongs to someone else. The statistician may, for example, help the client to organize a statistical unit to carry out the work. He may even help to find competent people to take charge of it. He may help to wash the cups after coffee, or to do some of the typing, maintaining his usual standard of error. He may, in a small organization, undertake to write the instructions for the coding, and to draw up the forms for the codes. As such work belongs to the subject-matter, he will insist on clearing every point with the client or with the experts in the subject-matter empowered by the client to assume the necessary responsibility.

For example, I have done a considerable amount of interviewing, under the regular supervision, and I count the experience as indispensable. I have helped to develop and to test many a questionnaire, but the client always knew that he alone was responsible for it. I wrote out the coding instructions and forms for a medical study in a small organization. The physician in charge was of course responsible for the codes and definitions, and I was careful, before putting them into use, to make sure that he made final revisions and understood his responsibility for this phase of the work.

The statistician's report or testimony. The statistician's final step in the survey is his report to the people that he works with. He will direct this report so that it will throw light on the questions that prompted the survey in the first place, thus to assist management or the substantive experts to draw conclusions concerning their problem. A client will in some engagements wish to include the statistician's report in claims to quality of performance or of product. Sometimes the statistician will draw up a report that may become legal evidence. Some examples of reports appear later in this chapter (p. 16).

The main contribution in the statistician's report will be estimates of the margins of sampling error for the results of chief importance, and the possible effects of the nonsampling errors. Ability to make statistical calculations, and to make comparisons and tests of significance, is what distinguishes a statistician from other professions. Evaluation of the nonsampling errors, however, is not purely mathematical: it requires the aid of expert knowledge of the subject-matter, and of the methods used in the field.

The statistician's report to management should not talk about probabilities. It will merely give outside margins of error for the results of chief importance. Administrators are not interested in probabilities; they need to know what figure they can use and within what range they may depend on it. They hire the statistician for his knowledge of probabilities, so that they themselves won't have to worry about such things. In my own presentation of the results of a sample, I use 3-sigma limits or their normal equivalent for practical upper and lower bounds of the possible margin of difference between the result of the sample and the results of the equal complete coverage, with no mention of probability.

The statistician's report will cover only the frame that was subjected to sampling. It will not attempt to generalize to other firms, people, cities, material, concentrations, levels, and conditions not covered by the survey or experiment, nor to other questionnaires and methods of test.

The statistician should not, in my judgment, give advice to the client to tell him what decision he ought to make (see Paragraph 9 on page 12). The statistician's job is to make useful statistical interpretations of the results of the investigation, so that management may understand them and use them intelligently. This is a big enough job. The minute a statistician steps into the position of the executive who must make decisions and defend them, the statistician ceases to be a statistician. He becomes just an ordinary citizen, as he has no special qualifications as an executive. His recommendation, if he were to make it, might well detract from his professional standing as a statistician.*

The statistician's report will declare his responsibility for the statistical design. It will state who was responsible for the completeness of the frame, for the questionnaire, for the method of test, for the field-work, for the coding, and for the computations. It will state that the figures in the report came from the client at the statistician's request for the results of the survey. It will explain the margin of sampling variation and the statistical significance of the results and of the comparisons that are important to the client. It will cite the amount of nonresponse, and of any other uncorrected operational blemishes detected. If there was a

* Borrowed heavily from David D. Rutstein, "Doctors and politics," *Atlantic Monthly*, vol. 198, Aug. 1956: pp. 32–35.

formal statistical control of the work (p. 39), the statistician's report will include evaluation of operational blemishes such as:

a. Failures to define or otherwise designate the sampling units in the manner prescribed.

b. Failures to reach and to cover sampling units that the random numbers selected.

c. Sampling units not intended for the sample but covered or partly covered by mistake.

d. Other slips and departures from the prescribed sampling procedure.

e. Errors and difficulties in reporting; wrong weights, wrong counts, wrong prices, and the like.

f. Nonresponse.

The possible effects of these blemishes will depend on what assumptions one makes with respect to them. These assumptions and their effects on uses of the results of the survey may involve expert knowledge of the subject-matter. This is especially likely to be true in the disposition of nonresponse. Hence, if the statistician's report describes the effect of the above blemishes, it will state not only the assumptions, but who is responsible for them. It will also describe the procedure that the statistician used in the calculations of the possible effects of the blemishes, so that the user of the results may try out the effect of other assumptions if he wishes to do so.

If there was no formal statistical control, the statistician's report will say so.

The client, if he prints or publishes the statistician's report, will print it in full, and will not omit any part of it without the consent of the statistician.

The statistician will prepare any statement that bears his name in connexion with the sampling procedure. Such writing is part of the engagement.

Besides the report to the client or to experts in the subject-matter, the statistician may wish to publish in a scientific journal any new theory or experience with respect to the sample design. The article may show actual figures from the survey unless there is some agreement or other reason to substitute fictitious figures.

The statistician will sometimes write a joint article in a scientific or professional journal, with an expert in the subject-matter as co-author. In accord with the principles stated above, writers will see to it that the statistical inferences in the article will be properly labeled as such: likewise substantive inferences and generalizations. Thus, if a medical survey be carried out in Chicago, the margin of sampling error and the tests of significance will clearly refer only to Chicago, or to whatever part of Chicago was in the frame and hence subjected to the sampling procedure. Such inferences are objective; they are not matters of opinion. However, any inference that the results will hold also in Denver is substantive: such generalizations require knowledge of medicine. No statistical theory in existence can carry the generalization from Chicago to Denver, as Denver was not in the frame.

The statistician is a logician and must insist that the inferences bear the proper labels. To fail to do this is to invite misuse of good data.

Details of the sampling plan are not helpful to the user. The margin of sampling variation between repeated samples carried out after the manner of the one being reported is calculable by theory, for a desired probability, from the results of the sample. (The fundamental requirement for a probability sample is that it be laid out so that this calculation is possible.) The margin of sampling error embraces the effect of any accidental mistakes whose expected value is 0, including variation in judgment, wherever judgment may enter into the collection of the information from any sampling unit chosen.

Then there is also, of equal importance, the evaluation of any persistent noncanceling errors of one kind or another; also the amount of non-response or the proportion of sampling units designated for the sample but not found (the so-called nonsampling errors, Chapter 5). This information comes from the audit or control, and sometimes also by the help of outside comparisons.

Details of the actual plan of sampling the material, such as the size of the sample, the modes of stratification if any, the choice and size of sampling unit, etc., all so important in a technical paper on sampling methods, are irrelevant to the user of the data; they add no new information and are of no help in an evaluation of the reliability of the data.

Incidentally, sample-surveys ought to be priced on the basis of the amount of information that they supply, and not on the number of interviews.

The practice of reporting full details concerning the sampling procedure, especially prevalent in reports from marketing-research agencies, undoubtedly had its origin in the use of judgment samples, for which full details are of course necessary.

Counsel in a legal case may, however, present the sampling plans as an exhibit, in order that either side in the case may probe the execution of the plans. The sampling plans may under such circumstances serve a purpose. At any rate, the statistician should prepare his plans and his testimony with this possibility in view.

Responsibilities of the client. The client should state in advance how he expects to use the results of the survey or experiment. He and not the statistician must assume the responsibility for those aspects of the problem that are substantive.   He must also declare whether a proposed frame and supplementation if any are sufficiently complete for his purpose; likewise the types, methods, and conditions to be sampled and studied. The criterion for decision in respect to a frame is whether a complete coverage thereof would suffice.

1. Specifically, the client must assume final responsibility for:

a. The type of information that the survey or the experiment is to elicit.

b. The methods of test, examination, questionnaire, or interview by which to elicit the information from a sampling unit.

c. The decision on whether the frame is sufficiently complete; likewise the methods, types, conditions, and environment that are subject to sampling and experiment (Ch. 3).

d. The classes and the areas of tabulation.

e. The actual work of preparation, training, testing, interviewing.

f. The supervision of this work.

g. The careful recoverage or retest of any units that the statistician specifies for audit or control.

h. The coding.

i. The tabulations and the computations.

2. The client will carry out the sampling procedure as specified, and will make no changes therein without authorization so long as the statistician's responsibility remains in force.

Examples of a statistician's report.   The 1st example is a brief one, as there was no formal statistical control of the field-work.   It refers to a sample of 2 counties in which the results displayed certain financial, economic, and social characteristics of the readers of a certain newspaper. The statistician's statement was included in a bulletin that the newspaper published to describe the purpose of the survey and to exhibit the results.

Example 1
Statistician's Statement in Regard to a Sample of Limestone and Cherokee Counties

The specifications of the sample for this survey followed generally accepted theory and principles of probability sampling.   The specifications if followed would yield results for the responses whose standard errors have the usual interpretation.

The standard error does not measure the effect of nonresponse nor of persistent omissions, inclusions, or departures from procedure.   It does include, however, the effects of variable performance of an interviewer, and also in this case the differences between interviewers.

I had no responsibility for the questionnaire.   The sampling plan in this case did not call for any formal statistical audit of the field-work, or of the tabulations or of the computations, nor do I take any responsibility therefor. I did satisfy myself that the firm O'Brien-Sherwood understood the sampling procedure, including the formation of the estimates and of the standard errors. I was on hand constantly to ask and to answer questions that came up in procedure.

I may say, however, in respect to coverage, that the sample gave an estimate of 470,000 dwelling units in the 2 counties combined, with a standard error of about 1.5%. The Census count, taken about a year previously, was 454,400. The difference is 2 standard errors, which could arise from sampling error, or from growth, or from some of both. The direction and magnitude of the difference appear to indicate successful coverage of the selected segments by the field-workers: incomplete coverage would have produced a deficit.*

The figures on which I base this estimate of the total number of dwelling units in the whole area, and the standard error thereof, came from O'Brien-Sherwood at my request.

The firm also informed me that the interviewers obtained responses in 87.3% of the households visited, and that the nonresponses were distributed amongst all interviewers, and in all areas, not being confined nor concentrated in any one class. My instructions asked the firm to make no adjustment for nonresponse, but to show in the tables the figures that came from the households that actually responded. I offer no adjustment for the nonresponses.

The 2d example is an excerpt from legal testimony, in which a telephone company had carried out an inspection of their various classes of telephone plant through the aid of sampling, to arrive at a figure for the overall per cent physical condition of the entire plant that was subjected to sampling and the inspection.

Example 2

Excerpt from Legal Testimony†

Q. Would you please explain the nature of your engagement with the Illinois Bell Telephone Company?

A. Mr. B., General Staff Engineer of the Company, informed me that he wished to make a survey to determine the overall physical condition of the Company's plant, and he asked me to draw up the proper sampling procedures.

Q. Did you accept this engagement?

A. Yes sir, I did.

---

* Even a simple statement like this requires substantive knowledge, specifically, some crude knowledge about field-work (pointed out to me by my colleague Leon Pritzker).

† The Illinois Commerce Commission, Docket No. 39126, 1951, and Docket No. 41606, 1954: the Illinois Bell Telephone Company in the matter of the proposed advance in rates. The passage printed here is testimony prepared in advance, and is not necessarily the same word for word in the record. Moreover, I have supplied some lines from several other subsequent dockets.

Q.   What was the scope of your engagement?

A.   To furnish sampling plans for the plant that Mr. B. asked me to sample; and to tell him, on the basis of figures that he would furnish to me as the result of applying the sampling procedures that I would furnish, and as the result of an audit that I would prescribe to examine the Company's performance, what was the reliability of the overall per cent condition derived from the sample, using as a norm a 100% inspection of every one of the millions of items on the records presented to me for sampling, carried out by the same definitions and methods of inspection as were to be used on the samples, and calculated in the same way.

Q.   Were there any special terms about your engagement?

A.   No, there was nothing unusual about it.   I accepted the engagement subject to my code of professional conduct, which binds me to complete technical responsibility for the sampling procedures, and which binds the Company to follow the plans in every detail; to make no departure without authorization from me.

Q.   Did you explain to Mr. B. what his responsibilities would be?

A.   I told him that he would be responsible for the completeness and the accuracy of the engineering records and other lists that he would present to me for sampling: that he would be responsible for the methods of inspection, for the inspection itself, and for the supervision thereof; for the translation of the inspectors' observations into numerical percentages; for the prices of the items in the sample, for the weights of the various classes of property, and for the form of computation.   He was also responsible for the accuracy of the computations.

Q.   Did Mr. B. inform you of the overall per cent condition of the plant calculated as the result of applying your sampling procedures?

A.   Yes sir, he did.   He informed me that he had estimated from the samples, following formulas specified in my sampling procedures, that the overall per cent condition was 82.3%.

Q.   What is the reliability of this figure?

A.   First of all, there is an allowance to make for the variation between repeated applications of my sampling procedures as Mr. B. actually carried them out.   If you were to select over and over again, using a prescribed set of rules, samples of the inspection records from a complete 100% inspection of all the millions of items included in the lists that Mr. B. presented to me for sampling, the overall per cent condition calculated from the samples would show variation from one sample to another, and variation from the 100% inspection.   There are standard methods by which to calculate, from the results of a sample that follows the procedure that I specified, the outside margin of variation between repeated samples.   The numerical value of this margin of variation in either direction turned out to be in this instance .5%, or 5 parts in 1000, based on figures that Mr. B. furnished to me at my request for the results of the inspection of the samples.   Repeated samples like this one, drawn from the same 100% inspection, would thus reproduce each other within very narrow limits. This statement is not a matter of opinion: it is a mathematical consequence of the sampling procedure and of the figures that Mr. B. furnished to me.

The mathematical formulas allow for accidental mistakes and departures from instructions, of the kind that may cancel each other.   For example, they allow for variation in the judgment of an inspector from one day to the next, and for varying appearances of items, wet and shiny one day, dry and older-looking the next.   They allow also for the canceling accidental effects of mistakes and blemishes in the identification of items in the office and in the field, such as failures to inspect an item that the random numbers had designated for inspection, or inspection of additional items, or inspection of the wrong item.   The effect of canceling accidental variations, from whatever source, is to widen the calculated margin of variation between the results of repeated samples.   As I said, the mathematical formulas automatically made and included the proper allowance for this kind of variation.

Q.   Is there any other source of uncertainty in the result derived from the sample?

A.   Yes sir, there is.   Omissions, substitutions, and the inclusion of extra items in the inspection will lead to error of a noncanceling nature in either a sample or a 100% inspection if the items omitted, substituted, or included additionally are predominantly in better condition or in worse condition than the average condition of the whole plant.   The mathematical formulas do not detect the noncanceling effects of mistakes.

Q.   What did you do about such possible mistakes?

A.   I asked Mr. B. to conduct an audit of the sampling, after the inspectors had done their work.   I prescribed for this purpose a sample of about 10% of the items that were in the main samples of the classes of plant of chief importance.   I asked Mr. B. to furnish me with figures for the number of items in the audit that the inspectors had found and inspected in exact accordance with my instructions, and for a record of any items in the audit in which the inspectors had departed in any way from my instructions.   I wished to know how many wrong items they had inspected, if any, either in place of items that the random numbers had designated for inspection, or additional to items so designated; and how many items they had omitted to inspect, if any.   I wished to compare the condition-classification of any item wrongly inspected with the per cent condition of the item that should have been inspected but was not; and I wished to know the per cent condition of any item omitted.   I needed this information in order that I might detect and measure the effect of any tendency to substitute or to omit items in good condition, or items in low condition.

I might add that an audit is as necessary for a complete 100% inspection as it is for a sample.   In fact, it is more important for 100% inspection because, as experience shows, the bigger the job, the more prominent become the departures from instructions.

Q.   Did Mr. B. furnish you with a report of the audit?

A.   He did.

Q.   What did it show?

A.   I calculated from his report that if the mistakes and blemishes that he detected had had any effect on his result for the overall per cent condition, the effect could be at most .62%, or 62 parts in 10,000, and that it could as well lie in one direction as the other.   I made this calculation by ascribing extreme upper and lower limits to the grades of items not found, and to items inspected by mistake, to see what possible effects might arise from these operational failures.   Mr. B. gave me the figures to use for the extreme upper and lower grades.

Q.   What is your conclusion about the accuracy of the per cent condition which Mr. B. determined by the sampling procedures?

A.   I told Mr. B. that he should accept the result 82.3 % for the overall per cent condition of the portion of the plant subjected to the sampling procedure, and contained on the lists that he furnished to me, as a figure whose quality he now knows, through evaluation of both (a) the margin of sampling variation, and (b) any uncertainty that could arise from blemishes in carrying out the sampling procedure specified.   That is, the inspection of the samples gave a determination of the overall per cent condition that is the same, within very narrow limits, as would have resulted from a complete 100 % inspection of every item in the plant, carried out by the same definitions and methods of inspection, and by the same inspectors, were they to exercise the same care on the 100 % inspection as they did on the sample.

Q.   Do you have anything further to add to your testimony in regard to the samples?

A.   It is important to understand that the result of the samples and the reliability that I have calculated therefor apply only to the records and to the methods of inspection that Mr. B. used.

Example 3

The 3d example is a report on the results of a sample that was used to estimate certain components of inventory-content, and the LIFO adjustment on the inventory.   The sampling procedure is described on pages 309 ff.

Memorandum to the Comptroller

ABC Company

Chicago 14

Statement in respect to the reliability of the estimates of the dollar-value of 1957 year-end corporate material, of prior-plant conversion-costs, and of unrealized earnings, for the portion of plant in the paint-and-glass-products pool subject to sampling.

This statement refers to the reliability of the results that you derived from a sample that I prescribed.   I understood from you that the lists that you presented to me for sampling were prepared from records maintained by the Corporation for purposes of production programing and inventory control.   They showed part-numbers and descriptions of items, and they met a fundamental requirement, namely, your assurance that processing all these part-numbers would constitute a 100 % evaluation of the problem.

My responsibility is limited to the statistical methodology—the procedure of selection, the procedure for forming the estimates that you required, along with the standard errors thereof, and their interpretation; statistical tests of compliance with the sampling procedure specified, and an audit to test the repeatability of the entire procedure; and finally, the statistical evaluation of the reliability of the results.   Your responsibility covers those aspects of the study that would be the same whether you used sampling or not.

I designed a sampling plan to apply to the lists that you provided. I worked from time to time with your people on the selection of the sample and on the selection of the sample for the audit. I worked with them on the forms, controls, and verifications to apply to the selection of the sample and to the arithmetic processing. I have confidence in their ability and desire to follow accurately the whole procedure. I have reason to believe that the numerical results of the sample are an accurate summary of the figures fed into the routine of processing.

According to figures that you furnished to me at my request, the results of the sampling are in the table herewith. The book-inventory came from the financial statement; the other figures came from ratios calculated from the sample.

Book-inventory, 1957 year-end (from financial statement)	$202,850,010
Corporate material	170,243,916
Prior-plant conversion-costs, active	19,995,647
Prior-plant conversion-costs, inactive	1,210,756
Unrealized earnings	11,399,691

The design of the sample made it possible to calculate objectively by standard methods, from the results of the sample itself, the tolerance to allow for the outside margin of difference between any of these results and the result that would have come from a complete processing of all the items on the lists that you provided. The outside margin of difference for material falls within $\frac{1}{2}$ of 1 % of the figures in the table. The outside margin of difference for the corporate conversion-costs, active, falls within 3 % of the figure in the table. The outside margin of difference for the corporate conversion-costs, inactive, falls within 12 % of the figure in the table. The outside margin of difference for the unrealized earnings falls within 5 % of the figure in the table. Theory and experience show that limits so calculated include the results that you would have gotten from a complete processing of all the items on the lists that you provided, were you to carry out the complete processing under the same rules and with the same care that you exercised on the samples.

The above tolerances include the possible effects of any accidental errors of a canceling nature that might have occurred in the pricing and in the processing, as well as the uncertainty that arises from sampling, but they do not detect nor evaluate the effect of any possible persistent error that there might have been in the pricing or in the processing. The sampling plan therefore called for an audit by which to detect persistence if any and to evaluate what effect it could have on the results of a complete pricing and processing of all the items on the lists that you presented to me for sampling, were you to carry out the complete pricing and processing with the same care that you exercised on the sample.

The audit consisted of a probe of a subsample of items drawn from the main sample. It called for repetition of the entire procedure for the items in the audit, by use of the original instructions, including recalculation, with other investigations that seemed warranted. Analysis of the differences found in the audit indicates the possibility of a small amount of persistence and that it could act in either direction to affect any of the figures in the above table. If there should be any persistence, it would affect the complete pricing in exactly the same way that it would affect the sample. With respect to the total inventory, the maximum overestimate that could arise from persistence, if there be an overestimate from this source, does not exceed 5 parts in 10,000. The maximum underestimate, if there be an underestimate, does not exceed 13 parts in 10,000. With respect to the material in the inventory, the maximum overestimate that could arise from persistence, if there be an overestimate from this source, does not exceed 2.4%. The maximum underestimate of the material, if there be an underestimate, does not exceed 9 parts in 1000. With respect to the conversion costs, active plus inactive, the maximum overestimate that could arise from persistence, if there be an overestimate from this source, does not exceed 2.6%. The maximum underestimate of the conversion costs, active plus inactive, if there be an underestimate, does not exceed 15%. With respect to the unrealized earnings, the maximum overestimate that could arise from persistence, if there be an overestimate from this source, does not exceed 11.7%. The maximum underestimate of the unrealized earnings, if there be an underestimate, does not exceed 6.6%.

I recommend that you accept the results of the sample as figures whose reliability is objectively evaluated in the statements contained above.

Example 4

For a 4th example, the reader may turn to page 162 in Chapter 9, for a statistician's statement to the comptroller of a company which took their inventory of materials in process by sampling methods, to replace the weighing or counting of every lot of material as they had done it heretofore. This example could become a legal document, as it is part of the financial statement of the company and is subject to challenge by any stockholder.

EXERCISES

1. *a.* What should one try to achieve in the design of a sample?

b. Could you hope to accomplish these aims without theory?

2. Give an example of a statement of a problem and the corresponding universe of study.

3. How do the tabulation plans help to clarify the definition of the universe?

CHAPTER 2

Some Remarks on the Theory of Sampling

It is noteworthy that the etymological root of the word *theatre* is the same as that of the word *theory*, namely a view. A theory offers us a better view.—Raymond J. Seeger, in his retiring address as president of the Philosophical Society of Washington, *Journal of the Washington Academy of Sciences*, volume 36, 1946: pages 285–293; page 286 in particular.

What is sampling? Sampling is the science that guides quantitative studies of content, behavior, performance, materials, and causes of differences. Our quantitative knowledge of nature is based on theory combined with studies of small batches of material, some of which may be good samples, some bad. Even if we study a complete census, or 100% of a crop in a field in 1960, or all the accounts for a year, or inspect 100% of a month's product, we must interpret the results as one of the samples that the cause system can and will produce, if we hope to reach sensible answers to our problems.

This is true of everyday life and of scientific research as well. As Cochran* says in his book, a person's opinion of an institution that conducts thousands of transactions every day is often determined by the one or two encounters which he has had with the institution in the course of several years. The traveler who spends 10 days in a foreign country and then proceeds to write a book to tell the inhabitants how to revive their industries, reform their political system, balance their budget, and improve the food in their hotels, is a familiar figure. In every branch of science we lack the resources to study more than a fragment of the phenomena that might advance our knowledge.

What is probability sampling?† The special feature of probability

---

* William G. Cochran, *Sampling Techniques* (Wiley, 1953): p. 1.

† The term probability sampling and its definition originated with the author in the book *Some Theory of Sampling* (Wiley, 1950): p. 10.

sampling is that it permits use of the theory of probability for the computation, from the sample itself, of probability limits of sampling variation in the estimates that come from repeated application of the prescribed sampling procedure to the same equal complete coverage (p. 50), or to the same cause system.   The theory of probability (called in such applications the theory of sampling) leads to explicit formulas for the expected value of an estimate and for its sampling variation.   The same theory enables us to select, from various feasible sample-designs, the one that will deliver the most information per dollar.   Then, it goes further and helps us to govern in advance the magnitude of sampling variation for a given probability level, by use of the sample-design selected.   The aim of this book is to assist study of the theory of probability sampling.

A probability sampling design specifies definite rules for the selection of the sample, invariably by the use of random numbers.   No substitution of one unit for another is permissible.   As we shall learn in Chapter 3, the sampling procedure also prescribes definite rules for the formation of the estimate from the results of the sampling, and rules for the formation of the margin of sampling variation of this estimate.   Departures from the prescribed procedure, if serious, will impair the usefulness of these calculations.

Sample-surveys and design of experiment by use of the theory of probability are reliable and indispensable tools for use in production, administration, accounting, marketing and consumer studies, census and government statistical series, and research in many fields.

Mixing and randomness.   If the units in a lot of material are all very much alike, then one unit will give almost the same result as another. If we stir the soup first, we may taste some off the top or from any other layer, because the taste, after some mixing, does not change much from one layer to another.   In other words, a crude sample is good enough.

The sample must be bigger if the units are unlike, although we are sometimes able to group them or to use prior information in such manner that the variation between units does not cause high variability in the estimate formed from the sample.   Theory gives us objective and usable measures of like and unlike.

Material will sometimes appear to be thoroughly mixed when it is not. Actually, in practice, material never comes ready mixed.   It comes in layers, by geographic position or by order of production.   A company that buys or sells methanol goes to a lot of trouble to agitate the methanol and to take samples from various layers of the tank.   This may astonish some people, who would suppose that a substance like commercial methanol would be uniform from top to bottom.   One might suppose that a sample of blood could come from ear, finger tip, or toe, but the fact is that the blood has different properties in different parts of the body.   It is important to know the purpose of the sample and where to get it from.

The statistician does not assume that material comes ready mixed. Instead, he turns into good profit, by use of zones or other natural strata, the very fact that the material is not mixed.   Moreover, when he wishes to mix the material within zones, or within strata, for random selections, he uses a table of random numbers, and thus achieves a standard degree of mixing (see the section on randomness, page 54).

Why study theory?   The aim in the design of surveys and experiments is to get useful information at the right time, and to hold the sampling error to an acceptable margin, to *avoid a wild shot* that will lead someone to a wrong conclusion or to a wrong decision.   These aims require the use of theory.

Theory teaches us (*a*) what we need to know about a material in order to sample it properly; and (*b*) about what precision to expect from a proposed sampling plan.   With the aid of theory, one can discover a number of sampling procedures, all valid and all able to deliver the required precision, but with differing costs.   A rational choice is then possible on the basis of cost and smoothness of operation under the existing conditions.

A theoretical statistician is one whose practice is guided by theory. Without theory, no rational choice of procedure is possible.   One can transfer theory and principles from one problem to another, but not procedures.

The statistical principles and even the formulas are remarkably the same in all types of applications.   It is only the material that is different. This means that the sampling units and other aspects of the procedure may be different.   Yet one does not need one theory for medical problems, another for physical materials, another for auditing, another for consumers' attitudes, and some other theories for census information.

Every new survey presents opportunity for a nicety of balance between the possible choices of boundary, size, number, and dispersion of the segments in a sampling unit.   Should the segments be confined to a county or to a small area, or dispersed without restriction?   Every new survey presents opportunity also for new ideas in stratification, in number of replications, in the choices of the procedure of estimation, and in the probe of the nonsampling errors (Ch. 5).   The precision aimed at must be sensible in view of the uses intended, and in view of the possible nonsampling errors.   A plan that calls for care and skills beyond the capability of the people who will work on the survey may run afoul of the instructions and yield nonsampling errors of embarassing magnitude.   The choice of plan must therefore receive even more regard for smoothness of operation than for efficiency.   Only knowledge of theory, experience, and knowledge of men and of materials, can lead to a suitable choice of plan.

Two types of error in design.   No sample-design is right or wrong except in terms of the uses of the data to be obtained.   Statistical theory helps to avoid 2 types of error:

1. More precision than the purpose requires;
2. Insufficient precision for the purpose.

In the 1st error, the survey will cost someone more money than was necessary.   In the 2d error, which may be worse, the survey is too small and fails to achieve significance.

Use of statistical theory holds the net economic losses from both errors to a minimum.

The responsibility for the precision to aim at is actually not statistical, as it belongs to the subject-matter (business, demography, agriculture, marketing, page 11).   However, the statistician must lend his services in this step, especially in the computation of the costs of various degrees of precision.   If a standard error of 15% is good enough, this should be the aim of the sample-design, as 10% may be considerably more costly without being more useful.

Some advantages of sampling:

a. Improvement of an entire statistical program, through clarification of aims and purposes.

b. Improved reliability.   Sampling possesses the possibility of better interviewing (or testing), more thorough investigation of missing, wrong, or suspicious information, better supervision, and better processing than is possible with a complete coverage.   (See page 163 for an example.)

c. Precision governable, tailored to the requirements.

d. Speed (and hence greater utility of the data).

e. Low cost (permitting expansion of the statistical program, and expanded usefulness).

f. Reduced burden of response.

Small samples are usually cheaper than a complete census or a complete inspection would be.   However, economy by itself, without improved quality of information, is a poor argument for sampling.   No self-respecting executive or scientist is interested in trading savings for reliability.

In most respects, the ingredients of quality are the same whether the survey be a complete coverage or a probability sample.   The necessity for a frame, the problems created by any gap between the frame and the universe (next chapter), the questionnaire, the procedure of interviewing or of testing, nonresponse, missing information, wrong information, the coding, the interpretation, and the judgment of the expert on the usefulness or the propriety of the data for the problem at hand, are the same for a sample as for an equal complete coverage.

The quality of the interviewing, supervision, coding, and other work in a sample-survey CAN be superior to the quality in a complete coverage, PROVIDED one takes advantage of the smaller size of the job.   Quality does not just happen: it is built in.   One should not boast too loud about the accuracy of a sample-survey, and of its superiority over a complete coverage, without giving reasons.   Some complete censuses are amazingly complete.   The Census of Japan, for example, has traditionally been good. There have been remarkable improvements in the censuses in the U. S. and in many other countries during the past 2 decades, owing especially to recognition of defects; then to improved training and better administration through the use of sampling.   With all the opportunity that sample-surveys have to be superior, the sober fact remains that few of them here measure up to the Census of Population in the U. S. with respect to completeness of enumeration of people of all ages.   This is amazing, as a complete census is an enormous undertaking.   In the U. S., it requires 220,000 workers, many of whom are holders of political appointments, who never took a Census before and never will again.   A stiff, skilful, uniform training of 3 days emphasizes the prime importance of finding dwelling units and counting people.

It is not unusual for a sample to turn up deficient in certain age and sex classes (e.g., mobile males, 18–25).   This does not necessarily indicate carelessness.   It means that few people know how difficult it is to count all the people.   Moreover, it is possible in a Census, one should remember, to count people that are not at home, and even to get much important Census-information about them, by asking the neighbors; whereas in a sample-survey, where the purpose is not Census-information but personal information NOT provided by the Census, it is usually necessary to find the person himself to get the answers.   It is of course not unusual to find that a sample measures up well against the Census-count of dwelling units in an area (pp. 17, 185, 208); and many samples even measure up well in the mobile age and sex classes. This is due to skill and perseverance.

Speed is important because the social order is not static but dynamic. Census and sample are therefore both to some extent out of date by the time the data are processed and published, but the sample has a distinct advantage, as its time-interval is measured in days, weeks, or months, against many months or even years for a complete census.   For this reason, in many types of enquiries, the complete census, by the time it is processed and tabulated, is often not as good a basis for action as the earlier returns of a small sample would have been.   To the extent that sampling methods produce quicker results, they therefore often produce more accurate results for purposes of action and for prediction.

It is remarkable that the effect of the introduction of sampling into the management of any business, or into a government statistical system, has invariably been immediate improvement in the operations, and more accurate and timely statistical information.

Some disadvantages of sampling.   Sampling does not give information on every person, nor on every dwelling unit, nor on every farm, business establishment, etc.   Sampling does not provide information for action with respect to individual account.

Sampling produces results that contain the errors of sampling.   This is a disadvantage if the error of sampling is too big for some purpose that one has in mind.   Sampling errors are not a disadvantage if they are small enough to be innocuous for the purpose.   A sample should be so designed in the planning stages that it is sufficiently precise for stated uses.

Clarification of some common misconceptions about sampling:

1. Probability sampling is NOT the substantive expert's selection of "representative" or of "typical" cases, areas, or farms, or of weeks or months from the year.   Instead, the selection of the sampling units is accomplished by means of a standard tool known as a table of random numbers.   When the selections are made by judgment, inferences may be made only by judgment, not by the theory of probability.   True, one sometimes sees unjustifiable calculations of standard errors and unjustifiable inferences in papers on medicine and agricultural science, in accounting, and in marketing and opinion research, when the samples are not random. Such calculations have no meaning; and if they lead to a correct answer, it is so only by luck.

2. Expert knowledge, judgment, sincerity, and honesty, are all necessary ingredients of any science, but they are not sufficient to make a sample. There is no substitute for the use of statistical theory.

3. The size of a sample is no criterion of its precision, nor of its accuracy, nor of its usefulness (Ch. 5).   The procedure of stratification, the choice of sampling unit, the formulas prescribed for the estimations, are more important than size in the determination of precision.   Once these features are fixed, then as we increase the size of a sample drawn with random numbers, we gain precision (though the point of diminishing returns comes rapidly).

4. One can not offset the hazard of nonprobability sampling by increasing the size of the sample.

5. The number of sampling units in the frame that the sample was drawn from conveys no information in regard to the precision of the sample. The proper size of sample for a universe of 100,000 accounts might be identical with the size for a universe of 1,000,000 similar accounts. I often prescribe the size of sample and its procedure before the company has furnished me with a count of the number of items in the frame whence the sample will come. The only exception occurs in cases where the sample is 10% or more of the frame; the precision of the sample is then improved by the small number of units in the frame (p. 387).

6. The quality of a statistical sample is built in. It is not luck.

7. A sample is NOT a last resort, to be used when a complete investigation is impossible. Rather, it is the first resort: it is the answer to the question, "What is the best way to do the job?" or "What kind of sample will do it, and how big must it be to deliver the information that is required?" Viewed in this way, sampling can be of inestimable help to research, management, and industry.

8. The uncertainty that arises from the use of modern sampling procedures in place of a complete examination of every item in the frame is an objective measure, and is not a matter of opinion nor of expert judgment.

9. Many uncertainties arise from failure of the expert in the subject-matter (accountant, labor expert, engineer, economist) to formulate questions or methods of test that are suitable to the solution of the problem, or to specify ways in which to procure the information, or to exhibit it, or to provide a suitable frame (list); and finally, from failure to interpret the results in accordance with the best knowledge of the subject (Ch. 5). The errors of sampling are but one of the sources of uncertainty in statistical data.

10. The theory and use of sampling apply as well to studies that involve subjective measurements and opinions, or visual judgment, as they do to studies where the measurements are quantitative. The type of measurement is a property of the equal complete coverage, being inherent in the questionnaire or method of test. It has no connexion with sampling. The symbols in the theory of sampling don't care how the observations are made.

11. Good agreement or poor agreement between the results of a sample and of a complete census taken at about the same time does not of itself establish or disprove the precision of the sample, nor the quality of the complete census.

Remark 1. On one occasion in a legal hearing, a well-known engineer gave it as his "considered opinion" that sampling would be entirely valid if the measurements on the items of the sample had been made by calipers and expressed in inches or in millimeters; but that if the test of an item is subjective or visual, based on appearance, sampling is not applicable.

Suppose that the sample had been 100%. Would his argument apply? No. Neither did it apply to a sample less than 100%.

One should regard a sample as a *sample of the labels* that might be tacked on to the items in a complete 100% examination. If the method of test or of acquiring the information or of passing judgment on a single item is not satisfactory, NO SAMPLE, NOT EVEN A COMPLETE COUNT, will give useful results.

Remark 2. A committee of top executives from a number of companies in a certain industry, well educated and able in their fields, chose recently to study the records in their respective companies for the month of July 1958 as a "typical month" to provide legal evidence concerning the amount and character of their business in 1958, in a case that threatened their industry. Was this a sample? Of what? Of 1958? There is no way to calculate limits of uncertainty, by use of probability methods, on any inference that one could draw concerning 1958, from a study of July. A judgment sample of 4 months of the year, January, April, July, and October would be little better, in spite of its size and cost. In contrast, a probability sample of half the size of July alone, drawn from all 12 months of 1958, produced data of known and demonstrable reliability.

Remark 3. In another instance an accountant wished to learn something about 100,000 transactions. He opened a textbook and drew a sample, and examined the transactions therein. He was disappointed because the sample contained no carloads of alumina, yet the company's records contained a few such transactions amongst the 100,000. The fault lay in the planning of the job. The accountant failed to state what classes he expected to get data on, and to design a sample that would deliver what he required. There is no cook book on sampling. The first step in the design should have been to enquire what kinds and sizes of transactions were important, and what were their relative frequencies of occurrence. If data on alumina were desired, these transactions should have been segregated and sampled separately. If the cost of such segregation is great, the cost is chargeable to the system of filing the records, and not to the sampling thereof.

Sample-design embraces a complete count as a special case. Choice of the proper sample-design demands consideration of all possible plans, including a complete coverage (a 100% sample). The choice settles on the plan that should deliver the required information with the required speed and accuracy, at the lowest total cost compatible with the human skills available. I have on several occasions, when consulted for a sample-design, recommended a complete coverage, which I believed to be the best way to get what was wanted. Examples are likely to be found when the frame to be covered is all the manufacturers of some special product, such as tires (which in the U. S. one may cover almost completely with 4 manufacturers), steel ingots, aluminum ingots; or the members of some small scientific society; or the advertisers in some journal. Another example sometimes occurs when one requires information for small areas or in fine classes, or for a small number of accounts.

Some remarks on nonprobability sampling.   Judgment samples.   This book deals with the methods of probability samples, but I believe that it is well to mention the existence and occasional utility of judgment samples, deliberately biased samples, and chunks.   A judgment sample is one in which an expert in the subject-matter makes a selection of "representative" or "typical" counties or other areas or business establishments.   For an evaluation of the reliability of such a survey, we must rely on the expert's judgment: we can not use the theory of probability.   In contrast, the precision of an estimate made from a probability sample is never in doubt, as the probabilities associated with any given margin of error one estimates by formulas directly from the sample itself.

There is another kind of judgment sample called a quota sample.   The instructions in a quota sample ask the interviewers to talk to a specified number of people of each sex and age, perhaps by section of the city, perhaps by economic level.   The report of the results usually boasts of good agreement between the sample and the census in respect to the classes specified, but what does this mean?   It means that the interviewers reported what they were supposed to report concerning these classes; it proves little or nothing with respect to the accuracy of the data that constitute the purpose of the study.

There is no way to compare the cost of a probability sample with the cost of a judgment sample, because the two types of sample are used for different purposes.   Cost has no meaning without a measure of quality, and there is no way to appraise objectively the quality of a judgment sample as there is with a probability sample.

Remark 4.   One will occasionally see judgment samples that are large and costly, under the excuse that only a rough answer is needed, and that a judgment sample is sufficient.   This may be a wasteful procedure, as a probability sample, tailored to low precision, may actually be cheaper, and it entails no risk of unknown precision.   It is wrong to assume that a probability sample is necessarily precise.   In good practice, a probability sample is tailored to the requirements, which may be high precision, or may be very low precision.   The essential point is that, be the precision high or low, it is *known*.

In a recent instance in my experience, a company discovered loss of $4,000,000 annually through wrong figures from a mass judgment sample. A small probability sample cost no more and discovered the discrepancy.

Remark 5.   If a sample must be confined to only 1, 6, or 10 units (blocks, tracts, cities, counties, farms, pieces of material), a judgment sample would be preferable to a probability sample.   In such very small samples, the errors of judgment are usually less than the random errors of a probability sample.

Remarks on nonprobability sampling, continued.   Deliberately biased samples. There are circumstances in which one wishes to discover a maximum or a minimum, or the worst possible conditions that his interviewers will meet, or the worst possible spots in a proposed frame. For such purposes, a biased selection may be preferable to a broader coverage by a probability sample.

I might cite an example. I was about to use a frame which consisted of pieces of property listed on the tax assessor's records. His records showed how many dwelling units were in each piece of property. There was good reason to suppose that these records were complete, except for new construction not yet taxed. A few tests showed that most buildings not yet completed were already on his list, and there seemed to be no omissions of ordinary existing property.

The question then arose whether some dilapidated properties, perhaps not now taxable but containing dwelling units, were still on his list.

To test this question, I brazenly assumed some knowledge of the subject-matter and suggested to the client what he thought of the idea of selecting several blighted areas in the city, and some shacks in the surrounding country, with the thought that if the assessor's records were deficient anywhere, they should be deficient in areas of this kind. The client agreed. Tests of a dozen of the worst areas showed no omissions, whereupon the client concluded that it was safe to use the assessor's records as a frame.

A deliberately biased selection is sometimes the answer to studies of special problems that experts in the subject-matter believe to be isolated in one area or another. For example, a company was having special trouble trying to sell its product in Phoenix, and they asked me to design a national sample of dealers. The proper procedure, it seemed to me, was first to study Phoenix, maybe by the aid of a probability sample of dealers in Phoenix. The point is that a probability sample of the whole country would not be a big help in that stage of the problem. If one wishes to study some of the sociological problems in a blighted area in Philadelphia, one should go to this area and study it, possibly by aid of a probability sample of dwelling units within the area. A probability sample of dwelling units of all Philadelphia might not be the best approach.

Remarks on nonprobability sampling, continued.   Chunks. A judgment sample is planned by an expert in the subject-matter, and one may hold him responsible. In contrast, there is another type of selection called a *chunk*,* which is a hazard for generalizations, but a refuge for exploratory

---

* The term chunk was coined by Dr. Philip M. Hauser in 1939.

studies in time of trouble. It is simply some portion of the universe that happens to be handy, such as—

The first 1000 questionnaires or forms that come in.

A certain city, selected mainly because the surveying organization has a field-force there, not yet disbanded from a previous survey.

Any group of people that happens to be handy for test (a class of students, for example).

Interviews of "average people" on street corners.

A low response from a mailed-questionnaire, or any other questionnaire.

The returns from a clipping in a journal.

The sacks of cement on the top of the pile.

We all take chances now and then on a chunk, and take the consequences. We look at a few apples or potatoes on top and decide whether to buy the box. If the grocer has put the big ones on top, he has given us a bad sample, and we settle our sampling problem by going elsewhere next time.

There was a difficult medical study to be carried out by sampling methods to cover a sample of a few hundred households over the State of New York. To test the first version of the questionnaire, I elected to work on a chunk of a dozen people not far from the headquarters at the Medical Center in the City of New York, to cut down the cost of travel and to increase the convenience of any recalls that would turn out to be necessary. In fact I did not even make the selections myself, but asked the girl who was constructing the frame to pick out 12 addresses not far away, 6 men and 6 women, 3 of each under 30, and 3 of each over 30. This chunk served a purpose.

It is remarkable and well to remember that some of the important gains in the arts of questioning and interviewing have been made with judgment samples, biased samples, and even with chunks. One can often explore the difficulties of a proposed questionnaire and field-procedure, and the problems of training, supervision, coding, and tabulating, in an area chosen by judgment for the express purpose of encountering typical conditions or extremely adverse conditions that will throw the spotlight on the deficiencies in the instructions and plans. The "dress rehearsal" for a full census is conducted in a county that is chosen because it will contain both urban and rural areas and will provide most of the problems, average and difficult, that the big census will present.

EXERCISES

1. Show that the following samples are judgment samples, not probability samples. Why aren't they probability samples?

a. A judgment selection of areas, followed by a random selection of dwelling units from within these areas.

b. A judgment selection of 60 stores, followed by a random selection of 20 of them.

c. A selection of areas by probability methods, followed by a quota selection of people within these areas.

2. What is the chief characteristic of a probability sample?

3. Name 2 faults that one may commit in the design of a sample. Can you avoid both of them? How does theory help you to minimize the net cost of both errors?

4. *a.* What are some of the advantages of sampling?

b. What are some of the disadvantages of sampling?

5. *a.* Is a sample a substitute in every respect for a complete count?

b. What advantage has sampling over a complete count?

c. What advantage has a complete count over a sample?

6. Why is it that careful studies of income, savings, purchases, mental retardation, employment and unemployment, opinions, attitudes, reading, listening, and many others, are carried out by samples and (except for questions on employment and unemployment) not in a complete census?

7. *a.* Is sampling used to count the number of people or of pigs in a large area? Name examples.

b. Is it economical to use sampling to count the number of people or of pigs in a very small area? Why is this?

8. Why are comparisons hazardous between samples, and between a sample and a census of several years ago?

9. Does the number of interviews in a survey give by itself a measure of the

a. Accuracy of a result?

b. Precision of a result?

c. Usefulness of a result?

Give your arguments.

10. A manufacturer of pharmaceuticals wished to consult the owners of all 2000 drug stores in a certain city with regard to their opinions on the most marketable sizes and prices of several medicines. The procedure was fortunately to visit the 400 big stores first; then to start in on the smaller ones. After the interviewers had visited the 400 big stores, and 465 smaller ones, so much time and money had already been consumed that the manufacturer decided to call the survey to a halt, and to examine the results thus far obtained to see what he had learned. It is important to remark that the interviewers went to the small stores at their convenience. Criticize this procedure.

Hints

1. Fortunately all the big stores were covered, wherefore it is possible to study their returns, to learn something about the big stores. However,

it is difficult to make comparisons with the small stores, or to draw any definite conclusions about the small stores, because they were not sampled properly.

2.   The money that was sunk into this particular survey netted little but the returns from the big stores.   The rest was mostly lost.

3.   For half as much money, or less, and with greater speed, the manufacturer could have taken a proper sample, consisting of perhaps 100 big stores and 100 smaller ones, both drawn at random with geographic stratification.

11. The following remarks occurred in a debate in the House of Representatives concerning the appropriation for the Census of Population and Agriculture, *The Congressional Record*, vol. 96, 21 April 1950, p. 5599 (House) 81st Congress—

> Mr.*.... Now if the Government needs this information for some important purpose it can be very conveniently obtained from the Bureau of Internal Revenue.   The information from the Bureau of Internal Revenue will incidentally be much more accurate.   As the distinguished chairman stated, they ask every fifth person.   This alone makes the information quite inaccurate.   For example, in our district there is a case where an enumerator thought you just asked each fifth person, so he went to a family that happened to have five people.   The fifth one happened to be a 4-year-old child.   So the enumerator marked that person down as, "Unemployed; no income."   In that particular case the head of the family had a pretty fair income. . . .

Wherein lies the error in the Congressman's reasoning?   How would you explain to him why it is necessary to stick to the sampling procedure? What would happen to the results of the sample if all Census enumerators changed the rules every time they encountered a family "with a pretty fair income," and juggled the order of enumeration so as to bring this income into the sample?   (*Note:* The Census-procedure in 1950 automatically selected 1 person in 5 for questions on income and for certain other characteristics not required of the other 4 out of 5.)

Why might it not be satisfactory to get information on personal incomes from the Bureau of Internal Revenue?   (Answer: The Bureau of Internal Revenue can not furnish information on employment, unemployment, seeking work, rent, and the like, nor on family-size, all of which are important in a study of income.   Moreover, below certain levels of income, there is no information at all in the Bureau of Internal Revenue.)

Why were Congressmen expressing views on a technical question?

12. The following questions arose in the design of a medical study, for which twins are known to be suited admirably,† as they supply for many purposes many times the genetic and environmental information that can be obtained from a random selection of people in general (not necessarily

* Name withheld.

† Franz J. Kallman, *Heredity in Health and Mental Disorder* (Norton, 1953).

twins).   There were 90 pairs of twins (some identical, some not) born of parents who were members of the Health Insurance Plan in New York in the year 1955.

a. Would these 90 twins serve as a sample for studies of differences between the characteristics of identical twins and nonidentical twins? The purpose of the study was medical, and was not to obtain proportions applicable to the population as a whole.

> My answer is that this decision must rest on the basis of medical knowledge.   It is not one that statistical theory alone can solve.   Statistical theory can, of course, make tests of a proposed frame and material outside the frame, with the aim of providing comparisons that will assist judgment.

b. (For discussion.)   How else would you find a sample of twins for such a study?

13. *a.* Does size alone determine the precision of a sample?

b. What else is important?

c. Is the number N of sampling units in the frame important?   (Hardly ever: see page 387.)

d. Under what conditions will an increase in the size of the sample yield more reliability?

14. A public utility company in a large city wished to know the proportions of various types of apparatus that they own on customers' premises. An official of the company proposed that the repairmen, when called to customers' premises to make a repair or change in installation, make an inventory of all the apparatus that he finds there.   A summary of all the inspectors' reports, at the end of a year, would, he thought, provide the required figures.   What was wrong with the plan?

15. An accounting firm decided to test out probability sampling for evaluating the inventory of a large manufacturing company.   They proposed to use 3 methods:

1. A complete count of all lots.
2. A judgment sample by their previous time-honored procedures.
3. A probability sample.

Then to choose, for their future practice, Method 2 or 3, depending on which came closer to the result of Method 1.

a. Explain why it is that the comparison would mean nothing.

b. If the 2 samples agreed, what could you conclude?   (Nothing.)

c. If the probability sample differed with the complete count by more than 3 standard errors, what would you conclude?   (Experience shows that 99% of such disagreements arise because the complete coverage suffered miscounts and different application of the definitions.)

d. Explain the fundamentals of sample-design by probability methods—
(i) statement of the acceptable sampling tolerance; (ii) design to meet it at
lowest cost; (iii) evaluation, from the results themselves, of the sampling
tolerance actually reached. Can you predict the performance of a
probability sample? (Yes.) Of a judgment sample? (No.) Then
show that the above proposal is incompatible with good business
administration.*

The best plan is to design a probability sample to meet the purpose, and
then to carry it out carefully, and to have confidence in the results. It is
poor administration and poor statistical practice to dissipate energy, skill,
and funds on tests and comparisons that will jeopardize both the complete
coverage and the sample.

* My colleague Professor Ernest Kurnow of New York University kindly contributed
this exercise.

CHAPTER 3

The Frame and the Elements of a Sampling Plan

> I've got a little list—I've got a little list . . .
> I've got *him* on the list . . .
> They'd none of 'em be missed.
> —Gilbert and Sullivan, *The Mikado*.

The five elements of a sampling plan. Once there is a decision to go ahead on a survey, and the aim is clear, and the frame also, then the sampling plan itself presents 5 elements to the theoretical statistician. The rest of this book is mainly an elaboration of these 5 elements.

1. The frame (the subject of this chapter).

2. The procedure for drawing the sample of sampling units from the frame (with or without stratification, but invariably with random numbers: see the definition for randomness on page 54).

3. The formula by which to calculate from the returns of the sample an estimate of the numerical value that the equal complete coverage (p. 50) of the whole frame would give for some statistical characteristic of the frame. This estimate, along with a measure of its reliability, is the aim of the survey.

An example of a formula for an estimate is $X = N\bar{x}$, which will give an unbiased estimate of the total x-population A of a frame, provided that $\bar{x}$ is an unbiased estimate of the average x-population per sampling unit. Another example of an estimate is $f = \bar{x}/\bar{y}$, the ratio of the x-population to the y-population. There will be many examples later on.

The procedure for the formation of an estimate is as essential to the sampling procedure as the method of selection. No sampling plan can be judged as biased or unbiased, adequate or inadequate, purely on consideration of the method for the selection of the sampling units from the frame.

4. The formula for calculating the standard error of any estimate; also the bias of any estimate (p. 425).

An example of an estimate of a standard error is $\hat{\sigma}_{\bar{x}} = \sqrt{[\Sigma(x_i - \bar{x})^2/(n-1)]}$, page 437.

38

5. An audit or control to evaluate the nonsampling errors and their possible effects on the final estimates.   See Chapter 5 for more details on the audit.   See Chapter 9 for an example.

Information on probabilities is not sufficient.   A statement concerning the probabilities of selection of the sampling units does not describe the procedure of selection, and does not supply the information that will enable one to calculate the margin of sampling variation.   A probability of .01, for example, could mean the selection (by random numbers between 01 and 00) of 1 person in every consecutive 100 names on a list, or it could mean the selection of 1 line in every consecutive 100 lines.   It could mean the selection of 1 segment from every 100 consecutive segments, or it could mean the selection of 2 segments from every 200 segments.   It could have many other meanings.   The probability of selection tells us nothing about the stratification that is built into the plan, nor whether the selection was carried out with or without replacement.

The probabilities of selection are important, but to compute the margin of sampling variation, and to compare one plan with another, WE MUST KNOW MORE ABOUT A PLAN THAN THE PROBABILITIES OF SELECTION. WE MUST KNOW ALSO THE PROCEDURE BY WHICH TO DRAW THE SAMPLING UNITS, AND THE FORMULA OR PROCEDURE BY WHICH TO CALCULATE THE ESTIMATE. We shall see, as we move through the book, that every sampling plan has its own formula by which to calculate the margin of sampling variation.

The frame.   A frame is a means of access to the universe or to a sufficient portion thereof.*   A frame is made up of sampling units, and this is true whether the survey is to be conducted by a complete coverage or by a small sample.   A sampling unit may contain dwelling units, people, firms, farms, business establishments, industrial or agricultural product, accounting records, or other material.   In the sampling of industrial product, the sampling unit may be a case, layer, or individual article, or the material between certain lines on a conveyor belt, at a certain random instant.   Every piece of material covered by the frame will belong to one definite sampling unit, or will have an ascertainable probability of belonging to any given sampling unit.   Without a frame there can be neither a complete coverage nor a probability sample, as there would be no way to lay out the work nor to know the probability of selection of any sampling unit.   The material in all the sampling units of the frame is the material presented for study, whether by a complete coverage or by a sample.   A sample of all the sampling units in the frame

* See the reference to Stephan on page 9.

is by definition a complete coverage of the frame (a 100% sample). This is true whether the frame covers all the universe or not.

A Census list of bounded areas is the usual starting point in a survey of human populations, and is an excellent example of a frame. A map, city directory, list of blocks (see Chapter 11 for use of the "block statistics"), list of segments prepared by an enumerator on the job (Ch. 12), are all examples, and one may use one or several such frames in one survey. A list of debtors to a department store, a card file, a list of parts used in an automobile, the manifest on a ship, a list of members of a trade association, a list of the members of the American Statistical Association, a list of business establishments which one can purchase, a list of drug stores in the classified pages of a telephone directory, a list of license plates issued to owners of automobiles, all the actual and possible telephone numbers in all the telephone offices owned by a telephone company for an inventory of telephone apparatus, are further examples.

Some of the frames just mentioned cover the universe completely, or practically so, the manifest on a ship being an example. Others leave a gap, a portion of the universe not covered, concerning which we have a word to say later on.

In any event, every sampling unit in a frame will bear a serial number, or will have a prescribed way of getting one. A random number thus selects a definite sampling unit, and leads to the investigation of all or of a sample of whatever material therein belongs to the universe.

The frame exists sometimes in the form of a set of rules for the creation and definition of a sampling unit, when and as needed. In surveys for the yield of a crop, the sampling unit may be a small plot of ground; 2 random numbers for the x- and y-coordinates locate a sample-plot of ground of specified size and shape on which to cut, harvest, and weigh the crop.

The one basic requirement of a frame, without which there is really no frame at all, is that it must show a definite location, address, boundary, or set of rules by which to delineate and to find any sampling unit that the random numbers draw. A frame is especially useful for sampling if it shows also for each sampling unit some further statistical information that will permit use of stratification or ratio-estimates. The reader may turn at this point to Exercise 1 on page 47 for an illustration of a list that was not a frame.

Some sampling units in the frame may be blanks, such as a line on a ledger which contains no entry of interest in the study, or a farm with no wheat in a study of the yield of wheat. If a random number draws a blank, we go on to the next random number, and make no substitution. Blanks create no problem unless they are so numerous that we spend most of our time and money delineating sampling units and knocking on doors only to find no member of the universe therein. This is one of the problems in sampling for rare populations.

A combination of frames is often useful. For example, in the sampling of business establishments, there might be 2 frames:

1. A list that contains most of the big firms, with addresses, in which the sampling unit is the firm.

2. A list of segments of area, to pick up a sample of firms not on the list.

The sample might be (e.g.) 100% of the list of big firms, plus a small sample of the segments of area. In the sampling of records, the frame might be the cards in a file, and the sampling unit might be a card or several successive cards. The frame might be the pages and lines of a ledger, and the sampling unit might be a line or a group of lines or a whole page in a ledger (p. 111).

The frame in the design of an experiment, or in the test of a product, or in a comparison of one product against another or against several, would be a list of the types of product to test, and in fact a list of a number of items of each type. In a manufacturing process the frame would include the levels of concentration of one or more chemicals, specified rates of flow to test, specified temperatures, and other experimental conditions.

In analytic studies, which constitute most studies of the consumer and of test of product and of tests of process and methods, the frame is never complete, because use of the results is in reference to future consumers, future products, future processes.

As we have learned already in Chapter 1, the frame is satisfactory if, when canvassed 100%, it provides results useful for inferences with respect to the universe. The decision whether the frame covers a sufficient portion of the universe is the client's responsibility (p. 16).

The application of the results of a complete count or of a survey that cover a given frame or set of frames must be carried by judgment to the remainder of the universe. There is no statistical theory that will objectively bridge this gap.

Careful statisticians have long distinguished between the universe and the frame, and they have implicitly concerned themselves with the gap between the frame and the universe. An explicit statement of the possible existence of the gap, and of the responsibility for the decision on whether a frame covers a sufficient portion of the universe, is, I believe, important for successful practice.

Fig. 1 is a schematic diagram that portrays the material that the universe covers, part or all of which lies in the frame. The portion of the universe that the frame fails to include, if any, is the gap between the frame and the universe.

It is often very difficult to find a suitable frame for a universe of people or of business firms that possess some rare characteristic, such as people who lead nearly normal lives but suffer from some rare disease.   Other examples are (1) people of high income; (2) dwelling units that contain a certain type of air-conditioning equipment; (3) men who make decisions for their companies on purchases of typewriters, *and* who read certain industrial or business journals.   An example that I met recently: how to find a frame or frames for a study to discover the problems of people who have been totally deaf from birth or from early childhood (about 1 person in 1200), and to estimate the number and proportions of the various subgroups of these people, and to study their problems of adjustment.   In

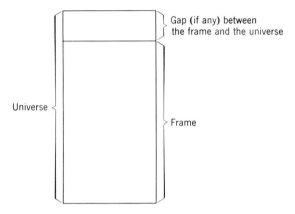

Universe

Frame

Gap (if any) between
the frame and the universe

Fig. 1. Schematic diagram to show the relation between the frame and the universe. The gap is the material in the universe that the frame fails to include.   There is sometimes no gap (see text).

this case, hospital records and pastors were very helpful; their lists and knowledge led to more lists and names.

Sweat, blood, prayer, and tears, plus bright ideas and some good luck, will usually produce results.   There are maps, lists, and specialized lists for sale.   The city engineer can often help.   The gas company has information on how many meters there are in each area.   Lists of voters and of registered voters sometimes serve specialized needs: so do the registrations of children in school.

Remark 1.   I listened to a learned lawyer and economist as he explained how difficult it is to formulate some of the problems of the Federal Trade Commission, especially in cases that involve alleged monopoly or restraint of trade, so that they become statistical—i.e., so that a statistical study (of people, records, sales, revenue) will provide information that will be helpful as evidence.   The trouble was mainly to define the universe, and to find a suitable frame therefor.   There were also problems in finding suitable questions.   One can well appreciate the difficulties that he described, but the title of his speech, "Problems of sampling in certain legal cases," was misleading.   His problems were the problems of the complete coverage, not of sampling.   Once the legal expert formulates a satisfactory complete coverage, the problems of sampling are in most cases relatively straight-forward to the statistician. (See also Exercise 2, page 47.)

The effect of indefinite boundaries of the sampling units.   Suppose that sampling units No. 17 and 18 contain dwelling units or pieces of material within the solid boundaries of Fig. 2.   However, suppose that owing to failure of the solid boundary to be clear, the field-worker or the inspector

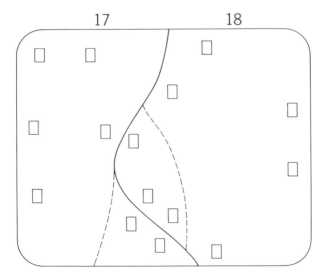

Fig. 2. Sampling units 17 and 18 are 2 adjacent sampling units defined by the solid boundaries. If the solid boundaries are not clear to the worker, he may follow a dotted boundary instead. Ambiguous boundaries, if numerous, will give rise to bias, and this bias does not diminish as the size of the sample increases.

can not be sure which pieces belong to No. 17 and which belong to No. 18. If the random numbers draw No. 17, he will almost certainly warp the boundary (*a*) to include the 3 pieces of No. 18 that lie within the dotted boundary in No. 18; or (*b*) to exclude the 2 pieces of No. 17 that lie within the dotted boundary in No. 17.   If the random numbers draw No. 18, he will exhibit a similar pattern of behavior.   Unclear boundaries will thus lead to noncompensating errors.   Most conscientious workers will include too many pieces, if there is any doubt about what to do.   Hence we see the need for unmistakable identification of every sampling unit that comes into the sample.

Sampling can not overcome the fundamental defects of a frame. SAMPLING is not a magic word which repeated 100 times will make everything right no matter what. Indeed, any hazard that arises from incompleteness in the frame or from any proposed method of covering a sampling unit will persist undiminished, no matter how big the sample, even if it be 100% of the sampling units. If this hazard is bad enough, the frame may not be suitable for either a complete coverage or a sample.

Fractional probability of inclusion in a sampling unit; linkage to 2 sampling units. The reader should not confuse indefinite boundaries with fractional probabilities of inclusion in a sampling unit. Thus, suppose that the 2 dwelling units included by the dotted boundary in sampling unit No. 17 in Fig. 2 are definitely bounded and have a probability of $\frac{1}{2}$ of being in sampling unit No. 17 and a like probability of being in sampling unit No. 18. (Heads or tails—an even random number could include them, and an odd random number could exclude them.) No weighting would be necessary, as their total probability of selection would be the same as the probability of any other dwelling unit in sampling units 17 and 18.

Now suppose that the probability of including the aforesaid 2 dwelling units in sampling unit No. 17 is $\frac{1}{2}$, and that the probability of including them in sampling unit No. 18 is 0. The 2 dwelling units will then come into the sample (1) if the random numbers draw sampling unit No. 17 from the frame, (2) if a 2d play with the random numbers turns up an even digit. Then, if we give to the interviews from these 2 dwelling units a weight of 2, as by duplicating their cards, or by making a carbon copy of their returns, there will be no bias.

The only requirement is that every piece of material must have an ascertainable probability of selection. This can not be if the boundaries of the sampling units are not clear, or if there is any doubt about what material to link with whole or fractional probability to any sampling unit.

We meet fractional probabilities when we select for interview 1 eligible person out of 2 or 3 eligible persons in a family. We make the selection with random numbers and we know the number of eligible persons that the 1 person is selected from: the probability of selection is therefore definitely known for every person interviewed. (See page 242 for instructions by which to select 1 person out of 2 or out of 3.)

Exercise. Suppose that (1) all the sampling units in the frame had ambiguous boundaries, so that some material could be in either of 2 sampling units, like the dwelling units within the dotted boundaries in Fig. 2; (2) all interviewers persistently either included or excluded the portions beyond the dotted boundaries, but that we know nothing about any such mistake; (3) every sampling unit contains 10 pieces, 2 of which are ambiguous.   Show that the expected value of the estimate of the total number of dwelling units, people, or pieces of any specified characteristic will be either too high or too low by 20%.

Remark 2.   This bias remains regardless of the size of the sample.   The only exception may occur in the case of a complete census of any human population, where people included once in the census will complain to an enumerator that comes later and attempts to enumerate them again.   In other words, overcoverage is mostly self-correcting in a complete census of a human population—not entirely self-correcting because the 2d interviewer may find a different respondent at home.   The contrary error of undercoverage is self-correcting only to the extent that people complain that no enumerator has yet visited them.

The reader will sense the great care taken in the instructions in various illustrations in the book to ensure that the sampling units are defined unmistakably.   See, for example, pages 111, 155, and 234, and elsewhere.

Decision on the sampling unit.   Once a frame meets approval of the client (page 16), the size, shape, and definition of a sampling unit are sometimes adjustable and subject to choice, in advance, at varying costs, and with varying degrees of statistical efficiency.   The optimum balance between cost and efficiency is a statistical problem, on which there exists a large body of theory and experience.   The decision on the choice of sampling unit is thus statistical and is the responsibility of the statistician, but he will make the choice not on the basis of statistical efficiency alone: he will concern himself also with the speed, the smoothness of preparation, and with the ability of the investigators to identify a sampling unit that they are sent to investigate.

Examples of various sampling units, together with reasons for the choice, will occur in Part II.   Some of the necessary theory for the choice of sampling unit and for the choice of estimate is in Chapter 19.   Two equally competent statisticians may make choices that differ in some details.   The 2 estimates will differ from each other, just as estimates made by replication of exactly the same sampling procedure will differ.

Use the frame before it is outmoded.   If the creation of a frame is going to occupy a long period of time, it may be smart to start using it while it is still in the process of creation; otherwise the early accessions may have died or moved on before you get any good out of it.   The drawing of a sample need not arrest further work on the frame.   One simply adds the new accessions to the frame, with continued serial numbers, and samples them at the same rate as the others.

Of course one must not start too soon to sample the frame.   If one starts too soon, the traveling costs in running back over the same area time after time to investigate new cases that have come into the frame and fallen into the sample may go out of bounds and exhaust the funds.   The plan requires careful calculation.

If the survey must refer to a particular date, then the frame will have the date in its definition.

Don't spend too much on the frame.   It is a statistician's delight to have a good frame, one with sampling units definitely bounded, complete, and with useful information concerning each sampling unit (for purposes of stratification and for ratio-estimates).   However, one must sometimes proceed into a survey with an incomplete frame, because a better one would cost too much.   In any case, in careful practice, one uses an incomplete frame only if it will give useful results.   A list of business establishments that one may procure (at a cost of say $4000) for a study of the past and intended future purchases of typewriters might cover 96% of the typewriters.   The other 4% may be purchases in small firms, schools, libraries, and other nonprofit institutions.   A possible additional frame for the remaining 4% might be a set of maps, on which one could create small segments of area as sampling units.   These maps might, however, be very expensive.   It might be better to go ahead now and use the list that is 96% complete, and learn how to construct a useful questionnaire, and how to use the data; then, later on, extend the frame if the usefulness of the data warrants the cost.

> The man who pays for the job is the one to make this decision, but the statistician should explain the alternatives, and their advantages and disadvantages (p. 10).

Special example of a frame.   One sometimes requires a bit of ingenuity to see just what he might use for a frame.   For example, the manager of a big department store wishes to discover the proportion of customers that come in during various periods of the day (9 to 10, 10 to 11, etc.) and look around and make no purchases, either because they had nothing in mind or because the sales assistants are occupied with other customers.   It is possible to observe a customer from the time she enters an aisle until she makes her exit therefrom, and to make a record of whatever she does with her time.

There is no list of customers.   What does one use for a frame?   The frame here is the second-hand of a watch.   Each customer is tied to an interval.   A random second will select a random customer.   An interval during which no one enters is a blank.   If 2 or more customers enter in one interval, one may contrive a quick rule by which to select 1 of them at random.   In case of such selection, one has the choice of making an unweighted estimate or a weighted estimate, where for weight one might use the average number of customers for that quarter hour or hourly period. One could tabulate the results by hour to isolate special difficulties. (See Chapter 17, page 398, for an example on the choices between weighted and unweighted estimates.)

EXERCISES

1. I was once called upon by a foreign government to assist in the design of a sample of agriculture. The governor of each prefecture had a list of the owners of big farms, and much statistical information in regard to each farm, but no addresses nor locations for any farm, only the names of the owners. Of course, the governor knew some of the owners personally, but by no means all. Moreover, sometimes several people had identical names. Show that such a list is not a frame, and that we could not use it for a complete coverage or for a sample.

2. Criticize the excerpt below, which is taken from the summary record of a meeting of the UN Statistical Commission. As this record was intended only for use of the participants, I have omitted the date and the name of the man that the report refers to. Moreover, I am not sure whether the confusion originated with the man that the report refers to, or with the reporter himself.

Mr. . . . . (of the UN Statistical Commission) congratulated the Sub-Commission on Statistical Sampling on its report but felt that greater stress should have been laid on the need for common sense in using sampling methods. There are many dangers involved in using random sampling methods. For example, if one took a random sample of farms for the purpose of finding out the number of pigs (for example), in Norway, the result would be inaccurate, he said, because some pigs live in town (the family pig). Thus, the statistician should always keep in touch with reality and should examine closely the conditions in the country concerned before deciding what his sampling unit should be.

Author's comment. The excerpt illustrates some common fallacies: (*a*) confusion of sampling errors with nonsampling errors; (*b*) confusion between the frame and the universe; (*c*) failure to appreciate the fact that the definition of the universe and the responsibility for the frame belonged, in this case, to the officials in the Ministry of Agriculture that needed the count of pigs. Common sense is as necessary for a complete count as it is for a sample. Failure to include the towns in the frame would of course result in failure to count the pigs that live in town. The wrong frame would be the wrong frame whether you take a sample or a complete census of pigs.

The reporter's use of the term sampling unit at the end of the excerpt is perplexing.

3. *a.* Name some examples of universes and of sampling units almost all of which would contain several members of the universe. (One example: a study of the purchases of bread by housewives. A sampling unit that contains on the average 5 dwelling units would usually contain several housewives who had bought bread within 2 days.)

b. Name some examples of universes and of sampling units almost all of which are empty—i.e., do not contain a member of the universe. (One example: a study of readers' preferences and comparisons of 2 magazines. A sampling unit of even 10 or 20 or even 30 dwelling units would contain perhaps 0, 1, or 2 people who could qualify for the universe. The expense of knocking on the other doors is an important share of the total cost of the survey; see Chapter 18.)

4. *a.* What is a frame?

b. What is the one essential function of a frame?

5. *a.* Is there sometimes a gap between the frame and the universe?

b. When is a frame a satisfactory coverage of the universe? Whose responsibility is it to say whether the frame is satisfactory from the standpoint of coverage of the universe?

6. *a.* May there be more than one possible frame for the same survey? For the same area?

b. May one frame be more desirable than another for sampling the same material?

c. What are some desirable characteristics of a frame?

d. Must there be a frame in order to carry out a complete count?

7. Cite 2 possible types of frames for an area in which the universe is housewives.

8. *a.* Must every sampling unit in a frame contain a member of the universe? (No; page 40.)

b. What problem is there to face when most of the sampling units in the frame contain no member of the universe?

9. What are the 5 statistical elements in a sampling procedure? (p. 38.)

10. Suppose that the method of selection gave equal probabilities to all the areas in the country. Is the sampling procedure necessarily unbiased?

11. You are told that all the sampling units in a region had equal probabilities in a sample. You are told also that an estimate $X = N\bar{x}$ was formed of the number of teen-agers that ate a certain brand of cereal last week, where $\bar{x}$ is the number of teen-agers in the sample that ate this cereal last week, divided by the number of sampling units in the sample, and N is the number of sampling units in the frame. Is there enough information here to enable you to say that X is an unbiased estimate of the x-population in the frame?

We shall see later on that X is indeed an unbiased estimate of A, the total x-population in the frame.

12. Suppose you knew only that the formula for an estimate was $X = N\bar{x}$, where $\bar{x}$ and N are defined as in the preceding question, and you knew nothing more about the sampling procedure.   Could you assert that X is an unbiased estimate?   (No.)

13. A man called me on the telephone to attempt to engage me to calculate for his company the number of retail grocery stores that would be required in each of several categories (chain stores, supermarkets, medium, small, delicatessen) in each of 20 metropolitan areas in order to produce figures for the sales of several important commodities of food. It would be fairly simple, he assured me.   Upon enquiry, I learned that he had no list of stores in any area, and had no intention of getting any. but that he had agents who could scour the country and find stores.   I discovered to my horror that what he had in mind was that his agents would search until they found as many stores of any category in any area as I would specify.

a. Criticize the suggestions from the standpoint of the 5 parts of a sampling plan.   (The plan is almost beyond criticism.)

b. Did he define the universe?   (Yes.)

c. Did he have a frame?   (No.)

d. Would his plan produce one?   (No.)

e. Was it possible to evaluate the probability that any specified store would come into the sample?   (No.)

f. Was there any formula for estimating a total or percentage?   (No, and there could not possibly be one for such a method of selection, as there are no probabilities.)

g. Would there be any possible way to compute the sampling error of any estimate that one might make from data collected from stores so selected.   (No, for several reasons.)

CHAPTER 4

Operational Definitions of Expected Value and of Standard Error

Of that there is no manner of doubt,
No probable, possible, shadow of doubt,
No possible doubt whatever.
—Gilbert and Sullivan, *The Gondoliers*.

Definition of the equal complete coverage. The concept of the equal complete coverage is fundamental to the use of samples. Every sample is a selected portion of the sampling units in the frame. The equal complete coverage is the result that would be obtained from examination of all the sampling units in the frame (segments of area, business establishments, accounts, manufactured articles) by the same field-workers or inspectors, using the same definitions and procedures, and exercising the same care as they exercised on the sample, and at about the same period of time.

A complete coverage may be conceptual or it may be actual. It is easy to point to examples of samples drawn from actual complete coverages. For example, in the Census of Population, there is a machine-card for every person enumerated in the Census, but many of the volumes of tables published by the Census are made, not from tabulations of all these cards, but from a sample thereof. In fact, sampling enlarges greatly the scope of publication in the Census. Many special studies, as of fertility, are made by the examination of a sample of families drawn by a prescribed rule from the original Census records. The Census is in such examples the equal complete coverage for the sample. Complete census and sample both contain the same proportion of careful responses, of careless responses and of nonresponse, of careful coverage and of careless coverage.

Operational definition of the error of sampling. The error of sampling in an estimate X is the difference between the estimate X obtained from a sample and the result A of the equal complete count.

Let us be formal about it. We have a frame, whose sampling units bear the serial numbers 1, 2, 3, and on to N. Let there be a complete coverage of this frame. That is, subject every sampling unit to the procedure of measurement supposedly agreed to. This measurement could be a test by eye or by instrument of some sort, applied to physical items, or it could be medical tests of people. It could be a count of the people that have one characteristic or another (e.g., a count of the males 20–29, employed). It could be a count of errors made, an evaluation of transactions of a certain type, or the yield of a crop. We refer to any such count or observation as a population of a sampling unit.

The testers or the interviewers have had a certain course of training, or maybe none at all. They may follow a certain ritual. They may be careful; they may be careless. They may fail to find all the dwelling units, all the people, all the material, whatever it be. They may report on nonexistent sampling units. They may make mistakes. Some respondents may misunderstand some questions. Some people may not be at home when the interviewer calls; some will refuse to be interviewed. There may be errors in the records that constitute the frame. In spite of possible errors, the complete coverage may be very useful, and likewise the sample thereof.

The results for the count of people or of the items of some characteristic in the N sampling units will be the N numerical values

$$a_1, a_2, a_3, \ldots, a_N$$

The symbols don't care what we measure, nor how. The sum of these N individual populations will be

$$A = a_1 + a_2 + a_3 + \cdots + a_N \tag{1}$$

This is the total x-population in the frame, as determined by this specific complete coverage.

The numbers a_1, a_2, etc., are not "true" values of the populations in the N sampling units; they are instead only the results of the complete coverage.

An operational definition* of the error of sampling is contained in the following experiment.

* An operational definition of an error of sampling or of anything else is a statement of tests or procedures that when carried out will tell us whether an item meets the definition. An operational definition will not get us into trouble with impossible words like true value, perfect questionnaire, perfect complete coverage, none of which, so far as I know, has meaning. The definitions given here are based on a complete coverage as actually carried out, imperfections and all. They make a clear separation between what is sampling and what is not. The operational definition given here for the error of sampling was described first by Morris H. Hansen and W. Edwards Deming, "On an important limitation to the use of data from samples," *Bull. Inst. Int. Statist.* Bern 1950, vol. xxxii, part 2: pp. 214–219.

1. Write each of the N population-values a_i on a card, and number the cards serially 1, 2, 3, etc., to N. These N cards show the complete coverage (Fig. 3).

2. Draw in the manner specified in the sampling plan a sample of n sampling units (Fig. 3 again).

For illustration here, the sampling plan may be to draw n cards without replacement by reading out n unduplicated random numbers between 1 and N. Let

x_1 be the x-population in the sampling unit drawn by the 1st random number
x_2 ,, ,,   ,,   ,, ,,   ,,   ,,   ,,   ,, ,, 2d   ,,   ,,
x_3 ,, ,,   ,,   ,, ,,   ,,   ,,   ,,   ,, ,, 3d   ,,   ,,
.
.
.
x_n ,, ,,   ,,   ,, ,,   ,,   ,,   ,,   ,, ,, nth   ,,   ,,

3. Form the estimate X by the formula specified in the sampling plan. The symbol X will be some function of the sample-populations (x_i).   To be specific, we may focus attention on one possible estimate, viz.,

$$X = \frac{N}{n}(x_1 + x_2 + x_3 \cdots + \cdots x_n) = N\bar{x} \qquad (2)$$

The difference between this one sample-result X and the result A of the equal complete count is

$$\Delta X = X - A \qquad (3)$$

ΔX or $X - A$ is the *error of sampling* for this one sample.   In practice, we usually do not know A, and can not compute ΔX for our sample. However, we may estimate probability limits for the range of ΔX, without knowing A.

Operational definitions of the expected value, standard error, and bias of a sampling procedure.   We continue our experiment.

4. Return the sample of n cards to their proper places in the complete coverage, and repeat Steps 2 and 3 to form a new estimate X.   Repeat these steps again and again, until you have about 10,000 estimates X.

5. Plot the distribution of X with any suitable class-interval.   Compute the mean and the standard deviation of this distribution.   The mean is an estimate of EX, and the standard deviation is an estimate of σ_X, both defined below.

Any one sample is a random selection from all the $N!/(N - n)!n!$ or $\binom{N}{n}$ possible samples of size n, all of which samples have the same probability. X is therefore a random variable, whereas A, the result of the complete coverage, is a constant for our present purpose. The $N!/$ $(N - n)!n!$ or $\binom{N}{n}$ possible values of X form the theoretical sampling distribution of X. The mean EX and the standard deviation σ_X of this

The frame		The sample	
Serial number and identification	Population or observed value	Order of appearance in sample	Population or observed value
1	a_1		
2	a_2		
3	a_3	1	x_1
4	a_4	2	x_2
.	.	3	x_3
.	.	.	.
.	.	.	.
.	.	.	.
N	a_N	n	x_n
Total	A		Σx_i
Average per sampling unit	a		$\bar{x}$
Standard deviation	σ		s

Fig. 3. Pictorial representation of the frame and the sample. x_1 in the sample is some one of the a_i in the complete coverage, drawn at random: likewise $x_2, x_3, \ldots, x_n$. Statistical theory enables us to make predictions about the sampling variation of the results from future samples, all drawn from the same complete coverage and processed according to the same rules.

distribution are of special interest. EX is the "expected" value of X. If

$$EX = A \qquad (4)$$

the *sampling procedure* is said to be unbiased. But if

$$EX = A + B \qquad (B \neq 0) \qquad (5)$$

the sampling procedure has the mathematical bias B.   The variance of the distribution of X is

$$\sigma_X{}^2 = E(X - EX)^2 \tag{6}$$

In other words, $\sigma_X{}^2$ is the expected value of the squared deviations measured from EX.   By definition, this is the *variance of the sampling procedure*, and its square root (σ_X) is the *standard error of the sampling procedure*. Thus, a sampling procedure has an expected value, a standard error, and possibly a bias.

It is an exciting fact that a single sample, provided it is laid out properly, will provide an estimate of the margin of sampling variation of all the estimates that one can form by repeatedly drawing samples from the given complete coverage, and processing them by the prescribed sampling procedure.   The same theory enables us to design a sample in advance that will deliver about the precision required.   This is the great contribution of modern statistical theory.

Randomness.   A random variable is the result of a random operation. A system of selection that depends on the use in a standard manner of an acceptable table of random numbers is acceptable as a random selection. Methods that depend on physical mixing or shuffling are hazardous and are not acceptable as random.   The words "thorough mixing," for example, have no standard meaning.   I should need to know who did the mixing, and much more.

An estimate or any statistical characteristic of a random sample is a random variable, and will vary with a definite probability from sample to sample.   The following properties of a probability sample are examples of random variables:

x_1—the x-population of the 1st sampling unit drawn into the sample.

x_2—the x-population of the 2d sampling unit drawn.

$\bar{x}$—the average x-population of the n sampling units in the sample.

s^2—the variance of the x-populations of the sample (p. 432).

$X = N\bar{x}$—an estimate of the x-population of the frame.   (N is the number of sampling units in the frame.)

$Y = N\bar{y}$—an estimate of the y-population of the frame.

$\hat{\sigma}_X$—an estimate of the standard error of X (p. 180).

$f = X : Y$—an estimate of the ratio of the x-population to the y-population in the frame.

Remark 1.   Random numbers are cheap, and interviewers and inspectors will use them correctly in the field, if they have proper instruction.   For use on the job, a man may tear pages out of a book of random numbers, and may carry one or two pages and use them until he exhausts them. Random numbers in sealed envelopes, prepared in advance for use on the spot, as soon as the worker finishes serializing certain sampling units, are simple to use (p. 175).

Remark 2. Devices that merely remove the choice of the sampling units from the judgment of the interviewer, without giving definite probabilities to each unit, do not constitute a random selection, and do not produce a probability sample.

Remark 3. Thus every characteristic estimated from a sample has its own standard error. If the reader will turn to page 182, he will see standard errors for the number of dwelling units, for the number of inhabitants, for the number of males, for the proportion male, and for still other characteristics estimated from the one survey.

The measure of sampling variation. The results (X) of repeated samples from the same complete coverage will distribute themselves as a random variable about EX. As we just saw (p. 54), the standard deviation of this distribution is called the standard error of the sampling plan for the x-characteristic. The maximum variation between the results of repeated samples all drawn from the same complete coverage, and following a prescribed sampling procedure, is usefully placed at 3 standard deviations $(3\sigma_X)$ in either direction from EX. This rule is a statistical standard long used in industry.* It gives the client or the user of an estimate what he needs to know about the sampling variation. One may wish to widen his estimate of 3 standards errors, to be conservative, when there are only a few degrees of freedom to work with (theory on page 449). One may also in rare instances wish to calculate the effect of extreme skewness or other departures from normality (p. 453).

Remark 4. A practice has grown up in some marketing-research companies of reporting 2 standard errors (or $\bar{x} \pm 2\hat{\sigma}_{\bar{x}}$) for a "confidence interval" of probability 19 : 1. One possible explanation is that 2 standard errors in a report give the impression to the unwary reader of greater precision than 3 standard errors would give. It is certainly good practice to report 1 standard error, with the degrees of freedom (p. 433) and interpretation.

Another possible explanation is confusion between an estimate and a test of a hypothesis, for which one should adjust the risks of the errors of the 1st and 2d kind to the costs of these errors.

I believe that what the user in commerce and in law needs is limits that represent practical certainty: that is, the 3-sigma limits.

Limitations of the standard error. The standard error of a result does

* See, for example: (1) Report of Committee on Standards of Probability Sampling for Legal Evidence, Current Business Studies (Society of Business Advisory Professions, New York University, March 1957). (2) Tentative recommended practice, "Acceptance of evidence based on the results of probability samples," E 141–59T (American Society for Testing Materials, 1916 Race Street, Philadelphia 3).

not measure the usefulness thereof. The standard error, however helpful in the use of data from samples, only gives us a measure of the variation between repeated samples. A small standard error of an estimate means: (1) that the variation between repeated samples must be small; hence also (2) that the accidental blemishes and variations from all sources must be small; and (3) that the result of the sample agrees well with the result of the equal complete coverage of the same frame. It does not mean that the persistent components of the nonsampling errors are small.

It is important, for such reasons, I believe, not to focus attention on the standard error alone. An ever-present question for consideration is whether the equal complete coverage would serve the purpose. In my own practice, I steadfastly refuse to compute or to discuss the interpretation of the standard error when large operational nonsampling errors are obviously present (Ch. 5). Once these mistakes are apprehended and corrected (the responsibility of the people that are doing the testing, pricing, computing), then one may usefully discuss the standard error.

The standard error estimated from the result of a single sample predicts the variability of repeated samples all carried out by the same sampling procedure, and coming from the same equal complete coverage. The standard error so estimated includes automatically not only the variability that arises from new selections in repetitions of the sampling procedure, but also the variability that arises from fluctuations in the investigator's judgment and performance, which may be different before and after lunch, and the variable effect of the weather and of other conditions that change the material or change the investigator's judgment over the period of the survey. The result of a complete coverage is also afflicted by these same variations, but we never know anything about them in a complete coverage unless we design it as a composite of interpenetrating subsamples.*

The standard error does not detect nor measure, however, the constant component of any persistent changes nor of any nonsampling errors that may be present; these one measures by the statistical audit or control, and by outside comparisons (next chapter).

Remark 5. It is possible for a result to be useful and still to possess a wide standard error. A result obtained by definitions and techniques that have been drawn up with care, and carried out, by excellent interviewing and supervision may have a wide standard error because the sample was small; yet such a result might well be preferable to one obtained with a bigger sample, with a smaller standard error, but whose definitions,

---

* I am indebted to my colleague Professor P. C. Mahalanobis, F.R.S., for pointing out this fact to me.

techniques, and interviewing were out of line with the best practice and knowledge of the subject matter.

Remark 6.   The results of a sample or of a complete coverage refer to the frame, not necessarily to the universe.   The objective inferences in regard to the precision of a sample refer to the frame.   If there is a gap between the frame and the universe, these inferences do not necessarily apply to the universe.   One fills in the gap (i.e., extends an inference from the frame to the universe) with aid of substantive knowledge, not by statistical theory.

Acceptability of a sample in legal evidence.   The actual difference between an estimate produced by a sample and by the equal complete coverage depends not only on the margin of sampling variation, but also on how rigidly the sample in question followed the procedure prescribed. The audit or statistical control of the actual sampling procedure (next chapter) enables one to calculate the bias attributable to persistent errors that may have been made in the sampling.   It also estimates how much operational error there is in the equal complete coverage.   (Some estimates also possess mathematical biases, but these are usually negligible if they exist at all, and they are easy to evaluate; see Chapters 16 and 17.)

It follows that the result of a sample has the same status in legal evidence, or in a scientific enquiry, as the result of THE equal complete coverage would have, PROVIDED that (a) the limit of sampling variation is sufficiently small to be innocuous; and (b) the persistent (noncanceling) effect of any mistakes in the selection of the sample and in the processing thereof is also innocuous.

Whether the equal complete coverage would be acceptable is thus the real problem, once the sampling procedure is certified, and if the standard error is not too big.   The acceptability of the equal complete coverage depends on answers to questions about (1) the content of the questionnaire or the method of test; (2) the frame; (3) the nonsampling errors, and how they affect the results.   These points are not to be settled by statistical theory, as they belong to the subject-matter, being the same whether one uses sampling or not.

Remark 7.   A masterful statement in regard to the objectivity of the standard error appeared in legal testimony delivered by Leslie F. Kish, of the Survey Research Center, University of Michigan, at the request of the Illinois Commerce Commission.*

The soundness of the statistical theory which underlies the computations of this average, and of its standard error, as well as the meaning of the confidence intervals formed from them, are universally accepted and incontrovertible.   By this I mean that there is no controversy; no personal

* Illinois Commerce Commission, Docket 41606, Chicago, 17 August 1954.   The lines that appear here may differ by a word or two from the reporter's record: I have only my own notes.

judgment enters into their meaning; they are well-established mathematical consequences.   Every trained statistician will make the same interpretation of the results of those computations. . . .

I have made a detailed, independent, and critical examination of the sampling plan, both in its theoretical and practical aspects.   I took it to be my duty to find any fault that might make the sampling plan invalid. It is my professional opinion that the sampling plan represents an excellent example of combining theoretical knowledge with sound understanding of the practical requirements to produce a valid, and reliable statistical sampling plan. . . .

The examination of the theoretical aspects of a sampling plan and of the formulas connected with it, is a relatively brief assignment for one who knows the theory of sampling. . . .

Efficiency, validity, and usefulness. A sample-design is *efficient* for a certain estimate if it produces at low cost a small standard error for this estimate (see the quantitative definition further on).   An estimate made from a sample is *valid* if it is unbiased or nearly so and if we can compute its margin of sampling error for a given probability.

A sample-design may be valid though inefficient.   It may be valid though useless, even though the standard error be small.   Validity and efficiency and usefulness are unrelated.   A standard error retains its validity and interpretation even though some other method of sampling, known today or to be discovered tomorrow, turns out to be more efficient than the plan actually used.

This point is not well understood outside professional statistical circles, and the reader may therefore wish to turn back to page 55 to the section, "Limitations of the standard error," which relates closely to this point. If we calculate the standard error of a result produced by a valid sampling procedure, we may interpret it with full confidence; and we gain no useful information for this purpose by enquiring whether the sampling procedure was efficient or inefficient.

Quantitative measure of efficiency. Fisher* gave in 1922 a very useful quantitative definition of efficiency, which is to compare the efficiencies of 2 sampling procedures A and B by the inverse of their variances for the same size (n) of sample.   In symbols (E for efficiency here),

$$E_A : E_B = \sigma_B{}^2 : \sigma_A{}^2 \qquad [n_A = n_B] \tag{7}$$

Or, equally, if we choose the sample-sizes n_A and n_B so that $\sigma_A = \sigma_B$, then

$$E_A : E_B = n_B : n_A \qquad [\sigma_A = \sigma_B] \tag{8}$$

* Sir Ronald Fisher, "On the mathematical foundations of theoretical statistics," *Phil. Trans.*, vol. 222A, 1922: pp. 309–368.   Variance is an imperfect measure of precision unless the distribution of the estimates is normal; Sir Ronald Fisher, *Statistical Methods and Scientific Inference* (Oliver and Boyd, 1956): p. 152.

A more useful comparison is in terms of cost, as Hansen first pointed out.* Let the costs be C_A and C_B for equal variances. Then

$$E_A : E_B = c_B : c_A \qquad [\sigma_A = \sigma_B] \qquad (9)$$

A slight complication may arise from the fact that a sample-design that is more efficient for one characteristic may be less efficient for another characteristic (see page 182).

EXERCISES

1. Define operationally an unbiased sampling procedure.

2. Define operationally the standard error of a sampling procedure.

3. *a.* Does an unbiased sampling procedure necessarily give a good estimate of a characteristic of the frame? (No.)

b. Of the universe? (No.)

4. When is a sampling procedure valid?

5. When is a sampling procedure efficient?

6. *a.* May a sampling procedure be valid but not as efficient as the best design possible?

b. Describe some conditions under which it might be wise to use an inefficient procedure.

7. What is a random variable? Name some random variables that are used in the theory of sampling.

8. What is an acceptable standard of random selection?

9. Why not use dice or cards, or some trick, to make a series of random selection?

10. *a.* Is an unbiased estimate necessarily a good estimate? Explain.

b. Are biased estimates useful? How is this?

11. Suppose that we draw a sample of size n from a frame of N sampling units by reading out n random numbers between 1 and N from a table of random numbers. Show that the order in which the N sampling units appear in the frame is immaterial.

12. *a.* Explain why the instructions for giving the serial numbers to the sampling units in the frame, either actually or by rule, and for reading out from a table the n random numbers for the sample, must specify that the 2 steps be independent.

b. Suppose that one were to read out the n random numbers for the sample first; then to arrange the order of the sampling units in the frame (i.e., to reassign the serial numbers). Would it be possible, by merely rearranging the order of the sampling units in the frame, to produce almost any desired result? (Yes.)

* Private communication to the author, about 1937. Mahalanobis was taking the same steps in Bengal about the same time.

c. Could we use the theory of probability to calculate the margin of error of such a result? (No.)

d. Would such a procedure be a probability sample?   (No.)

13. What use may one make of a standard error?   Why is the standard error so important?

14. *a.* Does the standard error of a result measure the usefulness of the result?

b. Does it measure the biases in the field-procedure?

c. If not, why is the standard error so important?

15. Does a small standard error mean that:

a. The data will be useful? (No.)

b. The questionnaire was designed skillfully? (No.)

c. The interviewers performed uniformly? (Yes.)

d. The interviewers performed well? (No.)

e. The supervision was good? (No.)

f. Why is the standard error so important?

16. X is an unbiased estimate formed from a sample-survey.   What does X estimate?   (The population of the universe?   The population of the frame?)

17. *a.* The result of a sample over a region gives:

$X = 792,000$ dwelling units having a particular characteristic
$\hat{\sigma}_X = 10,000$ dwelling units

What is your interpretation of the margin of error of X?   (Assume that the estimate of the standard error is pretty reliable.)

b. Does the estimate X refer directly to the number of dwelling units in the universe?   In the frame?

CHAPTER 5

Uncertainties not Attributable to Sampling

The fact that a general impression is more or less universal can not in itself be a guarantee of its validity.—P. C. Mahalanobis, The National Sample Survey, General Report No. 1, Ministry of Finance, Delhi, December 1952.

Reasons for studying all the sources of uncertainty. We shall learn in this chapter something about the various possible uncertainties in surveys, other than those that arise from the random selection of sampling units.

The user of the results of a survey has an interest in the possible effects of all the sources of uncertainty in data that he intends to use: otherwise he may fail to understand the limitations of the data, and may too easily draw wrong conclusions and incur losses. To the user of the data of a survey, it is only the total error that counts: he does not care whether it is a standard error or some other kind of error.* To the statistician, however, the causes and magnitudes of error are important, as his aim is to reduce error and to increase reliability. Statistics are a basis for action, and the more we know about the limitations of a figure, or of a procedure, the more useful it becomes.

Errors not connected with the selection of the sample belong to the equal complete coverage—i.e., they are errors, but not sampling errors. They belong also to the sample, as a good sample will deliver very nearly a proportionate share of the nonsampling errors that would afflict the equal complete coverage.

Economic balance of errors. It is well to keep in mind the triangle shown in Fig. 4. One leg of the triangle represents persistence. The

---

* One may recall here the famous remark made by Dr. Alfred N. Watson at the meeting of the American Statistical Association in Chicago 1942: "A standard error is just as bad as any other error."

For further reading, see the masterful work by Frederick F. Stephan and Philip J. McCarthy, *Sampling Opinions* (Wiley, 1958).

other leg represents the random errors, which include the sampling variation. The hypotenuse is the sum of the uncertainties, or the total error. When the non-sampling errors are large, it is uneconomical and ineffective to waste funds on a big sample, as a big sample will decrease the sampling error but leave the total error about the same. One must face the fact that the *overall usefulness and reliability of a survey may actually be enhanced by cutting down on the size of sample* and using the money so saved to reduce the nonsampling errors. In the sampling of records, this might mean tracing and correcting wrong and missing

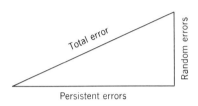

Fig. 4. The total error of a survey is the hypotenuse, which is the combined effect of the random errors and the persistent errors.

information. In a survey of human populations, this might mean more time and money on the questionnaire, hiring fewer and better interviewers, providing better training and better supervision in the field, and making more recalls on people not at home on a previous call.

The preferred technique and the working technique. We may speak of the definitions, the method of test, the questions, the methods of interviewing, the method of supervision, the treatment of nonresponse, etc., as the survey-technique. Whatever be the survey-technique, a complete coverage of all the sampling units in the frame will produce some result. Another complete coverage carried out with the same survey-technique, before changes have taken place, would give results slightly different from the results of the first coverage. This is so because people do not always give the same answer when you ask them a 2d time. Or, some other member of the household may answer on the 2d complete coverage. Some other survey-technique (different definitions, different questions, different procedures) would give still another figure. Neither figure is right or wrong, and neither of them is a true value. Physical measurements, and data transcribed from records, will also show differences from one complete coverage to another. There is no such thing as a true value.

Any result, whatever it be, is the result of applying some set of operations. Although there is no true value, we do have the liberty to define and to accept a specified set of operations as preferred, and the results thereof as a *master standard* (so-called by Harold F. Dodge).

There may be, by agreement of the experts in the subject-matter, for any desired property of the material, a *preferred* survey-technique. Unfortunately, it often happens that the preferred technique, usable on a laboratory-scale, is too expensive to apply in a full-scale survey, or it may be objectionable otherwise. The experts must then supply also a working technique. Thus, the preferred technique by which to define a person's age might be to compute the difference in time between today and the date shown on his birth-certificate. But some people don't have birth-certificates at all, and few people have them handy. Moreover, some people would not be happy with an interviewer who asked for birth-certificates. The Passport Division can ask for birth-certificates, but interviewers may ask the person how old he is, and record the result. This would be the working technique by which to measure age.

The preferred technique and the working technique will give different results. A working technique is acceptable to the experts if it gives results not too far, in their judgment, from the results of the preferred technique.

The difference in the 2 techniques, applied to a complete coverage of the frame, is the *bias* of the working technique. A working technique is said to be *accurate* if its bias is small.

Experts in subject-matter who are alive in their subjects are continually changing their definitions and methods of questioning. These changes arise partly from gains in knowledge of the subject, and also from the fact that changes in social and economic conditions require from time to time new kinds of information. No set of questions is static. The questions on the labor force in the Census of 1940 were entirely different from those used in the Census of 1930. The results were hardly comparable. Further changes have occurred almost continually during the years following 1940. This is as it should be.

Two experts may prefer different survey-techniques. Take watching television, for example. One expert may prefer to know the total length of time in a week that a television-set was tuned to this or that station. Another expert may prefer to enquire by telephone, at selected times, whether anyone is actually looking at the television, and if so, at what program. The 2 techniques measure different characteristics of television-viewing.

The bias and the accuracy of a working technique will thus change from time to time as one preferred technique is supplanted by a new one, with new results as a basis from which to measure the bias of the working technique.

Remark 1. It is important to remember that the bias of a working technique is not an error of sampling. The result of a sample will possess the bias of whatever technique is built into the survey; it will also possess sampling error. The sampling error will disappear as the size of the sample increases, but the bias of the working technique will remain fixed, independent of the size of the sample. The sampling error is calculable from the results of the sample. The bias of the technique used is measurable by a properly designed experiment, which will compare by the use of interpenetrating samples the results from the 2 techniques, the preferred technique, and the working technique.

Classification of uncertainties and deficiencies common to complete coverages and to samples.*

TYPE I. BUILT-IN DEFICIENCIES; MISSING THE POINT; MEASURING PROPERTIES OF THE MATERIAL NOT FULLY SUITED TO THE PROBLEM.

The distinguishing characteristic of this type of uncertainty is that it is built into the questionnaire or into the method of test. It does not arise from flaws in carrying out the specified survey-procedure: a recanvass (audit or control; *vide infra*) will not discover it. It is independent of the size of the sample.

Examples:

1. Failure to perceive what information would be useful; eliciting (perhaps accurately) information that is of little help on the problem. We encountered this deficiency in Chapter 1 (p. 16).

In the sampling of accounts, errors in source-documents will carry through into the final estimates, whether one covers the documents by a complete coverage or by a sample. Failure to know about these errors, or to correct them (best done in the sample) is an error of Type I.

2. Too big a gap between the frame and the universe. An example occurs when one applies an interpretation or forecast to domains and universes not covered by the frame used in this survey. (A gap is not mere mistakes of omission in the preparation of the frame.)

3. Ineffective rules for coding.

4. Ineffective tabulations.

5. Failure to recognize secular changes that take place in the universe before the results are written up and recommendations made.

6. Bias arising from bad curve-fitting; wrong weighting; incorrect adjustment.

7. Unwarranted deductions from the results, with a report that may lead to misunderstanding and to misuse of the survey. The report concerning the findings of the survey should make clear the limitations of the data. It should take into account the fact that the users of the figures

---

* This classification agrees well with a paper by D. B. Lahiri, "Observations on the use of interpenetrating samples in India," *Revue internationale de Statistique*, vol. 36, part 3, 1959; pp. 144–152.

may lack survey-experience, and be unable to comprehend uncertainty in a figure. The report should evaluate and interpret the margin of sampling error, and the possible effect of blemishes and blunders made in carrying out the survey-procedure (see the example on page 19). It should call attention especially to the possible misinterpretation that could arise from nonresponse, or from any gap between the frame and the universe.

Although a recanvass will not discover the existence of an uncertainty of Type I, an outside comparison may do so. Examples are well known. A survey, sample or complete coverage, carried out without operational blemish (i.e., in exact accordance with the procedures prescribed), to ask people how they will vote, or what they will buy, may turn out to disagree with how the people actually do vote, or what they do buy. The actual vote, or the actual purchases, furnish an outside comparison. Such disagreement may arise partly from built-in deficiencies (inability to elicit the relevant information for such predictions), although misuses of the results of the survey may also be responsible.

A superb illustration of errors, mostly of Type I, occurred as an editorial in the *Saturday Review of Literature*, vol. xxxii, 26 March 1949, under the title "Bogus best sellers." The article dealt with errors that book stores make in reporting books that have sold best during the past week. What prompted the author to write the article was that some publisher had laid claim to 1st place for a certain novel that had appeared in the 11th place on a list of best sellers. The author pointed out the following sources of error, most of which belong to Type I.

1. The book stores that are asked to name the best sellers are chosen "with little regard to statistical justice." (A bad sample he meant, I suppose.)

2. The book stores are not weighted with respect to volume of sales.

3. They do not have the correct geographical proportions (more bad sampling, I suppose).

4. Many book stores keep no records of sales, but report from memory.

5. A book-store owner sometimes reports the title for which he has the heaviest inventory, forgetting that heavy inventory may mean slow sales.

6. The latest figures that the book store has, if it has any, may be several weeks out of date.

7. Book stores report literary titles, for prestige.

8. Some best sellers were inadvertently omitted entirely from the list submitted.

9. Long-term sellers are usually ignored.

Note that a recanvass of the stores in the sample would not discover all these difficulties; nor would a complete coverage of all stores discover them.

TYPE II. BLEMISHES AND BLUNDERS MADE IN CARRYING OUT THE FIELD-WORK, THE TESTING, THE INTERVIEWING, THE CODING, THE COMPUTATIONS, AND OTHER WORK.

These errors have their origin in imperfect workmanship. They are discoverable and measurable by repetition or recanvass (called the audit or control) of a sample of the main sample.   All of them can occur in complete coverages as well as in samples.

8. Failure to find or to visit all the sampling units that were drawn into the sample.

9. Failure to provide definite boundaries or clear definition of a sampling unit.   As a result, or possibly through carelessness or by accident, the investigators may fail to test or to interview some part of a sampling unit, or may go out of bounds and test or interview units not intended for the sample, or not even in the frame.

10. Failure to cover a sampling unit completely, such as failure to find all the dwelling units or all the people therein.

11. Covering some material twice.

> One can avoid this error in either a complete coverage or a sample of physical material if the inspectors will mark any unit that they test, so that anybody can see that it has already been tested.   In a destructive test, a 2d test is impossible.   Human populations usually halt a 2d coverage (see Remark 2 in Chapter 3, page 45).

12. Failure to ask some of the questions, or to make all the tests prescribed.   Getting wrong answers.*   Asking questions not on the questionnaire.

13. Using the wrong test-instrument.   Errors in counting and in weighing.   Looking up the wrong price, or computing it incorrectly.

14. Nonresponse and refusal.   (See separate section on nonresponse.)

15. Mistakes in calculation and in transcription.

Persistent omission or inclusion of material above or below average value, or persistent mistakes in one direction, will cause biases.   The only way to evaluate them is by the audit or control (*quod vide, infra*), or with the help of outside sources of information.

If one were to write out a list of all possible uncertainties, of whatever type, he would of course include sampling variation, and this one might designate as uncertainty of Type III.   This chapter deals only with the uncertainties that do not originate in sampling; hence it has mentioned so far only Type I and Type II.   The rest of the book, from here on, deals almost entirely with sampling variation.

Nonresponse.   Nonresponse in a survey is devastating and discouraging. It is often difficult and expensive to find certain people at home, and some

---

* Morris H. Hansen, William N. Hurwitz, Harold Nisselson, and Joseph Steinberg, "The re-design of the Census Current Population Survey," *J. Amer. Statist. Ass.*, vol. 50, 1955: pp. 801–819.

Leslie F. Kish and Irene Hess, "On noncoverage of sample dwellings," *J. Amer. Statist. Ass.*, vol. 53, 1958: pp. 509–524.

people, though at home, refuse to talk. Nonresponse is often the main uncertainty of Type II. A problem arises in nonresponse because (1) the returns from a survey are heavily weighted with people who are always at home, and who are willing to talk; (2) many characteristics of these people are different from the characteristics of people who are rarely at home or who are unwilling to talk. The same problem often arises in enhanced degree in a mailed questionnaire: the solution is to interview a sample of people that did not respond by mail.*

There is no general safe limit to nonresponse. People sometimes enquire whether 50% response is good enough, or whether 80%, or 90%, or 95% is good enough, or just what do we consider to be good enough? The answer depends on the characteristic and how it is distributed. If half the people or firms with very high incomes, sales, employment, production, or inventories are nonrespondents, the error may be large, even though the nonresponse over all classes combined be but 5%.

Enlargement of the sample, or substitution of respondents in the same area and of similar characteristics does not solve the problem of non-response; such schemes merely overweight the sample further with people who have the characteristic of being at home most of the time.

The author recently evaluated by a probability mechanism† the possible bias from nonresponse under various assumptions. The possible ill effects cause one to ponder (next section). The only remedy is to try to eliminate the nonresponse, either by calling back again and again, or by use of the Politz plan.

The Politz plan enquires of each person interviewed whether he was at home the night before the interview, 2 nights before, 3 nights before, and perhaps 4 or more.‡ The respondents thus classify themselves into people that were at home not at all during the preceding nights, or once, or twice, etc. The plan continues by weighting a respondent's results inversely in proportion to the number of nights (including the night of the interview) that he was at home.

* Morris H. Hansen and William N. Hurwitz, "The problem of nonresponse in sample-surveys," *J. Amer. Statist. Ass.*, vol. 41, 1946: pp. 517–529.

Hansen, Hurwitz, and Madow, *Sample Survey Methods and Theory*, Vol. 1 (Wiley, 1953): p. 473.

M. A. El-Badry, "A sampling procedure for mailed questionnaires," *J. Amer. Statist. Ass.*, vol. 51, 1956: pp. 209–227.

† W. Edwards Deming, "On a probability mechanism to attain an economic balance between the resultant error of response and the bias of nonresponse," *J. Amer. Statist. Ass.*, vol. 48, 1953: pp. 743–772.

‡ Alfred Politz and Willard R. Simmons, "An attempt to get the not-at-homes into the sample without call-backs," *J. Amer. Statist. Ass.*, vol. 44, 1949: pp. 4–31.

Willard R. Simmons, "A plan to account for not-at-homes by combining weighting and callbacks," *J. Marketing*, vol. xix, 1954: pp. 42–54.

It is possible to show that, on the basis of certain reasonable assumptions, use of the Politz plan that enquires concerning 3 nights in addition to the 1, 2, or 3 nights when interviews were attempted, is the equivalent of 3 more random recalls.

There are circumstances in which recalls are impossible. For example, if one were to conduct a survey Monday evening concerning the content of yesterday's Sunday newspaper, he could hardly call back on people that he missed Monday night, because by Tuesday people have forgotten about Sunday. The Politz plan offers a solution. It has, in any case, the advantage of speed.

It is necessary in any survey, for administrative reasons, to fix a closing date beyond which any returns not yet in are declared nonresponses. If the survey is conducted by interviews, this date should permit time for a specified number of recalls. If the survey is done by mail, the closing date should usually permit 2 or 3 waves of mailing. There are 4 choices open for the nonresponses that remain.

1. Make vigorous fresh attempts by skilled interviewers, by mail, by telegram, or otherwise, on a sample of the nonresponses. Add the results, with the proper weights, to the responses. Success in this step transfers, in effect, the nonresponses to responses, and eliminates the problem. This is the best solution, when it is possible. An example appears in Chapter 8.

There is an optimum fraction to use in any survey for special pressure on the nonresponses. This fraction is sometimes a random 1 in 2, sometimes 1 in 3, sometimes 1 in 4, depending on costs.*   If the special interviews on the nonresponses are going to be especially difficult or costly compared with the original interviews, the optimum fraction may be as low as 1 in 5. I have seen the time when the special interviews were exceedingly difficult, and depended on help from people who would probably cooperate only indifferently (heads of fraternal organizations that some of the nonresponses belonged to). I reduced the ratio in this case to 1 in 10, with the hope that 1 in 10 would be better than a complete black-out of the nonresponses.

2. Calculate extreme effects of the nonresponse. Assume extreme skewness in the distribution of the nonresponses (or material omitted), and assume extreme values, high or low or both, in the various classes, and thus calculate upper and lower limits for the responses and

* See the footnotes on the preceding page; also William G. Cochran, *Sampling Techniques* (Wiley, 1953): p. 298.

nonresponses combined.   This procedure very often solves the problem.*
A report on this kind of calculation occurred in Chapter 1 on page 19.
   If the nonresponse is small, or if the extreme assumptions are not
greatly different, the calculated extremes may not be far apart.
   The results that one derives by use of extreme values depend on the
assumptions that he makes with respect to the nonresponses.   The results
are no better than the assumptions.   Assumptions, to be plausible and
acceptable, require knowledge and authority in the subject-matter.   The
statistician should therefore state the source or authority for the assumptions, and he should be careful not to take this responsibility himself
unless he is prepared to defend himself.
   3. Calculate plausible results for the responses and nonresponses.   This
choice may be midway between the extremes.   The same precautionary
warnings that we saw in the preceding paragraph apply here also.
   4. Merely report the results of the responses by class, and the number
and characteristics of the nonresponses, so far as possible, and leave
it to the user to carry it on from there.   An example appeared on
page 17.

   It is sometimes a fact that the results from the responses alone solve the
problem without any special action on the nonresponses.   This may be
the case when the decision to be taken on the basis of the results depends on
a minimum or a maximum number of some characteristic, such as yes or
no.   If the responses already contain so much or so little of this characteristic that they will overwhelm any possible results that might come from
the nonresponses, then there is no point going further.   This is a special
case of the use of extreme values for the nonresponses.

   It is not generally appreciated that a ratio (typified in later pages as
$f = x/y$), calculated from the responses alone, will remain unchanged if we
ascribe to the nonresponses the same average x- and y-values as the survey
showed for the responses, and add them to the responses.   This is an easy
way to bury the difficulty, but it is not a solution.

   Missing, illegible, or unacceptable information creates about the same
problem as nonresponse.   Fortunately, missing, illegible, and unacceptable
information are often but by no means always fairly well distributed amongst
all classes.

Futility of sheer size of sample to combat nonresponse.   The author's
calculations referred to in the footnote on page 67 show that certain
levels of accuracy are unattainable with any sample-size, however great,
unless effort is made through 3, 4, or even 6 recalls to find people at home
who were not at home at the first call.   An increase in the size of the sample
will decrease the variance of response, but it will have no effect on the

* See William G. Cochran, *Sampling Techniques* (Wiley, 1953): p. 294 for an example
and theory.

bias of nonresponse.   Successive recalls, on the other hand, to find people at home and to elicit response, are really effective in reducing the total error.   The following excerpts from the paper cited show some of the conclusions derived from one example of application of a probability mechanism.   The level of response at the initial call is assumed to be 62.5%.

1. With no recalls at all, the minimum relative total error attainable, under the assumptions made for the different classes of the population, is 11%.   No sample however big, not even a complete count, can penetrate below this minimum without recalls.

2. With one recall, the minimum total error drops to 7.6%.   No sample however big can penetrate below this minimum, with only one recall.

3. With 2 recalls, the minimum total error drops to 5.7%.   No sample however big can penetrate below this minimum, with only 2 recalls.

4. With 3 recalls, the minimum total error drops to 4.5%.   With 4, 5, and 6 recalls, the minimum total error drops to 3.7, 3.0, and 2.5%.

5. To attain a total error of (e.g.) $7\frac{1}{2}\%$, we may use 3 recalls, or 4, or 5, with the initial samples and costs shown in the accompanying table. Note that the more recalls, the lower the cost.

Number of recalls	Initial sample	Cost
6	345	$2290
5	378	2390
4	408	2450
3	512	2800

With only 0, 1, or 2 recalls we can not attain the prescribed total error ($7\frac{1}{2}\%$) with any sample however big.   This is why, in modern practice, a sample-design includes provision for a large number of recalls, or for use of the Politz plan (p. 67).

Remark 2.   At a lecture on sampling that I gave in Germany, someone said that it would be impossible to get answers from all the 3000 pig-owners in the sample, because some of them would not be at home when the interviewer called and recalled.   The sample would thus be in error.

I asked him if he had experienced any such trouble with a complete count of pigs, and he said oh yes, that the response in a complete count was about 90%; that for 90% nonresponse it was customary to multiply by $\frac{10}{9}$ the number of pigs actually counted.

My reply was that whatever you do about nonresponse in your complete count, do the same in your sample, and don't call it sampling error.   The uncertainty from nonresponse in the results of the sample will then be about the same as the uncertainty in the complete count.

The audit or control.   A statistical audit or control detects the existence of errors of Type II (p. 65), viz., blemishes and blunders made in carrying out the instructions for the survey, including the selection of the sample, the interviewing or the testing, the thoroughness of coverage, and the computations.   The purpose of the audit is to detect and to measure any persistent component of the blemishes and blunders of Type II, such as leanings toward high or toward low values, from whatever source.   The variable part of the blemishes and blunders is already absorbed into the standard error, but the mathematical formulas for the standard error can not evaluate persistence.   The way to detect and to measure persistent leanings toward high or toward low values is by a careful audit, although comparison with outside sources of information may also help at times.   For example, if every inspector or every interviewer added 1 to his count, the standard error would be unaffected, but a careful audit should detect the persistent addition.

The purpose of the audit is not to correct the main sample, but to provide information that will enable us to decide whether to accept the main sample or to redo it.   The standard error evaluates the possible sampling error; it enables us to say whether the sampling was sufficiently precise.   The audit evaluates the nonsampling error; it enables us to say whether the equal complete coverage would have been satisfactory.   The audit also helps us to understand better the meaning of the data in the main sample, and it teaches us ways in which to improve the next survey.

The procedure for the audit is to examine very carefully a small sample of the main sample (or of the complete census), to discover what went wrong.   It is a good idea to draw into the audit a portion of each interviewer's or of each inspector's work, in order to isolate the source of any trouble (Ch. 13).   The audit or control is the 5th part of a sampling plan (p. 39).   It is as necessary for a complete census as it is for a sample, and the method of conducting the audit is the same.*

* The first published example of the use of sampling to evaluate the errors in a complete census, and an example of *action* taken on the basis of the sample, was published by my friend G. Chevry of the Office of the Census in Paris, "Control of a general census by means of an areal sampling method," *J. Amer. Statist. Ass.*, vol. 44, 1949: pp. 373–379.   As a result of the resurvey of a sample of areas, the officials of the Census in Paris decided NOT to publish the results of the Census of Manufactures of France, 1946.   (Incidentally, M. Chevry used the word control in the proper sense, viz., count, inspection, evaluation, calibration.)

Meanwhile, Mahalanobis in India had for years been measuring the errors of sampling, and calibrating the work of observers, by means of interpenetrating samples: see his articles, "On large-scale sample surveys," *Phil. Trans.*, vol. 231B, 1944: pp. 329–451;

It is necessary that the audit be carried out by someone who is aware of his responsibility.   He should use a regular form or questionnaire that will supply the information needed (p. 73).   When he discovers a wrong result, or an omission or a substitution, then the best practice, I believe, is for him to report the flaw to the supervisor in charge of the interviewing or testing.   The supervisor will thereupon send the original interviewer or inspector to do the job as originally prescribed, and to bring in a revised figure.   There will thus be 2 results for any sampling unit in which the auditor discovers a blemish: the original result, and the revised result.

After the audit is complete, and the corrections made, one may then compare the results wherever there was a difference, and can evaluate the maximum numerical effect of the blemishes.   An example occurs in Chapter 9 (p. 159), and another one in Chapter 15.   This evaluation is no simple job.   It requires skill in the use of probability, and it requires also mature understanding of the nature of the errors committed by man and by machine.

Remark 3.   It is important to keep the audit or control small enough so that it remains a professional job and not mass production.   I should rather cover 50 or 100 sampling units carefully in the audit than to try to cover 1000 or more and fail to find half the blemishes.   One may extend the audit if the first results throw suspicion on the main sample.

Remark 4.*   The pattern of the 10 results, when one replicates the sample in 10 subsamples (next chapter), is often very helpful in apprehending mistakes of Type II.   In one instance, an error in the price of an item in an inventory produced an error of \$271,131 in Subsample 4.   This error knocked Subsample 4 far away from the other 9 subsamples and called attention to the mistake.

Remark 5.   The scheme for drawing the sample for the audit or control may be very simple.   Thus, if the main sample was selected with fresh random numbers in every zone, one may select, for the audit, every nth sampling unit of the main sample (e.g., the 15th and 30th thereafter). If the interviewers or inspectors were allotted to the sampling units of the

---

* The reader may wish to return to this paragraph after the use of 10 subsamples is more familiar.

"Recent experiments in statistical sampling in the Indian Statistical Institute," *J. Roy. Statist. Soc.*, vol. cix, 1946: pp. 325–378.

A further interesting example, out of scores that one could mention, is a paper by S. S. Zarkovic "Sampling control of literacy data," *J. Amer. Statist. Ass.*, vol. 49, 1954: pp. 510–519.   This is an evaluation, by a sample, of the data on literacy obtained in the complete Census of Yugoslavia, 1954.

The first large-scale regular and continuing use of an audit of censuses and sample-surveys was instituted by Hansen and colleagues during the decade 1940–1950.   The theory and procedure appear in Hansen, Hurwitz, and Madow's *Sample Survey Methods and Theory*, Vol. 2 (Wiley, 1953): Ch. 12.

main sample at random (Ch. 13) so as to measure the variance between the interviewers or inspectors, the selection of every nth item for the audit will automatically draw very nearly a proportionate sample from every interviewer or inspector.

Remark 6. I give below an example of the questions on an audit of a sample of aerial telephone property, inspected to estimate the physical condition of this property.

Question number	Question	Yes	No
1	Did the inspector inspect the right item?	60	0
2	Did he identify the item on his report so that anyone could definitely identify it later?	60	0
3	Did he inspect all the items in the sample?	58	2
	Note: (1) One inspector missed one pair of No. 14 hard-drawn covered wire in good condition. (2) Another inspector missed 16 pairs of B-rural distribution wire in excellent condition.		
4	Did the inspector use the random numbers correctly for subsampling the items within a sampling unit?	Does not apply to the aerial property, as there was no subsampling.	
5	Did he include in the inspection any item that he should not have included?	3	234
	Note: (1) One inspector reported 8 half-spans of No. 109 wire in excellent condition that should have been No. 104 copper-steel wire. (2) One inspector reported 12 half-span pairs of No. 6 B-rural distribution wire in excellent condition, where there were only 6. (3) Another inspector included a half-span of strand which was used as guy, not as strand to hold the cable. This strand was therefore not part of the regular aerial property.		
6	Did the inspector omit any complete unit?	0	60
7	Were there any departures from the instructions and procedure not noted above?	1	59
	Note: One inspector failed to sign and to date one of his records.		

Remark 7. Discovery of blemishes and blunders is empirical, and has not the objectivity that the sampling error has. These nonsampling errors must be. observed by human eyes. Anyone who is familiar with methods of inspection in factories knows that the number of defects in finish, or the number of defective items discovered, depends on how carefully one looks for trouble. Better light means that the inspectors will discover more defects. Even with good light, defects are often matters of opinion, and depend to some extent on whether the inspector has had his lunch. The discovery of dwelling units or of people missed in a segment of area depends on how carefully one beats the bushes to find them.

The sampling error, on the other hand, is of a completely different nature, being pretty closely predictable, and known objectively afterward, not as somebody's opinion, but as a mathematical theorem, provided the sampling plan was carried out correctly.

Remark 8. Another point is that the nonsampling errors that arise from the blemishes and blunders may be of almost any magnitude, depending on the care and supervision that goes into the survey. In a discussion of the sampling plans for an inventory, the auditor enquired, "What is the chance of a big error?" I replied that the chance of a sampling error as big as 3.5% was very small, and that a sampling error as big as 5% was for all practical purposes impossible: that, year after year, the sampling errors would be mostly within 3%. Oh, he was not worried about errors of only 3%: "What is the chance of an error of 20%?" (which would put his company out of business). I replied that any error beyond 5% could arise only from blunders such as careless counting or pricing, omissions, or counting some material twice; that it is the responsibility of the company to hold the blunders and the nonsampling errors to a respectable minimum, and not let them get out of hand.

Remark 9. Chapter 13 will describe a simple design for measuring the difference between interviewers. It is easy to extend the design to measure the difference between 2 forms of the questionnaire, or between 2 methods of test.

Use of sample-probes in administration. The same techniques that one uses to discover blemishes and blunders made in the execution of a sampling plan may be applied to discover flaws and waste in regular routine procedures, whether machine or office. Careful study by a proper sampling plan will often show that operations that are apparently doing well are actually full of flaws, producing waste and wrong results; that certain information being recorded on routine forms is incomplete and inaccurate. One may trace to their source a small sample of blemishes in hand work or in machine work. He may then decide whether to remove the source of the difficulty. Sampling provides a procedure by which one may generalize objectively to discover causes and percentages of departures and blunders. Sampling offers the possibility of achieving genuine creative administration.

Comparison of a survey with outside sources. In addition to the audit or control, one should not fail to compare the results of a sample with the results of any other sample or complete census that may possibly throw light on any deficiency in the present sample. Disagreement of a sample with a complete census or with another sample raises a question mark. Disagreement between the book-inventory and an estimate of the inventory taken by a sample also raises a question mark. It is a mistake, however, to jump to the conclusion that a sample is wrong because it disagrees with some other result. In my experience, it has usually been the complete census that was wrong, but not always.

Disagreement is a red light, on which the engineer slows down to await a green signal, to find out what is wrong. Deficiency in the total number of dwelling units estimated from a sample, when compared with the Census, may indicate that some of the areas are not yet tabulated. Or, it may indicate that the interviewers were not doing their work carefully. On the other hand, there may be nothing wrong at all. The shortage in dwelling units in some area may indicate change in a blighted area, as where a slum has turned into a park: it has happened. If a sample is deficient in certain age and sex classes, when compared with the Census, the deficiency may indicate lack of persistence on the part of the interviewers in the sample to enquire in every household the name, age, and sex of all the people therein. One should remember that a small proportion of honest differences in classification of residence and occupation are to be expected between main sample and audit, or between both of these and a complete census.

In the case of an inventory, disagreements often arise between (a) estimates of grand totals furnished by the main sample, and (b) supposedly comparable figures derived from the audit or from the accounting department, not from sampling error, nor from mistakes, but from different practices in accounting. For example, the figures on the quantities of material in an inventory obtained by the investigators that work on the main sample may be (by choice of the proper executive) the normal quantities of items supposedly in stock in the factory at any time, whereas the quantities that the accounting department shows are actual orders less material that has flowed outward through requisitions for use. A second possible choice of variation which one commonly sees is differences in prices. For example, at the books at the local plant, the cost of material may include freight, whereas the accounting department excludes freight. Changes in specification or in price, on record in one place and not in another, will lead to discrepancies. These and other types of nonuniformity often turn up in the comparison of the main sample with the audit and with the book-inventory. They are not sampling errors, but they are discrepancies that require systematic correction.   Discoveries of such nonuniform practices are often of great help in the administration of a company.

Will 2 samples agree?   Will 2 complete counts agree?   The precision of a sample is not established by comparison against a complete census UNLESS the complete census is THE equal complete coverage for this sample.   Simultaneous trials of complete count and sample, just to see whether sampling will give the same result, is in my opinion a woeful waste of funds.*   In my own practice, I have steadfastly refused to engage in such tests.   A simultaneous test, for the sake of comparison, to see if sampling will work, is almost sure, I believe, to impair the results of both the complete coverage and the sample.   The 2 results might still agree, of course.   Moreover, we know by theory, in advance, better than any number of comparisons could possibly establish, what the performance of a sampling procedure will be, *provided we really carry it out according to plan.*

The only circumstance in which a direct comparison is possible between 2 samples is when the 2 samples have been drawn from the same complete coverage, in which case the complete coverage is THE equal complete coverage for both samples.   Examples of direct comparisons abound in the censuses of most countries today, where most of the published tables are the results of samples drawn from the complete census (one of the many ways in which sampling is used in censuses).

The most common reasons for a significant difference between 2 surveys, complete or sample, is that the 2 surveys really do not cover the same frame, they do not have the same questions and definitions, one is much more extensive in questioning than the other, or they were carried out at different dates, or the equality of the 2 field-forces is questionable.   Another explanation is the existence of blemishes and blunders made in carrying out one or both of the 2 surveys.   This is why careful statisticians now prescribe for any survey an audit or test of the performance of the workers, to find out just what happened.   The explanation of a significant difference is not sampling error.

One might suppose that 2 people sent out to count dwelling units in an area would agree with each other exactly or nearly so.   It is a fact, though, that the counting of dwelling units or of anything else may be difficult.

* There are many examples, published and unpublished, but I shall be content to mention only two: (1) Winston C. Dalleck, "Inductive accounting: settling interline accounts by sampling methods," *Industr. Qual. Contr.*, vol. xiii, 1956: pp. 12–16. (2) W. E. Courtright and A. A. Procassini, "Inventory evaluation by sampling," *ibid.*, vol. xiv, 1958: pp. 16–20.   Both these examples, in spite of my criticism of carrying out both sample and complete coverage for comparison, led to the adoption of sampling methods to replace complete coverages.

Even with careful definitions of what constitutes a dwelling unit, one continually encounters questions because of unusual arrangements for living, cooking, and sleeping that the definitions simply do not cover. In any case, a careful count of dwelling units requires one to pause at every door that could lead to a dwelling unit, and to enquire into the family composition, cooking facilities, and economic dependence and independence of the people that live there. Nevertheless, under most urban conditions in this country, counts of dwelling units by 2 different people are often remarkably close, such as 88 and 88; 93 vs. 98; 25 vs. 24; etc. Occasional differences as great as 10% are to be expected.

EXERCISES

1. What are some of the kinds of errors and biases and of other uncertainties that occur in statistical data? Which ones are the same both for "complete counts" and for samples? Which ones are different?

2. Are we justified in assuming that the various kinds of errors in a statistical enquiry cancel each other? Explain.

3. Suppose that interviewers miss somehow or other 5% of the dwelling units in every area in a sample.

a. Does this have the effect of decreasing the size of the intended sample? (Yes.)

b. How much? (5% of the intended dwelling units.)

c. What biases may this 5% shortage introduce?

d. May these biases be serious? (Yes. Why?)

> **Remark 10.** If the interviewer consistently passes by any special type of dwelling unit (upstairs, rear, behind trees, off the main road), the result is not a mere random reduction in the size of the sample. It is instead the introduction of a bias, even though the special type of dwelling unit omitted be but 5% of all dwelling units. The omission may wipe out or shrink seriously certain special groups of people, with a resulting distortion of some income-level, of some rent-level, of some age-group, of the purchases of some products, or of some types of readers.

4. To obtain some item of information concerning a population of people in an area there are 2 alternatives: (*a*) a complete census; (*b*) a sample. In both studies, in 20% of the homes visited there will be either a refusal or no one at home who can answer the questions. In which method will the bias of nonresponse be the more serious; in the complete count, or in the sample?

5. In what ways do careful statisticians try to decrease the amount of nonresponse to the lowest level feasible?

6. How is it that one does not encounter nonresponse (not at home) in a quota method, and few refusals?

7. Why is it essential, in the planning of a survey, to be aware of the distinction between (1) errors in data that are common to complete censuses and samples alike, and (2) errors of sampling?

8. A company plans to take a survey, and a statistician has computed that 2400 interviews carried out by a prescribed method of sampling will provide the information with the precision required. They expect to encounter nonresponse on the first call of about 33%.

a. Will they solve the problem of nonresponse by boosting the size of the sample to 3600, so that there will be about 2400 responses at the first call? (No.)

b. What is the remedy for nonresponse?

9. Read the following passage.

> ... a bill which was introduced by Senator Pepper and which would enable the Secretary of the Interior to carry out a program of research and experiment with respect to natural sponges, to study the disease of natural sponges and to propagate and plant natural sponges.
>
> This paper is a further contribution to the subject ... and reports the results of a survey which was undertaken during the summer of 1950. Questionnaires were sent to distributors of natural and synthetic sponges, requesting information on preferences and purchasing habits. There are 91,124 retail outlets in the United States engaged in selling sponges. Of these, 33,221 handle hardware and 57,903 are drugstores. The questionnaire, prepared by the Fish and Wildlife Service, was distributed to representative stores in these groups through the National Association of Retail Druggists and the National Retail Hardware Association, each of which sent out 400 copies at its own expense. Both the preparation of the questionnaire and the stratification of the samples were done by use of modern methods of opinion research and by applying rules of procedure established by experienced poll technicians. There were 206 answers received. This is a 25.7% response; 20% is judged to be sufficient to be representative. The answers obtained by the questionnaire have provided a valuable basis for the evaluation of certain factors in the marketing of sponges.*

a. What methods do you suppose the National Association of Retail Druggists and the National Retail Hardware Association used for distributing the questionnaire?

b. Is 25.7% response sufficient? (No.) Is 20% response "representative"?

c. What figure on response would have been sufficient in this study?

d. Do you agree that "The answers obtained by the questionnaire have provided a valuable basis for the evaluation of certain factors in the marketing program?"

* Richard A. Kahn, *Proceedings of the Gulf and Caribbean Fisheries Institute, 3d Annual Session*, Nov. 1950.

e. What are "representative stores?"

f. Do the methods described correspond with "modern methods of opinion research?"

g. Do you think that stratification of the stores helped the results under the conditions of the survey? (I am doubtful.)

10. The sample for a survey contained 2200 dwelling units, and was to elicit information from the housewife. In all, there were 300 failures; 200 not at home after 4 recalls, and 100 refusals.

a. Why is it not satisfactory to substitute 300 housewives from neighboring dwelling units?

b. Why not just increase the size of the sample in the first place by 300 dwelling units or so, and stop worrying about the problem?

11. Criticize the following classification of bias (an excerpt from a worthy paper by my friend C. A. Moser, "Interview bias," *Review of the International Statistical Institute*, vol. 19, 1951: p. 1).*

... it may be worth while, at the outset, to distinguish the main categories of bias ... :

A. Bias in the sample—
    i.   Due to sampling from a frame which is inadequate, inaccurate, or incomplete.
   ii.   Due to sampling from a frame which contains some form of periodicity which is not taken into account.
  iii.   Due to incomplete achievement of the selected sample—on account of refusals, non-contacts, etc. (non-response in mail questionnaires falls into this category).
  iv.   Due to the use of a non-random method of selection. Purposive and quota sampling fall into this category. In quota sampling, the selection of respondents is entrusted—within set quotas—to the interviewers, and the possibility of biased selection arises.

Remark 11. The 1st error is not an error of sampling at all, as it is not a random variable. It would be present and in the same amount even if the sample were increased to 100% of the frame (a complete coverage). The error is in the frame, parallel to the wrong frame for the count of pigs in Norway (p. 47). Complete coverage or sample, it is necessary to cover the right area.

With respect to the 2d alleged error, periodicity in the frame is not a cause of bias. Periodicity may, however, cause loss of precision if one uses a patterned sample (e.g., a random start and every 20th unit thereafter)—a good reason to avoid patterned samples except with material known not to possess harmful periodicities.

The 3d error is as bad in a complete coverage as in a sample, as we have learned. It is not an error of sampling: it is a survey-error of Type II that afflicts complete coverages and samples equally.

* I wish to thank Dr. Moser for his generous consent to my comments.

12. In the magazine *Business Week* (New York) for 2 February 1952 occurred an article entitled "Industry's capital spending plans for 1952 through 1955." This is always an important subject, and the writers of such articles perform a real public service. The student will derive benefit by trying to answer the following questions, after he reads the excerpt below.

1. Is the description of the sampling plan adequate? Does it enable one to calculate the standard error of any result?

2. Would you judge, from the description, that the sample was a judgment sample?

3. Would you have confidence in such data if you were trying to make a forecast of industry's capital spending? (The 1st paragraph says that this is not a forecast. If it is not a forecast, what is it?)

4. If you had a complete list of all firms, and if the officers in each one answered the questionnaire to the best of their ability, would you be able to make a perfect forecast? (No.)

5. If you had the problem in hand, and wished to use a probability sample, what lists would you use for a frame?

6. Would a complete coverage of all firms, and careful answers to the questions, enable you to make a perfect forecast? (No.)

This is a report of the fifth annual survey of Business' Plans for New Plants and Equipment conducted by the McGraw-Hill Department of Economics. It is not a forecast, but a summary of plans industry now has for investment in capital goods.

The companies cooperating in this survey employ more than 60% of all workers in industries where capital investment is highest. That includes chemicals, oil, railroads, electrical machinery, autos, utilities, and steel. These industries account for two-thirds of all spending for capital goods. The companies included in the sample were mostly the bigger companies in these industries.

In other industries, coverage was not so complete. But the participating companies were carefully picked to make up a representative cross section.

In all, the sample includes companies employing over 5-million workers. That is about one-quarter of the total employment of all industry.

Physical capacity was measured by figures supplied by the cooperating companies themselves. All companies were asked to select their own measures of physical output. A steel company may use ingot tons, for example, while an aircraft engine maker compares the rated horsepower of the engine he produces. The results are the only available direct measures of the expansion in industry since 1939.

The figures on capital expenditures in this report are not directly comparable with those given in previous McGraw-Hill reports on capital spending. Instead, survey data for 1952 and succeeding years has been put on the same basis as the revised series published by the Department of Commerce for the years 1945–51. Figures on industries' plans are therefore comparable with the Commerce figures for back years. Figures on capacity are comparable with those published in last year's McGraw-Hill report, entitled "Industry Expands."

The McGraw-Hill study makes no allowances for the change in the value of the dollar. The totals reported by industry are compiled exactly as they are given. Therefore, to obtain the actual physical difference between one year's spending and another's, you must remember that the value of money has shrunk.

13. Explain the limitations in the conclusions drawn in the following excerpt, which came from a bulletin that was circulated in a clinic for tuberculosis in April 1953, and which came to me by chance. The figures refer to the results of a campaign in which the public was invited to come in for an X-ray of the chest, free of charge. In particular, what about the statement that in the District of Columbia tuberculosis is approximately 6 times as prevalent among persons 54 years of age or older as among those between 14 and 54? Is this a sample or a chunk? (A chunk.)

The authors comprised a special committee which conducted a two-year follow-up study of the community-wide chest X-ray survey in the District of Columbia from January 12 to June 30, 1948.

Among 349,988 adult residents X-rayed during the survey, evidence or suspected evidence of tuberculosis was found among 6159 persons. Of this number 2758 were new cases, that is, previously unknown to health authorities.

More than 40 per cent of the tuberculosis found in the survey, according to the report, was among persons 55 years of age and older, although this age group comprised only about 10 per cent of those examined. Only 39 per cent of these cases were previously known to health authorities in contrast to 79 per cent in the age group from 15 to 24. The report points out that in the District of Columbia tuberculosis is approximately six times as prevalent among persons 54 years of age or older as among those between 14 and 54. This, the report states, is of "practical importance."

14. Read the excerpt at the top of the next page.

(*a*) Do you think that 10,600 questionnaires were enough to send out for the purpose (not clearly stated)? (The amount and distribution of the nonresponse is more important then how many go out in the first place.)

(*b*) Do you know of any "general principles of sampling" that say that 20% is sufficient response?

(*c*) What proportion would be enough?

(*d*) Suppose that the response had been 100%; could you generalize for all of Chicago? (I wouldn't).

(*e*) For the United States? (No.)

(*f*) Is there any information in the 10,600 questionnaires? (I suppose so, but I know no statistical theory that will dig it out.)

PUBLIC OPINION SURVEY IN THE ELEVENTH ILLINOIS
DISTRICT, NORTHWEST SIDE OF CHICAGO

Extension of Remarks
of
HON. TIMOTHY P. SHEEHAN
of Illinois

In the House of Representatives
Monday, June 9, 1952

Mr. SHEEHAN.   Mr. Speaker, on April 17, 10,600 questionnaires were
sent to a representative cross section of the constituency of the Eleventh
District in the northwest side of Chicago, asking for their opinions.

Of this number, 2,097 questionnaires were completed and returned, and
these have been tallied on the basis of percentages of the "yes" and "no"
answers, eliminating the qualified answers and the absence of answers.
This represents a return of 19.8 percent which, under the general principles
of sampling, is considered a very good return.

The question concerning the policies of action in Korea are based upon
only the affirmative replies to the three parts of the question.

15. The following paragraph came from a report based upon a quota
survey.   I have modified the actual wording in order to conceal the
identity of the source, but I have retained the sense.

a. Do you think that the various segments of the sample (urban, rural,
white-collar workers, housewives, etc.) are "representative" of the same
groups in the universe?

b. What does the interviewing procedure guarantee?   (Only that the
"quotas" will be filled; nothing else.)

c. What if the quotas are outmoded by obsolete census data?

d. Does a probability sample depend on quotas?   (No.)

> The sample was a quota sample. . . .   Weights are then established to
> bring the sample frequencies up to the quotas as specified.   This balances
> all component segments of the sample, so that the resulting estimates for the
> universe are entirely proper.   The only assumption involved in this re-
> weighting is that the segments which are weighted are themselves represen-
> tative of the parent group in the universe that they represent.   This, of
> course, is guaranteed by the interviewing procedure itself.

16. I clipped the excerpt below from the *Journal Herald* (Dayton) on
19 May 1954, while at the airport in Dayton waiting for a connexion to
Indianapolis.

a. Suppose that some of these blanks came in to the Courthouse
Editor, filled out to show preferences, do you think that they would indicate
to him any definite course of action?

b. About how many people would see the questionnaire? (I don't know.)

c. About what proportion of these people would send it in? (I don't know in numbers, but perhaps 5% at the most.)

d. What could the results indicate? (I don't know.)

Journal Herald, Dayton, 19 May 1954, p. 1

CITIZENS ASKED VIEWS ON COURTHOUSE ISSUES

What should be done about Montgomery county's courthouse problem?
The Journal Herald is publishing a questionnaire to sample public opinion.

County commissioners are taking steps to place sale of the present "new" and "old" courthouse site and erection of a new courthouse and juvenile center on the November ballot.

Readers are invited to fill out the form below and return it to The Journal Herald by mail.   Names will not be published.

Courthouse Questionnaire

To: Courthouse Editor,
Journal Herald,
111 East Fourth Street,
Dayton 2, Ohio

DO YOU FAVOR:	Yes	No
Building a new courthouse and juvenile center	—	—
Selling property occupied by courthouse and jail	—	—
Moving the old courthouse to a new location ..	—	—
Tearing down the old courthouse............	—	—
Keeping old courthouse on its present location .	—	—

Name
Street
City

Are you a Registered Voter ....   —   —

PART II

Replicated Sampling Designs

CHAPTER 6

Some Simple Replicated Designs

How I envy the clarity of vision that comes to the travelling salesman in a railway buffet-car at the third highball! How simple the great problems become!—Clarence B. Randall, as reported in *Business Week* (New York), 12 June 1954: page 192.

A. INTRODUCTION TO REPLICATED SAMPLING

Purpose of this chapter. The purpose of this chapter is to introduce some simple replicated sampling designs, with some general remarks that will apply also in more complex situations. It is possible to replicate any sample-design merely by halving the sample and repeating the same sampling procedure to make 2 replications. This book will teach a special system of replication, which has a number of advantages. The distinguishing feature of the design is 2 or more subsamples, drawn and processed completely independent of each other. The chief advantage of replication is ease in estimation of the standard errors. Replication also renders easy the evaluation of the bias in the estimating procedure (p. 425).

The sampling procedure to be learned here draws directly the serial numbers of the sampling units for investigation (interview, test). Intermediate units come in with replacement; the units of the last stage come in without replacement (although we could, if we wished, draw with replacement). The theory to use for the total variance of an estimate is therefore the simple theory of the single stage. In our first applications, we shall use 10 replications in 10 independent subsamples.

This system of replicated sampling is now in regular use here and abroad in social and economic studies of many types, including estimates of acreage and of yield, in marketing research, and in studies of attitudes, in program-listening, in the appraisal of buildings and of other kinds of physical plant, in the testing of industrial materials, and in studies of accounting records. The footnote on page 186 gives some brief history.

Every student of statistics knows the formula $\sigma_{\bar{x}}^2 = \sigma^2/n$ (Ch. 17), for the variance of the mean of a random sample of size n. We may regard n as the number of replications of a sampling plan that draws 1 sampling unit in each replication. $\bar{x}$ is then simply the mean of the n estimates $x_1, x_2, \ldots, x_n$ obtained in the n replications.

Some characteristics of the system of replicated sampling to be learned here. *

1. The zones within a stratum will contain equal numbers of sampling units. Every sampling unit in the stratum will have the same probability as any other to come into the sample, a fact that simplifies greatly the instructions for the tabulations.

> Because the sampling units all have the same probability, so do all the dwelling units, all the people, cows, business establishments, unless we deliberately alter the probabilities within a sampling unit to select (e.g.) 1 person for interview in households where more than 1 person is eligible (p. 240). The special point here is that the system of drawing eliminates the series of multipliers that one must introduce for the varying probabilities of selection in multistage sampling if he draws intermediate units without replacement—a simplification that one can appreciate only if he has used both methods.

2. The zones provide fine geographic stratification, additional to any other stratification that is built into the design.

3. There is no need for special consideration and theory for the selection of primary units of different sizes or of extra-large size. The sampling procedure automatically gives to any primary area probability in proportion to the number of sampling units therein.

> It is thus a very simple matter to give to any primary unit (*a*) probability in proportion to the number of dwelling units therein, or (*b*) probability in proportion to the square root of the number of dwelling units therein, or (*c*) any other probability that we choose. This we do by fixing the number of sampling units in each primary unit. To lower the probability of any primary unit, as we may wish to do if the cost of creating segments is extra-high in this primary unit, we merely increase the size of the sampling units therein, so that there are fewer of them. However, we make no use of the probability of drawing any primary unit, as it does not appear in the estimates nor in their variances.

4. There is complete freedom in the basic design, such as in the size and distribution of the sampling unit, in the size of the segments that make

---

* The reader may prefer to return to this section after he has acquired some familiarity with the replicated method.

up the sampling unit, in the modes of stratification, and in the formula of estimation.

5. The same procedure of replication applies (with reduced efficiency, of course) to conditions where there is not reliable prior information nor reliable census data.

Some examples of sampling units.   The first thing to enquire into is the choice of sampling unit.   Some examples of sampling units appear below:

1 employee, or 4 or 8 or more employees in 1 factory	(Secs. B and C)
1 line in a ledger or a page, or 2 consecutive pages in a ledger	(Ch. 8, Sec. A)
Revenue on a sheet of paper	(Ch. 8, Sec. B)
1 segment of area containing households	(Chs. 10 and 11)
1 or 2 segments of area confined to a "block"	(Chs. 10 and 11)
1, 2, or more segments of area confined to a county or to some other area	(Ch. 11)
A business establishment listed on a card	(Ch. 7)
An acre of crop	
A small plot of crop in a field	
A dollar of investment	
A unit in a shipment of manufactured product	
1 lot of manufactured items shown on 1 line of a list	(Ch. 15, Sec. C)

The procedure of selection

1. Decide on the sampling unit.

2. Construct the frame, which will be a list of the sampling units, each with identification so that we can find it if the random numbers draw it (Ch. 3).   Give a serial number to every sampling unit, or construct a rule that will give a serial number to any sampling unit on demand.   (Our first example occurs on page 92.)

> The sampling units may come in convenient packages, called primary units, and it may be advantageous to list only the primary units.   If so, show for each primary unit the number of sampling units therein.   A primary unit may be a factory, a volume of records, a drawer, a cabinet of a file, a census area, a county, or some other natural and convenient subdivision of the material to be studied.
>
> Cumulate the serial numbers of the sampling units, primary unit by primary unit.   The cumulated totals give a serial number to every sampling unit (p. 92).

3. Decide the zoning interval Z (theory later).

4. Construct the sampling table in 2 or more subsamples by use of a table of random numbers.   This table selects the sampling units for the sample.   (Our first example is on page 94.)

The frame assimilates a long string of beads, each bead being a sampling unit.   Each bead bears a serial number.   Markers show for identification the end of one primary unit and the beginning of another.   Other markers, uniformly spaced, will divide the entire string of beads into zones.

The number of sampling units (beads) in a zone will be determined by the intended average number of people, accounts, dwelling units, or other items in a sampling unit, by the number of replications per zone, and by the total number of sampling units required in the sample.   The symbol Z will denote the number of sampling units in a zone.   The zone is merely a convenient subdivision of the frame, containing Z sampling units.

Replicated drawings.   Replication means the selection of 2 or more sampling units from every zone.   The 1st random number between 1 and Z (the zoning interval) draws a sampling unit for Subsample 1; a 2d random number between 1 and Z draws a sampling unit for Subsample 2; etc.   In some surveys there will be 10 subsamples in a zone (Sections B and C of this chapter, and elsewhere).   I prefer 10 subsamples unless there would be a distinct gain in statistical efficiency from use of fewer samples drawn from narrower zones.   This question is discussed further in Chapter 21.

The subsamples will bring forth different results—different numbers of yes and no, different numbers of dwelling units, and different numbers of males, females, and children.   These numbers will be random variables. It is the replication of the sampling design in every zone that creates the simplicity in the theory and in the preparation and in the computations.

The simple illustrations in the remainder of this chapter may help to clarify the foregoing introduction and to establish general principles of procedure for replicated sampling that the reader will find applicable to his own problems.

B.   A SAMPLE OF EMPLOYEES IN SEVERAL FACTORIES

Description of the problem, the universe, and the frame.   We may now turn to some simple sample-designs, all of which are drawn from actual experience.   In the first example, the management of a concern that owns a dozen factories wished to seek some information from their employees. The question was "Do you believe that the management studies the

suggestions offered by employees?"   There were some preliminary indications that the proportions of yes and no would be about equally divided.
A rather wide tolerance of 10% or 15% would suffice.   I prescribed a
sample of about 100 employees, to be drawn by the following plan.   The
justification for the size 100 was theory.

The universe was about 9600 employees in all the factories.   This
figure was only rough, and it is important to note that in the design of
the sample there was no need of an exact figure for any one factory nor
for the total: approximations introduce no bias, and no complication in
the procedure.

To commence the design of the sample, we ascribe to each factory a
number of sampling units, which will be approximately but not necessarily
equal to the number of employees therein.   Each sampling unit, in the
design just now under consideration, will usually contain 1 employee, but
sometimes 0, sometimes 2 (see Table 2).   We should not worry if an
occasional sampling unit contains 3 employees.   Inequalities in size do
not cause bias.   It is only necessary that every employee lie in some
sampling unit.   The cause of the nonuniformity in size is the roughness
in the figures on size, furnished in advance.   All sampling units will
nevertheless have equal probabilities; hence all employees will have equal
probabilities (Exercise 3, page 97).

The frame in this problem was the list of sampling units shown in
Table 1.   The definition of a sampling unit is in Table 2.

The intention in this survey was to use a self-filled questionnaire.   The
superintendent of each plant drew the sample by following instructions
mailed from headquarters.   The main expense involved was to pay the
superintendent while he drew the sample, and to pay the employees in the
sample for their time spent on the questionnaire.   In Section C, we shall
suppose that a supervisor will travel from one plant to another to draw
the sample and to administer the questionnaire.   It may then be better to
draw the employees in groups, so as to decrease the number of factories
in the sample, and to decrease the cost of travel.

Remark 1.   The number of sampling units once ascribed to a factory
in Table 1 remains fixed.   The number of employees in the factory may
change from day to day, but not the number of sampling units.

Formation of the sampling units in the factory.   The frame (Table 1)
gives 800 sampling units to Factory No. 1, 800 being an approximate
figure for the number of employees there.

The procedure by which to form a sampling unit in Factory No. 1 is
to give each employee a serial number in any order, from 1 on up.   Suppose

that this factory, when we go there to interview, has actually 812 employees, not 800.   Give the serial numbers 001 to 812 to the 812 employees.   Then form 800 sampling units on the spot in the manner illustrated in Table 2. Sampling units 001–012 contain 2 employees each; sampling units 013–800 contain 1 employee each.   Had the exact number been 790 on the day of the interviews, then sampling units 001–790 would contain 1 employee each, while sampling units 791–800 would be blanks.   Random number 793, for example, would then draw no sample.

TABLE 1

THE FRAME.   THE NUMBER OF SAMPLING UNITS IN EACH FACTORY
The number of sampling units in any factory is simply a
rough figure for the number of employees therein

Factory	Number of sampling units	Serial numbers of the sampling units
1	800	001– 800
2	1500	801–2300
3	1000	2301–3300
4	400	3301–3700
5	100	3701–3800
6	100	3801–3900
7	1000	3901–4900
8	1500	4901–6400
9	2000	6401–8400
10	600	8401–9000
11	500	9001–9500
12	100	9501–9600

Use of rough counts.   The origin of the figure 800 for the number of sampling units in Factory No. 1 is an approximation to the number of employees there.   It could come from any source, such as from last month's pay roll.   This example is our first encounter with a rough count. We shall meet many examples later.   A common example is the use of Census data, which are necessarily to some extent out of date by the time they are printed.   We often in practice make rough counts purposely, as by quick cruises over an area, or by quick eye-estimates of the number of terminals shown on a page, in a long series of pages.

The important point (which I may repeat from time to time) is that rough counts do not cause bias.   They merely fix the number of sampling units in an area, or on a page, or in a factory.   They fix the probability that a random number will fall within an area (county, block, factory), and the probability of selection of a sampling unit therein.

We use rough counts, instead of no counts at all, to enhance the efficiency of a sampling plan, for which purpose rough counts are almost as good as exact counts would be. Roughness may lead to some slight increase in the variance of an estimate of a total; this we shall evaluate in Chapter 18. The variance of our estimates, whatever they be, and from whatever source, will always be known.

TABLE 2

DEFINITION OF A SAMPLING UNIT IN FACTORY NO. 1

Sampling unit	Employees with these sampling numbers
1	1 and 801
2	2 and 802
3	3 and 803
.	.
.	.
.	.
12	12 and 812
13	13
.	.
.	.
.	.
800	800

Table 2 shows one way to form the exact number of sampling units required in Factory No. 1. The section entitled "The formation of a sampling unit in a factory," on page 100 shows several other ways. For the theory to be valid, it is only necessary that we form the specified number of sampling units, and that each one have a serial number. The serial number will run from 1 on up to the specified number of sampling units. The groups may be unequal; some may even be blank.

Not all ways of forming groups are equally efficient, however. Some ways will produce better estimates than others. The way that we formed the sampling units in Table 2 above is probably the quickest, easiest, and most efficient in that situation.

The method of selection. This sample will be laid out in 10 subsamples, as wide zones in this type of investigation are fully efficient. All sampling units will have the same probability. We note first that there will be about 10 employees per subsample, because the total sample is to be about 100. As there are about 9600 employees in total, 10 in each subsample, the zoning interval will be

$$Z = \frac{9600}{10} = 960 \tag{1}$$

Once we know the zoning interval, we are ready to draw up the sampling table, to show which sampling units of the frame are in the sample. A sampling table is always made with random numbers.

As the total number of sampling units was large (9600), and as there was no reason to suspect periodicities that might lead to high variances (Remark 3), I prescribed, for simplicity, subsamples to be formed by a systematic or patterned selection of sampling units, by the simple addition of 960 to every random start. Each subsample consisted, then, of a random unduplicated start between 1 and 960, and every 960th sampling unit thereafter. Table 3 is the sampling table, formed in this manner. Every sampling unit of the frame whose serial number falls under Sub-sample 1 of the sampling table is in Subsample 1, and similarly for Sub-samples 2, 3, and so on to 10.

Remark 2. The random starts in the top zone of Table 3 came from Kendall and Smith's *Random Sampling Numbers* (Cambridge University Press, Tracts for Computers No. 24, 1951), 65th Thousand; Columns 38, 39, 40; beginning in line 13; 001–960. This was the spot that I had marked in my own copy of this table at the completion of another sampling job—a procedure that I recommend as a methodical way to designate the starting-point for the next sampling table.

TABLE 3

THE SAMPLING TABLE, 10 RANDOM STARTS AND EVERY
960TH SAMPLING UNIT THEREAFTER
$Z = 960$

Zone	1	2	3	4	5	6	7	8	9	10
0001–0960	502	147	032	727	430	301	251	172	063	907
0961–1920	1462	1107	992	1687	1390	1261	1211	1132	1023	1867
1921–2880	2422	2067	1952	2647	2350	2221	2171	2092	1983	2827
2881–3840	3382	3027	2912	3607	3310	3181	3131	3052	2943	3787
.										
.										
.										
8641–9600	9142	8787	8672	9367	9070	8941	8891	8812	8703	9547

Remark 3. We shall not always form systematic or patterned subsamples, as we did here; we shall sometimes instead read out fresh random numbers in every zone; see page 105 for an example. Simple patterned subsamples are easiest and will give the same precision as fresh random numbers except possibly in the presence of periodicities in the frame. Periodicities or no, patterned subsamples will always give us a valid estimate of the standard error, so that there will never be any doubt about the precision actually reached.

Remark 4. Every factory contributed one or more employees in this sample. In Section C we shall decrease the dispersion of the sample by drawing the employees in groups of 8. Some factories will then contribute no employees to the sample, while some will contribute more than they do here.

Computations. The results are in Table 4. Every random number in Table 2 brings forth results from a sampling unit—0, 1, or 2 yes, or 0, 1, or 2 no, or a blank, as when a random number strikes a blank. There are 10 values (f_i) of the proportion yes, 1 for each subsample. The overall proportion yes in all 10 subsamples is

$$f = \frac{x}{y} = \frac{52}{101} = .52 \tag{2}$$

wherein x is the number of yes in all 10 subsamples combined, and y is the number of yes and no. The estimate of the standard error of f obtained by use of the range of 10 subsamples is

$$\hat{\sigma}_f = \frac{f_{\max} - f_{\min}}{10} \qquad \text{[P. 200]}$$

$$= \frac{.70 - .36}{10} = .034 \tag{3}$$

wherein w is the range of the f_i, the difference between the highest and the lowest. The circumflex (^) denotes an estimate. This figure gives $\pm.10$ for an estimate of the 3-sigma limits of sampling variation (p. 55).

Illustration of the usefulness of blanks. The following excerpt illustrates a common difficulty. I quote first a letter from a friend and well-known statistician in a far-off country; then my reply, calculated to show how easy it is to maintain equal probabilities with blanks.

Question that came in the mail. Assume that the proportion of areas to be chosen from each town has been decided in advance, and likewise the proportion of households that is drawn into the sample from each sample area. A difficulty arises because the stipulated proportions can not be maintained exactly. Thus, in a particular town, 1 area in 6 is to be selected for the sample; then 1 household in 4 from these areas that fell into the

TABLE 4

RESULTS

Zone	Subsample										1–10
	1	2	3	4	5	6	7	8	9	10	
1 Yes	1	1	0	1	0	0	1	0	0	0	4
No	1	0	1	0	1	1	0	1	1	1	7
2 Yes	0	0	0	0	1	0	1	1	1	0	4
No	1	1	1	1	0	1	0	0	0	1	6
3 Yes	1	1	1	0	1	1	0	0	1	1	7
No	0	0	0	1	0	0	1	2	0	0	4
4 Yes	1	0	1	1	0	1	1	0	1	0	6
No	0	1	0	0	1	0	0	1	0	1	4
5 Yes	0	1	0	0	1	1	0	0	1	1	5
No	1	0	1	1	0	0	1	1	0	0	5
6 Yes	0	0	0	0	1	0	B*	1	0	1	3
No	1	1	1	1	0	1	B	0	1	0	6
7 Yes	0	0	1	1	1	0	0	1	1	1	6
No	1	1	0	0	0	1	1	0	0	0	4
8 Yes	0	1	0	1	0	0	1	0	0	0	3
No	1	0	1	0	1	1	0	1	1	1	7
9 Yes	1	1	0	1	0	0	1	1	1	1	7
No	0	0	1	0	1	1	0	0	0	0	3
10 Yes	0	0	1	1	1	1	0	1	1	1	7
No	1	1	0	0	0	0	1	0	0	0	3
x_i (yes)	4	5	4	6	6	4	5	5	7	6	52
y_i (yes and no)	11	10	10	10	10	10	9	11	10	10	101
f_i $(x_i : y_i)$	.36	.50	.40	.60	.60	.40	.55	.45	.70	.60	.52

* B denotes a blank; no employee in the sampling unit.

sample.   Suppose that there are 23 areas in the town.   How can we draw 1 area out of 6?   Is it advisable to change the proportion of households so as to draw exactly $\frac{1}{24}$th of the households?   What should we do when an area contains 5 households, or 6, or 7, these numbers not being exact multiples of 4?

Reply.   Do not alter the proportions.   It is much more important to maintain constant probabilities over all areas and to avoid weights than it is to draw any designated size of sample.   It is easy to maintain constant probabilities by the device of adding blanks to fill up incomplete groups. Thus, here, where there are 23 areas, one may form 4 groups of areas whose serial numbers are 1–6, 7–12, 13–18, 19–24, the 24th being a blank.   Then with random numbers draw 1 area from each group.   If in the 4th group your random number draws area No. 24, you have no sample from that group.   If your random number draws No. 19, 20, 21, 22, or 23, you have real areas.

Likewise, within any area that you draw, form groups of 4 successive households, and complete the last group with blanks if necessary.   Draw one household from each group.   If you draw a blank from the last group, that group contributes no household to the sample.

You will maintain by this suggestion the constant probability of 1 in 24. No alteration or adjustment is then necessary in the estimates that you form, nor in the formulas for the variance.   Moreover, the field procedure is simplified, because the work in each town may proceed independently of the work in the others, as it is not necessary to tie an incomplete group in one town to the beginning of the next group in another, as is done sometimes to avoid incomplete groups.

EXERCISES

1. Show that one can form exactly 960 new sampling units, no more and no less, by starting with any of the original serial numbers between 1 and 960, and adding thereto every 960th serial number thereafter (a patterned or systematic formation).

This is so regardless of the number of sampling units in the frame.

The sampling numbers in Table 3 select at random and with equal probabilities 10 of these 960 sampling units.

2. Why are the 10 results obtained from the 10 subsamples formed by Table 3 random variables, even though each subsample is patterned or systematic?

3. *a.* Show that, before the sample is drawn, the probability that any sampling unit will fall into Subsample 2 is 1/960.

b. The probability that any employee will fall into Subsample 2 is also 1/960.   (Note that every employee belongs to 1 and only 1 sampling unit, and that the drawing is done without replacement.)

c. The probability that any employee will fall into any one or another of the 10 subsamples is 1/96.   (See also page 363 in Ch. 16.)

4. Show that the probability that any factory will contribute an employee to the sample is proportional to the number of sampling units in that factory.

5. One method of sampling, used much in previous years, by me as well as by others, was to take a random start and every kth sampling unit thereafter (a patterned or systematic sample).   Under what conditions:

a. Is this a probability sample?

b. Can you estimate the standard error from the sample itself?

Remark 5.   As there is no replication, there is no valid way to compute an unbiased estimate of the variance of an estimate made by this procedure. However, one may compute an approximation to the variance by assuming that 2 successive sampling units drawn by this procedure were drawn at random from a zone (stratum) of $2k$ successive sampling units.   There is unfortunately no way of knowing from the sample alone whether the expected value of this estimate is too high or too low.   However, in justification of the single random start, one may add that a long background of experience with a number of materials (including human populations, farms, and business establishments) shows that the estimate of the variance so produced is for many characteristics not far wrong.   Under such conditions, use of the single random start may be called a probability-sample.

The replicated method is so simple to apply that there is no point in taking a chance with an estimate that raises questions.

One must bear in mind, however, that whether we use the replicated method or any other, our estimates of variance are random variables, and any estimate of a variance may be too high or too low.   The theory for the variance of an estimate of a variance appears in Chapter 17.

C. CONCENTRATING THE SAMPLE IN GROUPS OF EMPLOYEES

Some considerations of cost.   If a supervisor from headquarters were to visit every factory in the sample of Section B, he would have to visit all 12 factories.   We don't like to train an interviewer and send him a long distance to supervise but 1 interview (as in Factory 12).   We therefore propose to redefine a sampling unit as several employees, specifically 8, which will be the minimum number of interviews in a factory.   We shall require a bigger sample in total to build up the precision to its former value, but the savings in travel will more than pay for these additional interviews.

Procedure for drawing employees in groups.   The problem and the method are general.   This illustration, though oversimplified, will teach the method.

Suppose that we return to the illustration of Section B and decide that we wish to draw a sample of about 160 employees in groups of 8.   Our new sample will consist of about 20 sampling units, and each sampling unit will contain about 8 employees. We shall suppose that the new sample drawn in groups of 8 (because of serial correlation) will deliver about the same precision as the sample of 100 employees drawn singly in Section B, but that it will cost less.

TABLE 5

THE NUMBER OF SAMPLING UNITS IN EACH FACTORY, WHERE
NOW EACH SAMPLING UNIT HAS AN INTENDED SIZE
OF 8 EMPLOYEES

Factory	Approximate number of employees	Number of sampling units	Serial number of the sampling units
1	800	100	001– 100
2	1500	187	101– 287
3	1000	125	288– 412
4	400	50	413– 462
5	100	12	463– 474
6	100	13	475– 487
7	1000	125	488– 612
8	1500	188	613– 800
9	2000	250	801–1050
10	600	75	1051–1125
11	500	62	1126–1187
12	100	13	1188–1200

We remake the frame, because each factory will have a reduced number of sampling units. Our new frame is Table 5, which will contain a total of 1200 sampling units (the last serial number in Table 5). The number 1200 comes from dividing 9600 by 8, 9600 being the advance figure on the approximate total number of employees. As we desire a total of 20 sampling units in the sample, 2 per subsample, the zoning interval will be

$$Z = \frac{1200}{2} = 600 \qquad (4)$$

The next step is to create Table 6, a new sampling table similar to Table 3. There will be 10 random starts in Zone 1 between 1 and 600. The sampling numbers in Zone 2 come by addition of the zoning interval $Z = 600$ to the numbers in Zone 1—systematic formation again.

One factory may contribute more than 1 sampling unit. The reader may observe that the random numbers in Table 6 strike Factory No. 2 in 5 places and strike Factory No. 9 in 5 places. For every strike, say for the number 220 under Subsample 1, we draw from Factory No. 2 one sampling unit (next section). These employees so drawn belong to Subsample 1. The number 178 under Subsample 2 calls for another and distinct sampling unit (as we draw without replacement) from the same factory; these employees belong to Subsample 2.

TABLE 6

The new sampling table for drawing employees in groups.
10 random starts between 1 and 600, and every
600th sampling unit thereafter

$(Z = 600)$

Zone	Subsample									
	1	2	3	4	5	6	7	8	9	10
0001–0600	220	178	554	274	271	044	160	329	447	039
0601–1200	820	778	1154	874	871	644	760	929	1047	639

The formation of a sampling unit in a factory. It is possible to form a sampling unit in several ways. Let us take Factory No. 2, for example. Table 5 gives it 187 sampling units (1500/8) for the sampling plan under consideration. This figure 187 remains fixed. Here are 3 ways to form the 187 sampling units. Method 3 is a little more trouble than the others, and I would not use it unless the cost of the interviews is high. Method 1 is simplest but least efficient because of possible clustering.

1. Divide the employees of this factory into 187 groups, preferably but not necessarily nearly equal. Give to each group a serial number 1, 2, 3, ..., 187. Each group is a sampling unit. A convenient way to do this, if each employee has a card in a file, is to place 187 separators in the file with the intention of including about 8 cards between separators. Then number the separators 1, 2, 3, ..., 187. A sampling unit is all the cards between one separator and the next.

2. Give to every employee a serial number. Define Sampling Unit No. 31 (e.g.) as Employee No. 31 plus every 187th thereafter.

3. Give to every employee a serial number.   Form blocks of 187 consecutive employees.   Complete the last block with blanks.   Form a sampling unit by drawing with random numbers 1 employee from every block of 187.   A blank drawn from the incomplete block at the end produces no sample.   Make no substitution.   Form another sampling unit by drawing another employee from every block.

Formation of the estimates and of the standard errors.   We form the estimates for this sample the same way as we did in Section B.   One adds up the number of yes and no for each subsample, and computes the ratio f_i, the proportion yes in Subsample i.   One computes also the overall ratio f for all 10 subsamples combined.   The standard error then follows as it did before.

> **Remark.**   The student should take note that the sampling procedure described above led to the omission of some of the smaller factories.   If there is suspicion, based on knowledge of conditions, that the small factories are greatly different from the big ones, and require separate figures, then it would be necessary to draw up a new plan.   Two immediate alternatives come to mind:
>
> 1. Draw and tabulate a separate sample from the small factories.   The small factories would constitute one frame and the big factories another.
>
> 2. Change the sampling unit to 3 or 4 employees in the small factories. Tabulate the results separately for small factories and for big ones.   (Smaller sampling units in a factory increase the chance that this factory will contribute a sampling unit to the sample.   They do not change the probability that any employee will come into the sample.)

EXERCISES

1. *a.* Show that, before the sample is drawn, the probability that any employee will fall into Subsample 2 of the sample just described is 1/600. This is so whichever way we form the sampling units within the factories, just so the number of sampling units in any factory is exactly the number shown in Table 5.

b. The probability that any employee will fall into the 10 subsamples combined is 1/60.

2. Show that if the factories have approximately the number of employees shown in Table 5, then each random number in Table 6 will bring about 8 employees into the sample, sometimes 7, sometimes 9.   (These inequalities do not cause bias.)

CHAPTER 7

A Survey of Business Establishments with Correction for Nonresponse

O time, thou must untangle this, not I.—Viola to Malvolio, in Shakespeare's *Twelfth Night*, Act II, Scene ii.

Purpose of the study. This study was carried out to discover whether the operators of the motels that were affiliated with the A.A.A.* were in favor of instituting some system by which they would make reservations for motorists in advance of their arrival. (A motel is an inn along the road; the word is a contraction of motor-hotel.) The questionnaire that went to the innkeepers, and the letter accompanying it, attempted to explain some possible systems of making reservations, and to enquire whether the innkeepers wished to have such a system.

The universe and the frame. The universe was all the "contract motels" that were affiliated with the A.A.A. The frame was a file of cards in the central office in Washington. The frame was complete, except for a short lag between admission of an inn as a member of the universe, and the typing and filing of the card, or a similar lag between revocation of affiliation (e.g., for failing to comply with the required standards), and the withdrawal of the card from the file.

The file consisted of 172 drawers. In each drawer were 64 cards with Celluloid name-plates. The rule adopted was to give each card a number 1, 2, 3, . . . , 64. The drawers were numbered serially, starting at the left and at the top. A random number between 1 and 64 for any one drawer would thus identify 1 and only 1 card. This card could denote any of a

* I am indebted to Mr. Elmer Jenkins, National Travel Director, and to Mr. Douglas Cochran, Assistant Travel Director, of the American Automobile Association, Washington, and to Mr. John Boddie, Consultant on Transportation, for the privilege of working with them on this study.

number of different kinds of establishments such as:

> A contract motel
> 1 to 10 rooms
> 11 to 24 rooms
> 25 rooms and over
> A hotel
> A restaurant
> A special attraction

Or it could be

> A blank card

Administrative restrictions and the general plan. The size of the sample was to be less than 1000 motels, so that the labor of recording and tabulation could be handled by the regular workers in the central office. The questionnaire went to the contract motels in the sample. A 2d notice went at the end of 10 days, and a 3d notice at the end of another week. Every questionnaire not in at the end of 24 days was declared a nonresponse. Traveling field-workers whose duties required them to move about continually to conduct inspections and to improve the service would then conduct face-to-face interviews with a sample of the non-respondents (see Table 4 for some results of these interviews).

The results were to be ready in about 8 weeks for use of the governing board. There was time for some pretesting of the questionnaire. This was done after conferences with innkeepers in several cities within 200 miles of Washington, and after trial by mail of a few questionnaires, followed by interviews to discover difficulties.

We were fortunate here to have a frame that was complete, or nearly so. The sampling of business establishments is not always so simple. One must sometimes spend many weeks or even months, in the sampling of business establishments, to build up a suitable frame or combination of frames. One may obtain lists from various sources. Some lists cover one area, and some another. Some lists cover one type of establishment, while other lists cover other types. Most lists are deficient in their coverage of small establishments and of nonprofit institutions: these one may pick up by a sample of areas.

The sampling procedure. The sample was drawn as 10 independent samples. The size of the sample was to be about 700 motels, based on theory, combined with surmises that members of the administration were willing to make with respect to some of the proportions to be encountered. However, because of wrong information on the number of contract motels, the actual size of the sample turned out to be 854 motels (the total in Table 3), which substituted into Eq. 2 (ahead) gave an estimate of 5978 motels instead of the alleged 5000. (Incidentally, the estimate of 5978 turned out to be almost exactly the correct figure, as a recount showed.)

As the intention was to take about 700 motels into each subsample, the zoning interval was computed as

$$Z = \frac{5000}{70} = 71 \tag{1}$$

then rounded to 70 for convenience. Each drawer became a zone, and any random number between 65 and 70 was a blank. A portion of the sampling table appears here as Table 1.

The probability of selection for any space and hence for any motel was thus 1/70 for any one subsample, or 1/7 for the entire sample. An estimate of the total x-population in the frame was then

$$X = 7x \tag{2}$$

where x was the x-population in the entire sample.

> There were 172 drawers in the file, yet the sampling table runs through 202 drawers (Table 1). Extra random numbers in the sampling table cost nothing and they are a wise safeguard against an undercount of the sampling units.

INSTRUCTIONS FOR DRAWING THE SAMPLE

1. The file consists of about 170 drawers. In each drawer are 64 Celluloid spaces.

2. The sampling unit will be a Celluloid space. Each space will bear a serial number between 1 and 64 in each drawer. No. 1 will be at the far end; No. 2 will be the next; No. 64 will be at the near end. Spaces 65–70 will be blanks.

3. The serial numbers of the drawers will start at the left tier and at the top, as you face the file.

4. The sampling table (Table 1) will show 10 random numbers in each drawer. Any Celluloid space that bears a serial number under Subsample 1 will belong to Subsample 1, and likewise for Subsamples 2, 3, etc.

> The method of construction of this sampling table (Table 1) was to read out fresh random numbers in every zone, in contrast with the patterned subsamples that we used in Chapter 6. (See page 95 in Chapter 6 for some notes on the choice between patterned samples and fresh random numbers in every zone.)

5. Send the questionnaire to every contract motel in the sample. Show on the questionnaire its drawer and serial number within the drawer.

6. The supervisor in charge of this work shall keep a record of the disposition of every sampling number. He will record B for a sampling number that strikes a blank, H for a motel, R for a restaurant, P for a

TABLE 1

THE SAMPLING TABLE FOR THE SELECTION OF CARDS
FROM THE DRAWERS FOR THE SAMPLE OF MOTELS
Zoning interval, 70 spaces.   Each drawer was a zone.

Drawer	Subsample									
	1	2	3	4	5	6	7	8	9	10
1	23	05	14	38	11	43	49	36	07	61
2	31	57	09	61	25	11	15	54	36	35
3	04	24	62	02	16	44	32	55	26	12
4	62	01	25	28	34	53	11	07	09	33
.										
.										
.										
199	39	51	53	59	63	62	43	03	28	23
200	31	17	41	13	38	25	15	48	61	63
201	41	53	49	63	14	62	33	60	48	36
202	56	18	47	45	49	53	01	05	52	62

TABLE 2

RESULTS OF THE TEST OF CONFORMANCE TO THE SAMPLING PROCEDURE

Subsample	Total number of sampling numbers	Contract motels Rooms			Hotel	Restaurant	Attraction	Vacant or blank
		1–11	12–24	25–				
1	172	26	45	18	27	16	1	39
2	172	21	46	19	29	13	1	43
3	172	23	36	23	25	21	1	43
4	172	30	27	21	37	18	5	34
5	172	20	48	16	27	14	2	45
6	172	19	47	22	31	15	0	38
7	172	29	41	20	24	12	1	45
8	172	25	43	22	23	14	2	43
9	172	19	43	16	44	13	1	36
10	172	26	48	15	21	12	5	45
All 10	1720	238	424	192	288	148	19	411
Average	172	23.8	42.4	19.2	28.8	14.8	1.9	41.1

point of interest, A for an attraction. These are all blanks of various kinds, and they will not receive a questionnaire: make no substitution.

7. He will summarize by subsample the results of applying the sampling procedure to the files, in the form supplied (Table 2).

8. This summary must have approval of the consulting statistician before the mailing takes place.

Test of conformance to the sampling procedure. Step 7 called for a test of conformance, in the form of a comparison between the counts of the various types of membership in the sample with the numbers in the accounting records. Comparisons of this kind will usually disclose any serious misunderstanding of the instructions. It is a good rule to require these comparisons and to scrutinize them before the next step takes place.

The test for conformance gave the results shown in Table 2. It was entirely satisfactory. It also discovered the error in the complete count mentioned on page 104.

You will observe that the 10 subsamples gave results that differed from one another well within the allowable limits, which the student will calculate in the exercises of Chapter 17, page 465. The number of "attractions" to be struck by any subsample is so small that we may treat it as a Poisson variate (Ch. 18).

The test for conformance is not a test of the behavior of the sampling procedure, but a test to discover flaws in the selections, and inconsistencies in information on file.

Reduction and correction of nonresponse. The sampling procedure prescribed the following steps for correction of the nonresponse.

9. At the 24th day, any reply not in shall be declared a nonresponse. Prepare on this date a list of the nonresponses.

10. The consulting statistician will draw a sample of 1 in 3 of the nonrespondents (1 at random out of every consecutive 3, all the way through all 10 subsamples, with balanced thinning, page 337).

11. Regular field-workers with automobiles, on duty to keep contact with members, will interview the sample of the nonrespondents, and will turn in the completed questionnaires. (Results in Table 4.)

12. Discard a questionnaire that comes in after the 24th day, unless it comes from a motel selected in Step 10 for the interviews of the nonresponses, in which case notify the field-worker to cancel his visit to that motel.

13. Call the interviewing to a halt at the end of the 14th day, and compile the results of the interviews (Table 4).

14. Multiply by 3 each figure so compiled, and add it to the corresponding figure in Table 3, to construct Table 5.

TABLE 3

ANSWERS TO THE QUESTION, "HOW FREQUENTLY DO PEOPLE
ASK YOU TO MAKE RESERVATIONS FOR THEM?"
Compiled at the 24th day, the cutoff for acceptance.   The
figures show the number of replies under each category

Subsample	Frequently	Rarely	Never	Ambiguous answer	No reply yet	Total
1	16	40	17	2	19	94
2	20	30	17	3	15	85
3	18	35	16	1	15	85
4	17	31	14	2	16	80
5	14	32	15	3	18	82
6	15	32	12	4	16	79
7	19	30	17	3	17	86
8	13	37	11	3	18	82
9	19	39	19	2	14	93
10	17	39	15	2	15	88
All 10	168	345	153	25	163	854

TABLE 4

RESULTS OF THE INTERVIEWS WITH A SAMPLE OF 1 IN 3 OF
THE INNKEEPERS WHO FAILED TO RESPOND BY MAIL
The same question as in Table 3

Subsample	Frequently	Rarely	Never	Temporarily closed (vacation, sick, etc.)	Total
1	1	2	2	1	6
2	1	2	1	1	5
3	2	2	0	1	5
4	2	1	2	0	5
5	1	3	1	2	7
6	2	2	0	1	5
7	1	3	1	1	6
8	1	2	1	2	6
9	2	2	1	0	5
10	1	2	0	2	5
All 10	14	21	9	11	55

An example of the results.   Table 5 is the final table.   The estimate of the number of contract motels comes from Table 3 as

$$X = 7x = 7 \times 854 = 5978 \tag{3}$$

A recount of the frame showed 5950, in good agreement.   Table 5 gives

$$X = 7 \times (856 - 33) = 7 \times 823 = 5761 \tag{4}$$

for the number of innkeepers that were able to reply.   The estimate of the number that replied with the words frequently, rarely, and never, combined, is

$$Y = 7 \times (210 + 408 + 180) = 7 \times 798 = 5586 \tag{5}$$

The proportion of these that are rarely or never asked to make a reservation is

$$f = \frac{408 + 180}{210 + 408 + 180} = 73.7\% \tag{6}$$

The highest ratio is 78.4% in Subsample 1; the lowest is 70.1% in Subsample 4.   An estimate of the standard error of the proportion f is therefore

$$\hat{\sigma}_f = \sqrt{1 - \tfrac{1}{7}}\, \frac{78.4 - 70.1}{10} = .77\% \quad \text{[P. 200]} \tag{7}$$

TABLE 5

ALL ENTRIES ARE NOW ADJUSTED BY MULTIPLYING BY 3 THE
RESULTS OF THE INTERVIEWS RECORDED IN TABLE 4, AND
THEN ADDING THEM TO TABLE 3
f in this table is the ratio of the rarely plus never to the
frequently plus rarely plus never

Sub-sample	Fre-quently	Rarely	Never	f	Ambiguous answer	Temporarily closed	Total
1	19	46	23	78.4	2	3	93
2	23	36	20	70.9	3	3	85
3	24	41	16	70.4	1	3	85
4	23	34	20	70.1	2	0	79
5	17	41	18	77.6	3	6	85
6	21	38	12	70.4	4	3	78
7	22	39	20	72.8	3	3	87
8	16	43	14	78.1	3	6	82
9	25	45	22	72.8	2	0	94
10	20	45	15	75.0	2	6	88
All 10	210	408	180	73.7	25	33	856

This figure gives 2.3% for an estimate of the 3-sigma limits of sampling variation.*

Remark 1. These limits met the specifications.   Narrower limits would have been no more useful (client's remark).

Remark 2.   The same survey furnished many other tables and other estimates with their standard errors.   The estimate $f = 73.7\%$, along with other estimates, with interpretation of the standard errors, constituted the report to management.

* I express here my indebtedness to my colleague Mr. Ralph Woodruff of the Census in Washington for pointing out an error in principle in the construction of Table 5 in an earlier manuscript.

CHAPTER 8

Examples in Sampling Accounts

A. QUICK STUDY OF ACCOUNTS IN LEDGERS*

You see, but you do not observe. The distinction is clear.—
Sherlock Holmes to Watson, in A. Conan Doyle's *A Scandal in Bohemia*.

Purpose of the study. This study is an investigation of the accounts in certain ledgers. There are 23 large volumes, and each contains perhaps 500 pages, some more, some less. There are 27 lines to a page. Names appear on many of the lines. These names constitute the catalog of listings. Some of the names show a money-entry (e.g., $752), which indicates that this person's account (tax) was unpaid on the date when it was posted.

Estimates of the number of money-entries by size was the aim of the sample. The purpose of the study was to provide a basis for deciding whether to try to collect any of the accounts due; and if so, which sizes might be most productive. The results are in the tables further on: they provided the information and the accuracy that the client required.

Administrative restrictions and choice of sampling unit. The sampling unit was defined as a pair of facing pages, as in Fig. 5. There were 27 lines to the page, and thus a sampling unit contained 54 lines and could contain therefore up to 54 names, and up to 54 money-entries. The money-entries constitute the universe. The list of sampling units (Table 1) is the frame.

The choice of sampling unit rested in this study on the limitations of time and available skill; not on efficiency. The ledgers were in constant use Monday through Friday. Moreover, it would have been unwise, in the judgment of the client, to run the risk of publicity. The work would therefore commence on a Friday evening and the ledgers would be back in place before Monday morning.

---

* I am indebted to my friend and colleague Dr. Melvin F. Wingersky, Attorney-at-law, who saw the need of the information and asked me to work with him.

There would be 3 helpers: one experienced and expert; one a novice in this kind of work; the third, a man who, the client told me, could number the sampling units in advance and who could move the heavy volumes from one table to another, as required.

It was essential, before settling on this choice of sampling unit, to be sure that no sampling unit in the entire 23 volumes could contain a high proportion of money-entries of Class A (a money-entry of $2000 or over; see Table 3), nor of Class B (a money-entry from $1000 to $1999).   Had this possibility existed, the results would have been impaired.   The names

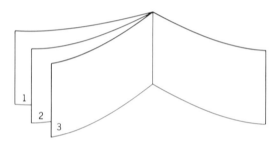

Fig. 5. Each sampling unit was a pair of facing pages.

were alphabetic by precinct, or by other small geographic area, and I ascertained from the man in charge of the posting that it was completely impossible that any sampling unit could contain more than a very few entries in Class A or in Class B.   Moreover, an examination beforehand of pages in every volume showed no indication of bunching of Class A or of Class B.

It would be easy to think of excellent alternative plans that would deliver good results in spite of bunching, but these alternative plans would all have required more skill and time than the administrative restrictions would permit.   One might, for example, draw a random number for the page, and then 2 or 3 random numbers to select lines thereon.   Another possible plan would have been to use stratified sampling, in which one might leaf through every page or a large sample of the 6000 or so pages, and record on a separate list the pages that had more than (e.g.) 4 entries of Class A; then to sample this list separately (an example of stratified sampling).   Any such plan, however inviting, would have brought down disaster, as there was neither time nor manpower to carry it out.

There was indeed some bunching, but not a great amount.   Exercise 2 (p. 122) asks the student to calculate the numerical effect of the bunching.

Size of sample, and the general plan.   Calculations showed that 200 sampling units would suffice.   The plan called for 10 subsamples.   The sample would have to detect the existence of accounts of Class A (accounts of \$2000 and over), and had to give a rough measure of their proportion amongst all accounts unpaid.   It was conjectured that the proportion of Class A might be as small as 1 %, yet it might be very important in dollars (it was; see Tables 8 and 9).   If the accounts of Class A were sprinkled throughout the lines like a Poisson variate with no bunching (Ch. 18), and if half the lines showed money-entries, the expected number of lines that belong to Class A would be $\frac{1}{2} \times 200 \times 54 \times .01 = 54$.   Now $1/\sqrt{54}$ is about 15 %, which would be the coefficient of variation to expect.   The results would still be useful even if through some bunching the standard error were double this amount.   (The coefficient of variation turned out to be 30 %;   Table 8.)

Steps in the procedure.   The following instructions went to the client.

1. Start with Volume 1 and write the serial number on the left-hand corner of each sampling unit.   Start with 1 in each volume.

Step 1 (the serialization of the sampling units) took place, under instructions sent by mail, during the 2 weeks before the actual sampling commenced. The omissions, skips, and duplications in serial numbers that the sample picked up gave eloquent testimony to the wisdom of using only the simplest procedures.   Fortunately, the skips and duplications caused no bias (Exercise 2, p. 122).

Write the numbers 1, 2, 3, etc. lightly in pencil at the lower left corner at the edge of the left-hand page of each sampling unit.   Skip even and odd faces that contain no name at all.

The supervisor may, at his discretion, omit from the frame a long series of consecutive pages that contain names but no money-entry at all.   These are the names of people who had paid after their names were posted.   The estimate formed later on from the sample for the total number of names in the 23 volumes refers to the frame only and must be increased by the number of names thus excluded, to form the estimated total of names in the universe of all 23 volumes (see Table 8).

In this instance, we excluded 1433 consecutive names that occurred in Volume 23.   As there was probably an unknown but small additional number of names excluded here and there, we shall in Step 20 and in Table 8 use the figure 1500 for the total number of names (paid listings) excluded from the frame in all volumes.

2. Make a list to show the number of sampling units in each volume, and accumulate the sums through the entire 23 volumes.   The accumulated sums give a serial number to every sampling unit.   This is the frame, Table 1.   Serial number 5380 was a blank at the end.

3. Divide the serial numbers into 20 zones.   The zoning interval will be $5380/20 = 269$.   Table 4 shows the boundaries of all 20 zones.

4. Read out 10 random numbers between 001 and 269 for the random starts in Zone 1.   The random starts appear in the top zone of Table 2.

TABLE 1

THE FRAME: LIST OF SAMPLING UNITS BY VOLUME

Volume	Number of sampling units	Serial numbers
1	217	001– 217
2	224	218– 441
3	289	442– 730
4	203	731– 933
5	226	934–1159
6	251	1160–1410
7	179	1411–1589
8	142	1590–1731
9	170	1732–1901
10	294	1902–2195
11	231	2196–2426
12	238	2427–2664
13	182	2665–2846
14	259	2847–3105
15	197	3106–3302
16	365	3303–3667
17	244	3668–3911
18	242	3912–4153
19	280	4154–4433
20	128	4434–4561
21	295	4562–4856
22	311	4857–5167
23	213	5168–5380

TABLE 2

THE SAMPLING TABLE (FURNISHED BY THE CONSULTING STATISTICIAN)

Zone	Subsample									
	1	2	3	4	5	6	7	8	9	10
0001–0269	137	110	268	261	009	125	191	132	016	195
0270–0538	406	379	537	530	278	394	460	401	285	464
0539–0807	675	648	806	799	547	663	729	670	554	733
0808–1076	944	917	1075	1068	816	932	998	939	823	1002
.										
.										
.										
5112–5380	5248	5221	5379	5372	5120	5236	5302	5243	5127	5306

They came from Kendall and Smith, 30th Thousand, Columns 2, 3, 4, beginning at line 1.

5. Now add 269 successively to each of the 10 random starts to form 10 systematic subsamples with a zoning interval of 269.   Continue the addition through the 20 zones for each subsample.   Enter these numbers, so formed, zone by zone, in the forms provided (Table 2).   These numbers are the "sampling numbers": any sampling unit that bears one of these numbers is in the sample.

6. By consulting the frame, translate each sampling number from a serial number into a volume number and a sampling unit within the volume.   Enter these numbers on the forms provided (Table 4).

7. Arrange the volumes for easy access.   Begin with Subsample 1 and study the first sampling unit therein that fell into the sample.   Record on the form provided the number of lines that belong to the various respective Classes A to I; also the total for all classes (see Table 4).   Continue this examination through the entire sample.

Prior to this step comes the definition of the classes.   These were defined by the client as shown in Table 3.   These classes would have been the same whether the study were a complete coverage or a sample of the accounts.

TABLE 3

DEFINITIONS OF THE CLASSES
(As specified by the client)

Class A	$2000 and over
B	$1000 to $1999.99
C	$400 to $999.99
D	$100 to $399.99
E	$50 to $99.99
F	$20 to $49.99
G	Under $20
H	A name with no money-entry (paid)
I	A line vacant

8. Proceed by recording first in the space provided the total number of lines in the sampling unit (right-hand column of Table 4).   Then record the number of items that you find in each of the other classes.   If 2 people work on the job, and if the books are all accessible simultaneously, 1 worker may well record first the total number of lines in the sampling unit, then the numbers in the Classes G, H, I.   The 2d worker may record subsequently the other classes, then form for each sampling unit the sum in the Classes A to I, to compare with the total that the 1st worker has

already recorded for that sampling unit.   Any discrepancy in the total will point to an error, which should of course be cleared up.

The number of lines in a sampling unit will usually be 54, which is twice 27.   However, a sampling unit will always begin with a designated left-hand page (which has already received a number in Step 1), and will extend up to but not include the next left-hand page that bears a number. This scheme is the half-open interval.   If the worker in Step 1 turned through inadvertence 2 pages instead of 1, he enlarged a sampling unit from 54 to 108 lines, but all sampling units and hence all names will nevertheless have equal probabilities.   The opposite error in Step 1 is to skip a number; if a skip occurs, some sampling unit is nonexistent, or a blank, in which case all classes therein will be 0.   (For an example, see Table 4, 2d zone; there was no sampling-unit number 406.)

9. When the 20 zones of a subsample are complete, show the sums for all the classes, and apply the usual corner check (see it in Table 4).   A discrepancy will point to an error.

10. Let

$$J = A + B + C + D + E + F + G \qquad (1)$$

$$K = J + H \qquad (2)$$

Record these sums in the vacant space at the bottom of the form.   Verify that $K + I$ equals the total number of lines.   Then compute for Subsample 1 the quotients

$$\frac{A}{J}, \frac{B}{J}, \ldots, \frac{G}{J}$$

and record them in percentages at the bottom of the form in Columns A to G.   These are the estimates of the proportions of unpaid listings in the various classes.   Their sum must be 100.0 to within a decimal.

11. Compute H/K and show it as a percentage in Column H, in a new row at the bottom.   This is an estimate of the proportion of names with accounts paid (uncorrected for the addition of names excluded in Step 1).

12. Compute the same quotients for the other 9 subsamples.   There will then be 10 estimates of each proportion called for.   The results are in Table 6.

13. Compute the average of the 10 estimates of each proportion.

14. Estimate from Subsample 1 the number of listings in each class. As there are 5380 sampling units in all, and as each subsample contains 20 sampling units, the number of listings of Class A estimated from Subsample 1 will be 5380A/20, where A is the number of listings of Class A in Subsample 1.   Similarly for the other classes, B, C, D, E, F, G, H.

15. Compute the same estimates for the other 9 subsamples.

16. Compute for each class the average of the 10 estimates just formed in the last two steps.   The results of this step are in Table 7.

17. Throw all 10 subsamples together to form the final estimates. First, form the totals in the various classes.   The lower part of the same form as used heretofore will serve well (Table 5).   Then compute from these totals the estimates called for in Steps 12 and 15.   Do this by treating the combined sample as an 11th sample, and proceeding exactly as you did with the 10 subsamples, but take 5380A/200 to estimate the number of listings, because there are 200 sampling units in all 10 subsamples combined.

> **Remark.**   The difference between the average of the 10 estimates and an estimate obtained by combining all 10 subsamples in Table 6 gives a measure of the bias in the ratio.   The small differences in Table 6 indicate that the bias is negligible.   See page 428 for numerical evaluation of bias.

18. Compute the standard errors of these averages, and record them. We use the range again to estimate the standard errors.   Let u be the final estimate calculated from all 10 subsamples combined.   Then

$$\hat{\sigma}_u = \frac{w}{10} \quad \text{[P. 200]} \tag{3}$$

for the estimated standard error of u, w being the range of the 10 estimates. Standard errors thus computed appear in Tables 6 and 8.

19. Or, still easier, the 10 estimates furnish a simple index of precision without any calculation at all.*   Thus, there is only 1 chance in $2^{10} = 1024$ that the highest of the 10 estimates will fall below the median of the result of a complete coverage (more precisely, the median of the results of all possible samples taken by the same procedure); and there is the same chance that the lowest of the 10 estimates will fall above the median. There is only 1 chance in 512 that the interval between the lowest and the highest of the 10 estimates will fail to include the median.   This inference is independent of the shape of the distribution in the frame.

20. Add the 1500 listings without money-entries that were excluded in Step 1, and revise the final estimates of the total number of names already paid (those without money-entries).   Revise also the proportions paid and unpaid (Table 8).   This step carries us from the frame to the universe of all listings.

21. Estimate for each class the total amount in dollars outstanding and unpaid.   Use the midpoint as the average in each class; use $10 as the average in Class G, and use $2500 as the average in Class A.   Let A, B,

* Called to my attention in 1946 by Professor P. C. Mahalanobis, F.R.S.

C, D, E, F, G be the final estimates of the number of names in each class. Then the total amount by class outstanding and unpaid will be:

$$2500 \times A$$
$$1500 \times B$$
$$700 \times C$$
Etc.

The results are in Table 9. By addition, we estimate $20,310,000 for the total amount outstanding and unpaid.

Conclusions

1. The total amount outstanding and unpaid is about $20,310,000.
2. 23% of the cases, or about 35,000 of them, are under $20 (Class G), and they amount to about $350,000.
3. 81% of the cases, or about 126,000, are under $100 (Classes G, F, E), and they amount to about $4,458,000.
4. 43.4% of the cases, or about 67,000 of them, lie between $20 and $50 (Class F), and they amount to about $2,350,000.
5. 13.1% of the cases, or about 20,000 of them, lie between $100 and $400 (Class D), and they amount to about $5,064,000.
6. 3.5% of the cases, or about 5400 of them, lie between $400 and $1000 (Class C), and they amount to about $3,785,000.
7. 1.2% of the cases, or about 1800, lie between $1000 and $2000 (Class B), and they amount to about $2,690,000.
8. 1.1% of the cases, or about 1700 of them, are over $2000 (Class A), and they amount to about $4,310,000.
9. Half the money owed lies in the 8900 cases in Classes A, B, C, over $400.
10. The other half lies in the 146,000 cases below $400 (Classes D, E, F, G).

Report to management. The estimates of the number of unpaid accounts and of the amount outstanding in each class (Tables 8 and 9), together with an explanation of the precision of the estimates, constituted the report to the client. This is what he paid for. This report should in my judgment not contain recommendations on what the client should do. It might be obvious, before or after the study, that the greatest yield would come from action on the large accounts. After the study was made, the client could estimate the size of the job and could decide rationally how to go about it. He might decide, on the basis of political expediency, to do nothing, in which case one may wonder why he spent his money for the statistical study. This is his privilege, however deplorable the waste may seem to the statistician (refer back to pages 12 and 13).

Date 24 Jan. 53

TABLE 4

THE RAW DATA OBTAINED FROM SUBSAMPLE 1

No.	Zone boundaries	Random number	Vol.	Sampling-unit number	Number of lines classed as									Total
					A	B	C	D	E	F	G	H	I	
1	1- 269	137	1	137	0	0	0	0	0	26	2	5	21	54
2	270- 538	406	2	189	0	0	0	0	0	0	0	0	0	see note
3	539- 807	675	3	234	0	0	0	2	0	0	0	2	266	270
4	808-1076	944	5	11	0	0	0	3	8	6	5	32	0	54
5	1077-1345	1213	6	54	0	0	0	1	1	5	1	23	77	108
6	1346-1614	1482	7	72	0	0	0	0	0	0	0	7	154	162
7	1615-1883	1751	9	20	0	0	0	1	0	15	9	2	28	54
8	1884-2152	2020	10	119	0	0	0	0	4	35	4	10	0	54
9	2153-2421	2289	11	94	0	0	1	13	5	19	9	6	2	54
10	2422-2690	2558	12	132	0	0	0	6	10	28	6	3	0	54
11	2691-2959	2827	13	163	0	0	0	0	4	17	1	9	23	54
12	2960-3228	3096	14	250	0	0	3	9	11	16	8	6	1	54
13	3229-3497	3365	16	63	0	0	0	0	6	26	3	18	0	54
14	3498-3766	3634	16	332	0	1	6	6	2	3	0	10	26	54
15	3767-4035	3903	17	236	2	5	5	5	3	1	1	32	0	54
16	4036-4304	4172	19	19	0	0	2	4	6	16	10	16	0	54
17	4305-4573	4441	20	8	0	1	0	3	12	24	9	5	0	54
18	4574-4842	4710	21	149	0	0	1	11	4	24	11	2	1	54
19	4843-5111	4979	22	123	0	0	0	3	1	17	0	3	30	54
20	5112-5380	5248	23	81	0	0	0	1	2	14	8	5	24	54
Sum					2	7	18	69	79	292	87	196	654	1404
Divide by J (%)					.4	1.3	3.2	12.4	14.3	52.7	15.7			100.0
Divide by K (%)												26.1		

$$J = A + B + \cdots + G = 554$$
$$K = J + H = 750$$
$$I = 654$$
$$K + I = 1404$$

Note: There was no sampling unit numbered 189 in this volume. The worker skipped this number in Step 1. Consequently this sampling unit is 0 in all classes.

TABLE 5

THE RAW DATA OBTAINED FROM ALL SUBSAMPLES: NUMBER OF LINES BY CLASS, BY SUBSAMPLE

Subsample	Total Unpaid J	Unpaid							Total Paid H	Blank lines I	Total all lines J + H + I
		A	B	C	D	E	F	G			
1	554	2	7	18	69	79	292	87	196	654	1404
2	656	4	8	20	75	117	306	126	198	226	1080
3	567	19	9	20	79	71	205	164	308	205	1080
4	530	3	4	27	87	68	180	161	292	258	1080
5	631	20	9	19	79	102	244	158	304	361	1296
6	516	1	2	19	78	90	241	85	207	465	1188
7	604	1	2	7	52	96	316	130	214	262	1080
8	542	2	10	25	79	79	249	98	225	313	1080
9	600	4	10	20	88	94	216	168	302	232	1134
10	557	8	6	26	67	76	250	124	254	431	1242
All 10 combined	5757	64	67	201	753	872	2499	1301	2500	3407	11664

TABLE 6

ESTIMATED PROPORTIONS, PAID AND UNPAID, BY CLASS, BY SUBSAMPLE

All figures are percentages

Subsample	All listings				Unpaid listings, by class						
	Paid plus Unpaid	Paid	Unpaid	Unpaid	A	B	C	D	E	F	G
1	100.0	26.1	73.9	100.0	.4	1.3	3.2	12.4	14.3	52.7	15.7
2	100.0	23.2	76.8	100.0	.6	1.2	3.0	11.4	17.9	46.7	19.2
3	100.0	35.2	64.8	100.0	3.4	1.6	3.5	13.9	12.5	36.2	28.9
4	100.0	35.5	64.5	100.0	.5	.8	5.1	16.4	12.8	34.0	30.4
5	100.0	32.5	67.5	100.0	3.2	1.4	3.0	12.5	16.2	38.7	25.0
6	100.0	28.6	71.4	100.0	.2	.4	3.7	15.1	17.4	46.7	16.5
7	100.0	26.2	73.8	100.0	.2	.3	1.2	8.6	15.9	52.3	21.5
8	100.0	29.3	70.7	100.0	.4	1.8	4.6	14.6	14.6	45.9	18.1
9	100.0	33.5	66.5	100.0	.7	1.7	3.3	14.7	15.6	36.0	28.0
10	100.0	31.3	68.7	100.0	1.4	1.1	4.7	12.0	13.6	44.9	22.3
Average	100.0	30.1	69.9	100.0	1.1	1.2	3.5	13.2	15.1	43.4	22.6
All 10 combined	100.0	30.3	69.7	100.0	1.1	1.2	3.5	13.1	15.1	43.4	22.6
Standard error	0	1.3	1.3	0	.3	.2	.4	.8	.6	1.9	1.5

The estimated proportions are computed as follows:

$J = A + B + C + D + E + F + G$ = total listings, paid plus unpaid; H, paid

$K = J + H$ = total listings, paid plus unpaid

$\dfrac{J}{K}$ = proportion of total listings unpaid

$\dfrac{H}{K}$ = proportion of total listings paid

$\dfrac{A}{J}, \dfrac{B}{J}$, etc. = proportions unpaid by class

TABLE 7

ESTIMATED NUMBER OF LISTINGS BY CLASS, BY SUBSAMPLE

Sub-sample	Paid plus Unpaid	Paid	Unpaid	Unpaid listings, by class						
				A	B	C	D	E	F	G
1	201 750	52 724	149 026	538	1 883	4 842	18 561	21 251	78 548	23 403
2	229 726	53 262	176 464	1 076	2 152	5 380	20 175	31 473	82 314	33 894
3	235 375	82 852	152 523	5 111	2 421	5 380	21 251	19 099	55 145	44 116
4	221 118	78 548	142 570	807	1 076	7 263	23 403	18 292	48 420	43 309
5	251 515	81 776	169 739	5 380	2 421	5 111	21 251	27 438	65 636	42 502
6	194 487	55 683	138 804	269	538	5 111	20 982	24 210	64 829	22 865
7	220 042	57 566	162 476	269	538	1 883	13 988	25 824	85 004	34 970
8	206 323	60 525	145 798	538	2 690	6 725	21 251	21 251	66 981	26 362
9	242 638	81 238	161 400	1 076	2 690	5 380	23 672	25 286	58 104	45 192
10	218 159	68 326	149 833	2 152	1 614	6 994	18 023	20 444	67 250	33 356
Average	222 114	67 250	154 864	1 722	1 802	5 407	20 256	23 457	67 223	34 997
All 10 combined	222 114	67 250	154 864	1 722	1 802	5 407	20 256	23 457	67 223	34 997

Note that the averages of the 10 estimates formed from the 10 subsamples agree everywhere with the estimates formed from the entire sample. There is thus no indication of bias (p. 428). The names excluded in Step 1 have not yet been added back into the totals.

TABLE 8

ESTIMATED NUMBER OF LISTINGS BY CLASS

Class	Number of cases (from Table 7)	As a per cent (Note 1)	
		of the total	of the unpaid
All listings, paid and unpaid	223 614	100.0 ± 0	· · ·
Paid (Note 2)	68 750	30.3 ± 1.3	· · ·
Unpaid	154 864	69.7 ± 1.3	100.0
Class A	1 722	· · ·	1.1 ± .3
B	1 802	· · ·	1.2 ± .2
C	5 407	· · ·	3.5 ± .4
D	20 256	· · ·	13.1 ± .8
E	23 457	· · ·	15.1 ± .6
F	67 223	· · ·	43.4 ± 1.9
G	34 997	· · ·	22.6 ± 1.5

Note 1: The figures in the right-hand column placed after the double sign ($\pm$) indicate the estimated standard errors, as in Table 6.

Note 2: The figure 68,750 is formed by adding 1500 to the figure 67,250 paid in Table 7.   The figure 1500 is the estimated number of paid listings excluded in Step 1.

TABLE 9

THE ESTIMATED AMOUNT OF MONEY IN EACH CLASS, COMPARED
WITH THE PROPORTION OF NAMES IN THAT CLASS

Class	Amount	Proportionate number of cases
A	$4 310 000	1.1 %
B	2 692 000	1.2
C	3 785 000	3.5
D	5 064 000	13.1
E	1 758 000	15.1
F	2 350 000	43.4
G	350 000	22.6
Total	20 309 000	100.0

EXERCISES

1. *a.* Why is it that a serial number missing from the frame introduces no bias?

b. Some pages with names received by accident no serial number. Some consecutive sampling units received the same number. Explain why it is that with the method of selection by which a sampling unit extends up to but does not include the next *different* serial number, the probabilities remain constant and there is no bias in the estimate.

This type of selection is known as the *half-open interval.* It is an extremely important invention by which to maintain equal probabilities.

Although missing serial numbers and varying sizes of sampling unit do not cause bias, they do increase slightly the standard errors of the results. Chapter 19 shows some calculations on the effect of varying size of sampling unit.

2. Compute whether there was any detectable bunching of the accounts in Classes A and B.

SOLUTION FOR CLASS A

The estimate of Class A as a proportion ($\hat{p}$) of the unpaid accounts was $1.1 \pm .3\%$. The total sample contained 5757 unpaid accounts. Had we drawn 1 name at a time from the entire list of unpaid accounts (estimated in Table 7 as 154,864 accounts), the estimate of the variance of the estimate $\hat{p}$ would have been $\hat{p}\hat{q}/n$ (p. 405), where $n = 5757$. This formula gives $\hat{\sigma}_{\hat{p}}^2 = .011 \times .989/5757$, whence $\hat{\sigma} = .0014$, which is about half of the standard error .003 as we estimated it from the 10 subsamples. We thus estimate that the efficiency of the sampling unit that we used (facing pages) was only about 1/4th as efficient as a single name would have been. See, however, the section on administrative restrictions, which explains the choice made (p. 110).

The advanced student may turn to Eq 25 on p. 484, which evaluates the effect of bunching in terms of the intraclass correlation, ρ. This equation reads, for our present purpose,

$$1 + (g - 1)\rho = \left(\frac{.0030}{.0014}\right)^2 = 4.5 \tag{4}$$

wherein $g = 54$. The solution is $(54 - 1)\rho = 3.5$, whence $\rho = 3.5/53 = .065$. This is our estimate of the intraclass correlation of entries of Class A, in sampling units of 54 consecutive lines.

3. A census is taken by writing the name of every person. The order of listing the members of a household is very strict, in order not to miss people—head of the family, wife, oldest child, second child, etc., married children living at home, servants living in, lodgers. A person living alone is a head. The sheets show 40 lines on one side and 40 on the other, numbered 1 to 80. Some lines are vacant, especially at the bottom, where the interviewer hesitates to start a new household, even though the instructions ask the interviewers to fill every line. Some lines contain not names, but notes of explanation. Some lines contain names subsequently canceled, owing to the fact that the interviewer discovered after he wrote the name that this person was a visitor and would be enumerated at his usual place of residence. The enumerator writes the letter H alongside the name of the head of a family, and there is therefore no difficulty in picking out the heads.

After the census is taken, there will be a study of families by means of a 5% sample of the census. To take a sample of families, it is only necessary to select a sample of heads, and to copy off on to a transcription sheet the information for the family whose head came into the sample.

Show that the following plan will yield an expected 5% sample of families, in 10 subsamples.

i. Serialize the sheets 1, 2, 3, etc., for any area or other work-unit.

ii. Now start reading from 2 columns of a table of random numbers. Record for each sheet 4 unduplicated random numbers between 01 and 80. These numbers designate line numbers for the sample. Record also a 3d digit from the table to indicate the subsample. (See the example in Table 10.)

iii. Record on the transcription sheet provided for the purpose the family-information for every line in the sample that shows the head of a family. The transcription sheet must show (a) for every line in the sample that is not a head the notation B for a blank line, or for a line containing a note; NH for a name, not a head; (*b*) the subsample number.   Make no substitution for B or NH.

iv. Tabulate by subsample and compute the final estimates and their standard errors by the procedures shown in earlier chapters.

TABLE 10

EXAMPLE

The random numbers came from Kendall and Smith's
Random Sampling Numbers, 33d Thousand, Columns 1, 2, 3; 5, 6, 7

Sheet	Line number	Subsample	Sheet	Line number	Subsample
1	61	3	4	16	8
	20	7		64	2
	35	2		58	3
	65	2		49	7
2	30	0	5	14	3
	62	6		56	7
	44	8		62	4
	70	3		55	3
3	53	3	6	57	2
	01	1		22	9
	59	5		37	1
	12	6		78	6

Remark 1.   The procedure just described does not balance the sub-samples.   To illustrate, Subsamples 4, 5, and 9 do not appear in the 1st 10 lines of Table 10, while Subsample 2 appears twice and Subsample 3 appears thrice.   The frequencies of the subsamples approach equality as a statistical limit, but there is an addition to the variance of any unbiased estimate of the form $X = N\bar{x}$, the addition being pq/n to the rel-variance* of X, where $p = .1$, $q = .9$, and n is the number of lines in the sample.

This additional variance is obviously trifling if n is large or even moderate, but one may wish to balance the subsamples if n is very small.   Instead of writing down the 3d digit just as it comes from the table, one may balance the subsamples by rejecting any digit that has already appeared in the same block of 10 sample-lines.   The procedure is to write down for any block 9 unduplicated digits; the 10th digit is automatic, being the subsample that has not yet appeared.   In the above example, if we use the 3d column for the subsamples, the order in the 1st block of 10 lines would be 3, 7, 2, 0, 6,

* Rel-variance is a handy term invented by Hansen, Hurwitz, and Madow for the square of the coefficient of variation.

8, 1, 5, 9, 4.   The order in the 2d block would be 8, 2, 3, 7, 4, 5, 9, 6, 0, 1. The reader may construct other ways to balance the subsamples.

Lack of balance causes only negligible increase in the variance of the estimate of a proportion, such as the number of children per family.

Remark 2. The reader will note that the above instructions call for fresh random numbers on every page. I would not use patterned subsamples here, because one may expect cyclic effects, owing to the fact that some enumerators will leave lines vacant toward the bottom of a sheet, whether or not the instructions say to fill up the sheet. Although our estimate of the standard error will always be valid, cycles or no cycles, and whether we use patterned subsamples or not, and regardless of whether patterned subsamples give a lower or a higher variance than fresh random numbers, there is no point in running the risk of a large variance from cycles, to save the small effort required to read out fresh random numbers for every page.*

Remark 3. Suppose that the size of the sample is not yet decided: it may be $2\frac{1}{2}\%$ or 5%, depending on time, costs, funds available, and variances. It may be wise, under such circumstances, first to complete a sample of $2\frac{1}{2}\%$; then possibly to proceed later into the other $2\frac{1}{2}\%$. A way to divide the sample into 2 equal random portions 1 and 2 is to copy a 4th digit along with each line number in the sample: the 4th digit if odd indicates that the line is in Portion 1; if even, that the line is in Portion 2. (The discussion of this topic continues in Chapter 14, page 271.)

This plan as described for dividing the sample will not maintain the balance of the subsamples. The reader may wish to construct a scheme that will divide the 5% sample into two $2\frac{1}{2}\%$ samples, both balanced. One way is to (1) balance the 5% sample; then (2) erase the designations of the subsamples in the 2d, 4th, 6th, and all other even blocks of 10. (3) Delete by random numbers 5 subsamples from the 1st (balanced) block of 10. (4) For the pattern in the 2d block, simply reverse the pattern in the 1st block. Likewise, for the 4th block, reverse the pattern for the 3d block; etc.

4. The files or lists that are to be sampled are sometimes located in several different places. An example is the sample of railway accounts in Section B of this chapter, where the files of the involved interline abstracts were located in 10 different offices, some of them 2000 miles apart. Suppose that a certain class of record (abstract, card, case, line, page) is to be identified and numbered serially, starting with the numeral 1 in each office, and suppose that the sampling plan will call for a proportionate

* The student who wishes to study the conditions under which patterned samples produce greater or less variance than fresh random numbers in every zone may turn to a paper by Werner Gautschi, "Some remarks on systematic sampling," *Ann. Math. Statist.*, vol. 28, 1957: pp. 385–394; also to W. G. Madow and Lillian H. Madow, "On the theory of systematic sampling," *Ann. Math. Statist.*, vol. 15, 1944: pp. 1–24 and W. G. Cochran, "Relative accuracy of systematic and stratified random samples for a certain class of populations," *Ann. Math. Statist.*, vol. 17, 1946: pp. 146–177, all of which are summarized in Cochran's *Sampling Techniques* (Wiley, 1953), Ch. 8.

sample from all offices.   The zoning interval is 200 and the 1st random number drawn from each zone will belong to Subsample 1; the 2d random number drawn from each zone will belong to Subsample 2; etc.

a. Is there any reason why it would be better to prescribe different random numbers for each office?

Different random numbers in every office would break up any possible cyclic effects that might arise from similarity in the method of filing in the different offices, and would produce a better estimate if cycles do exist. However, no bias would arise from use of the same random numbers in every office, even if there were cycles (see also page 95).

b. What is the probability that any one abstract will go into Subsample 1? (*Answer:* 1 in 200.)

c. If there are to be 10 subsamples, what is the probability that any one abstract will go into the whole sample? (*Answer:* 1 in 20.)

d. Suppose that in each sampling unit there are several cards or lines or spaces.   The entire sampling unit is to be processed.   What is the probability that any one card or any one line in the entire frame will come into Subsample 1? (*Answer:* 1 in 200.)

e. What is the probability that any one card or line will come into any of the 10 subsamples? (*Answer:* 1 in 20.)

f. Would you be worried if 1 of the files contained only 151 sampling units, and if the random numbers for some of the subsamples lay between 152 and 200?   Would this vacant space 152–200 destroy the probabilities?

The answer is no; any number that falls between 152 and 200 draws a blank, and there is no cause for worry.   Every sampling unit has the same probability of selection regardless of the size of the file.   The only danger is that the workers may not understand this and may put 10 sampling units into the sample, in violation of the procedure.   If they understand that blanks are entirely acceptable and are to be expected, and why they must not alter the procedure, they may not be tempted to depart from instructions.

Remark 4.   Although blanks do not cause bias, they do increase the variance of an estimate ($\bar{x}$ or X) of a mean or of a total; see Eqs. 9 and 10 in Exercise 9 below.   Blanks do not increase the variance of an estimate (f) of a ratio.

5.   A file contains 100,000 white cards, and an unknown number of red, green, and blue cards.   The colors are intermingled, but all the cards are filed alphabetic, by date, by city, independent of color.   Each color designates a special class.   (The file sampled in Chapter 7 was an example.) A sample of 500 white cards will provide the information that is needed for a proposed study.   The red, green, and blue cards will provide no information for this study.   The sample is to be drawn and processed in 10 subsamples.

a. Show that either of the 2 following plans will provide a sample of about 500 white cards.

b. Why is it that the colored cards do not decrease the expected size of the sample of white cards?

c. Show that if there were 100,000 colored cards in the file, mingled with the 100,000 white cards, there would be twice as many zones in Plan 1 as in Plan 2, and that about half the sample in Plan 1 would be blanks (colored cards).

Plan 1

Step 1.   Give a serial number to every card, in order, regardless of color.

Remark 4.   There are many ways to avoid the labor of numbering every card.   One may give serial numbers to the drawers, sections, and slots. The use of blanks (illustrated in several places in this book) to fill out vacant spaces in sections, so as to build up an incomplete section to an even 100, is also an important labor-saving device.   There are several illustrations in this book of accumulated totals, which serve to serialize the units.

Step 2.   Use for the zoning interval

$$Z = \frac{100,000}{500} = 200 \tag{4}$$

Remark 5.   Note that the zoning interval is based on the white cards alone, without any knowledge of the number of other cards.

Step 3. Construct a sampling table in 10 subsamples, with a zoning interval of 200.   One way, if there is no danger of loss of precision from periodicities, is to draw 10 random starts and take every 200th card thereafter.

Step 4.   Record the desired information for the white cards in Subsample 1; similarly for Subsample 2; etc.   Ignore the red, green, and blue cards, but count them if there is any desire to learn how many there are in the frame.

Plan 2

Step 1. Number the white cards 1, 2, 3, . . . and up to the last white card.   Ignore the others completely.

Step 2.   Use the zoning interval 200, computed as in Plan 1.

Step 3. As before.

Step 4. As before, but now no colored cards come into the sample at all.

6. Show that the following steps inserted before Step 4 in the instructions in Plan 2 in the preceding exercise will ensure that every white card shall have the same probability of selection as another, even though a few white cards, through inadvertence or otherwise, have no serial numbers on them, and even though 2 consecutive white cards received the same serial number.   These steps define a sampling unit by the half-open interval.

   a.   The search for skips.   Define a skip as a white card (a member of the universe) that for any of various reasons received no serial number.   To search for skips, search in the forward direction only, from every white card that fell into the sample, up to but not including the next different serial number (whether in the sample or not).   If this search discovers a white card with no number, give it the same number as the serial number that this search commenced with, but add the suffix X (to indicate that this is an extra card with the same number), and include it in the sample. For example, suppose that white card No. 10 fell into the sample.   We commence there our search for white cards without numbers, and we discover one lying between 10 and 11.   The serial number of the white card just discovered we make 10X.   A 2d white card between 10 and 11 could take the number 10Y.

   b.   The search for duplicate numbers.   Search in both directions, from every white card that fell into the sample, up to but not including the next white card with a different number (whether in the sample or not).   If this search discovers a duplicate number, give it the suffix X.   Thus, if we discover an extra serial number 10, we make it 10X.

   7. Why must one look only in one direction for skips (white cards with no number), and in both directions for duplicate numbers?   Why search only at the sample-points?

   Comparison of Plans 1 and 2 (blanks vs. no blanks).   In Plan 1, the user simply numbers all the cards in order 1, 2, 3, etc., regardless of color, draws a sample of cards, and takes what the random numbers give him.   A card drawn into the sample that is not a white card is a blank.   Plan 2 requires the user to find the white cards, and it is only the white cards that take serial numbers.   The sample in Plan 2 thus consists of white cards only.

   Plan 1, though easier at the 1st step, has the disadvantage of producing the higher variance for an estimate of a total, especially if the proportion of white cards is small.   We now derive the theory by which to compare the 2 variances for an estimate of a total.   We require, as a 1st step, Eq. 6, which I leave to the reader as an exercise.

   Given 2 distributions with means a and b, variances σ_a^2 and σ_b^2, number of sampling units in the respective proportions P and Q, show that:

   a. The mean of the 2 distributions combined is $Pa + Qb$.

b. The variance of the 2 distributions combined is

$$\sigma^2 = P\sigma_a^2 + Q\sigma_b^2 + PQ(b - a)^2 \tag{5}$$

c. Let every member of the *b*-distribution be 0.   Then $b = 0$ and $\sigma_b^2 = 0$.   The rel-variance of the combined distribution is then

$$C^2 = \frac{1}{P}(C_a^2 + Q) \tag{6}$$

where $C = \sigma(Pa + Qb) = \sigma/Pa$ in this case, and $C_a = \sigma_a/a$ is the coefficient of variation of the *a*-distribution by itself.   The variances of the 2 plans follow at once.

Let the intended size of sample be *n* white cards.   For Plan 2, we select the *n* cards by reading out *n* random numbers between 1 and the highest serial number of the white cards alone.   For Plan 1, we select *n/P* cards by reading out *n/P* random numbers 1 and the highest serial number of all colors of cards combined.   (The zone and the zoning interval computed in Exercise 5 will automatically adjust the sample-sizes.)   Compute $\bar{x}$ in either plan as the sum of the *x*-populations on the sample of white cards, divided by the number of white cards in the sample.   Then,

$$C_{\bar{x}}^2 = \frac{C^2}{n/P} = \frac{1}{n}(C_a^2 + Q) \qquad \text{[Plan 1]} \tag{7}$$

Strictly, there should be a factor $1 + Q/n$ on the right, because the size of sample in Plan 1 is not fixed, but is a random variable.   For practical purposes, we replace $1 + Q/n$ by unity.

$$C_{\bar{x}}^2 = \frac{C_a^2}{n} \qquad \text{[Plan 2]} \tag{8}$$

It would be wrong to conclude that Plan 2 is preferable, simply because it has the smaller variance for a sample of *n* white cards.   One must also take into consideration the cost of numbering the cards, and in Plan 2, the cost of studying every card to decide whether it is a member of the universe (a white card).   In Plan 1, we study only the cards that fall into the sample. An illustration with a cost-function appears in Section B of this chapter.

The above paragraphs deal with estimates of a mean ($\bar{x}$) or of a total (X). The variance of a ratio ($f = \bar{x}/\bar{y}$) is often in practice about the same by either plan, in which case the lower cost of Plan 1 will be attractive.

B. A LARGE-SCALE STUDY OF ACCOUNTS OF FREIGHT SHIPMENTS IN MULTIPLE LOCATIONS*

Purpose of the study. I shall now describe a sample of accounts in which the purpose of the study was to investigate the revenue that accrued to the carriers of freight-shipments of various commodities between different sections of the country. When a shipment traverses 2 or more railways, the railways that participate in the shipment divide the total revenue for the shipment amongst themselves according to a formula, which can at times be very complex. The reason for making this study was that certain railways were threatened in a lawsuit with a change in the formula that would have brought a reduction in the share of the revenue that they derived from shipments that traversed some portion of their roads. The aim of the study was to furnish evidence by which to predict, on the basis of past records, the future effects of the proposed reduction.

The frame. The records to be studied consisted of the "interline abstracts" in the offices of the railways. An interline abstract is a piece of paper, uniformly $8\frac{1}{2}$ by 11 inches for all railways, designed to show line by line, for a given month (e.g., July 1959), the serial numbers of all the cars that traversed the same route, with the same commodity, at the same rate. The abstract shows the commodity hauled, the route, the rate, the total revenue, and the calculations for the division of the total revenue amongst all the carriers that hauled the car or cars from origin to destination. Two cars can not appear on the same abstract unless they traversed the same route with the same commodity during the month. Most abstracts show but 1 car, although some abstracts show 2, 3, and even 10 or more cars. A solid train of 30 cars that moved together from New Orleans to Vancouver would appear on 30 lines of 1 abstract (continued to a 2d page if necessary). The revenue on 1 abstract might be only a few dollars for a single car that moved a short distance over 2 roads, or as high as $200,000 for a train-load of ore that had run from New Orleans to Vancouver. Some abstracts ran even higher.

The interline abstracts for any month are bound in volumes. These volumes may contain also other interline abstracts of shipments that moved over routes not subject to the lawsuit: they are not involved in this study.

The records of the Class I railways in the case were located in 17 different cities and towns, which accounts for the term "sampling in multiple locations" in the heading of this portion of the chapter. In addition,

---

* I am indebted to Mr. Howard Trienens of the law firm of Sidley, Austin, Burgess, and Smith of Chicago for the privilege of working with him on this sample.

there were 80 smaller railways (Class II), for which I prescribed simplified instructions (not shown here).

The sampling unit.   The sampling unit was any involved abstract with a serial number, plus any other involved abstract up to but not including the next abstract that bore a different serial number.   By this definition, the sample picked up its proportionate share of abstracts that the workers failed to serialize; also of those with duplicate numbers (see paragraph 13 of the instructions).   Every involved abstract in either stratum had thus the same probability as any other in that stratum.

The sample-design.   A little familiarity with the characteristics of the abstracts indicated that there should be at least 2 strata; also that the stratum of high value should be processed 100% because of the extreme variability in the revenue shown on the abstracts.   (Preceding paragraphs have already indicated that the revenue could run up to $200,000 and more.) A small sample from the stratum of lower value would then suffice.

The first step was a pilot study, to find flaws in the proposed instructions (written about 15 times), to learn the kinds of errors, difficulties, and misunderstandings that might arise.   Of equal importance, the pilot study provided an approximate picture of the distribution of revenue, which would enable one to make a rational decision on where to break the scale to form the 2 strata.   Should division between the 2 strata occur at $10,000, or at $7500, or at $15,000?

The results of the pilot study (carried out for the abstracts dated in the month of July 1957) gave the figures contained in Table 12 and shown roughly in Fig. 6.   It is clear from these figures that if the breaking point were at $10,000, then the upper stratum would contain about 21% of the revenue, but only 1% of the abstracts.   These abstracts, together with a 5% sample of those below $10,000, would carry 25% of the revenue, at a cost of processing only 6% of the abstracts.   Such a sample-design would focus with good efficiency the skill and effort to be expended on the study. The instructions further on show the procedure.

Another criterion for the breaking point which was helpful here, as it often is in the absence of specific information or experience with similar material, is given by Hansen, Hurwitz, and Madow.*   The rule, translated into the present problem, states that if Y represents the total amount of revenue on all abstracts, then one can not go far wrong by placing the breaking point at or near $Y/2n$.   With a 5% sample in the lower stratum, $n = .05N$ and $Y/2n$ reduces to $10\bar{Y}$, where $\bar{Y}$ is the average revenue per abstract.   The pilot study gave $\bar{Y}$ equal to about $1200, so the rule suggests $12,000 for consideration as a possible breaking point.   There is some

---

* Hansen, Hurwitz, and Madow, *Sample Survey Methods and Theory*, Vol. 1 (Wiley, 1953): p. 220.

TABLE 12

THE REVENUE AND THE NUMBER OF ABSTRACTS ABOVE
VARIOUS POSSIBLE BREAKING POINTS

Breaking point	Proportion of the revenue at and above this point	Proportion of the abstracts at and above this point
$ 0	100.0%	100.0%
500	95.0	76.2
1 000	69.6	30.5
1 500	56.5	16.5
2 500	44.0	7.9
3 500	37.3	4.9
5 000	31.7	3.2
7 500	24.4	1.6
10 000	20.8	1.0

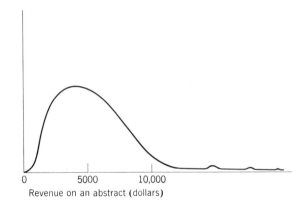

Fig. 6. The distribution of total revenue. Plotted roughly from the data in Table 12 and from other information. The ripples in the right-hand tail come from rare big accounts.

advantage in using $10,000 as the breaking point, as any abstract with 5 figures would then belong to the upper stratum. Moreover, it is better to set the breaking point too low than too high.

Decision on the plan.*   It was necessary to decide at this point whether to use Plan 1 or Plan 2 (just discussed at the end of Section A of this chapter)—i.e., whether to draw a sample of all abstracts below $10,000

* I am indebted to my colleague Dr. Morris H. Hansen for formulating this comparison between the 2 plans, and for pointing out the utility of Eq. 6 in the preceding section for such purposes.

and throw out those not involved; or to find, serialize, and sample only the involved abstracts. Fortunately, Eqs. 7 and 8 above provide the theory necessary for a decision. We now compute the costs of the 2 plans, and the information per unit cost.

TABLE 13

Costs of the 2 plans (blanks vs. no blanks)

Plan	Study the abstracts to decide whether they are involved		Number the abstracts		Compute and transcribe the information on the involved abstracts in the sample		Total cost ($N = 20n$)
	For 1 abstract	Total	For 1 abstract	Total	For 1 abstract	Total	
1	2	$2n/P$	1/15	$N/15P$	15	$15n$	$n(3.33/P + 15)$
2	2	$2N/P$	1/15	$N/15$	15	$15n$	$n(40/P + 16.33)$

To carry on, we must adopt some figures for the cost of the chief operations in the 2 plans. We shall express costs in minutes. Let us suppose that a man can number 15 abstracts in 1 minute. This cost includes time out for coffee, time out to listen to the ball game, time for revising the numbers when he makes a mistake, the cost of an index (p. 138), and the cost of supervision. The cost of numbering an abstract is then 1/15. There are N/P abstracts to number in Plan 1, and N in Plan 2.

Then there is the cost of studying an abstract to decide whether it is involved or not. We shall assume that this cost is 2 minutes. There are n/P abstracts to study in Plan 1, and N/P in Plan 2. Finally, there is the cost necessary to compute and to transcribe the information contained on the abstracts that fall into the sample. We shall assume that this operation requires on the average 15 minutes per abstract. The number of abstracts in this operation is n by either plan.

Table 13 shows the cost of the 2 plans, for a sample of n involved abstracts, under the assumption that $N = 20n$. The reader should satisfy himself that the costs of the various operations are not critical for the choice of plan. It is only necessary to introduce some approximate figure, as we have done.

TABLE 14
Comparison of Information per Unit Cost

Plan	$P = \frac{3}{4}$			$P = \frac{1}{2}$			$P = \frac{1}{4}$			$P = \frac{1}{10}$		
1 Cost/n	19.4			21.7			28.3			48.3		
2 Cost/n	69.7			96.3			176.3			416.3		
C_a^2	1	.4	.04	1	.4	.04	1	.4	.04	1	.4	.04
1 $nC_{\bar{y}}^2$ (Eq. 7)	1.25	.65	.29	1.50	.94	.54	1.75	1.15	.79	1.90	1.30	.94
2 $nC_{\bar{y}}^2$ (Eq. 8)	1.00	.40	.04	1.00	.40	.04	1.00	.40	.04	1.00	.40	.04
1 Information per minute expended	.041	.079	.178	.031	.049	.085	.020	.031	.045	.011	.016	.022
2 Information per minute expended	.014	.036	.359	.010	.026	.260	.0057	.014	.142	.0024	.0060	.060

We continue now into Table 14, which shows the costs and the variances of the 2 plans for some plausible values of P (the proportion of abstracts involved in this study), and for some plausible values of C_a^2, the rel-variance between abstracts, of whatever characteristic requires an estimate. The last 2 lines show the information per unit cost, the unit of cost being 1 man-minute.   We recall from Chapter 5 that information in an estimate is conveniently defined as the reciprocal of its variance.   We are concerned here with the cost of a unit of information; hence we divide the amount of information by the cost required to get it.   This means that the information per unit cost will be 1/variance × cost.

The tabulation plans were in a fluid state at the time the sampling commenced; hence it was necessary to prepare for many and various types of estimates.   It was possible that estimates of totals would be required for small segments of traffic (commodity, mileage-bracket), which would have small values of P and rel-variances (C_a^2) possibly intermediate between .4 and .04.   Plan 2 makes a good showing under these circumstances.

There were some persuasive attractions to Plan 2 not included in the mathematical formulation of the comparisons of variances.   For example, it was necessary to look at every interline abstract anyway, by either plan, to discover every involved abstract whose gross revenue amounted to $10,000 or over (Group a; *vide infra*).   This requirement would write off part of the expense of carrying out Plan 2.

Moreover, preliminary investigation showed that the number of cars, tons, and dollars of revenue were highly variable from one abstract to another.   Hence it seemed risky to prescribe a plan that was only capable of forming direct estimates of totals: there might be distinct gains with ratio-estimates of totals, for which the total number of cars on the involved abstracts, or the total weight, or the total revenue, might be excellent multipliers.   Plan 2 would identify the abstracts whence to draw the required totals (see Step 10 of the instructions further on): Plan 1 would not.   Such forces led to the adoption of Plan 2.   Without Table 14, we should have no idea whether we are paying a relatively high price for Plan 2 and for the advantages of having totals for ratio-estimates.

Instructions for the selection of the sample of interline abstracts.   The actual instructions follow.   The statistical reader will recognize the definition of the universe, the preparation of the frame, the designation of the sampling units by the half-open interval, their serialization, and finally, various searches to discover and to correct flaws in the preparation of the frame.   The following excerpt taken from a letter from the chief accountant of one of the railways to another chief accountant, written after a trial of the instructions, may be of interest:

We now appreciate our consulting statistician's advice to be sure of the quality of each step before we accept it, and even before we verify it, and to carry out one step at a time in the instructions.   We have profited much by this experience and advice.

INSTRUCTIONS FOR THE SELECTION OF THE SAMPLE OF INTERLINE ABSTRACTS

Definitions of the Groups

1. The universe of abstracts in this study will be abstracts declared by the Traffic Committee to be involved.*   For the purpose of selecting the sample, they will consist of 2 groups:

Group a: Any involved abstract that bears gross revenue of $10,000 or over.   These will be processed 100%.

Group b: Involved abstracts under $10,000.   A 5% sample of these abstracts will be selected and processed in the manner prescribed below.

Preparation of the Frame (Serialized Abstracts)

2. Process 1 month at a time.   Start with the forwarded† traffic for January 1959.   Verify the count of the volumes for the month to be sure that they are all there.   Find and mark all the involved abstracts in each volume.   Mark the abstracts of Group a in a manner that will distinguish them from the abstracts of Group b (for example, red marks on the a-abstracts; green marks on the b-abstracts).   The aim here is to find and to distinguish every involved abstract, and to make a clean separation between Groups a and b.

Do no work on any volume that consists entirely of abstracts not involved in this study.   The supervisor may, at his discretion, tie off and thereafter ignore sections that are solid LCL or otherwise not involved.‡

---

* The Traffic Committee was a committee appointed by the railways to define what abstracts would be relevant to this case.   The definitions were stated in terms used by the railways so that an experienced rate clerk could decide, upon inspection, whether an abstract was involved in this study (and hence a member of the universe) or was not involved (a blank).   It is not necessary for the statistician to learn the definitions.

† An interline abstract is a statement rendered by the terminating railway to other railways for the purpose of settlement, to divide the total revenue amongst the railways that hauled the shipment.   Forwarded and received are terms used by railways to designate different kinds of abstracts.   Railway A designates a shipment as forwarded if this railway carries the shipment from the point of origin to a connecting railway, which we may call Railway B.   The same shipment is then designated by Railway B as (a) received, if Railway B delivers the shipment to the point of destination, or as (b) intermediate, if Railway B carries the shipment onward to a 3d railway for further carriage.   There was no intermediate traffic in this case.   The meaning of these terms is not necessary to an understanding of the sampling procedure.

‡ LCL means less than a carload.   LCL shipments were by definition not involved in this study.

Some abstracts occupy more than 1 page. It is the practice of some roads to show the total gross revenue and division thereof on the last page. Other roads show the total and division thereof on the first page. The rule concerning which page to mark is simple: mark only the page that carries the total gross revenue and division thereof. The total gross revenue on this one page will determine which group the entire abstract belongs to. At least 1 road shows the total and division separately page by page. Here again, the rule is simple: mark every page. The total gross revenue on a page then determines which groups this one page belongs to, irrespective of any other page.

If a volume contains abstracts for 2 or more months, process the whole volume as a unit. The proper month will show on the transcription sheet (Step 17).

3. Now test the quality of the preceding step. The supervisor will appoint someone to look for (a) skips (abstracts involved but not yet marked); (b) abstracts not involved but marked in error; (c) abstracts misclassified a for b or conversely; (d) a page whose revenue carries forward to another page, and which was marked in error. He will keep a record volume by volume of all errors found, for review with the consulting statistician. He will correct all errors found.

The purpose of Step 3 is to furnish information by which to judge whether the work in Step 2 is acceptable. The purpose is not to try to raise the quality of Step 2 by finding and correcting what went wrong, although the supervisor will correct all errors found. It is impracticable to find all the flaws and to make all the corrections necessary in work that was not done right to start with. This is a well-known principle in industrial inspection and we shall apply it here. As Harold F. Dodge said, "You can not inspect quality into a product." If the quality of the work in Step 2 fails to satisfy a criterion* that the consulting statistician will set as an aid to the supervisor, the supervisor will make a fresh start on Step 2.

It is important to communicate directly and immediately with the consulting statistician in regard to any question or difficulty not explicitly covered here.

4. Once Step 3 is satisfied, begin with the 1st volume that you marked in Step 2, and give a serial number (1, 2, 3, and on) to every involved forwarded abstract regardless of whether it belongs to Group a or to Group b. Continue into the 2d volume without a break, then into the 3d, and so on. The serial numbers will run from 1 upward continuously through the last volume.

It matters not if the abstracts of 2 roads are mingled. Thus, there is no point in giving a separate series of numbers to abstracts of a subsidiary road, whether they be separate or mingled with the abstracts of the main road.

* The criterion was to accept 1 error in a volume (about 300 abstracts, involved and not involved), and to reject work on a volume that showed 2 errors or more. The acceptance number for misclassification of $20,000 or more was 0. See any book on acceptance sampling.

Local preference should dictate the choice.　The sampling plan will be valid regardless of how the abstracts are bound or mingled or numbered by class or by date, provided any information needed later shows on the transcription sheet when you prepare it in Step 17.

As blanks create no bias (see an indented paragraph below Step 6), it will be possible to arrange special dispensation in any location where, for administrative reasons, it is desirable to start 2 people at the job of numbering abstracts.　Call the consulting statistician on the matter.

5. Now test the serialization just completed in the preceding step.　Look for duplications and for jumps in the serial numbers.　Look for stray numbers out of order.　Give a serial number out of order to any abstract that was marked but not numbered.　Do the same to the 2d of 2 abstracts that bear the same serial number (a duplication).　A number jumped will henceforth be a blank: do not attempt to use it (see an indented paragraph below Step 6).

6. Make a catalog of all blanks and of all strays and other numbers assigned out of order, so that you can account for every number and can readily find any abstract numbered out of order.　One form of catalog is the index suggested below.

It is permissible at this point and in the next step to correct any errors wherever and however found.　For example, cross out the serial number on any abstract now discovered to be not involved.　Do not re-use this number; it will henceforth be a blank.　Give a serial number out of order to any abstract now discovered to be involved but not yet numbered, or to an abstract that bears a duplicate number.

Blanks create no bias, and there is no point in trying to use a blank number, once it occurs.　Use of a blank number would require entry in the catalog of abstracts numbered out of order, and extra work to find the abstract if this number falls into the sample.

The index suggested below accounts for every number and shows which numbers are out of order.　It also provides ready reference for review by the consulting statistician.　You may attach an index like this to each volume, or you may make a centralized catalog.

Volume_____Forwarded_____19____
(month)

4–1637	1642	1648	7–1653
a–1638	1643 (blank)	1649	7–1653 Skip
1639	1644	3–1650	etc.
1640	1645	3–1650 Dup	
1652*	a–1646	1651	
1641	1647	(1652)	

The asterisk (*) denotes that the number that appears here is out of order. The parentheses ( ) indicate that the number enclosed is not here but is elsewhere out of order. Look for it at an asterisk. Serial number 1638 belongs to Group a, and is so marked. Later on (Step 11) you will prefix the subsample number to the abstracts of Group b that fall into the sample. Thus, No. 1637 fell into Subsample 4, and No. 1650 fell into Subsample 3. The notation Dup signifies a duplicate found in Step 13. The notation Skip signifies a skip found in Step 16.

The serialized abstracts constitute the frame.

7. Count the abstracts marked in the volume and compare it with the count in the catalog or index. Compare this count with the opening and closing serial numbers therein, with allowance for blanks. The purpose of this count and comparison is to discover duplicates, skips, and jumps. The serial numbers are not yet frozen, and you may still correct any errors found, as by declaring a number blank if the abstract is not involved, and by giving a number out of order to any abstract missed so far, or found to bear a duplicate number. Correct also any mistakes in classification, though you must report to the consulting statistician, as in Step 3, all errors found.

8. Repeat the preceding steps for the abstracts for traffic received.

9. At the conclusion of the preceding step, the serial numbers will henceforth remain frozen: there will be no new numbers, in or out of order. (*Exceptions:* an entire volume that was out when the other volumes received their serial numbers, or new traffic brought in or taken out through new definitions.) Moreover, any abstract that you have designated as Group a will remain in Group a, even though you misclassified it. Similarly for any abstract in Group b, subject to the review of the consulting statistician.

10. Draw directly from all the serialized abstracts for the month designated, the number of abstracts, cars, and tons, and the total gross revenue, separately by Group a and by Group b, and separately by traffic forwarded and traffic received. This is a 100% job for all of Group a and for all of Group b. Record the results on lines 14 and 16 of Form S (Table 15, p. 144) and send this form to the consulting statistican. (*Note:* You will use these totals in Step 14, where they will furnish the base for certain comparisons that will be possible as tests of conformance with the sampling procedure.)

The order of working the 2 classes of freight is immaterial; you may vary it to suit your local conditions and preference. When you finish January 1959, or some designated set of volumes, commence another month, or another set.

Identification and Preparation of the Sample

11. The consulting statistician will furnish a set of 10 random starts for Group b for a given month as soon as the official in charge of the work for any road certifies that the serialization for that month is complete. There will be a set of starts for abstracts forwarded, and a set for abstracts received. There will be a new set of random starts every month, for every road. It is important to use the right set of numbers.

The forwarded abstracts whose serial numbers coincide with the random starts for Subsample 1 and with every 200th number thereafter will belong to Subsample 1 of the forwarded abstracts. Similarly for Subsamples 2, 3, etc.; and similarly for the 10 subsamples of abstracts received. Tag or mark the index for ready identification of every abstract of Group b whose serial number falls into the sample; also prefix the subsample number to the serial number for every one of these abstracts.

Thus, 4–1637 would indicate that abstract No. 1637 belongs to Subsample 4 in Group b. Any serial number without a prefix is neither in Group a nor in the sample of Group b.

Make no substitution for any sampling number that falls on an abstract of Group a or on a blank of any kind. The sample will bring in the proper proportion of every kind of abstract and of every kind of blank, provided you follow these instructions and do not substitute.

12. Measures have been built into the earlier steps to ensure the necessary quality, and to provide objective measures of the quality, whatever it be. It is nevertheless inevitable that a few mishaps will slip through. Some of them will be only honest differences of opinion on whether an abstract is really involved in this study. Steps 13, 15, and 16, if you follow them carefully and exactly as they specify, will detect and correct without bias the effect of mistakes heretofore undetected, provided a high level of quality was achieved in the early steps.

13. Search in both directions, forward and backward, as far as the next different serial number, from every abstract that was designated as Group a, and from every abstract now marked as belonging to a subsample of Group b, to see if the same serial number appears on 2 abstracts. Two abstracts that still bear the same serial number at this point will be called hereafter a doublet, and the 2d member of a doublet is a duplicate. The 2d member of a doublet, whether it be in Group a or in Group b, will carry the mark Dup after its serial number (see the index suggested in Step 6). The 2d and 3d members of a triplet will carry the marks Dup 1 and Dup 2. Mark the index to show any duplicates found. (An abstract that slipped through the previous steps without a number is a skip.)

Each member of a doublet will receive treatment according to its original classification, Group a or Group b. For example, if both members of a doublet had been marked for Group a, they will stay in Group a, and both of them will be in the 100% sample of Group a. If both members had been designated for Group b, they will stay in Group b, and they will go into the same subsample together.

It may happen that one abstract of a doublet was marked for Group a, while the other one was marked for Group b. The abstract of Group a will remain in Group a, and will be processed as such; while the abstract of Group b will go into a subsample in the regular manner IF its serial number is in the sampling table: the same serial number will then occur both in Group a and in Group b.

Each member of a doublet, if its number belongs to a subsample, will (like every other member of the sample of Group b) carry a prefix to its serial number to indicate its subsample number.

Record in a memorandum for the consulting statistician's review, full information concerning every doublet that you find, and the adjustment thereof.

Abstracts not in Group a nor in a subsample of Group b are not search points. Ignore them in this step and in Steps 15 and 16. The review prescribed here will be intensified and more effective than a 100% review. Moreover, the corrections for errors found here will be unbiased ONLY if you use the search points designated and only if you follow precisely the instructions given here.

If your index shows data for cars, tons, and revenue, AND if you drew your totals in Step 10 from this index, then you must draw the figures for the sample in Step 14 from this index. You will in this case postpone Step 13 until you have completed Step 14 and sent in your Form S complete through line 18 to the consulting statistician. Steps 13, 15, and 16 will then all fall under the "Review for correction of blemishes in preparation." Make the same transfer in case you are using a set of punched cards for the involved abstracts. Complete count in Step 10, and sample in Step 14, will then come from precisely the same list.

14. Count the number of abstracts, and draw the cars, tons, and the gross revenue for each subsample. Show these figures on Form S, separately by subsample, and separately by traffic forwarded and received. Form S will also require the corresponding figures that you drew in Step 10 from all the abstracts serialized.

Complete Form S through line 18 and send it to the consulting statistician. The comparisons thereon may indicate satisfactory compliance with these instructions, or they may indicate some misunderstanding or other difficulty. You will hear from the consulting statistician in respect to these comparisons. You may proceed meanwhile on the next steps, but only subject to rectification of any misunderstanding that the above comparisons indicate.

Review for Correction of Blemishes in Preparation

This sample is a probability sample, so laid out that the margin of uncertainty that one may ascribe to the use of sampling will be calculable and hence known from the results of the sample itself (Step 18).   Neither this calculation, however, nor the comparison in Step 14, will detect or measure the effect of persistent errors in the preparation of the frame, as these errors will affect equally the figures from the sample and from the total of all abstracts.

Careful search, carried out exactly as Steps 13, 15, and 16 prescribe, with disposition of errors found as directed therein, will effect correction without bias for all the skips, duplicates, and other errors of the kind that the search calls for, including those that remain in and between the abstracts not drawn into the sample.

The only bias left in the final results will arise from mishaps that remain undiscovered or unreported at the points where these instructions call for a search.   Reasons for diligence and for exact reporting in accord with these instructions are obvious.

15. Review every abstract in Group a, and every abstract in the sample of Group b, to discover if any of these designated abstracts was numbered in error, being actually not involved according to the definitions in force in the previous steps.   Look for a serial number on any page whose revenue was carried on another page.   Mark any of these abstracts or pages "Not involved," and add the word "blank" after the serial number in the index.   Mark the transcription sheet "Void" and hold it.   Keep a memorandum of these mishaps for attention of the consulting statistician.   As in Step 13, ignore the abstracts of Group b that did not fall into the sample.

16. Search for skips (an abstract involved but not numbered) on the far side, only up to the next different serial number, from every abstract of Group a and from every abstract of Group b that fell into the sample. As in Step 13, abstracts not in Group a nor in a subsample of Group b are not search points.   Some skips may already bear flags from Step 13. Ignore a skip on the near side.   If you discover a skip on the far side, give it the serial number of the preceding abstract and annex the notation Skip (Skip 1 and Skip 2 if there are 2 successive skips), and include it in Group a or in a subsample of Group b, according to the classification of the preceding abstract.   Show in the index every skip discovered.

17. For every abstract of Group a, and for every abstract in the sample of Group b, fill out a transcription sheet to show the information required. Carry out for this purpose any recalculations of revenue prescribed by the Traffic Committee, as for transit-shipments and combination rates.   Send the transcription sheets to Atlanta for punching and tabulation.

Communicate with the consulting statistician with respect to any errors that you find not covered by these instructions.

Formation of Estimates and Standard Errors

18. Form estimates and standard errors.

The procedure below follows a memorandum from the consulting statistician to the man in charge of the computer (IBM 705). The procedure calls for ratio-estimates, based on cars. The actual instructions were a little more complex, as they called for estimates of totals by month, to consolidate for the year.

x_{ij} the sum of the x-populations (e.g., revenue from hauling products of agriculture) on the abstracts of Subsample i, Railway j, northbound, forwarded

y_{ij} likewise for the y-population (e.g., revenue from hauling products of agriculture, forests, mining, and manufacturing)

c_{ij} the number of cars in Subsample i, Railway j, northbound, forwarded (lines 1–10, Table 15)

$x_{.j}, y_{.j}$ sums of x_{ij} and y_{ij} over all 10 subsamples

A_j the sum of the x-populations in Group a, Railway j, northbound, forwarded

B_j likewise for the y-population

C_j the number of cars in the complete count of Group b, northbound, forwarded (line 14, Table 15)

$c_{.j}$ the number of cars in all 10 subsamples, Railway j, northbound, forwarded (line 11, Table 15).

$X_{i.} = \sum_j \{(C_j/c_{ij})x_{ij} + A_j\}$ estimate of the total x-population furnished by Subsample i for all railways combined, northbound, forwarded   (1)

$Y_{i.} = \sum_j \{(C_j/c_{ij})y_{ij} + B_j\}$ estimate of the total y-population furnished by Subsample i for all railways combined, northbound, forwarded   (2)

$f_i = X_{i.} : Y_{i.}$ estimate of the ratio of x to y furnished by Subsample i for all railways combined, northbound, forwarded   (3)

$X_{..} = \sum_j \{(C_j/c_{.j})x_{.j} + A_j\}$ estimate of the total x-population furnished by the entire sample for all railways combined, northbound, forwarded   (4)

$Y_{..} = \sum_j \{(C_j/c_{.j})y_{.j} + B_j\}$ likewise for the y-population   (5)

$f = X_{..} : Y_{..}$ estimate of the ratio of x to y furnished by all 10 subsamples, for all railways combined, northbound, forwarded   (6)

TABLE 15

SUMMARY OF RESULTS BY SUBSAMPLE

*Continental Pacific Ry. Co. for the month of January 1959

Date of this report 17 Sept. 1959

Form S

Subsample	Forwarded				Received			
	Abs'ts	Cars	Tons	Revenue	Abs'ts	Cars	Tons	Revenue
1	17	23	638	12,384	14	24	550	9,570
2	18	33	1,045	18,738	14	17	447	7,983
3	17	23	652	12,613	13	25	959	7,978
4	17	20	552	10,203	14	26	928	17,726
5	17	29	1,106	14,547	14	18	627	8,731
6	17	44	1,152	22,860	14	25	879	13,635
7	18	22	597	10,184	14	21	613	11,157
8	17	36	1,029	19,641	14	17	472	7,716
9	17	47	1,961	28,629	14	19	523	9,945
10	18	24	715	14,530	14	16	424	7,547
11 Total sample	173	301	9,447	164,329	139	208	6,422	101,988
12 Average (11) ÷ (10)	17	30	945	16,433	14	21	642	10,199
13 Total × 20	3,460	6,020	188,940	3,286,580	2,780	4160	128,440	2,039,760
14 Group b by the complete count	3,507	5,974	194,088	3,254,411	2,756	4160	115,352	1,956,331
15 Difference	47	−46	5,148	−32,179	−24	0	−13,088	−83,429
16 Group a	24	514	24,606	346,204	18	354	9,588	268,632
17 Groups a and b by the complete count (14) + (16)	3,531	6,488	218,694	3,600,615	2,774	4514	124,940	2,224,963
18 Difference in per cent 100 (15) ÷ (17)	−1.3	.7	−2.3	.9	.8	0	10.5	3.8

There were similar formulas for traffic northbound received, and for traffic southbound forwarded and received.

Estimates of the standard errors come from the individual subsamples in the usual way, either by the sum of squares or by the range.

Unbiased estimates of totals may be formed by substituting 200 for C_j/c_{ij} and 20 for $C_j/c_{.j}$.

Comparison between the sample and the complete count. Form S (Step 14 and Table 15) enables one to compare the complete count of abstracts, cars, tons, and revenue with the number estimated from the sample. Any important omission or slip or misclassification that occurred during the processing may show up in this comparison, either by differences in lines 15 and 18, or by disagreements between the subsamples (lines 1–10). Small persistent (systematic) omissions and faults in classification that occur month after month will often be detectable by persistence of sign. A negative sign in line 15 or 18, or a positive sign, repeated in the same column 6 or 7 consecutive months, or even 5, no matter how small be the amount, indicates almost positively the existence of something wrong. (This happened on 2 occasions. Detection and removal of the mis-understanding brought a balance of plus and minus signs. The amounts involved were only a few hundredths of a per cent. The detection of so small an error naturally called forth amazement at the power of so sensitive a test.) Table 15 shows the comparison that one railway made for January 1959.

EXERCISES

1. Suppose that you discover in Step 13 a doublet (2 abstracts bearing the same serial number). Why not leave one member of the doublet numbered the way you found it, and assign to the other member a new serial number out of order, and let this one take its chance with the sampling table? Explain the bias that this suggestion would introduce, and explain why the procedure specified in the instructions introduces no change in probability.

2. Suppose that a random number in the sampling table turns out to be a jump (no abstract bears this serial number). Why not give this number to some abstract, as would be convenient to do if the search in Step 16 turns up an abstract that is actually involved, but not hitherto recognized as such, nor numbered? Explain the bias that this procedure would introduce.

3. Suppose that you find in Step 10 or at any time after Step 9 an abstract with total revenue $17,000 (i.e., above $10,000), but classed as Group b. Why not change it to Group a? Explain the bias that would result if you followed this procedure.

4. *a.* Why scrutinize in Step 15 only the abstracts that fell into the sample to discover and correct for the error of including abstracts in the frame that were not really involved?

b. What bias would you introduce if you searched in both directions for skips? (Compare this question and the next with Exercise 7, page 128.)

c. Why is it that if you look in only one direction for duplicates, you might fail to eliminate the bias arising from duplication?

5. Suppose that in the entire year the search called for in Step 6 turns up 49 skips—abstracts that should have been recognized in Steps 2 and 3 as being involved. Assume that these were all in Group b. There were 78,000 abstracts in the sample of Group b for the entire year.

a. What is your estimate of the total number of abstracts not recognized and numbered in Steps 2 and 3? (*Answer:* 20 × 49 or 980.)

b. What is the upper 3-sigma limit to this estimate?

SOLUTION

Use the square-root of a Poisson random variable (p. 461). $\sqrt{49} = 7$. $7 + 1.5 = 8.5$. $8.5^2 = 73$. $73 \times 20 = 1460$, which is the upper 3-sigma limit. You may assert that not more than 1460 abstracts in Group b escaped recognition. You may also assert that not more than the proportion 73/78,000 or about 1 in 1000 escaped recognition. Unbiased correction is automatic in Step 16.

6. A set of records of material purchased consists of 20,000 pages, 27 lines to the page which we may number 00 to 26 as required. Lines 00 and 26 are almost always blank. There are, in total, about 200,000 purchases listed on the 20,000 sheets. The desired size of sample is about 4500 purchases. Show that the following plan will yield a sample of about the required size, stratified proportionately in zones of 18 pages. (*Hint:* 200,000/18 × 25 = 445, the expected number in each subsample.)

Number the pages 1, 2, 3, etc.

10 subsamples. Zoning interval for the pages, 18.

The sampling unit will be a line, and the sample will be a sample of lines. Select for each subsample 1 line from the 18 pages in each successive zone. Follow the instructions in the note further on to build up the sampling table (Table 16). Use the random numbers for the lines in the following manner:

01 draws into the sample any entry on line 01, plus any entry on line 00.

02 ,, ,, ,, ,, ,, ,, ,, ,, 02.
03 ,, ,, ,, ,, ,, ,, ,, ,, 03.
.   .
.   .
.   .
24 ,, ,, ,, ,, ,, ,, ,, ,, 24.
25 ,, ,, ,, ,, ,, ,, ,, ,, 25, plus any entry on line 26.

A blank line produces no sample; make no substitution. Table 16 shows a portion of the sampling table.

Note on the random numbers. Start in Kendall and Smith's *Random Sampling Numbers*, 1st Thousand, Columns 9–12, line 13; use the columns in groups of 4 as printed. To select a page, use the 1st pair of columns. Add to the left-hand boundary of the zone the random number that you find therein between 00 and 17 in the 1st pair of columns. Ignore any number between 18 and 99. To select a line on this page, use the 2d pair of columns; translate the 2 digits modulo 25—i.e., record the remainder after division by 25; see the examples below.

Line	Random number	
13	60 06	ignore (too big)
14	99 10	,,
15	26 27	ignore
16	00 68	gives page 01 (see Table 16) line 18
17	18 40	ignore
18	70 43	,,
.		
.	(10 lines produce blanks)	
.		
4	05 17	gives page 06 line 17
5	75 53	ignore
6	33 93	,,
7	01 64	gives page 02 line 14
Etc.		

TABLE 16

EXAMPLE OF ACTUAL SAMPLING TABLE SUPPLIED TO THE COMPANY

Zone for the page	Subsample									
	1	2	3	4	5	6	7	8	9	10
00001–00018 Page	01	06	02	18	06	05	09	09	16	04
Line	18	17	14	11	09	20	20	22	19	07
00019–00036 Page	26	36	25	25	21	27	36	29	31	30
Line	19	18	12	05	07	12	12	03	08	14
Etc.										

7. Show that the zoning interval 36, with 2 random numbers between 01 and 25 for the selection of 2 lines on each page of the sample, would have produced the same size of sample as the size produced by the procedure of the last exercise.

The zoning interval of 36 would have shortened considerably the labor of preparing the sampling table, and it was a temptation to use it.   However, there was undoubtedly some serial correlation between the entries on a page, and about 16% of the pages would have contributed 2 entries to the sample, not blanks.   It would have been necessary in my judgment, because of this correlation, to increase the size of sample from 4500 to 4600, if we were to draw 2 lines per page.   Each entry in the sample would cost $12 to investigate.   It was cheaper to construct a lengthy sampling table, 60% blanks.

8. *a.* Calculate the average occupancy-rate of the pages in Exercise 6. (*Answer:* 40%.)

b. Suppose that 15,000 pages contained on the average 8 entries per page, and that the remaining 5000 pages contained on the average 14 lines per page.   Would this disparity in occupancy rates affect the expected size of the sample?   (*Answer:* No.)

c. Suppose that some pages are full, and some empty.   Would this distribution or any other distribution of occupancy affect the expected size of the sample?   (No.)

d. If you know neither the total number of accounts, nor the average occupancy rate, can you govern in advance the expected size of the sample? (*Answer:* No, not its size; it would only be possible to govern the expected proportion, such as 1/45th of the blanks, 1/45th of the entries, 1/45th of the illegible entries, etc.)

9. Show that the following plan of selection will produce an expected sample of 4000 registered voters.   The frame was 12,000,000 registered voters in 12 cities in India.

1. Ten subsamples.
2. Assume as maxima, 4 columns to the page, 50 lines to the page.

TABLE 17

Zone		Subsample				
		1	2	3	· · ·	10
001–150	Page	148	013	013		067
	Column	2	3	3		4
	Line	37	24	02		14
151–300	Page	162	284	261		248
	Column	1	4	2		4
	Line	09	11	29	· · ·	46
Etc.						

3. The sampling unit shall be a line in a column.   Three serial numbers will designate a sampling unit: 1 number for the page, 1 for the column, 1 for the line.

4. Number the pages in every binder or other holder, in every precinct, in all cities.

5. In each city, prepare an index by precinct and by volume or other holder that will give a serial number to every page throughout the city, so that (e.g.) random number 1623 will designate a certain page and none other in a certain binder in a certain precinct.

6. The successive zones for pages will be 001–150, 151–300, 301–450, etc. (150 pages to the zone.)   Read out 10 random numbers in every zone. Accept duplicate numbers.   (Why?)   These numbers designate the pages that will contain the sample-lines.   Do this for every city.

7. Read out, for every page so selected, a random number 1, 2, 3, 4 to designate the column; also a random number between 1 and 50 for the line.   Accept duplicate line-numbers in the same zone unless the line so designated is an exact duplicate—same page, same column, same line. The lines so selected constitute the sample.   For an example see Table 17.

8. For every line in the sample, transcribe the information required (name and address for interviewer).   A blank produces no sample; make no substitution.

> *Note:* The actual number of registered voters so selected turned out to
> be 4010.   This example shows how it was possible to design in New York
> a sample of voters of prescribed size, to be drawn on the other side of the
> earth, with no information about the form and number of pages, nor about
> the number of columns per page, nor the number of lines.   The only
> information available in New York was the figure 12,000,000 for the total
> number of registered voters.   The plan of selection gave every line, and hence
> every voter, a probability of $1/150 \times 4 \times 50 = 1/30,000$ of coming into
> Subsample i, and hence a probability of 1/3000 of coming into the sample.
> This probability meets the requirements, as $12,000,000 \times 1/3000 = 4000$.

CHAPTER 9

Evaluation of Inventory of Materials*

Sir,—In your issue of December 31 you quote Mr. B. S. Morris as saying that many people are disturbed because about half the children in the country are below average in reading ability. This is only one of many similarly disturbing facts. About half the church steeples in the country are below average height; about half our coal scuttles below average capacity, and about half our babies below average weight. The only remedy would seem to be to repeal the law of averages. Yours faithfully, etc.—Letter to the editor of the *Times*, London, January 1954.

The nature of the problem. The aim of the sampling plan to be described here was to estimate the value of the materials in the process of manufacture in a large manufacturing plant. The auditors that examine the company's books and inventories must have results whose accuracy they can defend in their statement to the stockholders. The material in the process of manufacture (the subject of this investigation) is only part of the total inventory. A 2d part is the raw materials, pieces, and assemblies bought outside, and held ready for the production line. A 3d part is the finished product on hand, ready to ship. The accounting department has figures for these other parts, so it will not be necessary to include them in the sampling procedure.

Sampling was introduced in the materials in process to improve the accuracy of the estimate therefor, and to realize some small savings in costs (something like $12,000 annually). The greatest gain came in flexibility and speed. It is not now necessary to shut down the plant

---

* I am indebted to Messrs. Paul H. Wernicke, Comptroller, and to Harold Gulde, of the Quality Control Department, of the Minneapolis-Honeywell Regulator Company for the privilege of working on this problem and for permission to use the illustrations. Examples of the instructions and forms appear in a paper by Harold Gulde, "Physical inventory by statistical sampling," *Transactions of the 13th Midwest Quality Control Conference*, Kansas City, November 1958.

5 or 10 days for inventory; the company may now take the inventory by sampling methods on any Saturday and have the results in a few days. In contrast, there had previously been a delay of several months to process the complete inventory.*

Suppose (a) that the total inventory of all 3 parts combined was to be correct within 2% (a decision of the management and of the auditors); (b) that the 2d and 3d parts of the inventory were to be evaluated by the accounting department; (c) that the total number of dollars in all 3 parts combined is about $14,000,000; and (d) that the value of the materials in process is about $4,500,000.   Then

$$2\% \text{ of } \$14,000,000 = \$280,000 \tag{1}$$

This is the allowable 3-sigma error in dollars.   The allowable standard error is then about

$$\sigma_X = \$90,000 \tag{2}$$

The allowable coefficient of variation of X for the materials in process is

$$C_X = \frac{90,000}{4,500,000} = 2.0\% \tag{3}$$

Decision on the type of sampling plan.   The inventory of the materials in the process of manufacture will take place in 15 departments, in 4 buildings.   Lots of raw material (rubber, sheet metal, wire, bits of silver and gold, stampings, piece-parts rough and finished) go in at one end of a department and pass through one operation after another. The parts, materials, and assemblies in manufacture may be grouped into "lots" (in baskets, trays, pans, compartments, or in piles) along the production line, now at a standstill on Saturday.   Parts of the conveyor overhead are to be cleaned off and combined with similar parts below.   At each operation, the part, material, or assembly is more valuable than before.   There are, altogether, about 43,400 lots of materials of many kinds, and the lots range in value from perhaps $10 to $9000.

The decision on the type of sampling plan (arrived at after much discussion, and on the basis of theory and experience) was to form 2 strata:

Class 1. Lots that in the judgment of the foreman in charge of the department are worth K dollars or more.   Count (evaluate) these lots 100%.   (See page 153 for the value of K.)

---

* An account of the administrative advantages of sampling to estimate this inventory from the standpoint of the comptroller, is the article by Allan L. Rudell, "Sampling doubles accuracy of inventory; halves cost," *Bull. Nat. Ass. Accountants*, vol. xxxix, Oct. 1957: pp. 5–11.

Class 2. All other lots.   Draw 2 lots at random from every zone, 1 lot for Subsample 1, 1 lot for Subsample 2.   The zoning interval will be calculated later (bottom of next page.)

Remark 1.   I have already mentioned several times the technical and psychological advantages of this type of plan, by which items of high value go into the sample with certainty.   Another example appeared in Section B of the preceding chapter.

Remark 2.   There are 2 reasons for choosing 2 subsamples per zone and not 10.   First, simplicity.   Second, wider zones might have increased the variance (Ch. 21).

The next question is (a) What will be the best cutoff point K?   And then next: (b) How big a sample should one take in Class 2?   The theory by which to answer these questions follows.

Decision on the cutoff and on the size of the sample.   We collect here the theory and the symbols that we shall need; also in Table 1 the empirical data furnished by the company on the basis of some rough samples of last year's records of a 100% inventory.

Class 1   lots whose value the foreman declares to be above the cutoff

Class 2   lots that he declares to be below the cutoff

P_1 the proportion of lots in Class 1

P_2 the proportion of lots in Class 2

a the mean value of all the lots

a_1, a_2 the mean values of the lots in Class 1 and in Class 2

x an estimate of a

x_1 an estimate of a_1

x_2 an estimate of a_2

$$a = P_1 a_1 + P_2 a_2$$
$$x = P_1 x_1 + P_2 x_2$$

$$\sigma_x^2 = \left(1 - \frac{n_1}{N_1}\right)\frac{P_1^2 \sigma_1^2}{n_1} + \frac{P_2^2 \sigma_2^2}{n_2}$$ [The finite multiplier $1 - n_2/N_2$ for the right-hand term is replaced here by 1]

$$= \frac{0 + P_2^2 \sigma_2^2}{n_2}$$ (if we sample Class 1 completely)   (4)

n_1 = number of lots in the sample from Class 1   (here, $n_1 = N_1$)

n_2 = number of lots in the sample from Class 2

$n = n_1 + n_2 = N_1 + n_2$

Then from Eq. 4,

$$n_2 = \left(\frac{P_2 \sigma_2}{\sigma_x}\right)^2 = \left(\frac{P_2 \sigma_2}{a C_x}\right)^2$$ (5)

where $C_x = \sigma_x/a$ is the coefficient of variation of the sampling plan. In a rough sample of last year's 100% inventory, 2843 lots had a total value of $414,025, whence we adopt

$$a = \frac{\$414,025}{2843} = \$145 \text{ per lot} \tag{6}$$

We must remember that if a foreman has the job of separating the lots into 2 classes with a cutoff at (e.g.) $500, some lots that he places in Class 1 will have a value under $500, while some lots in Class 2 will run above $500. The distribution in Class 2 will be roughly triangular, tapering off to 0 at or near the cutoff K. For the variance of Class 2, we turn to the triangle in Panel C of Fig. 16 on page 260 for which

$$\sigma^2 = \frac{h^2}{18} \tag{7}$$

I shall assume that the lots of Class 2 run from 0 to

$$h = 1.4K \tag{8}$$

where h is the actual cutoff in Class 2, and K the intended cutoff. The coefficient 1.4 allows the foreman an error of 40%. (See Remark 3.)

Now if we set σ_2 equal to $\sqrt{h^2/18} = \sqrt{1.4^2K^2/18}$, $n_1 = N_1$, $a = \$145$, and $C_X = 1.8\%$ (shaded downward from 2.0% for safety), Eq. 5 reduces to $n_2 = (1.4KP_2/\sqrt{18} \times 145 \times .018)^2 = .016(KP_2)^2$. We may look in Table 1 to find P_2 for any proposed cutoff K, whereupon we are ready to calculate n_2. This I did for $K = \$400, \$500, \$600, \$700, \$1000$. The results are in Table 2.

Every pair of sample-sizes n_1 and n_2 in Table 2, when used with the corresponding cutoff K, and with the sampling procedure specified, will yield the desired precision, but with different cost. Which cutoff is most economical? The answer comes from the costs in the right-hand column. These costs came from the company, after I calculated the sample-sizes for the various cutoffs. They include only the variable part of the cost. The minimum cost appears to occur at about $500.

The exact location of the best cutoff is not critical. The cost of the sample is already small compared with the cost of the 100% inventory in earlier years, and we have yet to count the gain in flexibility. We thus set $K = \$500$ and proceed. h in Eq. 7 is then $700.

The calculated sample-size for the minimum cost is the figure $n_2 = 3760$ in Table 2. The ratio n_2/N_2 is then about $3760/(43,400 - 1300)$, which is about 1/11, or 2/22, which indicates a zoning interval of 22 in Class 2, with 2 drawings per zone. For simplicity, we specify 20 for the zoning interval.

TABLE 1

THE PROPORTION OF LOTS ABOVE AND BELOW VARIOUS DIVISION POINTS

P_1 and P_2 came from a sample of last year's inventory. $N_1 = NP_1$ where $N = 43,400$, the estimated number of lots this year. P_1 and P_2 are the proportions above and below the division point

Division point	$N_1 = NP_1$	P_1	P_2
\$ 1 and over	43 400	100.0%	0 %
100	11 600	26.8	73.2
200	6 900	15.9	84.1
300	4 687	10.8	89.2
400	1 950	4.5	95.5
500	1 300	3.0	97.0
600	1 040	2.4	97.6
700	868	2.0	98.0
800	739	1.7	98.3
900	650	1.5	98.5
1000	564	1.3	98.7
1100	477	1.1	98.9
1200	434	1.0	99.0

TABLE 2

THE SAMPLE-SIZES AND THE TOTAL COST OF INSPECTION FOR VARIOUS CUTOFFS

The sample-sizes n_1 and n_2, for any cutoff K, will yield the desired precision.   The costs came from the company

$$n_2 = .016(KP_2)^2 \qquad \text{(see text)}$$

Cutoff, K	$n_1 = N_1$	P_2 from Table 1	n_2 (calculated)	Cost
\$ 400	1950	.955	2 340	\$5060
500	1300	.970	3 760	4704 (min.)
600	1040	.976	5 498	4965
700	868	.980	7 530	5587
1000	564	.987	15 590	9436

One of the steps in preparation is to place 2 random numbers in every one of the 2085 sealed envelopes required for Step 3. These were fresh random numbers in every zone. An example would be this:

Sample 1   17
Sample 2   9

Further details of the plan. A lot is by definition a group of identical parts all contained in 1 box, pan, tray, basket, or compartment, or in 2 or more such containers that are in physical contact. The only safe way to bound a lot that consists of a number of trays or compartments is with generous use of white sticking tape. The steps on the floor are these.

1. Every foreman of a department will arrive several hours ahead of time in order to form and identify the lots that he believes have a value of $500 or more (Class 1). He will affix red tags to these lots, and number them 1, 2, 3, and onward. He will do this before the workers arrive for Step 2.

> **Remark 3.** The foreman made a pretty clean separation, although some lots did turn up in Class 2 with values up to $600. One turned up with a value of $900.

> **Remark 4.** The foreman's failure to make a clean separation does not introduce bias into the estimate. It does increase the variance of Class 2, but the variance of the estimate, whatever it is, will be calculated from the sample itself (see Eq. 20 later on for Var X).

2. Workers in pairs will group the remaining material into lots, 20 lots to a zone. The lots in a zone will receive the serial numbers 1, 2, 3, to 20. The serialization is simple with pads of note-paper, 20 pages to a pad, preprinted with the serial numbers 1 to 20. The 20 pages of a pad bear also the number of the zone. Each pair of workers receives a packet of pads, and they will use 1 pad to a zone. The serial numbers, once placed on the lots, must remain there.

> However, it is still permissible at this time for the workers to regroup their material into more convenient lots, or to call the attention of the foreman to any material that they believe should form a lot valued at $500 or more. After this step, no alteration of the lots nor of serial numbers is permissible. This step creates the frame. A lot is a sampling unit.

3. Open the sealed envelope specifically marked for each zone. This sealed envelope will contain 2 random numbers between 1 and 20. The lots that bear these serial numbers are the "sample-lots" in this zone. One number will be designated for Sample 1, the other for Sample 2.

4. Count the items in the sample-lots. Show each count separately on the proper card, along with the stock number, the unit cost (both from

the stock-book), the serial number of the zone and the serial number of the sample-lot, and the sample number (1 or 2).   Be sure to use the right card (green border).

5. Count the items in the lots of Class 1 (red tags) that you encounter in or near this zone.   Show for each lot the stock number, the unit cost, and the serial number of the lot.   Be sure to use the right card (red border).

> The "count" of a lot, if the lot contains not more than a few dozen pieces, is an actual count of the number of pieces therein.   For bigger numbers, a multiplying-balance is safer.   (A skilled operator will be on duty at each multiplying-balance.)   The 2 workers will carry their lots to the balance, or will ask for help if the lot is heavy.
>
> Defective parts are to be counted in a lot: they are part of the lot until they are officially removed and designated as scrap.
>
> The number of pieces multiplied by the unit cost gives the value of the lot in dollars and cents (e.g., 1284 pieces at 27¢ gives $1284 \times \$.27 = \$346.68$).   A worker shows the count and the unit cost on his report for the lot.   Later on, these numbers are punched into a card, and the multiplication (value of the lot) is punched automatically into other columns of the card.

The estimating procedure.   The formation of the estimates and of their standard errors follows the methods of previous chapters, but there are some new features.

There are 15 departments.   There will be for Department i the figure A_i for the total value of the lots of Class 1.   There will also be x_{i1} and x_{i2} for the sums of the dollars in Samples 1 and 2 from Class 2.   Now let

$$X_1 = 20\Sigma x_{i1} \quad \text{[The estimate from Sample 1 of the total dollars in Class 2]} \tag{9}$$

$$X_2 = 20\Sigma x_{i2} \quad \text{[The same for Sample 2]} \tag{10}$$

$$A = \Sigma A_i \quad \text{[The total dollars in Class 1]} \tag{11}$$

All the sums (Σ) run over all 15 departments.   The figures for the sums come from Table 3.

Then the final estimate of the total value of all the lots in both Class 1 and Class 2 in all 15 departments will be

$$X = A + \tfrac{1}{2}(X_1 + X_2) \tag{12}$$

and

$$\text{Var } X = 0 + 100\Sigma w_i{}^2 \quad \text{[From Eq. 2, page 197]} \tag{13}$$

where

$$w_i = |x_{i1} - x_{i2}| \quad \text{[Department } i] \tag{14}$$

The calculations go rapidly. The figures in Table 3 show that

$$X_1 = 20 \times \$127,678 = \$2,553,560 \quad \text{[By Eq. 9]} \quad (15)$$

$$X_2 = 20 \times \$131,857 = \$2,637,140 \quad \text{[By Eq. 10]} \quad (16)$$

$$\tfrac{1}{2}(X_1 + X_2) = \$2,595,350 \quad (17)$$

The lots that the foremen placed in Class 1 had the total value of

$$A = \$2,178,727 \quad (18)$$

whence the total inventory of the materials in process is

$$X = \$2,178,727 + \$2,595,350 \quad \text{[By Eq. 12]}$$

$$= \$4,774,077 \quad (19)$$

Table 3 gives

$$\text{Var } X = 100 \times 55,723,396 \quad \text{[By Eq. 13]} \quad (20)$$

$$\hat{\sigma}_X = 10\sqrt{55,723,396} = \$75,000 \quad (21)$$

The standard error aimed at was \$90,000 (Eq. 2). The correspondence between the precision aimed at and the precision attained shows how one may study in advance the statistical characteristics of the material to be sampled, with allowances for human mistakes in classification, and reach just about the precision desired.

Remark 6. Any estimate of a standard error is itself a random variable with a standard error. The number of degrees of freedom in the estimate of Var X is about 11, so the standard error of $\hat{\sigma}_X$ is about 21% (p. 439). The advanced student may wish to turn to the appendix in Chapter 11 (p. 225) to see a calculation of the number of degrees of freedom in this example.

Remark 7. As the estimate $\hat{\sigma}_X$ has a standard error of about 21%, we may not assert that the standard error of X is exactly \$75,000. We are safe in saying, however, that the precision attained is close enough to the precision aimed at.

There was also in the inventory certain additional material: in the form of steel and copper in sheets; paints, oils, and greases; and other non-productive items. Altogether, the figure furnished by the company for this additional material brought the total estimated inventory up to \$12,098,000, of which only the portion sampled is subject to sampling error. The coefficient of variation, based on the entire inventory, was therefore

$$\hat{C}_X = \frac{74,648}{12,098,000} = .62\% \quad (22)$$

which is excellent precision, according to the client.

Comparison with the accounting records. The final estimate of the inventory of materials in process is ready for comparison with the number of dollars charged to this account by the accounting department. This comparison and any other comparison that is possible should never be omitted. The reasons have already been stated in Chapter 5. It is all too simple to omit part of the inventory, or to evaluate some materials in a way different from the way they are charged by the Accounting Department. The comparison with the books, the 1st year that sampling was introduced by this company, showed a difference of only \$25, or about 1 part in 150,000. This fantastic agreement is of course sheer coincidence that no sample or complete count could hope to duplicate. It had nevertheless a powerful psychological effect in favor of sampling. A statistician is entitled to a coincidence like this once in a while. Another one occurred in the Census of Japan (p. 267).

TABLE 3

THE RESULTS OF THE SAMPLES

(Figures altered)

Department, i	Number of zones	Sample 1	Sample 2	Difference	(Difference)^2
1	24	\$ 2 973	\$ 2 279	\$ 694	482 108
2	25	1 737	2 008	− 271	73 224
3	80	7 630	7 068	562	315 574
4	27	1 270	1 371	− 101	10 124
5	133	10 208	10 297	− 90	8 055
6	131	7 428	8 225	− 797	635 448
7	185	9 875	7 819	2057	4 230 755
8	210	14 175	15 257	−1082	1 169 772
9	112	5 871	4 707	1164	1 354 012
10	289	13 477	16 155	−2677	7 168 631
11	238	15 824	19 752	−3928	15 430 520
12	209	13 562	12 847	715	511 196
13	231	10 089	13 817	−3728	13 900 445
14	39	4 278	4 204	74	5 540
15	152	9 280	6 051	3229	10 427 991
Total	2085	127 678	131 857	−4178	55 723 396

The audit to estimate nonsampling errors. An audit of the inventory took place according to a prescribed plan (see Ch. 5, page 71). A corps of 20 people regularly engaged in statistical quality control re-examined a total of 440 lots, drawn by random numbers from the main sample of the inventory, both classes. The search for material missed was

100%.   The auditors used a regular form printed for this purpose.   The audit searched for blemishes of the following types:

1. Material not in any lot (100% audit).
2. Fuzzy boundaries of lots.   How much material could belong to either of 2 lots?
3. Wrong count of the items in a lot.
4. Wrong stock-number; wrong operation-number.
5. Mixed parts on 1 ticket.
6. Discrepancies between (a) the sampling numbers on the lots counted and (b) the original list of random numbers.

The following memorandum on the nonsampling errors is based on figures furnished by the audit.   The reader will perceive that (a) the evaluation of the nonsampling errors depends on a probability model, and that (b) there is some latitude for the exercise of judgment.   There is no unique theory for the nonsampling errors as there is for the sampling errors.   The reader might perhaps treat some of the errors differently, but I doubt if the final net result would be substantially different.   In case of doubt about how to handle a certain error, one may choose the easiest course, which is to be as severe as possible.

MEMORANDUM ON THE NONSAMPLING ERRORS OBSERVED
IN THE AUDIT OF THE INVENTORY*

TABLE 4

SUMMARY OF THE ESTIMATES OF THE EFFECTS OF
THE NONSAMPLING ERRORS

Type of defect	Remark	Estimate of the error caused
Lots missed	The audit discovered 2 lots missing. These lots were thereupon put into the inventory. The coverage of this part of the audit was total, so I assume that there were no more lots missed.	0
Wrong counts Wrong part number Wrong operation number Mixed parts	I lump these into 1 broad type of error; see separate section for treatment. The over-all effect appears to be an overestimate.	The net result of these errors may be an overestimate; see separate section.

* This section and the next 2 came from my memorandum on the case.

Treatment of Wrong Counts, Wrong Part Number,
Wrong Operation Number, and Mixed Parts

I throw these errors together, as they all have, I believe, common roots in bad luck, incorrect entries, fatigue, misunderstanding, and possible carelessness.

There were 440 lots in the audit, 15 of which disclosed mistakes. As the work was of the same nature in both the high-value lots (declared by the foremen to be $500 or over) and in the low-value (sample) lots, I make no distinction in what follows. The tabulation of the 15 errors is in Table 5. The net bias from the above sources observed in the audit of 440 lots was $287.63. The average net bias is therefore $287.63/440 or about 65.37¢ per lot. The total number of lots in the regular inventory was about 43,370; hence the estimate of the bias is 43,370 × 65.37¢ = $28,350. This figure is positive, and it indicates that the nonsampling errors may have caused an overestimate. We now calculate the limits of the bias.

$$\hat{\sigma}_1^2 = \frac{124,172}{14} = 8869 \qquad [\text{From Table 5}]$$

$$P_1 = \frac{15}{440} = .0341, \text{ the proportion of lots in the audit with error}$$

$$P_2 = \frac{425}{440} = .9659, \text{ the proportion of lots in the audit with no error}$$

$$\sigma^2 = P_1\sigma_1^2 + P_2\sigma_2^2 + P_1P_2(a_1 - a_2)^2 = .0341 \times 8869 + 0 + 0$$

$$= 302.34$$

$$\sigma = \$17.39$$

Note that σ_1 and σ_2 here have not the meaning that they had in the early part of this chapter. They refer now to the 2 portions of the 440 lots in the audit, the 15 lots with errors, and the 425 lots without error. Similarly for P_1 and P_2.

The standard deviation of the errors in the 440 observations in the quality-audit is thus about $17.39. I shall assume that errors of the kind dealt with here can fall in either direction, and that $17.39 is the standard deviation of the process of taking an inventory.

There were 5843 lots in the sample of the regular inventory, and this figure includes the high-value lots (Class 1). The standard deviation of the error in a sample is $\sigma\sqrt{n}$, where n is the number of lots in the sample (Ch. 17). In this case, our estimate of the standard error of the bias that arose from the above nonsampling errors in the sample of 5843 lots is

$$\sigma\sqrt{5843} = \$17.39\sqrt{5843} = \$1330$$

I do not attempt to divide these dollars between the lots of low value and the lots of high value, as both kinds of lots are subject to the same kinds of error, and in approximately the same amounts. Moreover, they contribute about equally to the total inventory. I shall therefore merely expand the estimated standard error $1330 of the sample by the factor 43,370/5843 to find

$$\$1330 \times \frac{43,370}{5843} = \$9872$$

This is the estimate of the standard deviation of the estimate of the bias in the total estimated inventory that arose from the nonsampling errors. Three standard errors is $30,000. The maximum overestimate attributable to the nonsampling errors is then $28,350 + $30,000 = $58,350. The maximum underestimate is $28,350 − $30,000 = −$1650.

The standard deviation of the combination of sampling error plus the estimate of the nonsampling errors is $\sqrt{\{75,000^2 + 10,000^2\}} = 75,650$. Three times this number is $227,000. The maximum overestimate is thus $28,350 + $227,000 = $255,300, or 2.1% of $12,098,069 (furnished by the company for the total inventory). The maximum underestimate is $28,350 − $230,000 = −$201,650, or 1.7% of $12,098,069.

TABLE 5

CALCULATION OF THE AVERAGE VALUE AND OF THE VARIANCE

OF THE NONSAMPLING ERRORS

The figures for x_i came from the audit

Type of error	Error x_i (+ high, − low)	$x_i - \bar{x}$	$(x_i - \bar{x})^2$
1. Error in count	23.29	4·11	
2. „ „ „	−55.62	−74·80	
3. „ „ „	2.01	−17·17	
4. „ „ „	−.55	−19·73	
5. „ „ „	.22	−18·96	
6. „ „ „	−.17	−19·35	
7. „ „ „	0	−19.18	
8. Error in part number	−27.33	−46.51	
9. „ „ „ „	353.89	334.71	
10. „ „ „ „	0	−19.18	
11. „ „ „ „	0	−19.18	
12. Wrong operation number	−.22	−19.40	
13. „ „ „	0	−19.18	
14. „ „ „	0	−19.18	
15. Mixed parts	−7.89	−27.07	
Sum	287.63	−.07	124,172.39
Average (divide by 15)	$\bar{x} = 19.18$		xxx

Statement from the Consulting Statistician
to the Comptroller of the Company*

This statement is predicated on figures and other information furnished to me by your company, on the assumption that your people followed correctly my sampling procedures. I may point out that the method of counting, the pricing, the extensions, and the verification of the existence of the inventory, including the existence of the materials in process, are outside my province, and I take no responsibility on these aspects of the inventory nor for anything other than the statistical methodology and the interpretation of the results that you have furnished to me.

The sampling plan that I designed for your inventory provided procedures for (1) the selection of lots for the sample; (2) the formation of an estimate of the aggregate inventory of the materials in process; (3) the calculation of the margin of sampling error in this estimate; (4) a probe of a subsample of the main sample to evaluate some of the nonsampling errors.

I shall deal first with the margin of error of the sampling itself. In my opinion, the results that your company obtained for the inventory of the materials in process in June 1957 falls within a maximum sampling tolerance of $224,000 in either direction from what your company would have obtained had you counted and physically processed every lot of the designated inventory of the materials in process with the same care and with the same degree of skill that you exercised in applying the sampling procedures. The maximum sampling tolerance, $224,000, is 1.9% of $12,098,069, this being the figure that your company furnished to me for the estimated total regular inventory, including the materials in process and additional items.

I turn my attention now to the nonsampling errors, which are dependent on human observation and have not the objectivity of the calculation of a sampling tolerance. The sampling plan contained within itself a systematic probe for the evaluation of certain nonsampling errors, viz.: lots missed; wrong count of parts; wrong part number, wrong name for the part; wrong operation number; missing operation number; mixed parts on 1 ticket not designated as mixed parts.

The error of sampling, mentioned above, includes the effect of the variable part of the nonsampling errors, such as wrong counts, wrong part number, wrong operation number, mixed parts. It does not include the constant or systematic part of the nonsampling errors, such as a persistent tendency to overcount or to undercount.

The probe for lots missed and for lots counted twice was total. It detected no lot counted twice, and only 2 lots missed, out of the 47,370 or so lots in the regular inventory. This flaw was corrected, so it should lead to no error whatever, and I shall make this assumption.

I have evaluated the other nonsampling errors with the aid of a probability mechanism (previous section), with figures furnished by you. The results indicate a possible overestimate. The maximum overestimate, if there is an overestimate, can hardly exceed $58,000. It is possible that there is no overestimate at all, as the probability mechanism gives $1650 as the limit of any underestimate attributable to the nonsampling errors.

The limits of error from the combination of the sampling and the

---

* Figures altered.

nonsampling errors are in my opinion a maximum overestimate of $255,300, and a maximum underestimate of $199,000, these figures being respectively 2.1% and 1.7% of $12,098,069. The actual magnitude of the possible overestimate or of the possible underestimate lies, in my opinion, well inside these two extremes.

Why the sampling method gives a better inventory than the complete count had given heretofore.   The sampling method, for reasons like these, gives greater accuracy than the complete inventory and given heretofore:

1. Fewer people did the work (about 1/5th as many as on the complete inventory); better selection possible; better supervision. Less fatigue; better concentration on the work.

2. Training more thorough.   This statement applies to the instruction of the foreman, as well as of the workers.

> The slides, the printed material, the instruction itself, and the test-runs were much improved over those used previously for the complete inventory. There had been nothing to hold back such improvements in the complete inventory, but this experience was the usual one in which sampling proced-ures carefully conducted require and achieve quality of performance not previously achieved.

3. Greater success in including every piece of material in 1 and only 1 lot.

4. More care in counting (or weighing) the pieces in a lot.

> More care in writing down the correct stock number.   More inspection of stock numbers.
> More care in writing down the correct unit price.
> Proportionately greater amount of supervision and probe.

5. Speed.

> The complete inventory had required in past years 2 weeks for the inventory itself, plus something like 4 months for processing.   Most of this processing was done on overtime, with the usual number of mistakes.
> In contrast, this sample-inventory required 1 day, plus a few days for the processing.

6. Flexibility.

> Because of the speed of the sample, the inventory can now be taken and entirely finished in any week of the year.   It is thus possible now to adjust the time of the inventory to the convenience of the company and of the auditors.

7. Smaller proportion of illegible and missing entries.

> The effect was (*a*) greater accuracy; (*b*) less time lost in the punching of the cards, as the punchers spent relatively less time trying to decipher illegible entries and track down duplicates to try to fill in the missing figures.

8. More knowledge about the accuracy of the results.

There was not only greater accuracy in the sample, as claimed above, but also BETTER KNOWLEDGE OF JUST WHAT THE ACCURACY WAS.

The audit of the performance was carried out more thoroughly than is usual with a complete inventory.   This is the part of the sampling plan in which provision is made to measure the effect of all the errors that can occur (p. 61).   Inspectors made random selections of a subsample of lots, according to a plan laid out in advance; and they probed into the identification and counting of the lot, the stock number, the unit price.   They wrote up a report on every lot in their subsample.   This sort of verification was impossible because of sheer size when the inventory was complete; consequently no one ever really knew how good or how bad it was.   Now, with sampling, the quality of the job is known.

Exercise.   An inspector engaged in the audit of the sampling procedure spied an error in a sample-lot that was not designated for the audit. Should this count as a defect against the department?

No; but a report thereof should go to the foreman so that he can refer it to the original inspector for correction.

CHAPTER 10

Exercise in a Replicated Survey
of a Small Urban Area

Alas, my lord, your wisdom is consumed in confidence.—
Calpurnia to Julius Caesar, Shakespeare's *Julius Caesar*, Act II,
Scene ii.

The aim of this chapter. Although this exercise will be a small survey, and done on paper, it will present many of the practical problems that occur in actual surveys, big and little. The student will (1) break up into segments the area in the map in Fig. 7 on page 168. He will then (2) draw a sample of segments; (3) collect information about the people and the dwelling units in these segments; (4) from the information so collected, compute estimates and the standard errors of certain characteristics of the entire area. This exercise will prepare the student for study of the next chapter, which contains the general theory and procedure for use of the replicated method in larger areas, and under varied conditions.*

Instead of ringing door-bells, the interviewer will consult a complete census of the area, already taken. The complete census will be the equal complete coverage for the sample; so in this instance, we may compare the results of the sample with the equal complete census. A sample is a sample of the labels that the equal complete coverage would attach to every person, to every dwelling unit, to every farm, cow, and business establishment in the area. If the complete census contains nonresponse, then the sample will pick up its share of this nonresponse. If the complete census contains wrong answers through misunderstanding or inadvertence, then the sample will pick up its share of these same mistakes.

The student's first task, before he begins to carry out the steps of the sample-survey, will be to make up a complete census of the area. Table 1 shows the start of a complete census, which the student should finish. This census covered the map in Fig. 7.

* A paper that would be helpful to the student at this point is Leslie H. Kish, "A two-stage sample of a city," *Amer. Sociol. Rev.*, vol. 17, 1952: pp. 761–769.

Remark 1. The instructor may elect to pool the efforts of the class to compile one master complete census. The students would then all draw their samples from the same complete census. Their results and their standard errors would provide a set of interesting comparisons amongst themselves and with the complete census.

TABLE 1

THE COMPLETE CENSUS

Dwelling unit		Address		Number of		Children under 10	Rent or rental value	Owned or rented
no.	Block	Street	No.	M	F			
1	1	Stirling St.	36	2	2	1	$100	O
2			34	...	...	...	90	V-S
3		Fourier St.	3	...	...	...	110	V-S
4		Wallis Sq.	11	2	3	2	110	O
5			9	3	1	2	100	O
6			7	1	4	2	120	O
7		Cauchy St.	6	1	3	1	90	O
8	2	Stirling St.	30	1	2	...	90	O
9			28	2	2	1	80	O
10			26	2	3	2	90	O
etc.								

R = rented by the people that live in the dwelling unit
O = owned by the people that live in the dwelling unit
V = vacant

V-H = vacant, held for occupancy
V-S = vacant, for sale
V-R = vacant, for rent

The complete census is summarized in Table 7.

Remark 2. Another alternative for teaching is to start off with a real complete census of some area nearby. Such a project would show up problems of organization, of interviewing, and the perplexing problems of refusals and of people not at home, people calling the police, etc. It would also give some elementary ideas of costs, and of the variability of costs, because some interviewers will finish early and some late. In such a case this chapter would provide the procedures for drawing samples and compiling estimates and standard errors.

The sampling unit.   The sampling unit will be 1 or more segments in a designated area (called a block in this chapter).   The intended size of a sampling unit in this survey will be 3 d. us. (dwelling units), based on the rough count (Table 2).   The actual sizes of the segments in any block

TABLE 2.   THE FRAME

LIST OF SAMPLING UNITS BY BLOCK

Block	Number of dwelling units (rough count)	Number of sampling units	Serial number of the sampling units
1	6	2	1–2
2	13	4	3–6
3	33	11	7–17
4	10	3	18–20
5	14	4	21–24
6	6	2	25–26
7	6	2	27–28
8	9	3	29–31
9	7	2	32–33
10	9	3	34–36
11	11	4	37–40
12	25	8	41–48
13	7	3	49–51
14	9	3	52–54
15	7	2	55–56
16	10	3	57–59
17	9	3	60–62
18	9	3	63–65
19	7	2	66–67
20	8	3	68–70
21	8	3	71–73
22	7	2	74–75
23	4	1	76
24	10	3	77–79

will vary above and below the average, as the prime requisite is definite boundaries, rather than equality of size, or any particular size (see Ch. 12).   The zoning interval will be 8, for a 25% sample.

If the number of segments in any block exceeds the number of sampling units ascribed in Table 1 to that block, there is a chance that the random numbers in Step 5 will select 2 segments for interview in that block (see Block 18 in Table 5).   Regardless of the size of segment, every person of every age and of every characteristic, and every dwelling unit, has a

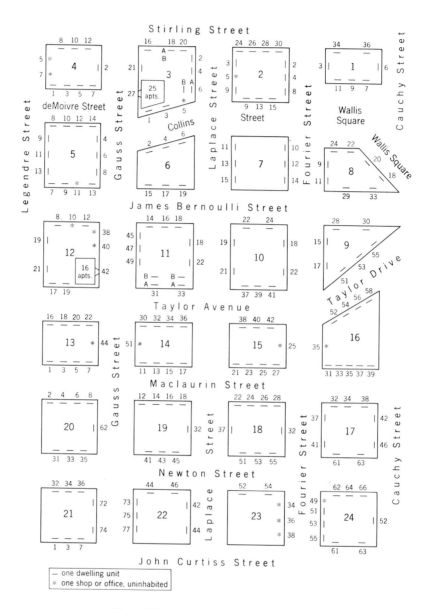

Fig. 7. Map to use for the complete census.

probability of 1 in 8 of getting into Subsample 1, and the same probability of getting into Subsample 2.

The frame. The frame will be a list of sampling units by block. Before we can decide on the number of sampling units in each block, we need first a rough count of the number of dwelling units block by block. The field-workers will carry out the following steps:

1. Make a quick trip over the area designated for the survey (Fig. 7), and record a rough figure for the number of dwelling units in each block. (Table 2 provides an illustration.)   Every block will have a serial number, and will be identifiable on the map, or otherwise.   This list is the frame.

   Remark 3.   The map was supposedly made some time ago, and is not necessarily accurate now.   The student should assume, in making up his complete census, and in carrying out Step 1, that there have been some changes, such as new dwelling units or new structures here and there, and an occasional removal of a structure.   Further, in accordance with reality, the student will occasionally mistake shops for dwelling units, and the converse.   He intends to include in his count of dwelling units any structure that could be inhabited.   He must therefore allow for the fact that a shop or an office building could have quarters for the owner or for the janitor.   He will include vacant dwelling units in the segments because we wish to count them, and because they may be occupied later when the actual interviewing takes place.   In practice, 2 people may differ by 10% in their rough counts of a block.

   Remark 4.   In practice, the "block statistics" furnished by the last census provide the figures for Step 1 in the "block cities" in the U.S. (see the next chapter for an illustration, and for the definition of a block city). In the "nonblock cities"' and in the urban areas outside the block cities, or in a country where the census does not show figures for very small areas, one may often purchase or procure otherwise a map and then carry out Step 1 by a quick cruise over the area on foot or with an automobile or with a bicycle, or by aid of the parish priest, or of the police, or of other officials (I have used them all), or by giving each small area an average size (see the sample of Mexico City in the next chapter).   Aerial maps are often very helpful in rural areas.

   Remark 5.   The field-worker will usually have instructions not to stop to ring door-bells in Step 1.   The only enquiry necessary will be an occasional question to the superintendent of an apartment house to learn how many flats there are inside, or how many on each floor.   The work should proceed rapidly, as the cost must be kept low, and there is no great advantage in having exact counts in Step 1.   Fortunately, we need only rough approximations at this stage.   Rough counts will cause no bias, as I have said before.

2. Show in Table 2 for each block the number of sampling units, and form the accumulated sums.   The number of sampling units in a block will in this survey be $\frac{1}{3}$d of the number of dwelling units recorded in

Step 1.   Note that the accumulated sums give a serial number to every sampling unit.

Remark 6.   It is important to note that the number of sampling units ascribed to any block in Step 2 and Table 1 remains fixed for the whole survey.

The sample.

3. Create zones with a zoning interval of 8 sampling units (Table 3).

TABLE 3

Selection of the Sample*

Zone	Zone boundaries	†Random numbers between 1 and 8 without replacement		Random numbers translated to the serial number of the sampling unit		Block that contains the selected sampling unit	
		1st	2d	1st	2d	1st	2d
1	1–8	3	1	3	1	2	1
2	9–16	2	7	10	15	3	3
3	17–24	7	5	23	21	5	5
4	25–32	8	5	32	29	9	8
5	33–40	4	8	36	40	10	11
6	41–48	2	5	42	45	12	12
7	49–56	6	5	54	53	14	14
8	57–64	8	7	64	63	18	18
9	65–72	3	8	67	72	19	21
10	73–80	8	5	80	77	blank	24

* This table and all the others in this chapter came from a survey carried out by Miss Josephine D. Cunningham.

† From Kendall and Smith's *Random Sampling Numbers* (Cambridge University Press, 1951).

4. Draw 2 random numbers in each zone, one number for the 1st subsample, and another random number for the 2d subsample. The sampling unit whose serial numbers fall into the 1st sample will belong to the Subsample 1, and those whose serial numbers fall in the 2d sample will belong to the Subsample 2.   Each student will draw his own random numbers and insert them in his own construction of Table 3.

The sampling plan draws 2 sampling units from every consecutive 8. Every segment, every dwelling unit, and every person, will have a probability of 1 : 4 of coming into the sample.

5. Put random numbers in a sealed envelope, 1 envelope for every block in the sample, to direct the interviewer, when he is ready for Step 7, to the segment (or segments) that he is to cover.

Remark 7. The numbers in the envelopes are recorded in Table 5. They select, in any block, 1 segment in every m segments, where m is the number of sampling units shown in Table 2 *for that block*. If the number of segments that the interviewer created in a block were never more than the number (m) of sampling units ascribed in Table 2 to that block, we should need only 1 random number between 1 and m. But there are sometimes more segments than sampling units in a block, as will happen (a) when the interviewer, in the interest of definite boundaries, creates segments smaller than the intended size (3 d. us., in this case), or (b) when the area has grown since the rough count was made. To play safe, we place in each sealed envelope 3 random numbers—one between 1 and m, another between $m + 1$ and $2m$, another between $2m + 1$ and $3m$.

The number of segments in which interviews will take place in any block may be 0, 1, 2, or possibly 3. Any random number in the envelope that falls beyond the actual number of segments in the block is a blank. The asterisks in Table 5 show the segments in which there were interviews. There were 2 instances where there were interviews in more than 1 segment in a block.

Step 5 should in practice be independent of the creation of the segments in the blocks. Why? (See Exercise 3 at the end of the chapter.)

The creation and selection of the segments. It is possible to ascertain from Table 2 which block any sampling units belong to, even though the segments are not yet created. The interviewer will now create segments in the blocks that contain the sampling units whose serial numbers fell into the sample (Table 3). He will not create segments in the remaining blocks, where no random number struck. The creation of the segments and the interviewing may take place in one trip. The steps follow:

6. The interviewer will go to each block that contains a random number in Table 3, and will create segments therein, counterclockwise from the northeast corner. Every segment will have a serial number within the block (Table 4).

Define a segment by the half-open interval; for example, from and including Stirling Street No 36 up to but NOT including Wallis Square No. 11. To define a segment, it will usually only be necessary to record the address that begins the segment. However, if there is any possibility of confusion later on, you should prepare notes or a rough map of the segment

TABLE 4

IDENTIFICATION OF THE SEGMENTS IN THE BLOCKS
IN WHICH THE SAMPLE FELL

Block	Number of sampling units in the block (from Table 1)	Serial number of the seg- ment (Step 6) within the block	From and including the given address, up to but not in- cluding the next address	Approximate number of dwelling units in the seg- ment	Remarks
1	2	1	Stirling St. 36	3	
		2	Wallis Sq. 11	4	
2	4	1	Stirling St. 30	4	
		2	Laplace St. 3	2	
		3	Collins St. 9	3	
		4	Fourier St. 8	3	
3	11	1	Stirling St. 20	5	
		2	Gauss St. 27 Basement	4	See Note A
		3	Gauss St. 27-D	3	2d floor
		4	,, G	3	3d floor
		5	,, J	3	4th floor
		6	,, M	3	5th floor
		7	,, P	3	6th floor
		8	,, S	3	7th floor
		9	,, V	3	8th floor, remainder of 27 Gauss St.
		10	Collins St. 1	2	
		11	Laplace St. 6	2	See Note B
		12	,, 4	2	
5	4	1	DeMoivre St. 14	4	
		2	Legendre St. 9	3	
		3	Bernoulli St. 7	3	
		4	Gauss St. 8	3	
8	3	1	Wallis Sq. 22	4	
		2	Bernoulli St. 29	2	
		3	Wallis Sq. 18	3	See Note C
9	2	1	Bernoulli St. 30	4	
		2	Taylor Ave. 51	3	
10	3	1	Bernoulli St. 24	4	
		2	Taylor Ave. 37	3	
		3	Fourier St. 22	2	
11	4	1	Bernoulli St. 18	3	
		2	Gauss St. 45	3	
		3	Taylor Ave. 31	2	See Note D
		4	Taylor Ave. 33	2	See Note E
		5	Laplace St. 22	2	

TABLE 4 (CONTINUED)

Block	Number of sampling units in the block (from Table 1)	Serial number of the seg- ment (Step 6) within the block	From and including the given address, up to but not in- cluding the next address	Approximate number of dwelling units in the seg- ment	Remarks
12	8	1	Bernoulli St. 12	2	
		2	Legendre St. 19	2	
		3	Taylor Ave. 17	2	
		4	Gauss St. 42-A	4	
		5	,, E	3	
		6	,, H	3	
		7	,, K	3	
		8	,, N	3	
14	3	1	Taylor Ave. 36	4	
		2	Maclaurin St. 11	4	
18	3	1	Maclaurin St. 28	4	
		2	Laplace St. 1	3	See Note F
		3	Laplace St. 4	3	
		4	Laplace St. 7	3	
		5	Laplace St. 10	2	
		6	Newton St. 51	4	
19	2	1	Maclaurin St. 18	4	
		2	Newton St. 41	4	
21	3	1	Newton St. 36	3	
		2	Curtiss St. 1	3	
		3	Gauss St. 74	2	
24	3	1	Newton St. 66	3	
		2	Fourier St. 49	3	
		3	Curtiss St. 61	3	

Note A: At 27 Gauss Street there is a superintendent's unlettered apartment in the basement. This is included in Segment 8 along with all the apartments on the 1st floor of which there are 3, lettered respectively A, B, and C.

Note B: This segment is composed of the entire structure at No. 6 Laplace Street. There is only 1 entrance but there are 3 name cards and 2 bells.

Note C: There are 2 structures both bearing the number 20 Wallis Square. Segment 3 in Block 8 includes both these structures plus 1 d.u. at 18 Wallis Square.

Note D: This Segment No. 3 is composed of the entire structure at 31 Taylor Avenue. There are apparently 2, 3, or 4 apartments in it as there appear to be 2 side and 1 rear entrance.

Note E: This Segment No. 4 is composed of the entire structure at 33 Taylor Avenue. The structure is physically similar to the one next door at 31 Taylor Avenue (Segment 3); it has multiple entrances and may contain several dwelling units.

Note F: The structure at 37 Laplace Street on the map has been destroyed and is now the entrance to a new low-cost housing development consisting of 11 houses numbered 1 to 11 Laplace St.

TABLE 5

RANDOM NUMBERS FOR THE SELECTION OF THE SEGMENTS FOR INTERVIEWS

Zone	Subsample 1				Subsample 2			
	Block	Number of segments in the block (from Table 2)	3 random numbers between 1 and the number of segments (Note 1)	Serial numbers of the segments for interview (in sealed envelope; Notes 2 and 3)	Block	Number of segments in the block (from Table 2)	3 random numbers between 1 and the number of segments (Note 1)	Serial numbers of the segments selected for interview in sealed envelope; see Notes 2 and 3
1	2	4	3, 3, 3	*3, 7, 11	1	2	2, 1, 2	*2, 3, 6
2	3	11	4, 2, 7	*4, 13, 29	3	11	6, 3, 11	*6, 14, 33
3	5	4	1, 1, 3	*1, 5, 11	5	4	3, 4, 1	*3, 8, 9
4	9	2	2, 1, 1	*2, 3, 5	8	3	3, 3, 3	*3, 6, 9
5	10	3	3, 1, 1	*3, 4, 7	11	4	4, 2, 1	*4, 6, 9
6	12	8	4, 2, 4	*4, 10, 20	12	8	1, 8, 2	*1, 16, 18
7	14	3	2, 2, 1	*2, 5, 7	14	3	3, 1, 3	*3, 4, 9
8	18	3	1, 1, 2	*1, *4, 8	18	3	2, 3, 3	*2, *6, 9
9	19	2	2, 1, 2	*2, 3, 6	21	3	1, 2, 1	*1, 5, 7
10	blank	blank	blank	blank	24	3	2, 3, 1	*2, 6, 7

Note 1: From Kendall and Smith's *Random Numbers*, 18th Thousand, Column 1, line 1, with replacement. In drawing the 2d sample, do not admit any random number that would draw a segment that is already in the 1st sample. In practice, with smaller sampling ratios (*n/N*), duplicated random numbers occur only rarely.

Note 2: These are the numbers in the sealed envelope in Step 7.

Note 3: The asterisk denotes a segment in which interviews took place.

or of the whole block to show the direction of motion from one end of any difficult segment to the other end.

Aim at a size of 3 d. us. to a segment, but never hesitate to alter the size of a segment to gain clarity in its definition. Obscurity about the boundary of a segment, where it begins or where it ends, and what dwelling units compose the segment, will alter to some unknown extent the probability of some of the dwelling units in this segment, and may cause some small but unknown bias. In contrast, inequalities that you introduce in the sizes of segments in order to obtain clear and unmistakable boundaries cause no bias.

Chapter 12 contains more details on field-procedure.

The interviews.

7. Open the sealed envelope to learn which segment or segments you will interview in (Table 5). The identification of the segments by number should be obvious. (If it isn't, your work in Steps 5 and 6 was not satisfactory.)

8. Carry out the interviews in the segments selected by the random numbers in the sealed envelope.

The interviewer will consult his complete census and will record on a form provided for this purpose the information required for each dwelling unit in his segments. (The form is not shown here; it will resemble the form in Table 1.)

Segments not used in this survey will serve for subsequent surveys.

Remark 8. The importance of complete coverage of a segment can not be overestimated. The omission of rear and side entrances, or of houses hidden by trees, or of apartments difficult to gain access to, may cause serious biases (not just a shrinkage in the size of the sample, which would by itself be unimportant).

Remark 9. In some surveys (not illustrated here*), the segment will be oversize, and the interviewer will conduct an interview for information in only every 2d or 3d dwelling unit. However, a short interview in every dwelling unit in a segment is still necessary in order to find all the dwelling units.

The tabulations.

9. Summarize by thick zone for Subsample 1 and for Subsample 2 the information in your interviews for the following characteristics of the

---

* Hansen, Hurwitz, and Madow, *Sample Survey Methods and Theory* Vol. I (Wiley, 1953): Ch. 12.

TABLE 6

SUMMARY OF THE RESULTS BY THICK ZONE

A thick zone is here defined as equal to 2 of the original zones

A.   Dwelling units, vacant, owned, and rented

Thick zone	Dwelling units (all)			Vacant			Owned by occupant			Rented by occupant		
	Subsample		$w_i(*)$	Subsample		$w_i(*)$	Subsample		$w_i(*)$	Subsample		$w_i(*)$
	1	2		1	2		1	2		1	2	
$i = 1$	6	7	−1	1	0	1	3	4	−1	2	3	−1
2	6	7	−1	0	0	0	6	6	0	0	1	−1
3	6	4	2	1	1	0	1	3	−2	4	0	4
4	11	7	4	0	1	−1	3	0	3	8	6	2
5	4	6	−2	1	0	1	1	1	0	2	5	−3
Sum	33	31	2	3	2	1	14	14	0	16	15	1
$†\overline{w}$	xxx	xxx	2.0	xxx	xxx	0.6	xxx	xxx	1.2	xxx	xxx	2.2
R^2	xxx	xxx	26	xxx	xxx	3	xxx	xxx	14	xxx	xxx	31

TABLE 6
(CONTINUED)

B. Male, female, and children under 10

Thick zone	Male and female			Male			Female			Children under 10		
	Subsample		$w_i(*)$	Subsample		$w_i(*)$	Subsample		$w_i(*)$	Subsample		$w_i(*)$
	1	2		1	2		1	2		1	2	
$i = 1$	21	29	-8	13	13	0	8	16	-8	5	9	-4
2	30	23	7	15	9	6	15	14	1	6	5	1
3	14	11	3	9	5	4	5	6	-1	3	2	1
4	48	26	22	25	12	13	23	14	9	13	9	4
5	14	26	-12	7	11	-4	7	15	-8	5	6	-1
Sum	127	115	12	69	50	19	58	65	-7	32	31	1
$\bar{w}$	xxx	xxx	10.4	xxx	xxx	5.4	xxx	xxx	5.4	xxx	xxx	2.2
R^2	xxx	xxx	762			540			211			35

* I have carried the signs in these columns to facilitate verification of the arithmetic.
† $\bar{w}$ is the average difference, taken without regard to sign. My secretary, Edith Del Peschio, compiled this table and the next one from data supplied by Miss Josephine Cunningham.

TABLE 6

(Continued)

C.   Rent or rental value (all d. us.)

| Thick zone | Rent or rental value | | |
| | Subsample | | w_i (*) |
	1	2	
$i = 1$	550	700	-150
2	640	550	90
3	460	320	140
4	830	480	350
5	240	300	-60
Sum	2720	2350	370
$\bar{w}$	xxx	xxx	158

people and of the dwelling units.   (See Table 6 for an illustration.)

    i.   The number of dwelling units, total.
    ii.   ,,   ,,   ,,   ,,   ,,   vacant.
    iii.   ,,   ,,   ,,   ,,   ,,   owned by occupant.
    iv.   ,,   ,,   ,,   ,,   ,,   rented by occupant.
    v.   ,,   ,,   ,, inhabitants (male plus female).
    vi.   ,,   ,,   ,, males.
    vii.   ,,   ,,   ,, females.
    viii.   ,,   ,,   ,, children under 10.
    ix.   The proportion male.
    x.   The average rent or rental value of all the occupied dwelling units.
(*Note*: the rental value of a dwelling unit occupied by the owner is by definition 1 % of the selling price of the dwelling unit.)
    xi.   The proportion of the occupied dwelling units owned by the occupants.

Compute the difference (range) w_i between the 2 subsamples for Thick Zone i.   Compute (as in Table 6) the average range $\bar{w}$ without regard to sign, and compute T^2, the sum of the squares of the ranges.

A thick zone in this chapter will be 2 successive thin zones.   A thick zone is only a convenient subtotal.   The purpose of the thick zones is to reduce the labor of tabulation, yet to provide sufficiently good estimates of the standard errors. In this chapter there are only 10 thin zones, and the labor of tabulation to tabulate them all would not be prohibitive. However, to gain experience, and to provide illustration, and incidentally to reduce the labor, we shall form 5 thick zones.

10. Summarize the above characteristics for your complete census, for comparison (Table 7).

TABLE 7

SUMMARY OF THE COMPLETE CENSUS COMMENCED IN TABLE 1

Characteristic	Total or average
i. The number of dwelling units, total	249
ii. The number of dwelling units, vacant	24
iii. The number of dwelling units, owned by occupant	123
iv. The number of dwelling units, rented by occupant	102
v. The number of inhabitants (male plus female)	920
vi. The number of males	456
vii. The number of females	464
viii. The number of children under 10	225 (plus a number unknown at 27 Gauss St.)
ix. The proportion male	.496
x. The average rent or rental value of all the occupied dwelling units (not including vacant units)*	$81.93 $\left(= \dfrac{18,425}{225}\right)$
xi. The proportion of occupied dwelling units owned by the occupants	.494 $\left(= \dfrac{123}{225}\right)$

* The rental value of a home occupied by the owner is by definition 1% of the selling price of the home.

The estimates and their standard errors. We are now ready to compute the entries in Table 9, which will summarize our survey.

11. Form estimates of the totals and proportions listed in Step 9. The figures that we need from the survey are in Table 6. The new formulas that we need we borrow from the next chapter. The roman numerals in the paragraph below correspond to lines of equal number in Table 9.

i. THE NUMBER OF DWELLING UNITS, TOTAL. Table 6, Part A shows

$x = 34 + 30 = 64$ [The number of dwelling units in the sample]

$T^2 = 26$ [Sum of squares]

Eq. 1 on page 197 gives the estimate.

$$X = \tfrac{1}{2}Zx = 4x = 256 \qquad [Z = 8 \text{ here}] \tag{1}$$

for the total number of dwelling units.   This compares with 249 from the complete census (see Table 7).   For the standard error of X we draw on the formulas in the next chapter.   As the sample here is 25 %, we introduce the finite multiplier.   Its general form, from Chapter 17, for an estimate of a variance, is $(N - n)/N$, which takes here the specific form $(Z - 2)/Z$, numerically 6/8.   The finite multiplier for an estimate of a standard error here will then be $\sqrt{(6/8)}$, or .87.   An estimate of the standard error of the number of dwelling units will then be

$$\hat{\sigma}_X = .87 \times \tfrac{1}{2}ZR \qquad [\text{Eq. 4 on page 197}]$$
$$= .87 \times 4R = 3.46R \qquad [R^2 = 26 \text{ from Table 6, Part A}]$$
$$= 3.46\sqrt{26} = 18 \tag{2}$$

ii. The number of dwelling units vacant.   Table 6, Part A shows

$$x = 3 + 2 = 5$$
$$R^2 = 3$$

We reuse the constant parts in the estimates that we have just calculated, now to find

$$X = 4x = 4 \times 5 = 20 \qquad [\text{As in Eq. 1}] \tag{3}$$

which compares with 24 for the complete census (Table 7 or 9).   For the standard error by the sum of squares,

$$\hat{\sigma}_X = 3.46R \qquad [\text{As in Eq. 2 above}]$$
$$= 3.46\sqrt{3} = 6 \tag{4}$$

iii–viii.   The student may now perform the calculations for these characteristics, by use of Eqs. 1 and 2 above.

ix. The proportion male.   From Table 6, Part B,

$$x = 69 + 50 = 119 \text{ male}$$

$$y = 127 + 115 = 242 \text{ male plus female}$$

$$f = \frac{x}{y} = \frac{119}{242} = .492, \text{ the proportion male in the sample}$$

For the standard error of f, we calculate in Table 8 to find $S^2 = 36.2$.

TABLE 8

CALCULATION OF THE SUM OF SQUARES S^2 FOR USE IN THE
STANDARD ERROR OF THE ESTIMATE OF THE RATIO OF
THE PROPORTION MALE ($f = .492$)

w_{xi}	w_{yi}	$w_{xi} - fw_{yi}$	$(w_{xi} - fw_{yi})^2$
(from Table 6, Part B)			
0	-8	3.9	15.4
6	7	2.5	6.2
4	3	2.5	6.2
13	22	2.2	4.8
-4	-12	1.9	3.6
Average 5.4	10.4	xxx	$S^2 = 36.2$

We insert the same finite multiplier .87 that we used before, whereupon

$$\hat{\sigma}_f = .87 \frac{S}{y} \quad \text{[Eq. 10, on page 198]}$$

$$= .87 \frac{\sqrt{36.2}}{242} = .023 \tag{5}$$

x. The proportion female is the complement of the proportion male, and has the same standard error (see Exercise 4 at the end of the chapter).

xi. The average rent or rental value of all the occupied dwelling units (owned plus rented):

$x = \$2720 + \$2350 = \$5070$ total rent or rental value

$y = 14 + 14 + 16 + 15 = 59$ d. us. occupied (owned plus rented)

$f = \dfrac{\$5070}{59} = \$85.93$ an estimate of the average rent or rental value of the occupied dwelling units

The reader may calculate the standard error of this estimate of the average rental value of all d. us. by use of those formulas used for the proportion male. The data and the ranges are in Tables 6A and 6C.

xii. The proportion of the occupied dwelling units owned by the occupants; also the standard error of this proportion (left to the reader as an exercise).

It is interesting to note from Table 9 that no estimate made from the sample departs as much as 2 standard errors from the complete census. Further, the standard errors of the proportions are much smaller, relatively, than the standard errors of the totals.   Why?

TABLE 9

THE ESTIMATES AND THEIR STANDARD ERRORS: AND
COMPARISON WITH THE COMPLETE CENSUS
Refusals ignored

Characteristic	Estimated from the sample	Standard error estimated from the sample	Complete census (Table 7)
i. Dwelling units total	256	18	249
ii. Dwelling units vacant	20	6	24
iii. Dwelling units owned	112	13	123
iv. Dwelling units rented	124	19	102
v. Male plus female	968	84	920
vi. Male	476	80	456
vii. Female	492	23	464
viii. Children under 10	252	20	225
ix. Proportion male	0.492	.023	0.496
x. Proportion female	0.508	.023	0.504
xi. Average rent of the occupied dwelling units	$85.93	$5.30	$81.93
xii. Proportion of occupied dwelling units owned by occupants	0.475	.045	0.494

Possible biases from nonresponse.   The reader will observe that there were a few refusals, although neighbors supplied some of the missing information.   There is no cure for nonresponse in either complete censuses or samples (Ch. 5).   To expedite the tabulations, it is necessary to adopt some rule that will make immediate allowance for the nonresponses. One simple rule for an estimate of a total is to duplicate the information from the preceding household.   For characteristics that are highly correlated door to door, this rule is not bad.   For characteristics that are uncorrelated, this rule is merely a quick way to bury the problem.

For an estimate of a proportion, or of an average of any kind, one simple rule is to ignore the nonresponse.   This is the equivalent of ascribing to the nonresponses the same average values that the responses gave.

Thus, if one were to apply the above rules in the present exercise, the proportions (characteristics ix, x, and xi) would be unaffected, but the totals for both the complete census and the sample would be affected. As an exercise the student may apply the above rules for characteristics i–viii, noting however that where the information was obtained from a neighbor, there is no need to make allowance for nonresponse, because we have the information.

EXERCISES

1. Explain why it is, once we fix the zoning interval as 8 sampling units, and decide to draw 2 sampling units from each zone, that the probability of any person's and of any dwelling unit's coming into Subsample 1 is exactly 1/8; likewise for Subsample 2; and that the probability of coming into both subsamples combined is 1/4; also that these probabilities are exact (*a*) regardless of the intended size of the segments; (*b*) regardless of the actual sizes of the segments (*c*) regardless of how rough be the rough counts that we started with; (*d*) even though the random numbers select 0, 1, or 2, or even 3 or more segments for canvass in 1 block, provided we follow the rules in putting the random numbers into the sealed envelopes.

Chapter 19 shows the theory and computation of the increase in variance that arises from inexactness in the rough counts, and from variation in the size and number of segments per sampling unit.

2. *a.* Under what circumstances could the random numbers in the sealed envelope fail to select any segment for interview in some block? (When the number of segments created is smaller than the lowest random number.)

b. If this happens, would you interview in 1 segment anyhow? (NO.)

3. Why is it a good idea for another person (other than the interviewer) to draw the random numbers for the sealed envelopes?

The person who creates the segments and numbers them 1, 2, 3, etc., if he knows which segments have been selected for canvass, may unconsciously and with the best of intentions create and number them so that his canvass will avoid unusual segments. It is easy, by proper arrangement of the work, to serialize the sampling units independently of the random numbers.

4. Prove that the standard error of the proportion of males and the standard error of the proportion of females will always be equal, under any plan of sampling.

5. The number of dwelling units found in a sample of New York, carried out in October 1955, are shown in Table 10.   Calculate (*a*) an estimate of the number of dwelling units in New York by Eq. 1 on page 180; and (*b*) the coefficient of variation of this estimate.   The zoning interval was 2000 sampling units.

TABLE 10

DWELLING UNITS IN NEW YORK (ALL 5 BOROUGHS);
OCTOBER 1955

Thick zone	Subsample 1	Subsample 2	Range w	w^2
1	65	93	28	784
2	81	81	0	0
3	84	80	4	16
4	119	74	45	2025
5	83	62	21	441
6	78	87	9	81
7	69	82	13	169
8	75	65	10	100
9	88	71	17	289
10	69	75	6	36
11	74	89	15	225
12	75	77	2	4
13	89	88	1	1
14	74	112	38	1444
15	99	59	40	1600
16	24	20	4	16
Total	1246	1215	253	7231
Average	77.9	75.9	15.8	451.9

ANSWERS

a. $X = \frac{1}{2} \times 2000(1246 + 1215) = 2,461,000$ d. us.   [Eq. 1, page 197]

b.   By the sum of squares, $\hat{C}_X = \dfrac{\sqrt{7231}}{1246 + 1215} = 3.4\%$ [Eq. 5, page 197]

Remark 10.   The Census of 1950 showed 2,434,000 d. us.   Some years later, as I understand it, the authorities of the City of New York, in the hope of demonstrating a sizable increase in the number of inhabitants, and the right to claim a greater cash allotment to the City from the State, ordered a new census of population, which took place in 1957.   This census turned out to be an expensive disappointment, as it yielded 2,487,000 d. us., only 2% above the Census of 1950, and incidentally 1 standard error higher than the sample of October 1955, calculated above. This sample or any other sample properly designed could have supplied to the City authorities, at a cost of less than 1% of the cost of the census, preliminary information that would have predicted little or no increase in the number of inhabitants, which is exactly what the census of 1957 confirmed.

Remark 11 (for information). The intended size of sampling unit was 2 segments of 5 dwelling units each, drawn from opposite halves of a block. The probability of selection was exactly 1 in 2000, regardless of the actual number and size of segment created in the field, and regardless of any growth or loss in population.*

A further calculation on page 449 evaluates the variance of $\hat{C}_X$.

* I am indebted to the firm O'Brien-Sherwood of New York for the privilege of working with them on this survey.

CHAPTER 11

General Theory and Procedure for Replicated Sampling of a Large Area*

No skill in pedagogy, and no lustre of personality, can atone for teaching errors instead of truth. Errors are very likely to be taught by those who do no research, and then, the more skillful the pedagogic indoctrination, the greater the harm.—Harold Hotelling, "On the teaching of statistics," a paper read at a meeting of the Institute of Mathematical Statistics at Dartmouth College, September 1945: *Annals of Mathematical Statistics*, vol. xix, 1948: pp. 95–115; page 106 in particular.

A. GENERAL STATEMENT ON PREPARATION

New problems in a sample of a large area. We carried out in the previous chapter a survey of a small urban area. The aim there was to learn some basic routine of preparation and of calculation. We shall now extend the procedure to large areas, and we shall learn some possible alternative variations, and when to use them. What additional problems do we now face?

---

* This chapter comes largely from my paper "On simplifications of sampling design through replication with equal probabilities and without stages," *J. Amer. Statist. Ass.*, vol. 51, 1956: pp. 24–53. Replication originated with Mahalanobis in 1936, and it is a pleasure to express my appreciation for the privilege of studying his methods in India, in 1946 and again in 1951 and 1952. He uses the term interpenetrating subsamples, which has much merit. Discussions with Mr. D. B. Lahiri of the Indian Statistical Institute in Calcutta, and with Messrs. William N. Hurwitz and Max Bershad of the Bureau of the Census in Washington, have helped in many ways. For detailed accounts of the Indian experience, see P. C. Mahalanobis, "On large-scale sample surveys," *Phil. Trans.*, vol. 231B, 1944: pp. 329–451; "Recent experiments in statistical sampling in the Indian Statistical Institute," *J. Roy. Statist. Soc.*, vol. cix, 1946: pp. 325–378; D. B. Lahiri, "Technical paper No. 5 on the National Sample Survey" (The Department

1. The best size of segment to use. The size of segment in the last chapter was 3 d. us., for illustration and ease of computation. Segments should sometimes be much bigger. On the other hand, they may sometimes have an intended size of 1 d. u., as in samples that I am engaged in working on at this moment for studies of conditions of housing in Turkey, where the door-to-door correlation in the characteristic of special interest (number of rooms, fuel for cooking, fuel for heating, etc.) is very high.

2. The number of segments per block. This number will often be 1, but it may be good to use 2 segments from opposite halves of a block, in districts where opposite sides of a block are sufficiently dissimilar in many characteristics.

3. The composition of a sampling unit. A sampling unit will sometimes be a minimum load of 50 or 60 or more d. us., broken into segments of 1, 5, 10, 20, or more d. us., all confined within a primary area such as a county (pp. 213 ff.).

4. The frame may now include rural areas and suburban areas along with urban areas. There is nothing fundamentally new about rural areas, except that definable areas of various irregular shapes and irregular density of population will replace the more regular urban blocks, and that the maps of the rural areas that we may wish to sample may be of varying quality and difficult to find.

5. There will be a question of the best order for listing the census areas in the frame, to facilitate tabulation and to achieve the best statistical efficiency. We shall return shortly to this question.

6. There will be, in a large-scale survey, operational problems in the attainment of quality and uniformity in the training, hiring, supervision, and performance of the interviewers. These problems we do not treat in this book, although Chapter 13 will present some new statistical aids to supervision.

The primary frame. The primary frame will be a list of areas with definite boundaries (p. 207). These areas will preferably be areas for which we have Census-figures, although one must sometimes draw up plans where there are no Census-figures, or where they are out of date (Remark 1 and Section F of this chapter). If we have Census-figures, or any other recent information, we use them, and show for each area the

of Economic Affairs, Ministry of Finance, New Delhi, March 1954)—published also in *Sankhya*, vol. 14, 1954: pp. 264–316.

Replicated sampling went under the name of the Tukey plan in my earlier papers and in my book *Some Theory of Sampling* (Wiley, 1950), in respect to my friend Professor John W. Tukey of Princeton, who in 1948 took the trouble to persuade me to use 10 interpenetrating subsamples in a certain application, to eliminate the labor of computing the standard errors. I have used no other method since.

number of sampling units therein.   In any event, the areas in the primary frame are primary areas.

A primary area may be struck by more than 1 random number, and thus contribute to 2 or more subsamples.   Indeed, if a primary area covers 2 zones, 2 random numbers in each subsample will strike it.   The simplicity of the calculation of variance by the replicated method lies in the possibility that a primary area may be struck by more than 1 random number, even in the same subsample (pointed out to me privately by my friend and colleague Professor John W. Tukey).   If 2 random numbers strike a primary area, we simply repeat the procedure of drawing a sampling unit therefrom, whatever be the procedure specified.   In other words, the larger units come into the sample with replacement.   It is only at the last stage, where we draw our final sample of segments or of other items to test or to interview, that we may draw without replacement.

A random number will fall within 1 of the identifiable primary areas. This primary area may be very small, like an enumeration district (Sections C, D, E of this chapter), in which case the sampling unit may be a segment of area or a pair of segments.   It may be a county or a group of counties (Section E of this chapter), and the sampling unit therein may be several segments, in pairs or independent of each other.   For example, a sampling unit that is to be 64 d. us. (as in an actual survey that followed the description in Section E) could be 1 segment of 64 d. us., 2 segments of 32 d. us., 4 segments of 16 d. us., 4 pairs of segments with 8 d. us. to the segment, etc. The segments may all be confined to a random portion of the primary unit, or they may be permitted to fall anywhere within the boundaries of the entire primary unit (p. 215).

In any case, we try to obtain maps and Census-figures with which to create a secondary frame consisting of secondary areas within the primary area struck.   If the sampling unit is to consist of (e.g.) 2 pairs of segments, there will be 2 drawings from the secondary frame.   Each drawing will fall in some secondary area, such as a block, or some other type of area. If our Census-figures are still about right, and if the interviewer creates segments of about the size intended, there will be, with rare exceptions, 2 segments on opposite sides of the block.   The size of the sampling unit, the size of the segment, and whether there should be 1 big segment or 2 segments of smaller size on opposite sides of the block, and how they are to be linked or distributed, should be chosen in any proposed survey to achieve efficiency and smooth performance.

Some remarks on the order of listing the census areas in the frame.   The order in which the primary areas appear in the frame should serve 2 requirements.   First, a primary area that is smaller than a zone (e.g., a block, tract, enumeration district) should, if convenient, go into the same zone with other areas that have similar characteristics (e.g., average income, occupation), in order to achieve possible gains from stratification. In other words, each zone should be as homogeneous as possible.   Within a zone, however, the order in which the census areas appear is immaterial. No problem of stratification arises for an area whose size is as big as a zone, since it will be in the sample no matter where it appears in the frame.

Second, the order in which the areas appear in the frame should facilitate tabulation.   For example, in a sample of a metropolitan area, the frame might commence with a list of the tracts in the big city or cities;   then go on to areas in the smaller cities;   and then to enumeration districts or other areas in the rural parts, in some significant order.   This order would permit separate tabulation for the central portions of the big cities, and separate tabulations for the part outside these cities.   Within a city, or within any other area, excellent natural stratification is inherent in the frame if the order in which the area appear is geographic, as is usually the case in a census-table.   It is important to note that sampling by zones captures all the possible gain in precision that is possible from the natural stratification in the frame.

Measure of size; the Cdu.   We need an elementary measure of size for an area.   It is convenient for this purpose to invent the symbol Cdus for for dwelling units contained in an area according to the last census or according to any later information.   Thus, an area that contained 317 d. us. at the last census, and for which we have no new information, contains now 317 Cdus, even though the actual (unknown) number of dwelling units in the area today be different.   If our sampling units are to have an intended size of (e.g.) 10 Cdus, this area will appear in the frame with 32 sampling units.   If there has been a 10% growth, a sampling unit of 10 Cdus will have today an average size of about 11 actual dwelling units.

Some rules on the number of thick zones.   We had, in the preceding chapter, an introduction to the use of thick zones, but with no discussion of the principles that should govern their thickness.   The reason for tabulating by thick zone, rather than by thin zone, is to gain economy and speed in tabulation, to normalize the estimates, and to increase the validity of the estimates of the standard errors computed by the range. The size of the thick zones should if possible meet the following requirements.

1. The number of thick zones should be enough to provide the minimum number of degrees of freedom that we require in our estimates of variance. If there are 2 subsamples, each thick zone yields 1 degree of freedom (p. 443).

Remark 1.   One will hold the number of thick zones to some reasonable minimum, to avoid unnecessary cost and delay in tabulation.

Remark 2.   In a continuing series of samples, month after month, on the same kind of material (studies of the labor force, rail traffic) for which the standard error remains fixed or changes only slowly, the cumulative estimate thereof becomes firm in a few months, even with only 4 or 5 subsamples and 1 thick zone.

2. An estimate of a total (typified in this book by X) is unbiased no matter how rare be the population estimated.   The estimate of Var X by the sum of squares is also unbiased.   However, any estimate of a variance for a rare population characteristic may be subject to high sampling variance.   For rare populations that are evenly distributed over the region to be sampled, one may avoid difficulty by use of segments that are big enough to contain, on the average, 1 or 2 or 3 members of the rare population.   A rule of thumb is that a thick zone need contain 5 or more members of a rare population to permit use of the range in the estimate of the standard error of X.   Special problems exist in the sampling of spotty characteristics, like the occupations of mining and fishing.   Chapter 17 shows some theory that is helpful in respect to the choice of thick zones, with a numerical example (p. 446).

3. An estimate of sampling error of a ratio (typified in this book by $f = x/y$) may require some care if the y-population is rare, but there will be no difficulty if the population in the denominator in each thick zone meets the requirement explained in the preceding paragraph.

Remark 3.   Thus, we see 2 forces pushing us toward big segments, and toward a large number of thin zones in each thick zone.   At the same time, the 1st requirement points out the need of a sufficient number of thick zones. Hence, in a survey of a small region, where there may be only 20 or 30 thin zones, requirements 1, 2, and 3 may be competitive.

Remark 4.   The thick zones retain the statistical efficiency of the initial zones while they reduce the number of tabulations.

4. The thick zones need not be uniform in their construction.   In fact, in the national sample described in Section E, a thick zone may be an entire stratum, such as the northeast metropolitan counties.   This is a convenient arrangement, as it is customary and in fact almost necessary as a routine procedure to tabulate in subtotals by stratum anyhow.

5. One should be aware, though, of the possibility of a loss in the number of degrees of freedom if the variances of 1 or 2 thick zones dominate the others in the formation of a total.   The estimates of a total (X) are additive, and so are the variances (except in some instances of stratified sampling where the samples are not independent from stratum to stratum,

Chapter 15), but the degrees of freedom are not necessarily additive. The appendix to this chapter treats the degrees of freedom.

6. The simplest rule to apply in the circumstance of high variability in the variances within thin zones is to form the thick zones by a systematic selection. Thus, if we wish to tabulate in 10 thick zones, we may form thick zone No. 1 by combining thin zones No. 1, 11, 21, 31, etc.; and thick zone No. 2 by combining thin zones No. 2, 12, 22, 32, etc.; and similarly for the other 8 thick zones. In fact, it is the effectiveness of the stratification that causes the variances within the thin zones to vary greatly from one thin zone to another, which in turn causes loss in degrees of freedom. A systematic selection of thin zones mixes the material and pretty well equalizes the variances of the thick zones.

One may also form the thick zones by a random ordering of the thin zones.* The expected effect of this random ordering appears in the exercise in the appendix to this chapter, page 222.

> **Remark 5.**   No matter how we form the thick zones, we introduce no bias, either into the estimate of X or into Var X. The problem is merely one of getting our money's worth of degrees of freedom, and of knowing how many degrees of freedom we have.

7. In connexion with rare characteristics, it is important to note from some recent work by Howard L. Jones† that the interpenetrating sub-samples will still give valid estimates and valid standard errors of a ratio $x : y$ so long as both numerator and denominator of at least 2 subsamples pick up members of the universe that possess the specified x and y characteristics. Of course, if some populations are rare or absent in 1 or more subsamples, the interpretation of the standard error must be made with the aid of the proper theory for skewed distributions.

Incomplete zones.   Unless we take steps to avoid an incomplete zone, an area of tabulation may begin or may end in the middle of a zone. When this happens we may, for purposes of tabulations, treat the rest of the zone as blanks, even though a new area begins within the zone where another one leaves off. The only reason to be unhappy about an incomplete zone is the inconvenience of apportioning the population in the zone to the 2 areas. There are several ways to avoid an incomplete zone, and to keep the tabulation clean.

---

* This suggestion came from Mr. Thomas Hayton of the Bell Telephone Company of Canada, Montreal.

† Howard L. Jones, "Investigating the properties of a sample mean by employing random subsample means," *J. Amer. Statist. Ass.*, vol. 51, 1956: pp. 54–83.

Sizes of Sampling Units Adjustable

1. Adjust upward or downward* the size of the sampling unit (e.g., the number of Cdus per sampling unit) so that the accumulated number of sampling units over the area is a multiple of the zoning interval Z.   There will be an accumulation of the losses and gains of rounding, and it will usually be necessary to add or to subtract a sampling unit or 2 in 1 or more of the biggest primary areas, to fill the last zone with no remainder.

2. Increase or decrease the number of sampling units in a few big primary areas only, to force the total number of sampling units in the area of tabulation to an exact multiple of Z.   This plan is equivalent to altering slightly the size of the sampling unit in the areas so reworked.

Either way, there is no change in the probability of selection.

Sizes of Sampling Units Not Adjustable

3. *If the last zone is nearly full.*   Fill up the zone with blanks, in which case a random number may strike a blank and draw no sample.

4. *If the last zone is nearly empty.*   Add the sampling units in the incomplete zone one by one to the sampling units in the preceding zone. For example, if the zoning interval were 2000 sampling units, and if 108 sampling units were left over in the last zone, we could simply double the size of the first 108 sampling units in the last complete zone, and eliminate the incomplete zone.   There will be some loss in precision for a total, but insignificant loss for a ratio x/y.

Incomplete thick zones.   A thick zone at the end of an area of tabulation may contain fewer than the desired number of thin zones.   No adjustment is necessary.   The calculation of the estimate X or f from the whole sample is unaffected.   The estimate of a standard error by use of the range may in principle be slightly disturbed because of the unequal weighting, though I have not seen a case in practice where I thought that unequal weighting made any difference.   One is always safe in using a sum of squares (R, T, or S in Eqs. 3–23 ahead), as the squares are self-weighting.

Complete freedom in the basic design.   Any sample-design that is valid can be laid out in a replicated design.   One not only has complete liberty in the basic design, but must exercise it.   Thus, one may use any mode of stratification that he deems to be efficient, by specifying the order for listing the census areas in the main frame.   One may, if he wishes, use

---

* The adjustment is usually only very slight.   In an actual example, the adjustment was from the prescribed size of 64 Cdus to 63.36.

intermediate stratification of a preliminary sample, followed by Neyman allocation for the main sample (p. 277).

One may meet subsequent demands for some other size of sample in some district. To decrease the size of the sample to half the original size, one may delete 1 sampling unit out of every 2 consecutive sampling units in one subsample, and delete the opposite sampling units in the other subsample. (See page 337 for balanced thinning.)

It is also easy to increase the size of sample. An additional sampling unit per zone will produce a 50% increase, with 3 subsamples. In a sample of New York, where the zoning interval was 2000 sampling units and where it was necessary to produce a 25% supplementation in the boroughs of Brooklyn and Queens, I drew for each subsample 1 supplemental random number between 0001 and 8000 for every consecutive 4 zones throughout these 2 boroughs. The interval 8000 was convenient because this number was to be the size of the thick zone for tabulation.

The intended sizes of the sampling units, and of the segments as well, need not be constant within any primary or secondary area. It may be important, for economy, to decrease the number of segments in areas that contain a very large number of Cdus, or which are difficult to carve, or which will be costly for the field-workers to reach. This decrease in the number of segments in an area is easy to accomplish, as the text has just explained. A rule that is often close to optimum is to make the size of the segment proportional to the square root of the number of Cdus in the area. Two or 3 size-classes will suffice.*

One may feel free to choose any formula that appears to be efficient for the estimation of any total, proportion, or other characteristic of the frame. Whatever be the choice of the form of the estimate, the computation of the standard error thereof will be, through replication, rapid and valid.

One may, as I said above, confine the sampling unit to a county, and scatter it in any prescribed manner, over the whole county or over only a portion of the county (Section E).†

Triplicate and quadruplicate drawings from each zone may at times be

---

* Morris H. Hansen, "The sampling of human populations," *Proceedings of the International Statistical Institute*, Washington, 1947.

Hansen, Hurwitz, and Madow, *Sample Survey Methods and Theory*, Vol. I (Wiley, 1953): p. 359. See also W. Edwards Deming, *Some Theory of Sampling* (Wiley 1950): p. 238.

† The theory for the optimum number of segments in a sampling unit, and for the optimum number of sampling units in a cluster, is in Hansen, Hurwitz, and Madow, *Sample Survey Methods and Theory*, Vol. I (Wiley, 1953): p. 291.

desirable, or even 10 drawings per zone, as we have seen in previous chapters. Formulas for multiple drawings appear later (p. 198).

It is not difficult to lay out a plan by which to measure the variance between interviewers, or the difference between 2 methods of training, or between 2 questionnaires, or both, by balanced random assignments in successive zones; see Chapter 13.

For extra pressure on (e.g.) 1/3d of the nonresponses, one may select at random 1 sampling unit from every successive 3 sampling units in Subsample 1, and likewise for Subsample 2 and for the other subsamples if any; then weight by 3 the results of the successful interviewers of the nonresponse, and add them subsample by subsample to the initial results (p. 106).

Heavy spotty growth that has taken place here and there since the last census will cause the same trouble in this sampling procedure that it causes in any other, and it calls for the same treatment.*

Replication calculates the total variance, not the components thereof. The simplicity that the replicated method provides for the calculation of the standard error of an estimate is an advantage that can hardly be over-estimated. Be it noted, though, that the replicated method measures the total variance. There are occasions, especially in the redesign of a sample, when one needs to know the components of the total variance—e.g., How much of the total variance came from the variability of the characteristic from dwelling unit to dwelling unit within the segment? How much came from the variation in the size of the segment? How much from the variation in the number of segments per sampling unit? From the variance between the means of the segments? It is important also in the design of a sample to have some knowledge of the costs of travel and of the average cost of creating segments and interviewing in 1 segment.

This book contains very little on procedures for the actual measurement of the various components of the total variance. The reader whose chief interest is research in design must therefore turn to other works, which would include most of the books on sampling that have appeared to date.

Two-way stratification with forced selection of heterogeneous sampling units.† One important variation in design is a 2-way stratification which

---

* See Hansen, Hurwitz, and Madow, *Sample Survey Methods and Theory*, Vol. I (Wiley, 1953): p. 35. See also Chapter 14 in this book, page 256.

† The plan offered here is a simplified description of a plan used by Roe Goodman and Leslie F. Kish at the Survey Research Center in Ann Arbor; see their article, "Controlled selection, a technique in probability sampling," *J. Amer. Statist. Ass.*, vol. 45, 1950: pp. 350–372.

forces areas of unlike characteristics to fall together into the sample, a device that in some kinds of studies has shown notable increases in efficiency. Thus, one may force areas of heavy industry to fall into the sample along with areas of light industry, urban areas to fall with rural areas, center with fringe, high-rent with low-rent, etc. One accomplishes this forcing by use of 2 frames, where we used only 1 heretofore. The 2 frames will cover some region of the domain of study (a province, or a region, or a city), but they will commence with census areas that are opposite in character. Thus, one frame might commence with the central part of the biggest cities of predominantly heavy industry, and the other frame might commence with open country sparsely settled.

In studies of public opinion in Germany, a separation of communities into 2 lists on the basis of religion was effective. One list commenced with communities that were practically 100% Protestant: the other list commenced with communities that were practically 100% Roman Catholic. Both lists merged at the bottom with communities that were about equally divided.

The sampling units in both lists will commence with serial number 1. The 2 zoning intervals will be equal, and each list will contain the same number of sampling units, with blanks at the end if necessary to fill out the last zone. However, the average sizes of the sampling units need not be equal in the 2 lists, nor need the segments be equal; in fact their average sizes may vary intentionally from one zone to another within the same list to accommodate variable costs, as mentioned earlier.

One random number z draws sampling-unit number z from one list; also sampling-unit number z from the other list. Segments drawn as prescribed from each of these 2 sampling units form the 1st sample. Together they form 1 sampling unit, which consists of a dumbbell, with opposite characteristics at the 2 ends. A 2d random number in the same zone draws 2 more sampling units, 1 from each list.

The sampling units are thus made heterogeneous. The gain in precision is usually slightly better than the gain of stratification alone, although neither may show much gain.

One may form a separate estimate for any region covered by either of the 2 frames alone, or for any region covered by both frames.

Further gains may accrue from the use of Masuyama's zigzag interval,*

* Motosaburo Masuyama, "Recent advances in sampling surveys in Japan," *Bull. Int. Statist. Inst.*, vol. xxxiii, part II, 1951: pp. 147–152, p. 149 in particular.

The use of a heterogeneous primary sampling unit under 1 supervisor has been basic since 1940 in the sample for the *Monthly Report on the Labor Force*, designed by Morris H. Hansen and colleagues in the Census; see Chapter 12 in the book by Hansen, Hurwitz, and Madow, *Sample Survey Methods and Theory*, Vol. 1 (Wiley, 1953).

whereby random number z between 1 and Z' draws also sampling unit number $Z' - z$, where $Z' = 2Z$.

EXERCISES

1. Explain the statement that the aim of sample-design is to increase the variance between the sampling units in the sample, which must necessarily decrease the variance between the results of successive samples, which is the characteristic desired of a sample-design.

2. Show that, whether we have figures from the Census or not, and regardless of the growth or loss that has taken place since the last Census, (a) the probability that a random number will strike a primary area is proportional to the number of sampling units ascribed to that area in the primary frame; and (b) the probability that any segment will fall into any subsample is $1/Z$.

3. Show that the probability of selection is not altered by an incomplete zone at the end.

B. FORMULAS*

Procedure for computing the estimates and their standard errors. For illustrative purposes we shall deal with 2 drawings per zone. If we use the subscripts 1 and 2 for the 2 samples in Zone i, then the results of the interviewers may be summarized as

x_{i1}, x_{i2} for the 2 x-populations. E.g., the x-population might be the number of packages of a certain item of food (e.g., coffee) that the families in the sample bought last week.

y_{i1}, y_{i2} for the 2 y-populations. E.g., the y-population might be the number of families that bought any food last week.

Usually we need estimates of:

    A the total x-population in the entire frame
    B the total y-population in the entire frame
    ϕ the ratio A/B

In the above example, A would be the total number of packages of coffee that all the families in the frame bought last week, while B would be the total number of families in the frame that bought food of any kind. The symbol ϕ would denote the ratio A/B, the average number of packages of coffee purchased last week per family that bought food.

Calculation of the variance of a direct estimate (2 subsamples). The direct estimate X of the total x-population A in the entire frame will be

$$X = \tfrac{1}{2}Zx \qquad \text{[There being 2 subsamples, as in Chapter 10]} \qquad (1)$$

Z being, as before, the number of sampling units per thin zone. We may then estimate the variance of X by the formula,

$$\sigma_X{}^2 = \tfrac{1}{4} Z^2 R^2 \qquad \text{[2 subsamples]} \qquad (2)$$

where

$$R^2 = \Sigma w_i{}^2 \qquad (3)$$

is the sum of the squares of the ranges $|x_{i2} - x_{i1}|$ in the m thick zones ($i = 1, 2, \ldots, m$). The summation (Σ) runs over all m thick zones. If there were only 1 thick zone, R would be the range $|x_2 - x_1|$. The circumflex ($\wedge$) denotes an estimate, as before. Out of this estimate of the variance of X come estimates of the standard error of X and of the coefficient of variation of X in the form

$$\hat{\sigma}_X = \tfrac{1}{2} ZR \qquad \text{[2 subsamples]} \qquad (4)$$

$$\hat{C}_X = \frac{R}{x} \qquad \text{[2 subsamples]} \qquad (5)$$

wherein x is the total x-population in the entire sample. The symbol y, later on, will be the total y-population in the sample.

The theory for estimates of variances is in Chapter 17.

We should introduce on the right of all the above equations for estimates of the standard error, whenever Z is small, the finite multiplier $\sqrt{(Z - 2)/Z}$ for 2 subsamples. The reader may turn back to the preceding chapter for illustration.

Calculation of the variance of a ratio (2 subsamples). For an estimate of the ratio ϕ we may take

$$f = \frac{\text{The } x\text{-population in the sample}}{\text{The } y\text{-population in the sample}} = \frac{x}{y} \qquad (6)$$

For an estimate of the variance of f we first define and calculate for (thick) Zone i,

$$\left. \begin{aligned} w_{xi} &= x_{i1} - x_{i2} \\ w_{yi} &= y_{i1} - y_{i2} \end{aligned} \right\} \qquad (7)$$

then calculate

$$h_i = w_{xi} - f w_{yi} \qquad \begin{aligned} &\text{[We here retain the algebraic} \\ &\text{signs of } w_{xi} \text{ and } w_{yi}\text{]} \end{aligned} \qquad (8)$$

whereupon

$$\hat{\sigma}_f{}^2 = \frac{S^2}{y^2} \qquad \text{[2 subsamples]} \tag{9}$$

$$\hat{\sigma}_f = \frac{S}{y} \qquad \text{[2 subsamples]} \tag{10}$$

$$\hat{C}_f = \frac{S}{x} \qquad \text{[2 subsamples]} \tag{11}$$

wherein

$$S^2 = \Sigma h_i{}^2 = \Sigma(w_{xi} - f w_{yi})^2 \tag{12}$$

This formula for estimating Var f comes from Eq. 16a on page 439.

Multiple drawings per zone (k subsamples). Triplicate drawings in a zone 50 % wider than the zone required for duplicate drawings will produce the same size of sample but will give a better estimate of the standard error, because there will be 33 % more degrees of freedom. Quadruple drawings with a zone twice as wide as the zone required for the duplicate drawings will yield 50 % more degrees of freedom (*vide* Table 11 in the appendix of this chapter). However, the wider zones reduce the amount of stratification and may under some conditions cause a slight loss in the efficiency of the estimates.

When there are k drawings per zone, the estimate of a total population will be

$$X = \frac{1}{k} Z x \tag{13}$$

where x is, as in previous chapters, the total x-population in the entire sample. The factor $1/k$ was $\frac{1}{2}$ in Chapter 10, where there were 2 drawings per zone, and it was $1/10$ in Chapters 6, 7, 8, and 9, where there were 10 drawings per zone.

An estimate of Var X by the sum of squares is

$$\hat{\sigma}_X{}^2 = \frac{Z^2 T^2}{k(k-1)} \qquad \text{[k subsamples]} \tag{14}$$

whereupon

$$\hat{\sigma}_X = \frac{ZT}{\sqrt{k(k-1)}} \qquad \text{[k subsamples]} \tag{15}$$

$$\hat{C}_X = \sqrt{\frac{k}{k-1}} \frac{T}{x} \qquad \text{[k subsamples]} \tag{16}$$

wherein

$$T^2 = \sum_i \sum_j (x_{ij} - \bar{x}_i)^2 \tag{17}$$

T^2 reduces to $\frac{1}{2}R^2$ when $k = 2$, and the last 3 equations reduce to Eqs. 2, 4, 5.

An estimate of the variance of $f = x/y$, as defined in Eq. 6, is

$$\hat{\sigma}_f{}^2 = \frac{k}{k-1} \frac{U^2}{y^2} \qquad [k \text{ subsamples}] \tag{18}$$

whereupon

$$\hat{\sigma}_f = \sqrt{\frac{k}{k-1}} \frac{U}{y} \qquad [k \text{ subsamples}] \tag{19}$$

$$\hat{C}_f = \sqrt{\frac{k}{k-1}} \frac{U}{x} \qquad [k \text{ subsamples}] \tag{20}$$

wherein

$$U^2 = \sum_i \sum_j \{(x_{ij} - \bar{x}_i) - f(y_{ij} - \bar{y}_i)\}^2 \tag{21}$$

This formula for estimating Var f, like Eq. 9, comes from Eq. 16a on page 439. The sums for both T and U run over all k subsamples and over all m thick zones. When $k = 2$, U^2 reduces to $\frac{1}{2}S^2$, and Eqs. 18, 19, 20 reduce to Eqs. 9, 10, 11.

Use of the average range in the following formula* will simplify even further the computation of the standard error of X:

$$\hat{C}_X = \frac{\bar{w}_x}{k\bar{x}\sqrt{m}} \qquad [2 < k \le 10] \tag{22}$$

In thick zone i there are the x-populations $x_{i1}, x_{i2}, \ldots, x_{ik}$. If w_{xi} is the difference (range) between the highest and the lowest without regard to sign, then $\bar{w}_x$ is the average of all the m ranges. $\bar{x}$ is the average x-population per thick zone per subsample.

Remark 1. Eq. 22 gives good results if the difference between the highest and the lowest of the m ranges in the m thick zones amounts to no more than the average range ($\bar{w}$) itself. This rule will restrict use of the range to cases where the rel-variance between the m ranges is $1/12$ or less (see Panel B in Fig. 16 on page 260), and this in turn guarantees that any bias toward an underestimate of the standard error does not exceed 6%. These remarks do not affect Eq. 23 below, as it applies to a single thick zone.

* See the reference to Nathan Mantel on page 437.

We rewrite now some of the above formulas for the case of only 1 thick zone and 10 subsamples. Eq. 21 then reduces to

$$U^2 = \sum_1^{10} (x_i - fy_i)^2 \qquad (23)$$

which is ready for use in Eqs. 18, 19, 20 with $k = 10$. Eq. 22 reduces, under the same conditions, to

$$\hat{C}_X = \frac{w_x}{10\bar{x}} \qquad \begin{array}{l} [w_x = |x_{\max} - x_{\min}| \,; \\ 10 \text{ subsamples; 1 thick zone;} \\ \bar{x} = \tfrac{1}{10} \Sigma x_i] \end{array} \qquad (24)$$

If further, the y_i are not highly variable from 1 subsample to another, and not correlated with $f_i = x_i/y_i$, the range in f_i will give a quick and useful estimate of the coefficient of variation of f by the formula

$$\hat{C}_f = \frac{w_f}{10f} \qquad \begin{array}{l} [w_f = |f_{\max} - f_{\min}| \,; \\ 10 \text{ subsamples; 1 thick zone}] \end{array} \qquad (25)$$

The reader may recall use of these estimates in previous chapters.

Remark 2.   The finite multiplier to use in the above estimates of the variance of X or of f is $(Z - k)/Z$. The finite multiplier in an estimate of a standard error or of a coefficient of variation is $\sqrt{(Z - k)/Z}$.

Remark 3.   To gain experience with a new material, one may in the 1st survey use duplicate drawings, and compare the average variance between the 2 samples in adjacent thin zones with the average variance within thin zones. If one can be sure that this ratio lies close to 1, he may say that there would be but little loss in using more subsamples.

A general plan for the variance of any estimate. *   For an estimate other than a ratio (x/y) or a total (X), we need a more general theorem.   If u_1 and u_2 are the 2 estimates of any characteristic, u_1 obtained from one random half of the full sample, u_2 from the remaining half, and u from both halves combined, then

$$\text{Var } u_1 = 2 \text{ Var } u \qquad (26)$$

and it is a fact that

$$\text{Var } u = \text{Var } (u_1 - u) \qquad (27)$$

except for the finite multipliers.   Hence, to estimate Var u one need only draw 1 of the 2 sampling units at random from every zone, and compute the estimate u_1 therefrom and from $(u_1 - u)^2$; then repeat the estimate by drawing another random half, then another, and another.   Subsample 1

---

* I am indebted to my colleagues William N. Hurwitz and Max Bershad of the Census in Washington for the privilege of publishing the method described in this section.

constitutes one random half; Subsample 2 constitutes another, but yields no new information. Subsequent halves thus require random drawings from the full sample already at hand. The successive values of $(u_1 - u)^2$ are correlated, but the cumulated average value of $(u_1 - u)^2$ will soon settle down to some number which we may accept as Var u. If there is not much change in the variance of u from zone to zone, the number of degrees of freedom in this estimate of the variance will be m, the number of thick zones.

More generally, when there are k sampling units per zone, we may draw at random 1 sampling unit per zone and use the accumulated average of $(u_1 - u)^2$ as an estimate of $(k - 1)$ Var u.

C. ILLUSTRATIVE EXAMPLE

INSTRUCTIONS FOR A SAMPLE OF THE CINCINNATI AREA*

Preliminary calculations. The district for the first illustration will be the Cincinnati area, composed of the cities of Cincinnati and Covington, and the remainder of Hamilton and Kenton counties, plus Campbell County adjacent. The aim of the study was to compare readers' opinions and comparisons of 2 special Sunday features of a certain newspaper. The sampling plan will be in 2 parts, the 1st part for the "block cities,"† and the 2d part for the remainder of the area. The reason for the split is that the firm that carried out the study could carry out the 1st part in their own office, as the only materials required therefor were the block statistics published by the Census for the 2 block cities in the survey, plus maps or directories as further aids, all of which one can purchase readily. In contrast, the sampling units in the 2d part were enumeration districts, figures and maps for which only the Census could supply.

The area to be surveyed contained 276,600 Cdus (1950), of which 179,139 Cdus or 64.8% were in the block cities Cincinnati and Covington. The number of dwelling units in the sample was to be about 900. A sampling unit was defined as 2 segments of 5 Cdus each. Thus, 45 zones

* I am indebted to the firm O'Brien-Sherwood of New York for the privilege of working with them on this survey of Cincinnati, and for the illustrative figures.

† Block city is a term used in the United States for a city for which the Bureau of the Census publishes "block statistics." To be eligible, the city must have contained 50,000 or more inhabitants in the preceding Census. A block in America is the smallest area that is bounded by streets. The block statistics show for every block city the number of occupied dwelling units and many other useful figures for every block that had occupied dwelling units in the last Census. In this illustration, the block cities in the sample were Cincinnati and Covington.

with 2 subsamples per zone should yield a sample of about 900 Cdus; or, owing to growth since 1950, something over 900 occupied dwelling units today.   To decide the zoning interval, we note that 276,000 Cdus will give 27,600 sampling units, and that $27,600/45 = 613$.   The zoning interval actually adopted in the face of possible growth since the Census of 1950 was 630 sampling units.

> I used 2 subsamples here; also in the last chapter.   I recommend nevertheless, for most surveys that cover only a county or 2 or 3 counties, the use of 10 subsamples, with a single thick zone of tabulation, the simplicity of which we have already seen in Chapters 6, 7, 8, 9.

Instructions for the part in the block cities.*   The following instructions apply to the block cities Cincinnati and Covington.

1. Prepare a list of the tracts in the order shown in the Census statistics for tracts, and show for each tract the number of sampling units therein. Form the accumulated total sampling units tract by tract.   The accumulated totals will ascribe a serial number to every sampling unit in the 2 block cities (Table 3).

2. Draw 2 random numbers for each of the 29 zones within the block cities, 1 for the 1st subsample, and 1 for the 2d subsample.   The zoning interval will be 630 sampling units.   Record these numbers in 2 columns in the order drawn (Table 4), Subsample 1, Subsample 2.   Every random number will identify by serial number a certain sampling unit for the sample; also the tract and the block in which it lies.

3. Prepare a list of the blocks in each tract struck by the random numbers.   The order of the blocks within a tract shall be the order shown by the Census block-statistics.   Show for each block the number of Cdus therein and allot to each block a number of sampling units.

> *a.*   First, tie any block of less than 20 Cdus to another adjacent, and allot a size to the pair.†   It may occasionally be necessary to tie 3 or more blocks together.   Tie likewise blocks that had size 0, as they may now be occupied (*vide* Table 5 for an example).
> *b.*   Force the total number of sampling units accumulated for the blocks within a tract to agree with the size ascribed to that tract in Step 1.   Do the forcing by adding or subtracting a sampling unit from the biggest block or combination.

* These are the actual instructions to the firm.   They are printed here only for illustration of the theory and principles, and not as patterns suitable without modification for other surveys.

† Step 3a provided enough segments for a 2d survey, to be taken a few months later. One may omit this step if there will be no further surveys in the area.   On the other hand, if the firm wishes to prepare for a number of surveys during the next year or two, the minimum size for a combination of blocks may well be 50 or more d. us.

4. Draw a random number between 1 and the total number of sampling units in the tract, to determine the block or combination in which the sampling unit falls.   Make a list of these blocks, and show how many sampling units were ascribed to each one in Step 3*b*.

In a big tract, it may save time to form groups of 5 successive blocks, and to show the detail block by block only for the group struck by the random number.

5. In each block or combination so drawn, create segments by the prescribed rules.*

6. If *c* is the number of sampling units allotted to a block, then draw 1 segment at random from every successive *c* segments in this block, and canvass it.   The random numbers will ordinarily be in a sealed envelope, ready for use (p. 110).   Tables 3 and 4 on page 207 and 208 illustrate the above steps.

Remark.   If we create segments of 10 d. us., and draw for interview 1 segment in *c* segments, there will be interviews in 1 segment in the block. If we create segments of 5 d. us., there will be interviews in 2 segments, usually on opposite sides of the block.   These numbers may change, of course, if there has been a significant change in the number of dwelling units since the census was taken; or where the interviewer, for the sake of definite boundaries, created segments smaller or larger than the intended size.

The variance contributed by rough counts and by varying size of segment is a subject for treatment on pages 479 ff.

Instructions for the area outside the block cities.   This part of the work was carried out by the Census, down through Step 10.   The instructions to the Census follow (numbered continuously from the previous steps, for convenience in reference).

7. Prepare a list of the enumeration districts† (hereafter E. DD.) in the

---

* See Chapter 12 for examples of procedures.

† The abbreviation E. D. denotes enumeration district.   The plural is E. DD.   An E. D. is 1 of the small administrative areas into which the entire country is divided for the purpose of taking a Census.   Every part of the country lies in 1 but only 1 E. D. An E. D. is delineated with the intention that it shall contain a population no greater than 1 enumerator can cover within the enumeration period.   It is further required, however, that no E. D. shall cross the boundary of any incorporated place, city ward, census tract (in a tracted city), township, or other minor civil division, no matter how small.   The usual range in the population of an E. D. is between 500 and 1500, though some were much larger, and others, because of the 2d of the 2 requirements just noted, contained populations much smaller than 500—some even as low as 1 or 2 persons, or none at all.   A small E. D. is usually assigned to an enumerator who is to cover another one nearby, with the consequence that there are more E. DD. than enumerators. As a matter of fact, in the 1960 Census there were 239,000 E. DD., but only 160,000 enumerators.

3 counties outside the block cities, in any order convenient.   Show for each E. D. the number of sampling units therein, and form the accumulated totals.   Tie an E. D. of less than 40 sampling units to an E. D. that is nearby on the Census list, and ascribe a number of sampling units to the combination.   It may occasionally be necessary to tie 3 E. DD. in one combination.   The accumulated totals will ascribe a serial number to every sampling unit.

8. In areas where there have been special censuses since 1950, without change of boundary, the Census will use the new figures.

9. Draw 2 random numbers for each of the 16 zones outside the block cities, 1 for the 1st subsample, and 1 for the 2d subsample.   The zoning interval will again be 630 sampling units.   Record these numbers in 2 columns in the order drawn.   Every random number will identify by serial number a certain sampling unit; also the E. D. (or a combination of 2 or of 3 E. DD.) in which this sampling unit falls.   The Census will furnish a map and a description of each of these E. DD. or combinations, together with the figure that in Step 7 prescribed the number of sampling units therein.

10. As an alternative to Step 7 the Census may, for economy, use 2 stages.   The 1st stage might be to list groups of 5 or 10 E. DD. and to show for each group the number of sampling units.   Within any group that is struck by a random number, it will then be necessary to ascribe a number of sampling units to every E. D. therein, and then to force the total number of sampling units for the group to agree with the original measure of size assigned to it.   Do the forcing by adding or subtracting a sampling unit from the biggest E. D. or combination.

11 and 12. The firm will now delineate segments in the E. DD. in which the sampling units fell, and will draw segments by random numbers for canvass.   The work follows Steps 5 and 6 and there is no need to write it out here.

Splitting a big block or other area into portions on the spot to save labor in the creation of segments.   If the number of dwelling units in a selected area is huge, the time required to create segments over the whole area will be much greater than the time required to interview in the sample of segments that will be drawn from the area.   It therefore seems desirable to put an upper limit on the cost of creating segments.   One way is to increase the size of the segments in the big areas, to reduce their probabilities of selection (p. 193).   There is still another way.

Imagine that we had been in possession of information by which we could break the big block into portions, and that we had showed each portion separately when we listed the blocks in Step 3.   Then, the random sampling unit would fall in a distinct portion, and we should create segments in only this portion, ignoring the other portions.

We may accomplish the same end *after* a sampling unit falls in such a block, provided the information comes into our possession before we have gone too far. A good arrangement is to ask the interviewer, before

Town: Middletown
E. D. No. 21–50, Block No. 21

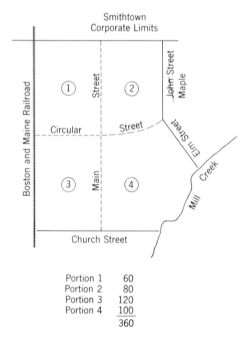

Portion 1	60
Portion 2	80
Portion 3	120
Portion 4	100
	360

Fig. 8. The main office furnished to the field-worker the boundary shown by the solid line. The field-worker broke the area into portions per these instructions by sketching in the dotted lines. She also changed John Street to Maple Street.

she creates segments, (*a*) to look into the possibility of breaking up any block that contains more than 100 or 150 d. us.; then, (*b*) if she succeeds, to call in by telephone or to send the information by mail. She must of course appreciate the importance of distinct boundaries for the portions. (See Appendix D of Chapter 12 for an example of instructions to the interviewers on this point.)

Suppose that a block that came into the sample had 360 Cdus, and that 36 sampling units were ascribed thereto. The interviewer goes to this block and finds that she is able to break it into 4 portions, with unmistakable boundaries.   She draws the map shown in Fig. 8 and furnishes by mail or by telephone the 2 columns in Table 1.   Someone in the office constructs Table 2 and in so doing adjusts the total number of sampling units to the required 36, and draws a random number between 1 and 36. Let the random number be 17.   As this number designates a sampling unit in Portion 3, instructions now go to the interviewer to create segments in Portion 3.   The random numbers that will tell her which of these segments she is to canvass will come to her by mail or by telephone.* They appear in a note to Table 2.   The 1st random number lies between 01 and 12; the 2d between 13 and 24; the 3d between 25 and 36.   (Why was the segment-interval 12 here, and not 36?)

TABLE 1		TABLE 2	
REPORTED BY THE INTERVIEWER		DONE IN THE OFFICE	
Portion	Approximate number of dwelling units	Number of sampling units	Serial numbers
1	60	6	1– 6
2	80	8	7–14
3	120	12	15–26*
4	100	10	27–36
Total	360	36†	

* Random number between 1 and 36, 17 (struck Portion 3).

† Size 36 is the number of sampling units ascribed to this block as it appeared in the list by tract (an example is on the next page).

The segment-interval in Portion 3 is 12.   Segments for interview: 02, 13, 27.

Some numerical results for Cincinnati.   Table 3 shows the number of Cdus in the tracts of Cincinnati and of Covington, the number of sampling units ascribed to each tract, and the serial numbers of these sampling units.

Table 4 shows the random numbers for these cities (Step 2).   The sampling units that bear these serial numbers belong to the sample.   There was a similar set of random numbers for the area outside the block cities (Step 9).

* The whole transaction, random numbers and all, can be handled in 2 telephone calls of 3 minutes each.

TABLE 3

SERIAL NUMBERS OF THE SAMPLING UNITS BY TRACT
IN THE BLOCK CITIES CINCINNATI AND COVINGTON

City	Tract	Cdus (1950)	Number of sampling units	Serial numbers of the sampling units
Cincinnati	1	2019	202	1– 202
	2	2334	233	203– 435
	3	2729	273	436– 708
	4	2861	286	709– 994
	5	4577	456	995– 1450
	6	1461	146	1451– 1596
	7	1705	170	1597– 1766
	8	1473	147	1767– 1913
	9	2724	272	1914– 2185
	10	2221	222	2186– 2407
	.			
	.			
	.			
	107	750	75	15754–15828
	108	175⎫		
	109	87⎭	26	15829–15854
	110	486	49	15855–15903
	Ward			
Covington	1	2316	232	15904–16135
	2	1829	183	16136–16318
	3	2080	208	16319–16526
	4	2066	207	16527–16733
	5	6222	622	16734–17355
	6	5508	551	17356–17906

Comparison with Table 3 shows which tracts contained the sampling units in the sample.   For example, sampling unit No. 1820 lies in tract No. 8.   Next comes Step 3.   One may turn to Hansen, Hurwitz, and Madow's *Sample Survey Methods and Theory*, Vol. I (Wiley, 1953), pages 248–252, to see an example that is similar; nevertheless, I give in Table 5 the detail in Tract 8, for the convenience of the reader.   The sum of the sampling units block by block turned out to be 147, without any forcing.   In Step 4 we draw a random number between 1 and 147: this

TABLE 4

THE SAMPLE IN THE BLOCK CITIES
CINCINNATI AND COVINGTON

Drawn by use of Kendall and Smith's *Random Numbers*, 4th
Thousand, Columns 5, 6, 7, line 15.   Each random number
from the book lies between 000 and 629, and is translated
by addition to the left-hand boundary of the zone

Zone (thin)	Boundaries of the zones	Random drawings			
		From the book		Translated	
		Subsample 1	Subsample 2	Subsample 1	Subsample 2
1	00001–00630	401	211	402	212
2	00631–01260	122	087	753	718
3	01261–01890	222	559	1483	1820
4	01891–02520	457	424	2348	2315
5	02521–03150	211	012	2732	2533
Etc.					

random number turned out to be 33, which drew Sampling Unit No.
$1799 = 1766 + 33$, which draws the combination of blocks 5 and 6.
Steps 5 and 6 need no description here.

The results for several characteristics for the 2 samples over the entire
Cincinnati area are in Table 6.   From this table we see that the total
number of households in the sample is $449 + 454$, which substituted into
Eq. 1 gives

$$X = \tfrac{1}{2} \times 630(449 + 454) = 284{,}445 \qquad (28)$$

for an estimate of the total number of households in the area in June 1954.
This figure compares with the number 276,000 in the Census of 1950.
For the coefficient of variation of this estimate, we note from Table 7 that
$R^2 = 833$, whence Eq. 5 gives

$$\hat{C}_X = \frac{R}{x} = \frac{\sqrt{833}}{449 + 454} = 3.2\% \qquad (29)$$

for the coefficient of variation of the estimate X.   The standard errors of
any other characteristics are computable likewise, but I shall not show a
further example here, except to add that the time required for the

TABLE 5

A random drawing between 1 and 147 selected a sampling
unit from the combination of blocks 5 and 6 for the sample

Tract	Block	Cdus	Number of sampling units	Serial numbers of the sampling units
8	1	24⎫	10	1767–1776
	2	78⎭		
	3	101	10	1777–1786
	4	121	12	1787–1798
	5	77⎫	9	1799–1807
	6	10⎭		
	7	85	8	1808–1815
	8	118	12	1816–1827
	9	91	9	1828–1836
	10	23⎫	6	1837–1842
	11	40⎭		
	12	114	11	1843–1853
	13	55	6	1854–1859
	14	98⎫	10	1860–1869
	15, 16	0⎭		
	17	99	10	1870–1879
	18	61	6	1880–1885
	19	96⎫	10	1886–1895
	20	0⎭		
	21	26⎫	5	1896–1900
	23	24⎭		
	22	132	13	1901–1913
Total		1473	147	

computation of 6 standard errors was about 15 minutes with a slide rule, once the figures by thick zones came to hand.

There was no randomization of interviewers here; hence the differences between the 1st and 2d samples in Tables 6 and 7, and also the standard errors computed above, are not purely errors of sampling, but are a combination of the uncertainty introduced by sampling and of the differences between interviewers, a topic to appear again in Chapters 12 and 13.

TABLE 6

SUMMARY FOR THE 2 SAMPLES OVER THE ENTIRE CINCINNATI AREA

Characteristic	Subsample 1	Subsample 2
Number of dwelling units encountered*	449	454
Number of individual persons		
Selected for interview†	452	460
Refused	41	42
On vacation	31	19
Not found;‡ deaf, sick	34	37
Households not qualified (do not receive		
the newspaper studied)	104	98
Households qualified	242	264

* This is the x-population used in Eqs. 28 and 29.

† These are the interviews attempted.   The rule was to interview all the people in families of 1 person; 1 at random from each family of 2 persons; 1 at random from every 3 persons in families of 3 or more persons (p. 242).   The final results for any household were then weighted inversely by the number of persons interviewed therein.

‡ Not found in 6 recalls.

TABLE 7

TOTAL DWELLING UNITS BY THICK ZONES BY SUBSAMPLE,
IN THE SURVEY OF CINCINNATI
The differences and their squares are for use in Eqs. 3 and 5
for the standard errors

Thick zone	Subsample 1	Subsample 2	Difference w_i	w_i^2
1– 5	51	45	6	36
6–10	55	40	15	225
11–15	50	47	3	9
16–20	37	56	19	361
21–25	49	60	11	121
26–30	40	42	2	4
31–35	52	48	4	16
36–40	46	52	6	36
41–45	69	64	5	25
Total	449	454	71	833
Average per thick zone	49.9	50.4	7.9	92.6

EXERCISES

1. Suppose that the figure from the Census on which we based the number of sampling units in a block in Table 5 is still accurate today. Prove that if the size of the segments is actually half the size of a sampling unit, then (*a*) the random selection of segments will yield 2 segments per block; (*b*) the 2 segments will lie in different halves of the block.

2. Suppose that the block has now 20% more dwelling units than it did at the time of the Census. Prove that if the size of the segments is actually half the size of a sampling unit, then the average number of segments randomly selected per block will be 2.4—i.e., sometimes 2 segments per block, sometimes 3.

D. A NATIONAL SAMPLE WITH SAMPLING UNITS UNRESTRICTED*

Preliminary calculations. The number of dwelling units in a national sample was to be about 5500, in 500 sampling units of 10 Cdus (1950), which because of growth, would yield on the average about 11 d. us. today. The number of zones would be 250, with 2 subsamples per zone. The number of Cdus in the entire country from the Census of 1950 was about 43,000,000. A convenient zoning interval was 17,500, derived from 4,300,000/250. There was to be no restriction on a sampling unit: it could fall in any county, regardless of whether any other sampling unit fell far away or near. Each sampling unit could be 1 segment of average size 11 d. us., or 2 segments of 5 or 6 d. us.

The size of 500 sampling units is the basic size. Actually, this sample, like most others, is capable of extension and of contraction. Thus, to double the size of the sample one need only draw 2 segments out of every segment-interval instead of 1. To decrease the size to half, one may knock out 1 of every consecutive pair of sampling units in Subsample 1, and retain the sampling unit in Subsample 2 in the zone that lost Subsample 1. Other sizes are also easy to derive.

Instructions for the block cities.†

13. Prepare a list to show the block cities by geographic region (e.g., New England). List the cities in geographic order within each region,

---

* I am indebted to the firm O'Brien-Sherwood of New York for the privilege of working with them on this survey.

† For convenience in reference, the numbers of the steps in this section are continuous with those of the previous steps.

east to west, south to north within size-groups (over a million inhabitants, 250,000 to a million, under 250,000).* Show opposite each city the number of sampling units therein. The accumulated totals will give a serial number to every sampling unit. Continue the accumulation from one region to another.

14. Draw 2 random numbers in every zone. Make a list of the random numbers and the cities that they fall in. Maintain identification by subsample.

The identifying random number and the subsample will appear on every questionnaire, and will follow it through the tabulations.

15. Each random number will fall in some city and will bring in a sampling unit, which will be a segment of 10 Cdus, or a pair of segments of 5 Cdus each in the same block (whichever way is prescribed). To locate these segments, accumulate tract by tract, or by any other convenient subdivision, the number of sampling units in the block statistics. Force the total number of sampling units at the end of the accumulation to agree with the number that you allotted to this city in Step 13. Various tables in the book illustrate how accumulated totals give a serial number to every sampling unit (see Table 1 on page 113, and Tables 2, 3, 5, 8, and 10 in this chapter).

16. Draw a random number between 1 and the total number of sampling units in this city to see which tract or other subdivision will contribute a sampling unit to the sample.

17. Accumulate block by block, or by groups of blocks, the number of sampling units from beginning to end of the tract or other subdivision that the random number struck in the preceding step. Force the total number of sampling units at the end of the accumulation to agree with the number that you allotted to this tract or subdivision in Step 15. Then draw a random number between 1 and the total number of sampling units in the tract or subdivision. This random number will fall in a certain block or a group of blocks. This is the block or group of blocks that will contribute the segment or segments to the sample.

18 and 19.   Same as Steps 5 and 6.

Repeat Steps 15–19 for every random number that struck the city. A big city or a big county may contribute a number of sampling units. (See Table 9, where Boston, in another sample, contributed 4 sampling units,

---

* This order will give good statistical efficiency for most purposes, and it will facilitate tabulations by size of city. However, this stratification is not offered here as a recommendation for other surveys. The stratification, like other parts of the basic design, should be altered to achieve convenience and statistical efficiency for the purpose at hand.

and New York contributed 17.) There will be concentration of sample wherever the population is heavy. No segment can fall on an uninhabited mountain top or swamp.

Instructions for the area outside the block cities. As in the sample for Cincinnati, this part was carried out by the Census. The strata were those defined by the Census for the *Monthly Report on the Labor Force.** The steps were otherwise similar to Steps 7–12 for Cincinnati.

Dispersion of the sample. Although the 500 sampling units were unrestricted, only 14 of them required new interviewers, not already in correspondence with the firm. This fact may help to dispel fears that a large portion of an unrestricted sample will scatter out of reach. The sample falls where there are people. There is no sample where there are no people. The unrestricted sample has great efficiency. We describe in the next section a sample in which the sampling units are restricted to a county.

E. A NATIONAL SAMPLE WITH SAMPLING UNITS RESTRICTED†

Preliminary calculations. The number of dwelling units in this national sample was to be about 14,000, in 200 sampling units of about 64 Cdus (1950), which, because of growth, would yield on the average about 71 d. us. today. The number of zones would be 100, with 2 subsamples per zone. The number of Cdus in the entire country was again 43,000,000, as in Section E. The primary units here are counties or groups of counties. Each sampling unit will be restricted to a primary unit (county or group of counties), so that no segment will be out of reach of a supervisor.

The sampling unit. Each sampling unit will split into 4 subunits of 16 Cdus, which will in turn split into 2 segments of 8 Cdus, usually on opposite sides of an area.

The primary units. Each new survey will require new segments (unless it is a resurvey of the same segments). One uses, if feasible, segments from the supply of segments already created. When this supply of segments is exhausted, it is necessary to move into new blocks (or other areas) within the primary unit and to create a new supply of segments. In a county

---

* Had there not been already in existence a suitable mode of stratification, it would have been necessary to prescribe one.

† I am indebted to Alfred Politz Research, Inc., of New York, and to my friend and colleague Mr. Lester R. Frankel, formerly Vice President and Technical Director thereof, for the privilege of working on this survey.

that is small, a firm that uses a national sample every few weeks will soon find itself with no place to go.   Moreover, it is bad practice to become conspicuous by interviewing a large fraction of the population in a county. The use of combinations of counties will eliminate this hazard, and it will avoid the complex theory of replacing, by a new drawing, a small county that is worn out.

Nearly 500 of the 3000 counties in the U.S. contained fewer than 10,000 inhabitants in 1950: one county contained only 265 people; one actually only 63—almost a family affair.   In Nebraska and in South Dakota, it was necessary to combine 7 counties to build up a combination that contained 10,000 people.   In Texas, it was necessary to make combinations of 3, 4, and 5 counties, and some of these combinations covered more than 10,000 inhabitants) joined to a big one adjacent, or preferably, it may be the small counties (fewer than 10,000 inhabitants), it was known in advance that only about 6 random numbers would strike the small counties or combinations thereof.   Yet the convenience and simplicity in the use of the combinations of counties is worth the effort of forming the combinations (about 8 or 10 man-days).

Rules for combining counties. *

1. If a county contained fewer than 10,000 inhabitants in 1950, it is a candidate for combination.   If it contained fewer than 8500 inhabitants, combination is imperative.

2. A combination of counties may be a small county (one of fewer than 10,000 inhabitants) joined to a big one adjacent, or preferably, it may be a pair of adjacent small counties.   It may be 2 small counties on opposite sides of a big one, and the only feasible combination may be to put all 3 counties together.   A combination may contain 3 or more counties.

3. One may seek either of 2 kinds of combination—

a. A combination of unlike characteristics (half heavy industry, half rural), with a common border, in a compact mass covering preferably 1000 square miles or less, although more area is permissible.   We refer to this as a small heterogeneous combination.   One interviewer can go to a segment anywhere inside such a combination.

b. When such a combination is imbossible, we seek a combination of like characteristics, in which there is a monotony of farming, industry, income, and scenery, from one end of the combination to the other.   The total area may run into thousands of square miles, as some small counties

---

* These are a set of rules that I have used.   I present them here as an example, not as a recommendation.   Every sampling plan presents individual problems, and there is no general procedure that is always the right one.

in Texas cover singly several thousand square miles. This happens to be so because areas of sparse population are also usually areas that show little change in character over hundreds of miles.

A common boundary is not necessary here. It is permissible and in fact desirable to join counties from different substrata within a stratum, if to do so will build a small heterogeneous combination. However, one must not combine 2 counties that lie in different areas of tabulation.

4. Sometimes one can not meet any of the above desired qualities of a combination, and must form some other kind of combination, merely meeting the requirement for the minimum number of inhabitants.

Method of distributing the sampling units within a rural county or combination. If a rural county (or combination) is heterogeneous, and not too expansive, it may pay to disperse the sampling units in segments over the county. On the other hand, if a county is homogeneous, it will be about as good to save the cost of long trips and to confine a survey to a random small portion of the county. That is, one may consume the county like a loaf of bread, a little at a time, as other parts of the county will give information but little different.

For a combination of homogeneous characteristics, there is no need of a common boundary between 2 counties of the combination. The argument that the same interviewer should be able to go into both counties breaks down because it may be necessary to change interviewers anyhow when in the course of time the sample moves far away from the place. where it started.

The procedure by which to draw a segment or a sample of segments from a primary area will not be described here in detail. The methods of Chapter 10 and of Sections C and E of this chapter are applicable, but there are many other ways. Aerial photos and maps are in wide use in the United States by concerns that are continually drawing samples. The advantage is that the capital investment in the materials (photos and maps) gives freedom and flexibility to meet a dead-line for the preparation of the sample: one is then not waiting in line, dependent on a second concern to provide the secondary frames.

For references to the use of aerial photos and maps, one may consult Hansen, Hurwitz, and Madow, *Sample Survey Methods and Theory*, Vol. I (Wiley, 1953), Ch. 6; also Leslie F. Kish, "A 2-stage sample of a city," *Amer. Sociol. Rev.*, vol. 17, 1952: pp. 761–769. Also, John Monroe and A. L. Finkner, *Area Sampling* (Chilton Company, Book Division, 1959).

The strata. There were 8 strata in this sample, the primary units being counties or combinations of counties as already described. The 8 strata were as follows:

TABLE 8

THE PRIMARY FRAME

Northeast, metropolitan areas

Primary Unit	Cdus	Number of sampling units	Serial numbers
Boston	791 804	12 497	000 001–012 497
Hartford	152 391	2 405	012 498– 14 902
New Haven	155 815	2 459	14 903– 17 361
Bridgeport	143 872	2 271	17 362– 19 632
Brockton	55 961	883	19 633– 20 515
Worcester	151 502	2 391	20 516– 22 906
Fall River	110 688	1 747	22 907– 24 653
Providence	197 385	3 115	24 654– 27 768
Manchester	46 274	730	27 769– 28 498
Springfield	126 104	1 990	28 499– 30 488
Portland	48 444	765	30 489– 31 253
Pittsfield	38 026	600	31 254– 31 853
New York	3 777 081	59 623	31 854– 91 476
Philadelphia	1 019 115	16 085	91 477–107 561
Pittsburgh	614 378	9 697	107 562–117 256
Altoona	39 684	626	117 257–117 884
Wilkes-Barre	106 658	1 683	117 885–119 567
Scranton	71 818	1 134	119 568–120 701
Johnstown	76 366	1 205	120 702–121 906
Binghamton	53 039	837	121 907–122 743
Erie	61 798	975	122 744–123 718
Buffalo	306 236	4 833	123 719–128 551
Reading	73 264	1 156	128 552–129 707
Allentown	122 632	1 936	129 708–131 643
Trenton	59 338	937	131 644–132 580
Utica	80 418	1 269	132 581–133 849
Rochester	143 605	2 267	133 850–136 116
Syracuse	97 223	1 534	136 117–137 650
Albany	154 213	2 434	137 651–140 084
Harrisburg	83 807	1 323	140 085–141 407
York	58 930	930	141 408–142 337
Lancaster	65 227	1 029	142 338–143 366
Atlantic City	40 175	634	143 367–144 000
Total	9 123 271		

1. Northeast States, metropolitan counties* (Table 8).
2. Northeast States, townships not in Stratum 1.
3. North Central States, metropolitan counties.
4. North Central States, counties not in Stratum 3.
5. Southern States, metropolitan counties.
6. Southern States, counties not in Stratum 5.
7. Western States, metropolitan counties.
8. Western States, counties not in Stratum 7.

The above strata were also areas of tabulation—i.e., the sample was to be summarized separately for each of these 8 strata. These strata were also the thick zones—i.e., there were no intermediate summaries within the strata. The thick zones were accordingly unequal in size, the only disadvantage being some decrease in the effective degrees of freedom in the estimates of the variances.

Steps in preparation.†

20. Make a list of the counties that had in 1950 fewer than 10,000 inhabitants.
21. Make combinations of these counties as described earlier.
22. Prepare the primary frames, 1 for each stratum (Table 8).
23. Read out 2 random numbers in each zone (Table 9).
24. Compare the tables prepared in the last 2 steps, to see which primary areas will contribute to the sample. Make a list of these areas. Show for each area—

The random number or numbers that struck it.
The subsample (1 or 2) of each strike.

   These identifying numbers will appear as identification on every questionnaire.

25. Proceed to draw a sample of 64 Cdus from each area for every random number that falls therein.

   Thus Boston, with 12,497 sampling units in the frame (Table 8), was struck by 4 random numbers, 2 from each subsample. Each random number will draw 4 pairs of segments, 16 Cdus to the pair (p. 213). Hence,

---

* A metropolitan area is a county in which there was at the last census a city of 50,000 inhabitants. An adjacent county is included if it is essentially metropolitan in character, and closely related socially and economically with the central city. Certain modifications are necessary in New England (Census of Population, 1950; Series PA1 and Series PB1).

† For convenience in reference, the numbers of the steps in this section are continuous with those of the steps in Section E.

TABLE 9

THE RANDOM NUMBERS AND THE AREAS THAT THEY FALL IN

1. Northeast, metropolitan areas

Zone	Subsample 1		Subsample 2	
	Random number	City	Random number	City
0 001– 7 200	4 130	Boston	6 250	Boston
7 201– 14 400	10 391	,,	8 433	,,
14 401– 21 600	21 184	Worcester	16 575	New Haven
21 601– 28 800	22 165	,,	22 781	Worcester
28 801– 36 000	34 180	New York	30 203	Springfield
36 001– 43 200	38 861	,,	39 855	New York
43 201– 50 400	49 532	,,	45 940	,,
50 401– 57 600	57 263	,,	50 565	,,
57 601– 64 800	63 362	,,	60 245	,,
64 801– 72 000	71 670	,,	66 977	,,
72 001– 79 200	75 250	,,	77 750	,,
79 201– 86 400	81 775	,,	82 357	,,
86 401– 93 600	90 221	,,	88 924	,,
93 601–100 800	97 979	Philadelphia	100 538	Philadelphia
100 801–108 000	106 480	,,	101 358	,,
108 001–115 200	111 683	Pittsburgh	108 542	Pittsburgh
115 201–122 400	116 910	,,	116 932	,,
122 401–129 600	128 241	Buffalo	125 070	Buffalo
129 601–136 800	130 138	Allentown	130 628	Allentown
136 801–144 000	142 586	Lancaster	139 628	Albany

2. South, metropolitan areas

Zone	Random number	City	Random number	City
0 001– 7 200	2 828	Washington	4 189	Washington
7 201– 14 400	12 448	Baltimore	9 358	Baltimore
14 401– 21 600	17 032	Louisville	21 023	Charleston, W. Va.
21 601– 28 800	22 459	Richmond	22 978	Norfolk
28 801– 36 000	35 312	Greenville	30 758	Columbia
36 001– 43 200	40 474	Miami	41 599	Miami
43 201– 50 400	43 876	Augusta	48 715	Mobile
50 401– 57 600	57 242	Little Rock	54 128	New Orleans
57 601– 64 800	64 696	Fort Worth	61 987	Dallas
64 801– 72 000	67 413	Houston	70 783	Austin
72 001– 79 200	78 898	Wichita Falls	78 868	Wichita Falls

we cumulate by tract within Boston the number of units of size 16 Cdus; then draw 1 unit from every 7200 such units, being careful to maintain the identity of the subsample. Each unit will be an expected pair of segments of 8 Cdus each (p. 189).

26. Keep the 2 subsamples identified. Form the desired estimates for the 2 subsamples combined. Form estimates separately by subsample for any characteristic whose standard error is needed.

F. SAMPLING WITHOUT STATISTICS
FOR SMALL AREAS

Census-figures desirable but not necessary. If there are no figures from the Census for the areas in a frame, we ascribe to each area, by our best judgment, some number of sampling units. This number might be the average number of sampling units over all the areas. This is the same thing as drawing the areas with equal probabilities, and then interviewing in 1 segment out of n.

The absence of figures from the Census does not render sampling impossible: it only increases the variability of the sampling units and compels one to use, for a stated degree of precision, a bigger sample than he would require if Census-data existed.

> **Remark.** The reader may wish to convince himself, as an example, that whether we have Census-figures or not, (a) the procedures of this chapter will give equal probabilities to all people, and to all dwelling units; and (b) the probability of any area being struck by a random number is proportionate to the number of sampling units ascribed thereto.

A sample of the Federal District of Mexico. * This area showed 604,000 d. us. in the Census of 1950. There are totals for each section, but not for the blocks within the section. The feature of special interest here, different from the previous illustration, is that within any section we assign to each block the average number of dwelling units. This plan draws the blocks within a section with equal probabilities, with replacement.

Another feature is that the city has been growing at the rate of 6% annually; hence, it seemed desirable to increase arbitrarily by 50% the sizes of the blocks of some of the outlying sections wherein the growth appears to be most prominent. This decision illustrates the fact that the number of Cdus in an area is not necessarily a figure published by the Census; it may instead represent one's best information.

---

* I am indebted to Señorita Ana María Flores, Chief of the Department of Sampling in the Census of Mexico, for the data in Table 10 and for the privilege of working with her on this survey.

Lack of census information for small areas does not alter the probability of selection or introduce bias; it only subtracts from the efficiency that would be possible otherwise, and it adds a few problems to the field-work, owing to the fact that the segments assigned for interview will usually be more variable.

This study was to be an investigation on the cost of living.   An interviewer could cover an average of 5 or 6 families in 1 day.   The possible adjustments in the tasks of the interviewers that sometimes arise from inequalities in the sizes of the sampling units would not be serious here, because a small sampling unit in any area would likely be counterbalanced by a big one not far away.   It appeared advisable, therefore, in the interest of simplicity of preparation and administration, to assign to each block within a section a designated number of sampling units, this number so chosen that it produced an average size of 5 d. us. per sampling unit over the whole section.   The plan for Mexico thus conforms otherwise to the plan for Cincinnati.   Table 10 shows the start of the assignment of serial numbers to the sampling units in Estrata 1, a portion of Mexico City.

TABLE 10

THE COMMENCEMENT OF THE TABLE OF SERIAL NUMBERS
FOR THE SAMPLE OF SAMPLING UNITS IN MEXICO CITY

The frame for Estrata 1

Section	Number of blocks	Average number of Cdus per block (1951)	Number of sampling units Per block	In the whole section	Serial numbers of these sampling units
1	4	140	28	112	0001–0112
2	11	127	25	275	0113–0387
3	16	126	25	400	0388–0787
4	17	100	20	340	0788–1127
5	8	145	29	232	1128–1359
.					.
.					.
.					.

One may also, in the absence of census data for small areas (such as blocks), make a quick tour of a city or other area and estimate very roughly the number of dwelling units in each small area, then proceed according to the instructions for the survey of Cincinnati.

APPENDIX: THE NUMBER OF DEGREES OF FREEDOM IN AN ESTIMATE OF A VARIANCE MADE FROM A COMBINATION OF ZONES OF UNEQUAL VARIANCE

Satterthwaite's formula. (For advanced students.) We give now a formula for the approximate number of degrees of freedom in an estimate of a variance, such as Var X. Let there be k subsamples in every zone. The formula that we shall use is*

$$K = (k-1)\frac{(\Sigma\sigma_i^2)^2}{\Sigma\sigma_i^4} \qquad (1)$$

where K is an approximation to the number of degrees of freedom, and σ_i^2 is the variance of thick zone i. The summation runs over all m thick zones. Let us see what this formula leads to in terms of the variability of the variances σ_i^2. Let

$$\sigma_i^2 = S(1 + s_i) \qquad (2)$$

where S is the average of the m values of σ_i^2, and s_i is the relative departure of thick zone i from this average. Then $\Sigma s_i = 0$ by definition, and

$$K = (k-1)\frac{\{\Sigma S(1 + s_i)\}^2}{\Sigma S^2(1 + s_i)^2}$$

$$= (k-1)\frac{\{\Sigma(1 + s_i)\}^2}{\Sigma(1 + s_i)^2} = (k-1)\frac{m^2}{\Sigma(1 + s_i + s_i^2)}$$

$$= m(k-1)\frac{m}{m + 0 + \Sigma s_i^2} = m(k-1)\frac{1}{1 + C_s^2} \qquad (3)$$

where $C_s^2 = (1/m)\Sigma s_i^2$, this being the rel-variance of the m values of σ_i^2.

Note that if the m values of σ_i^2 varied uniformly over a range of $\pm 100\%$ above and below the average, C_s^2 would be $\frac{1}{3}$ (Panel B in Fig. 16 on page 260), and the loss in degrees of freedom would be only 25%. More erratic variation will of course bring bigger losses.

* This formula in a more general form for any number of strata (or thick zones), and for any number of degrees of freedom in each stratum, was first given explicitly by F. E. Satterthwaite, "An approximate distribution of estimates of variance components," *Biometrics*, vol. 2, 1946: pp. 110–114. It was given later for 2 strata by Alice Aspin in her paper, "Tables for use in comparisons whose accuracy involves two variances, separately estimated," *Biometrika*, vol. 36, 1949: pp. 290–293, with an appendix by B. L. Welch, pp. 293–296.
    I am indebted to my colleague Howard L. Jones of the Illinois Bell Telephone Company for calling my attention to these papers, and for many conversations on this subject; I am also indebted to his paper, "Investigating the properties of a sample-mean by employing random subsample means," *J. Amer. Statist. Ass.*, vol. 51, 1956: pp. 54–83.

EXERCISES

1.* Suppose that we form our m thick zones by a random ordering of the M thin zones. Show that Satterthwaite's formula (Eq. 1) for the number of degrees of freedom in the estimate of Var X is $m(k-1)/\left\{1 + \dfrac{m-1}{M-1}\,C^2\right\}$, where k is the number of subsamples and C^2 is the rel-variance of the variances within the M thin zones.

SOLUTION

A thick zone is here composed of M/m thin zones drawn at random, and the variance of a thick zone is the random sum of the M/m variances of the thin zones therein. The rel-variance of the variances of the m thick zones formed at random is then merely the rel-variance† of the sum of M/m random variables whose rel-variance is C^2 (defined above). The rel-variance of the m thick zones is therefore

$$\frac{M - M/m}{M - 1}\,\frac{C^2}{M/m} = \frac{m-1}{M-1}\,C^2 \qquad \text{[Apply Eq. 20 on page 383]}$$

We only need to substitute this rel-variance into Eq. 3 already derived. The result for the random formation of thick zones is the approximation

$$K = m(k-1)\,\frac{1}{1 + \dfrac{m-1}{M-1}\,C^2} \tag{4}$$

2. Show that the formula just derived reduces to $K = k - 1$ if we tabulate in 1 thick zone, as we did in Chapter 6 and elsewhere.

3. Derive the number of degrees of freedom per tabulation shown in Table 11, which is based on the assumption that the variances of the m thick zones are all equal.

Decision on the best plan will depend on convenience, and partly on this comparison, with consideration of possible losses from wide strata.

Formulas for 2 thick zones, 10 subsamples. Suppose that we desire more than the 9 degrees of freedom that come from 1 thick zone with 10 subsamples. We may gain degrees of freedom by forming 2 thick zones instead of only 1. The 10 subsamples in one thick zone will furnish

$$x_1^{(1)} \quad x_1^{(2)} \quad \cdots \quad x_1^{(10)} \qquad \bar{x}_1$$

whence, following Eqs. 13 and 14 (p. 198), we form

$$X_1 = \frac{1}{10}\,Z \sum x_1^{(1)} \qquad \hat{\text{V}}\text{ar } X_1 = \frac{Z^2}{10 \times 9}\,\Sigma(x_1^{(i)} - \bar{x}_1)^2$$

---

* For advanced students, familiar with Chapter 17. I am indebted to my friend Robert J. Brousseau of the American Telephone and Telegraph Co., New York, for Eq. 4.

† Rel-variance means the square of the coefficient of variation, as already defined on page 124.

TABLE 11

THE NUMBER OF DEGREES OF FREEDOM AND THE NUMBER OF
TABULATIONS REQUIRED FOR VARIOUS NUMBERS OF
SUBSAMPLES

Number of thick zones	Number of subsamples	Number of tabulations required	Degrees of freedom (df)	df : tab
10	2	20	10	.50
7	3	21	14	.67
5	4	20	15	.75
1	k	k	$k - 1$	$1 - 1/k$

Likewise, the 10 subsamples in the other thick zone will furnish

$$x_2^{(1)} \quad x_2^{(2)} \quad \cdots \quad x_2^{(10)} \quad \bar{x}_2$$

$$X_2 = \frac{1}{10} Z \, \Sigma x_2^{(i)} \qquad \hat{\text{V}}\text{ar } X_2 = \frac{Z^2}{10 \times 9} \Sigma (x_2^{(i)} - \bar{x}_2)^2$$

The final estimates for the total frame will be

$$X = X_1 + X_2$$
$$\hat{\text{V}}\text{ar } X = \hat{\text{V}}\text{ar } X_1 + \hat{\text{V}}\text{ar } X_2$$

For a ratio, there will be also a y-population, the final estimate Y, and the final estimate

$$f = \frac{X}{Y} = \frac{x}{y}$$

x and y being the total x- and y-populations in the sample. Then by Eq. 14 on page 198,

$$\hat{\text{V}}\text{ar } f = \frac{10}{9} \frac{1}{y^2} \left\{ \sum_i [(x_1^{(i)} - \bar{x}_1) - f(y_1^{(i)} - \bar{y}_1)]^2 \right.$$
$$\left. + \sum_i [(x_2^{(i)} - \bar{x}_2) - f(y_2^{(i)} - \bar{y}_2)]^2 \right\} \tag{5}$$

If the variance between sampling units is decidedly different in the upper half of the frame from what it is in the lower half, it may be desirable to form the 2 thick zones systematically, odd thin zones to Thick Zone 1, even thin zones to Thick Zone 2, to equalize the variances within the thick zones. Otherwise the degrees of freedom in the final estimate of Var X will not be $2 \times 9 = 18$, but something less.

An example in calculation of degrees of freedom. We now make an approximate calculation of the number of degrees of freedom in the

TABLE 12

CALCULATION OF THE NUMBER OF DEGREES OF FREEDOM IN THE ESTIMATE OF VAR X MADE IN EQ. 20 IN CHAPTER 9

Department	Number of zones m	Sample			x^2	x^2/m	$(x^2/m)^2$
		1	2	1 and 2 combined x			
1	24	30×10^2	23×10^2	53×10^2	$2\ 809 \times 10^4$	117×10^4	$13\ 689 \times 10^8$
2	25	17	20	37	1 369	55	3 025
3	80	76	71	147	21 609	270	72 900
4	27	13	14	27	729	27	729
5	133	102	103	205	42 025	316	99 856
6	131	74	82	156	24 336	186	34 596
7	185	99	78	177	31 329	169	28 561
8	210	142	153	295	87 025	414	171 396
9	112	59	47	106	11 236	100	10 000
10	289	135	162	297	88 209	305	93 025
11	238	158	198	356	126 736	533	284 089
12	209	136	128	264	69 696	333	110 889
13	231	101	138	239	57 121	247	61 009
14	39	43	42	85	7 225	185	34 225
15	152	93	61	154	23 716	156	24 336
Sum						3413	1 042 325

$$K = \frac{3413^2}{1\ 042\ 325} = 11.2 \qquad \text{[Eq. 1]}$$

estimate of the variance of the inventory of materials in process, studied in Chapter 9. Eq. 21 on page 157 estimated the standard error to be $75,000. How many degrees of freedom are there in this estimate? There were 15 departments, each being 1 thick zone, 2 subsamples, 1 degree of freedom.

The variances of the 15 departments were obviously very different from one another, as one may see by looking at the squares of the differences in the right-hand column of Table 3 on page 158; hence we should expect the number of degrees of freedom in Var X to suffer considerable reduction below 15. One could use, for the calculation of the degrees of freedom, the squares of the differences as estimates of the variances, department by department, to substitute into Eq. 1 of this appendix; but as each department had only 1 degree of freedom, I propose that we might obtain a slightly better result for the calculation now in process by assuming that the average variance between the lots in thin zones in any department is proportional to the average number of dollars per sample-lot in that department. In Department 1, for example, the 2 subsamples contained respectively $3000 and $2300 worth of material. The average of these 2 figures multiplied by 20 gives an estimate of the number of dollars in Department 1 (Class 2 only). As the factor of proportionality will cancel from the numerator and denominator of Eq. 1, I omit it, and form in Table 12 the sum of the 2 subsamples, and call this sum x. If M denotes for a moment the number of zones in a department, then x^2/M will be proportional to the variance between lots within the thin zones in that department. Table 12 shows the 15 values of x, x^2/M, $(x^2/M)^2$. The sums at the bottom are ready to substitute into Eq. 1 with $k = 2$, which gives

$$K = (2 - 1)\frac{3413^2}{1,042,325} = 11.2 \tag{6}$$

for the approximate number of degrees of freedom in Var X, which we calculated in Eq. 20 in Chapter 9.

As an exercise, the reader may wish to show that if we would break up Departments 8, 10, and 11 each into 2 equal subdepartments, the number of degrees of freedom in Var X in Chapter 9 would increase by 4 to about 15.2.

As a further exercise, show that the estimate of the number of degrees of freedom is about the same if one assumes that the average standard deviation (instead of the average variance) between the lots in the thin zones in any department is proportional to the average number of dollars per sample-lot in that department. In other words, our calculation of the degrees of freedom in Var X is not very sensitive to our assumptions.

CHAPTER 12

Field Procedure for the Creation of Segments and for the Selection of People within Families

There is no knowledge of external reality without the anticipation of future experience. . . . what the concept denotes has always some temporal spread and must be identified by some orderly sequence in experience. . . . There is no knowledge without inter- pretation. . . . Thus, if there is any knowledge at all, some knowl- edge must be a priori.—C. I. Lewis, *Mind and the World-Order* (Charles Scribner's Sons, 1929), page 157. (*Author's note:* knowl- edge a priori means theoretical description—the statistician's probability model.)

Purpose of the chapter. The purpose here is to learn (*a*) how to create segments in a designated area; (*b*) how to make a frame of the segments in this area; (*c*) how to select on the spot from this frame a sample of segments for immediate interview; (*d*) how to select at a later date another sample of segments from the same frame; (*e*) how to select, when desirable, 1 person from 2 or more that qualify in the same family as members of the universe.

We have learned in the last 2 chapters how to use the results from a replicated sample of segments. We did not worry much about the inter- viewer who must create the segments and draw a sample of them. We merely assumed that he could somehow do the work. We now put our- selves in his place. We shall try to write instructions that will be clear to him, and not too difficult to carry out. Unmistakable boundaries of segments are of course vital, and every square foot of a designated area must be in one segment or another.

General description of a field-procedure for segments. A field-worker goes to the designated area for the purpose of creating segments, following the instructions in the appendixes of this chapter. The desired size of a segment may be 20 d. us., 10 d. us., 5, 3, or possibly even 1, depending on circumstances. Detailed maps, directories, and aerial photographs are helpful where the delineation of the segments is to be done in the office.

Whatever be the intended average size of a segment, there will usually be considerable variation in the sizes of the segments, because the primary necessity in delineation is clarity of boundary, not uniformity of size. Fortunately, segments need not contain equal numbers of people nor of dwelling units. Inequality and variability in the sizes of segments introduce no change in the probabilities of selection of the people or of the dwelling units therein, and no bias. Substantial inequalities in size will usually cause only negligible decrease in the precision of a ratio, but an estimate of a total or of an average will suffer some loss in precision (see pages 479 and 480 for examples). In any case, in a probability sample, the standard errors are calculable, and in good practice they accompany the estimates of the results of chief importance so that there need be no doubt what the precision really is.

The prime principle in the creation of segments is that every dwelling unit must be in one segment or another. The field-worker must leave no possible doubt about the bounds of a segment. Ambiguity in a boundary will cloud the probabilities of the dwelling units near the boundary, and will lead to bias. Every dwelling unit must be in 1 and in only 1 segment. Each segment takes a serial number 1, 2, 3, and on to the last segment in the designated area.

The best time to interview in a sample of these segments is right now, while the field-worker is on the spot. She requires no extra transportation, and the bounds of the segments are fresh in her mind. Random numbers in a sealed envelope, prepared independently in advance, which she will open only after she has defined and numbered the segments, will prescribe which segments to interview in. There will be an envelope for each designated area.

The first step, if the designated area is big, is to try to break it up into 2, 3, 4, or more portions. The purpose is to try to confine the segments to only 1 portion (which the supervisor will select by random numbers), thus to avoid the relatively excessive cost of creating segments over a big area, in comparison with the cost of interviewing.

The random numbers in the envelope will usually draw, from the segments created in a block, or in a selected portion thereof, only 2 segments for interview, no matter how big or how small be the block, and no matter how unequal be the portions. An exception is to be expected if the number of dwelling units in the block is nowhere near the number shown in the frame, or if the interviewer grossly miscounts the dwelling units in the portions, or if she creates segments persistently smaller or larger than the intended size. In any case, the exceptions create no bias, and only slight impairment of the precision (Ch. 19, pp. 477 and 479).

A sample of field-instructions for the creation of portions appears in

Appendix D of this chapter (p. 241). The procedure for the office appeared in Chapter 11 (p. 204).

There should be an instruction to the field-worker to halt her proceedings and to call or write for special advice if she has more segments than random numbers in the sealed envelope, for unless her segments are abnormally small, this condition may indicate abnormal growth, which may require special treatment (p. 256).

The procedures described in this chapter go smoothly in the field, with simplicity, economy, and statistical efficiency. They possess the following advantages: (1) The field-worker need not adjust the sizes of the segments in order to produce exactly any designated number of segments. Instead, the actual sizes of the segments may be dictated chiefly by the necessity for clarity of boundary and completeness of coverage, and only secondarily by statistical efficiency. (2) If the random numbers select more than 1 segment from an area, the segments selected will be scattered over the area, and there will thus be some small gain from the stratification so enforced. (3) The creation of the segments and the interviewing in a sample of the segments may take place in 1 visit, except for recalls where there was nonresponse. (4) The same list of segments will serve several surveys.

The appendixes to this chapter contain instructions for consideration. If the intended average size of segment is other than 5, the instructions will of course require modification.

For a subsequent survey, one uses the same list of segments, and selects, in the office, with random numbers, a sample of segments not hitherto used.

The selection of segments. Every block or other primary area will be listed in the frame, and the frame will ascribe to each block a certain number of sampling units, which we arrive at by consulting the figures from the Census, and perhaps other sources as well (Step 2 in Ch. 10, page 167). Random numbers strike some of these blocks (or other areas), and in these areas we need to create segments, and then to interview in a random selection of these segments. The only point that we need to understand here is that the segment-interval in any block is equal to the number of sampling units ascribed to that block, and that this interval will stand fixed regardless of the actual number of dwelling units in the block today, and regardless of the sizes of the segments. For example, if the number of sampling units ascribed to a block was 9, then a random number between 1 and 9 would select a segment to interview in, and another random number between 10 and 18 would select another, and another between 19 and 27 would select another, and so on (p. 203). The random selection of segments in the sealed envelope was, in an actual survey, 1, 14, 22, 29, 45, 52.

Random numbers beyond the number of segments actually created are blanks and draw no sample. The interviewer should be instructed to halt and to call or write for instructions if her supply of random numbers in any envelope is insufficient (see Appendix A, page 237), as this event may denote heavy growth, and the need of special treatment (p. 256).

Procedure for subsequent selection of a sample of segments. The list of segments that the interviewer makes will in most areas remain usable for some time, perhaps for years. New dwelling units built in between others, or carved out of existing dwelling units, will have the same probability as any others of coming into the sample. This statement does not apply, of course, to a whole area that was never carved into segments. People may move in and out, buy and sell; dwelling units may burn down, but so long as every square foot of an area was ever definitely in one segment or another, the probability of all the dwelling units will remain equal.

For a survey next month, or a year hence, it is then only necessary (except where there has been heavy growth) to draw segments from the list that the interviewer turns in today. Segments that have been in one survey will usually not be admitted in another one, unless the survey is the kind in which we wish by design to interrogate the same people after an interval, to measure a change, as in the Monthly Report on the Labor Force in the U. S. and in Canada.* The list of segments thus gradually wears out, if we use it repeatedly, until there are finally no segments left unsurveyed, and it is time for a new list of segments in a new set of areas.

> **Remark 1.** A sample of segments remains a valid sample of the population for many years. Movements and redistributions of the people into new areas, and heavy growth in old areas, gradually raise the variance of the estimates derived from a sample of segments until we finally require a completely new sample of areas, and new segments in these areas.

> **Remark 2.** We should nevertheless not attempt to use an old list of segments in an area that shows extremely heavy growth.

> **Remark 3.** A firm that finds use for a list of segments in subsequent surveys is fortunate because they may distribute the original cost of the preparation of the sample amongst all the surveys that use the original list.

Principles for selecting 1 member of the family for interview. If one could assert that every person (*a*) who lives in a segment that comes into

---

* See Hansen, Hurwitz, and Madow, *Sample Survey Methods and Theory*, Vol. I (Wiley, 1953), Ch. 12.

the sample and (b) who qualifies by age and sex and by perhaps other characteristics as a member of the universe will actually be interviewed, then every one of these persons would have the same chance as a segment has of coming into the sample.

Unless we plan and instruct the interviewers thoroughly, the actual chance of an interview may, however, unfortunately, be different from the probability of the selection of the segment. The sampling procedure must therefore be one that will equalize these probabilities, or at least determine them. Thus, if some man is never at home in the evening, then no matter how many times the interviewer calls in the evening, the man will have no chance of an interview. The interviewer must therefore make an appointment to see the man in the forenoon or at some time agreed upon. Women are at home oftener than men are: older people and young children are at home oftener than youths. Equalization of the probability of interview is usually attempted by repeated recalls. Another procedure is for the interviewer to collect information that will determine the relative probability of people's being at home, and to multiply the interviews accordingly by the Politz plan (q.v., Ch. 5).

If the job is only to collect information about the dwelling unit—the number of rooms and baths, year built, rent, whether there is a telephone, how many extensions, whether there is an electric drier or a television, the number of people, perhaps by age and sex, then the interviewer needs to talk to only 1 person. Anyone that has the information will do. For example, in a census of population, it is only necessary to find a responsible person at home who knows the facts.

But for information about an individual that only he can give, such as his attitudes and opinions and reasons and personal purchases, there is only 1 way—find him (or weight the interviews of his class by the Politz plan). No one else will do. It is not permissible to draw a good sample of segments and then to interview for opinions, attitudes, and other personal characteristics just whoever comes to the door and is willing to talk: we might just as well start with a quota method and save the trouble of drawing a sample of areas.

Even if everyone were at home and eager to talk when the interviewer called, it might nevertheless be desirable to select a sample of people for interview. There are many questions that are contagious. This is especially true in studies of habits, food, attitudes, and opinions. More precisely, the answers are contagious—i.e., if 1 person in the family expresses an opinion or a claim, the other members say the same, or the opposite. Sometimes the opinions, habits, and other characteristics of the members of a family are actually the same, whether influenced or not by the first interview. Under such circumstances, big families or even

2-person families are not efficient sampling units.   It is often customary, under such circumstances, to instruct the interviewer to select 1 person from amongst the members of the family that qualify.   A simple procedure for making the selection is in Appendix C, page 240.   If the person selected is not at home, the interviewer must take the usual steps to find him and accomplish the interview.

APPENDIX A.   EXAMPLE OF INSTRUCTIONS FOR THE CREATION OF SEGMENTS IN THE FIELD*

These instructions embrace 3 parts:

1. Work in the right area.
2. Create segments in this area.
3. Select (by random numbers in a sealed envelope) the segments in which you will interview.

WORK IN THE RIGHT AREA

First, you must understand how important it is to work in the right area. The map shows roads or streets that we believe will enable you to find the area and to decide definitely whether you are in the right place (Fig. 9).   If you work in the wrong area, you will create a lot of trouble.

You will work only inside the area that lies within the boundaries.   Do not concern yourself with any dwelling units outside the designated area. A house across the road outside your area is definitely NOT a part of your job.

It is entirely possible that some of the roads or streets have different names from those that the map shows.   The map that we used may be out of date a few months, maybe a few years, and changes may have occurred meanwhile.   It may be that the area now contains streets that the map does not show.   Please alter the map to make it correct.   You need not do a first-class draftsman's job, but do try to make your alterations neat and legible.   The first steps are therefore as follows:

1. Make sure that you understand what area you are to work in.
2. Test your map and alter it where it needs alteration.

* These were instructions that I used for a survey carried out by the firm O'Brien-Sherwood of New York.   They are specific, and will not necessarily be correct in other surveys.   I present them, not as models to copy, but to emphasize responsibility for definite boundaries in the creation of segments in the field.   Use of the simple indicative and imperative moods for instructions may be of interest.

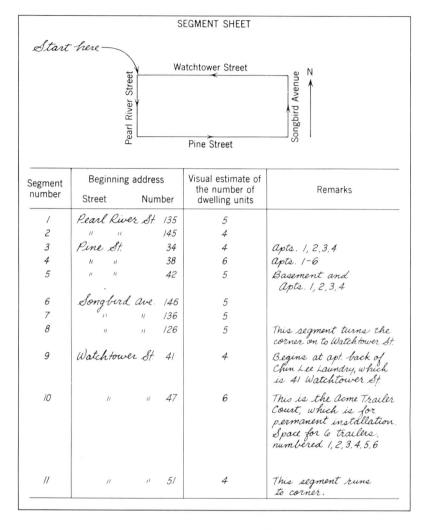

Segment number	Beginning address		Visual estimate of the number of dwelling units	Remarks
	Street	Number		
1	Pearl River St.	135	5	
2	" "	145	4	
3	Pine St.	34	4	Apts. 1, 2, 3, 4
4	" "	38	6	Apts. 1-6
5	" "	42	5	Basement and Apts. 1, 2, 3, 4
6	Songbird ave.	146	5	
7	" "	136	5	
8	" "	126	5	This segment turns the corner on to Watchtower St.
9	Watchtower St.	41	4	Begins at apt. back of Chin Lee Laundry, which is 41 Watchtower St.
10	" "	47	6	This is the Acme Trailer Court, which is for permanent installation. Space for 6 trailers, numbered 1, 2, 3, 4, 5, 6
11	" "	51	4	This segment runs to corner.

Fig. 9. An interviewer's segment-sheet.   Each address begins a segment, which includes this address and every dwelling unit up to but not including the next address.   The map was drawn in the office before the interviewer commenced work.

To remove a street that is no longer there, just draw a wiggly line through the street on the map.   If the name of a street has changed, draw a line neatly through the name on the map and print in the new name.   Show the name of a street that has no name on the map.

CREATE SEGMENTS IN THIS AREA

a. Dwelling Units

It is first necessary that you understand what a dwelling unit is. Very simply stated, it is a place where a family lives or could live. In the simplest case, a dwelling unit consists of a structure which is occupied or intended for occupancy by a single family. In other circumstances, most of which will be immediately apparent to you from external evidence, you may find structures that have 2 or more mail boxes, door bells, electric or gas meters, etc. Where such evidence indicates to you that more than 1 family occupies a single structure, simply count it as many times as there appear to be families within.

We would stress, however, that the criteria for determining whether a single house should be counted as a single dwelling unit or more than a single dwelling unit rests entirely upon your visual inspection. Unless there is *external* evidence to you that more than 1 family resides in a single house, you will count the house as a single dwelling unit. You must be very alert, however, in watching for side and rear entrances that lead to separate dwelling units, upstairs or basement apartments, dwelling units over stores, etc.

You may occasionally encounter a small structure, an apartment door, or some other entrance, that leaves you in doubt. Is it a dwelling unit, or only an extra side-entrance to a dwelling unit that you have already counted? Or is it a storage room, a shop uninhabited, a dentist's office? If it is definitely not a dwelling unit where people live or could live, do not count it. But if you have any doubt on the matter, include it in a segment. (If it turns out later not to be a dwelling unit, you have done no harm: in effect, you have only reduced the size of your segment by 1 unit, which you will see is not serious.)

There is 1 additional point. You will encounter occasional dwelling units not occupied at the time of your inspection. A dwelling may be vacant. Or, you may encounter a house which is not yet fully constructed, or though fully constructed, not yet occupied. These are all different kinds of vacant units. You will count the vacant units in a segment, just as you do the occupied units. One reason is that a vacant unit may be occupied weeks or months later, when this segment is selected for interview. Another reason is that your count will be compared with other counts.

We have been emphasizing what to include as a dwelling unit. You will *exclude* from your survey:

1. Institutions.
2. Dormitories.

3. Transient hotels.

4. Transient tourist courts, transient trailer courts, motels.

You will find special rules later on for apartment houses.

b. Segments

Segments are small identifiable groups of dwelling units. (You will see, as we go on, how important this word *identifiable* is.) These identifiable segments will contain approximately 5 d. us. each, and each segment will always lie wholly within one block or other area.

The Segment Sheet that you will use shows line numbers 1, 2, 3, and onward. The address that you show on line 1 begins the first segment. The address that you show on line 2 begins the 2d segment, and so on. A segment begins with an address that you will record, and it runs up to but does not include the next address that you record. The last segment begins with the last address that you record and runs up to but does not include the address that you wrote on line 1. Use a new Segment Sheet for every block or other area designated for you to work in.

Show in the proper column of your Segment Sheet your visual estimate of the number of dwelling units that each segment contains. Keep in mind the following points:

1. Every address that you record must be one that you or anyone else can find easily and unmistakably. Therefore, leave no doubt on your record about where any segment begins, and in unusual cases, what it contains.

2. The ideal size of segment in this survey is 5 d. us. However, it is of much more importance that each segment that you create have a definite and unmistakable beginning than it is for all of your segments to contain 5 units. Many of your segments may well contain 4 units. Some of your segments may for good reasons be as small as 2 or 3 units, while it may be more practicable for others to be as big as 7 units, or even 8 or 9 units. When you deviate from the ideal segment-size of 5, we prefer that you *form small segments* rather than big ones.

3. Explain any unusual situations with a rough diagram. Show on your diagram any new streets that now appear within the designated area.

One hard and fast rule is that EVERY DWELLING UNIT MUST LIE IN ONE SEGMENT OR ANOTHER. The only question is which segment shall any one dwelling unit belong to.

Now let us look closer at the job of creating segments and see just what to do. These instructions include a SEGMENT SHEET for each area wherein you will create segments. In the heading of the sheet will be a rough map, which will usually show the roads or streets or other boundaries

that enclose the area.   One corner will be marked "Start here."   Turn
to the segment sheet (p. 232) for an illustration.

1. Go to the corner marked "Start here."   Start walking around the
block counterclockwise.   As you proceed around the block, you will
count only the dwelling units on your left.

2. Record on the Segment Sheet the address of the first unit on Pearl
River Street.   This address begins segment No. 1.   The numeral 1 appears
in the left-hand column of the Segment Sheet.   Count 4 more units, and
the segment is complete.   Record in the space provided the number of
dwelling units in Segment No. 1 (5 dwelling units in this example).

3. Segment No. 2 will begin with the *next* dwelling unit.   *Record* this
address.   Count 4 more units, and segment No. 2 is complete.   Record the
number of units in this segment (5 again).

4. Follow this procedure throughout the entire block.

In rural areas and in towns where there are not streets and addresses,
you will describe the first house in a segment, not by number, but by the
name of the owner, by location, size and description, distance from main
roads.   The color of a house is risky, as people sometimes paint their
houses, and an interviewer coming at a later date might have difficulty.

Note that what you are doing is to record the beginning-addresses.
These beginning-addresses define or create the segments, as we said earlier.

As you might suspect, it will not always be so simple to break a block
into segments.   Here are some of the possibilities that will require deviation from the foregoing ideal situation.   As noted before, the beginning
point of each segment must be definite and unmistakable.   Consequently,
you should never begin a new segment in the middle of a small structure
that has multiple entrances (apartment houses are different and will come
later).   It is better to let the whole structure be 1 segment, even if it
contains 6, 7, or 8 d. us.   Break it into 2 segments only if you can do so
unmistakably.   Remember that there must be no doubt subsequently
which is the beginning-address of any segment.

Another point that requires consideration in this work is what to do
about the corners of blocks.   This can probably be explained best by
going back to our example.   Assume that you were working along Pine
Street and that you had started a new segment just before you reached the
corner of Pine Street and Songbird Avenue.   If there are any dwelling
units on Songbird Avenue, you may go around the corner to complete the
segment, in which case you will be sure to note in the "Remarks" column
that the "Segment turns the corner."

If, on the other hand, you find that there are no dwelling units on
Songbird Avenue, you will face 1 of 2 situations.

1. It may be that the segment that you started to form at the end of Pine Street contains 3, 4, or 5 d. us. (without the additional dwelling units that you might have expected to find on Songbird Avenue). The best thing to do here is to let these dwelling units at the end of Pine Street stand as a full segment, just as they are, and your next segment will begin on Watchtower Street at the corner of Songbird Avenue.

2. It may be that the segment that you started to form at the end of Pine Street contains only 1 or 2 d. us.   In this case, you may erase the last address that you wrote on Pine Street, and consolidate these 1 or 2 d. us. with the preceding segment.   Or, another thing that you can do, especially if this consolidation will form a segment as big as 8 or 9 d. us., is to break up this big segment into 2 segments of nearly equal size, provided you can do so quickly and unmistakably.

3. You will see that we have here introduced another rule; namely, that a segment that begins on Pine Street will not run around 2 corners on to Watchtower Street.   In other words, if there are no dwelling units on Songbird Avenue, the last segment on Pine Street will end definitely on Pine Street, and the next one will begin definitely on Watchtower Street.

If you work in a cluster of blocks, the serial numbers that you assign to the segments in one block will carry over to the next block.   For example, assume that the first block that you worked yielded segments 1, 2, 3, and 4.   The first segment on the next block that you would work would then be No. 5, the next segment would be No. 6, etc.   Keep in mind that you will never split a segment between 2 blocks: a segment must lie all in 1 block.   Confusion would arise if someone tried to interview later in a segment that started in one block and ended in another.

c. *Apartment Houses*

As we mentioned previously, apartment houses will require some special but simple rules.   If there are as many as 8, 9, or 10 apartments per floor, it is advisable to create 2 segments on each floor.   For example, apartments A, B, C, D, E could form one segment while apartments F, G, H, I, J could form the other segment.   List the actual apartment numbers or letters that the segment embraces in the column for "Remarks" on the Segment Sheet.   In other apartment buildings that have 7 or fewer apartments per floor, it is better to let the entire floor stand as a segment. Sometimes it might even be necessary to combine 2 floors to form a segment.

At any rate, whenever your segment includes apartments, always list the apartment numbers or letters in the "Remarks" column.   Should the apartments not have numbers or letters, list the names of the occupants, and describe the location clearly, as "Ground floor left." Ordinarily you can find all the information that you need for apartment houses from the mail boxes or door bells. It is advisable, however, that you probe a floor or two just to ascertain that no confusion could develop later in your methods.

Be sure to include the apartment where the superintendent lives, or the janitor. Watch for living quarters in stores, machine shops, etc.

SELECT THE SEGMENTS IN WHICH YOU WILL INTERVIEW

You have now finished the creation of segments in your assigned area. On your Segment Sheet you have recorded the address of the 1st dwelling unit of each segment, and they are numbered 1, 2, 3, 4, etc. You have of course given additional pertinent data in the "Remarks," and have even made a rough diagram where necessary. You are now ready to select the segments in which you will interview.

Actually, this selection of segments has already been made for you. All you have to do is to open the sealed envelope and find listed there the numbers of segments that have been chosen for you to interview in.

IT IS OF THE UTMOST IMPORTANCE THAT YOU FINISH YOUR JOB OF CREATING SEGMENTS BEFORE YOU OPEN THIS ENVELOPE. TO OPEN THE ENVELOPE PREMATURELY WOULD PROFIT YOU NOT IN THE LEAST, BUT IT WOULD SERIOUSLY IMPAIR YOUR WORK.

The envelope will contain 7 numbers. You will interview in every segment that bears 1 of these numbers. Do not worry if there are numbers for which you have no segments. For example, your envelope may contain the numbers 3, 18, 19, 35, 45, 51, whereas you had created only 19 segments in your area. You would then disregard the numbers 35, 45, 51. The supply of numbers is purposely more than you can usually use.

If, however, you have more segments than random numbers, the explanation can only be 1 of 3 possibilities:

1. This area has grown a great deal since the map that we used was made.

2. Your segments are mostly smaller than the intended size (choice of definite boundaries may have forced you into this).

3. You are working in the wrong area, in spite of the precaution that you took.

If you are in the wrong area, discard the work that you have done, and go to the right area and proceed in the regular manner. If this is not the explanation, however, please halt your work; send to the home office your revision of the map, and your list of segments, and any possible explanation in regard to growth. Await further instructions.

To proceed, it is highly important that you INTERVIEW IN EVERY OCCUPIED DWELLING UNIT within every segment that the random numbers select.   Search every selected segment for households in the basement, in the rear, side, top, behind trees.   Enquire into the possibility of another dwelling unit within a dwelling unit.   Interview in a trailer in the yard of any address in a selected segment, unless it is transient.

It is equally important that you should NOT GO OUTSIDE a segment. This brings up a possible question that may occur to you.   You may have showed in the "Remarks" column 6 d. us. for a segment.   You were told to do this by visual inspection.   But now, when you come to interview in this segment, you find that there are really only 4 households.   This will happen if 2 of the dwelling units are vacant; and it may happen also by sheer mistake, as when a door or an entrance that looked like a possible dwelling unit only leads you to an office, or to a closet, or to a room that houses tools or elevator machinery.

You will *in no case* substitute one segment for another, nor renumber or revise your segment.

APPENDIX B. FURTHER EXAMPLE OF INSTRUCTIONS RECORD OF INTERVIEWS AND VISITS

You will fill out 1 of these forms for every segment that the random numbers select for interview.   Please fill out this form carefully.   The figures at the top of the sheet are to be filled out after you have completed your work in the segment and are preparing to send your work to us.

In most instances, the "Number of possible interviews in segment" will be exactly equal to the "Number of dwelling units in segment."   The only time these figures will differ will be in cases where you have to conduct more than 1 interview in the same family.   When this occurs, the "Number of possible interviews" will be bigger than the "Number of dwelling units."

Fill out on the form the address for each dwelling unit as you begin the interview.   As you complete the interview, write down also the name of the respondent whom you interviewed or whom you were supposed to interview.   In homes where you are to interview 2 persons, or 3, record the names and addresses separately.

In cases where you are unable after 6 calls to reach any person at all in some dwelling unit, record the family name by looking at the bell or the mail box, or by enquiry.

If you obtain an interview on your 1st call, record the date and time of day when you completed the call and under the "Remarks" column record the work "completed."

If some person whom you are to interview is not at home, try to ascertain when he will return, and when you could see him. Make an appointment if possible. Record under the "Remarks" column the date and time of your attempt, and include any information that may help you

RECORD OF INTERVIEWS AND VISITS

Segment number __6__ (From sealed envelope)

Number of possible interviews in segment __5__
Number of dwelling units in segment __4__
Number of completed interviews __4__

Instructions: Fill in date and time of each call. Under Remarks fill in V for vacant; NPE for no person eligible in this household; AWAY if all the selected persons are away for an extended period; REF (with details) if any selected person refused. If the person to be interviewed is not at home, learn when he will be likely to be home for your 2d call.

Name and address		Date and time	Remarks
Mary Jones	1st Call	8 Feb 59 3 p.m.	Completed
123 Main St	2d Call		
New York 67	3d Call		
John Jones	1st Call	8 Feb 59 3 p.m.	Call back after 6. He works
123 Main St	2d Call	9 Feb 59 6:30 p.m.	Not home
New York 67	3d Call	11 Feb 59 6:30 p.m.	Completed
Charles Smith	1st Call	9 Feb 59 3:30 p.m.	On vacation
125 Main St	2d Call		Will not be back
New York 67	3d Call		until 12 Feb 59
Jane Doren	1st Call	9 Feb 59 3:40 p.m.	Not home today Call tomorrow after 6.
127 Main St	2d Call	10 Feb 59 7:10 p.m.	Completed
New York 67	3d Call		
Judy Trew	1st Call	10 Feb 59 4 p.m.	Completed
129 Main St	2d Call		
New York 67	3d Call		

Fig. 10. Sample of proper method of completing the record of interviews and visits.

to make your recall at the proper time, such as "Works in day; call back after 6 p.m." There must be a notation under "Remarks" for every dwelling unit in the segment. If you show "Away," state the reason, such as vacation, and length of expected absence from home. You are to make 5 recalls after your initial call before you give up an interview with any person.

APPENDIX C. SELECTION OF ONE PERSON FOR INTERVIEW WITHIN A FAMILY*

To Be Used When More than One Person in the Family
Qualifies, but Only One Is to Be Interviewed

Many of the households that you will visit will have only 1 person that qualifies for interview. In these cases, you interview him. In other instances, there may be 2, 3, 4, or even 5 persons that qualify within a family. You are to interview, in this survey, only 1 person in the family, and this instruction tells you how to make the selection.

Write the names of the people that qualify on the lines designated on page 1 of the questionnaire. These lines are numbered 1, 2, 3, 4, 5, 6. Enter on line 1 the oldest female, on line 2 the next female, on line 3 the next, and so on. When you have listed all the females, then write the name of the oldest male (usually the head of the household). Then, the

Tape

1. _____

2. _____

3. _____

4. _____

5. _____

6. _____

Lift here

After you have
listed all the names,
lift the tape until you
uncover an X with
a name. Interview
this person.

* This illustration is written up for a survey in which any person 21 years old or over in a family qualifies as a member of the universe, but in which there will be only 1 interview in any family. The age limit, and the definition of the universe, will of course vary from one survey to another. In one survey, for example, on use of lipstick and other aids to female beauty, females 16 years and over constituted the universe. In another (p. 398), a man or wife who read a certain newspaper constituted the universe.

next oldest male, the next, and so on.*   When you have listed all the females that qualify, and all the males, then (and not before) lift the tape at the right from the bottom and interview the person on the lowest line that shows an X.

NOTE TO APPENDIX C†

The arrangement of the X-pattern is as follows:

An X on line 1 of every questionnaire.
An X on line 2 of a random 1 in 2 of every 2 questionnaires.
An X on line 3 of a random 1 in 3 of every 3 questionnaires.
Etc.

As an exercise, the reader may wish to prove that (a) the probability is unity that there will be an X opposite the name in a case in which only 1 person in a family qualifies; (b) the probability of selection is $\frac{1}{2}$ for each of both persons in a family in which 2 persons qualify; (c) the probability of selection is $\frac{1}{3}$ for each of 3 persons in a family in which 3 persons qualify; etc.

It is simple to increase the probabilities of the later lines, to increase the probability of drawing males into the sample.

APPENDIX D. SPECIAL INSTRUCTIONS TO CONCENTRATE THE SEGMENTS INTO A SINGLE PORTION OF A LARGE BLOCK‡

It will be possible to shorten the job in a large block, by confining the creation of segments to only a portion of the block.   The instructions below will apply to any block that you can break up into readily identifiable portions, although we shall usually not apply them unless the block appears to contain more than about 60 d. us.

First, look around the block to discover how you could break it up into 2 or more portions.   One portion might be a large apartment building, or a part of a large building.   Another portion might be a smaller building or a group of small buildings, or a group of single dwelling units or duplexes.   The portions will be satisfactory if they are all definitely identifiable with unmistakable boundaries, and if no portion

* The reason for listing the females first, which departs from the usual Census order, is to be able to give greater probabilities to the males, as one often wishes to do because males are more mobile than females, and are less often found at home.   This scheme does not solve the problem of nonresponse, but with the proper X-pattern, it will deliver about equal numbers of interviews of males and females.

† I learned the use of this X-pattern at the Alfred Politz Research, Inc., New York, about 1946.

‡ An example of a map and list of portions, with selection of 1 portion, appeared in Chapter 11, page 204, so I omit them at this point.

has fewer than about 30 d. us.   (There is no use to create a very small portion.)

All that you need do, when you encounter a block that you believe could profitably be broken up, is to (1) make a list of the portions, on the form provided; (2) estimate for each portion the approximate number of dwelling units; (3) send in your map and estimates to the main office in New York.   (Use the telephone if you are within 300 miles of New York.) Someone in the main office will designate one of the portions for you to work in.   You will create segments only in this portion.   You will receive from the main office a new sealed envelope that will contain the numbers of the segments that you will interview in.

Exercise. Work out a scheme, for use by the interviewer on the spot, that will designate for interview every person in a family where only 1 person qualifies as a member of the universe, and which will give a balanced random selection of:

1 person in a family where exactly 2 qualify.
1 person in a family where exactly 3 qualify.
1 of the 3 fourth members of 3 successive families where exactly 4 qualify.
1 of the 3 fifth members of 3 successive families where exactly 5 qualify.
1 of the 3 sixth members of 3 successive families where exactly 6 qualify.

Hint: Instruct the interviewer to list, in Census order 1, 2, 3, etc., in space that you will provide on the questionnaire, the members of the family that qualify.   Then let him refer to the specific selection-sheet that you attach to the questionnaire, to learn whom to interview.   An example of a selection sheet appears below.   The asterisk in the column for 1 person will appear on every sheet.   The remaining asterisks will be randomized from one questionnaire to another by balanced thinning (p. 337).   (Hint for 4 or more persons: add blanks to fill each block of 3;   then balance the selection in blocks of 3, blanks and all.)

Interview every person with an asterisk

Order on list	Number of persons qualified in the family					
	1	2	3	4	5	6
1	*			*		*
2		*	*		*	
3						
4				*		
5						
6						*

CHAPTER 13

A Statistical Aid to Supervision

For whoso despiseth wisdom and nurture, he is miserable, and
his hope is in vain, his labours unfruitful, and his works unprofitable.
—The Wisdom of Solomon, iii: 11.

Limitations of a statistical tool. The aim of a statistical tool as an aid
to supervision is to achieve uniformity of performance, and performance
at the right level. As we learned in Chapters 1 and 5, statistical theory
will not of itself determine what is the right question, nor what is the
right answer. It will not determine what is the correct definition of age,
occupation, education, passed, failed, good, or bad. This chapter will
therefore not tell anyone whether his interviewers or his inspectors are
getting the right answers, except by comparison with a standard. What
statistical theory does is to help us to measure differences in performance,
and hence to isolate and to regulate the causes of variation and of departures from a standard. The aim of this chapter is to provide aid toward
the achievement of uniformity of performance, and to bring the performance into line with a standard measure when any exists.

General description of the tool. The statistical tool to be described
here is simple and useful, both in the interviewing of human populations,
and in the inspection of physical materials. The principle to apply is
that if there were no difference between the investigators, they would, in
effect, all draw chips out of the same bowl, and the differences between
them would follow the same distribution as the variance of samples of
the same size all drawn by the same interviewer.

The Mosteller-Tukey probability paper enables us to compare rapidly
the performance of investigators and to discover if any retraining is
necessary. (References to the Mosteller-Tukey paper are on page 407.)

A strict criterion by which to decide whether a point is significantly far
away would require knowledge of the theory of multiple comparisons.
We shall be satisfied here with a simpler tool.

243

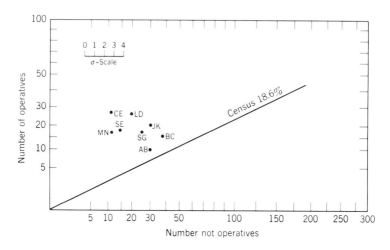

Fig. 11. The number of operatives and the number of not operatives in the occupations recorded by the various interviewers during the first 2 weeks of a survey in Wilmington in 1952, compared with the Census of 1950. The points all fall above the Census, and this fact indicates definitely that the reason for the difference between the sample and the Census arose from causes common to all interviewers. Retraining removed the cause.

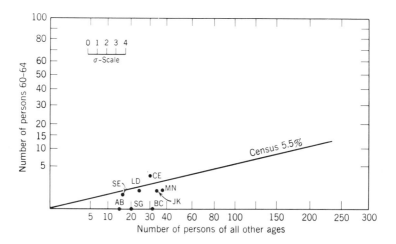

Fig. 12. The number of persons 60–64 and the number not over 64 in the ages recorded by the various interviewers during the first 2 weeks of a survey in Wilmington in 1952, compared with the Census of 1950. The chart indicated the possible existence of common causes, which turned out to be failure to understand the importance of thorough coverage and correct reporting of age.

The scales on the Mosteller-Tukey probability paper are built on the assumption that the variance of the distribution of any one investigator's work is binomial.   Actually, the distribution of one investigator's work is sometimes greater than binomial, yet we shall nearly always find a way to use this tool without strict knowledge of what his sampling variance actually is.   An example appears in Fig. 11, where the points, in relation to the Census line, definitely establish their own criterion for a common cause. Another example occurs in Fig. 12, where the pattern of the points on the horizontal line, and the number of zeros, indicate lack of uniformity. Sometimes most of the points will form a fairly compact cluster, which establishes an allowable tolerance, pointing a finger pretty definitely at some point that is out on a limb, far away from the others.   I had occasion to mention the same type of comparison in the discovery of gross blunders on page 72.

When we use the Mosteller-Tukey probability paper, we plot for each interviewer on the vertical axis his count of some specified population, and on the other axis his count of the remainder of the population.   If no point stands out from the crowd, there is no indication of nonuniformity, although all the points may be in error through a common cause, when compared with a standard.   A common cause will throw all the interviewers in one direction or another, or will affect the degree of scatter.   If 1 point falls far away from the crowd, then this investigator is either out of line, or he has by chance met unusual material.   The cause to look for is in this circumstance special; i.e., specific to him or to the materials that he tested.

Remark 1.   To summarize, this statistical aid to supervision separates out the special causes from the common causes.   An interviewer whose point lies far away from the bulk of the others exhibits a *special cause*—a cause not common to all the interviewers, but specific to him.   He may require retraining; or, if his point indicates extra-high quality (like EM's no refusals in Fig. 13), the problem is first of all to discover whether the quality is really as good as it looks, and if it is, to try to find out how he does it, and to train the others likewise.

On the other hand, when the bulk of the interviewers differ from some outside result (Fig. 11), there is indication of a *common cause* that afflicts all the interviewers, or the outside comparison.   A common cause may lie in the instructions, or in the training, or in the supervision.

Remark 2.   A usual mistake in administration is to attribute *any* difference to a special cause, to blame some interviewer for not doing his work, when there is actually no significant departure, and when no special cause exists.   The confusion between common causes and special causes is costly, as treatment for a special cause, when none exists, will actually bring more variation between interviewers.

Success in administration depends heavily on a judicious choice of problems and clever separation of the responsibilities for solving them. A statistical tool that aids the separation of special causes from common causes is therefore a heavy contribution to a science of administration, to better surveys, to increased output, and to accuracy.

What this supervisory tool does is to direct the efforts of supervision to the place where these efforts can be most effective. It helps the supervisor to avoid dissipating his energy and causing trouble where none exists.

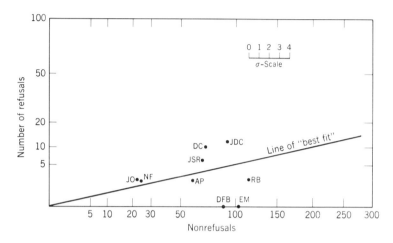

Fig. 13. Refusals and nonrefusals for 9 interviewers at the end of 4 weeks. Interviewers DFB and EM are significantly superior, or else there is something wrong with the records.

Example of common causes. I give first an example of a survey in which 1 of the questions was occupation. The survey covered a large city and several adjacent counties. There were 8 interviewers, and therefore 8 points. Each interviewer worked in a valid sample of the whole area. The allotments to the interviewers followed Appendix A. The initials beside the points are initials of the interviewers. The Census of 1950 had preceded the survey by 2 years. Shortly after the survey commenced, I plotted in Fig. 11 (p. 244) the proportions of operatives against all other occupations, in the male population 14 and over, and compared these figures with the results of the Census. The scale on the vertical axis shows the number of operatives encountered by each interviewer, and the horizontal axis shows the number of people in other occupations. The universe was all people in the labor force (at work, with a job, or seeking work). The line shows the proportion of operatives counted in the Census.

Although some interviewers came close to the Census, it was obvious that the 8 interviewers as a whole in the sample-survey were finding too many operatives compared with the Census. This we conclude because the points are not widely dispersed, and because they all lie above the Census line. The chart thus indicates the need to retrain *all* the interviewers, and not just to retrain 1 or 2 interviewers to bring them into uniformity. The trouble was attributed to common causes (*vide supra*), arising from faulty training. The next step was therefore to retrain the interviewers. The points on a new chart (not shown here), based on further interviews, fell above and below the Census line, and indicated successful retraining.

> **Remark 3.** The reader should note that the chart by itself, in common with any statistical method, does not indicate which group of workers is wrong. It might have been the Census, or it might have been these 8 interviewers in the sample-survey. In fact, the word right or wrong has no meaning, just as true value has no meaning. Statistical methods can only observe that there is a significant difference. Both groups of workers can't both be right, unless there was an intentional change in definition, which there was not. In view of the excellent training given to the workers in the Census, and the long course in occupational coding given to the people that code occupations in the Census Office in Washington, one could only accept the Census as a master standard (p. 62).

Another example of common causes. Fig. 12 (p. 244) shows the number of people of age 60–64 that various interviewers encountered in a survey of a certain city. There were 8 interviewers, and there are therefore 8 points. The scale on the vertical axis shows the number of people of age 60–64 in the interviews accomplished at the time the chart was plotted, and the horizontal axis shows the number of people of all other ages 10 and over. The line shows the proportion of age 60–64 in the Census which was taken 3 years earlier.

This chart is another good example of an exhibition of common causes of difference. Even though some interviewers came fairly close to the Census, the points raise a question mark and indicate the possibility of common causes. The chart led to a search for the reasons, which turned out to be failure to understand the importance of thorough coverage.

Examples of special causes. In Fig. 13 the number of refusals that 9 interviewers encountered are on the vertical axis, nonrefusals on the horizontal. This chart exhibits something remarkable—DFB interviewed 87 households without a refusal, and EM interviewed 104 without a refusal. In contrast, JDC encountered 12 refusals and 93 nonrefusals. The difference is significant. The action indicated is this: (*a*) find out,

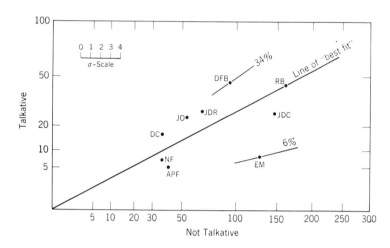

Fig. 14. The interviewers' subjective classification of the respondents as talkative or not talkative.   Strangely, DFB and EM are at opposite extremes, although neither had a refusal from these same respondents (see the preceding chart).

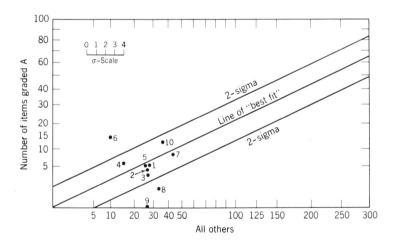

Fig. 15. The number of items graded A versus the number graded not A by the same inspectors.   The chart indicated nonuniformity of performance, and the need of further training, after which the performance improved.

if possible, how DFB and EM do it; and (*b*) teach the others, especially JDC, to follow the same techniques. (This is easier said than done, because one person can hardly copy all of another's personality and approach to a family.)

There is in this example no line of reference like the proportion determined by the Census in the 2 preceding examples. Here we are merely concerned with the scatter of the points.

Other examples of special causes.   The next chart, Fig. 14, shows the interviewers' subjective ratings of the people that they had interviewed, talkative on the vertical, not talkative on the horizontal.   The differences are significant. EM, who had no refusals, finds that only 6% of the people that she interviewed were very talkative.   My guess is that she did her interviewing in a businesslike manner and went on her way without giving people much chance to talk.   In contrast, DFB, equally successful in having no refusals, found 34% of the people to be very talkative.   JDC fell between.   All this indicates that the subjective classification of people as talkative or not talkative would in this survey be regarded as a characteristic of the interviewer, rather than of the people interviewed.

Another chart not shown was a plot of incomes above $5000 against incomes less than $5000.   The differences were significant.   Interviewers DC, JDC, and RB were getting figures, it turned out, only on income from wages or salaries, and not the total income of the family (which would include interest, sale of property and of stocks, and other transactions).

Fig. 15 shows some results obtained in the inspection of physical equipment, in this case, one part of a telephone instrument.   There were 10 inspectors in this group.   Each inspector was to grade a component part of a telephone instrument as A, B, C, or D, where the letters A, B, C, D had specified meanings, ranging from new or good as new down to worn out and in need of replacement.   I selected this chart out of a hundred or more* because it indicates the existence of special causes—that is, nonuniformity amongst the inspectors, and the need for further training and for careful supervision to discover the reasons for the nonuniformity.

Measurement of the variance between interviewers.†   The statistical tool to be described here requires the readiness of each interviewer to work in

---

* I am indebted to Mr. J. E. May, Inventory and Costs Engineer of the Southern Bell Telephone Company in the State of Mississippi, for this chart.   Mr. May plotted charts daily during the work, to be used as a supervisory tool toward the achievement of uniform performance.

† The reader who wishes to go further will enjoy the treatment by Walter Hendricks, *The Mathematical Theory of Sampling* (The Scarecrow Press, New Brunswick, N.J., 1956), Ch. xi.   Also, Hansen, Hurwitz, and Madow, *Sample Survey Methods and Theory*, Vol. II (Wiley, 1953), Ch. 12.

any area that the random numbers designate for him, at least within a prescribed area.

Let us see first what statistical design is necessary if we wish to compare interviewers. Take a design in 2 subsamples. Suppose that Interviewer A worked only in Zone 1, and that Interviewer B worked only in Zone 2. The diagram of their work appears below.

Zone	Sample 1	Sample 2
1	A	A
2	B	B

We could never know here whether the difference between the results of A and B could be attributed to the interviewers or to the fact that they worked in different areas. Nor could we be sure that Zone 1 differs from Zone 2: the difference could be attributed to the interviewers. I dare say that more than one apparent sociological or economic difference between 2 areas has given rise to ingenious explanations, when the real difference lay in the interviewers.

Anyway, the next scheme will not have that defect, provided the corners where A will work are decided by a random number, and not just because A likes to work there. The difference between the results of A and B *may* arise from differences between their areas, but not from any difference between the zones, as they will both work in both zones. Moreover, for the same reason, the difference between the 2 zones can not be attributed to the interviewers. (The possibility of interaction between an interviewer and an area is a real one, but is beyond the scope of this chapter, and I proceed on the assumption of no interaction.)

Zone	Sample 1	Sample 2
1	A	B
2	B	A

Now if in any survey we put 2 interviewers in every 2 zones, and if we let random numbers decide which corners 1 of them is to work in, we shall be able (*a*) to compute the variance between interviewers, and (*b*) to compute the pure sampling error (pure in the sense of being free of the differences between the interviewers).

It is the pure variance of sampling that we are entitled to know, I believe, because even a complete count is afflicted with the differences between interviewers, and by the error of sampling I mean the uncertainty that arises purely from the use of sampling in place of the equal complete coverage (Ch. 1).

We can go on, and discover ways to place and to overlap any number of interviewers into 3 subsamples, 4 subsamples, 10 subsamples.

Appendix A to this chapter shows an allotment of interviewers to 10 subsamples anywhere within an area.   This scheme gives good estimates of the differences between the interviewers, but it is expensive, as it forces all the interviewers to travel over the whole area.

Appendix B shows 2 possible ways to allot interviewers when they can not go so far from home.

The scheme in either appendix permits comparisons between interviewers, although the comparisons in Appendix B are weaker than the comparison in Appendix A.

Note that in both of these appendixes, all the interviewers work in all subsamples, and equally, except that 1 interviewer at each end may not have his full share.   The variance between the subsamples will thus be practically unaffected by differences between the interviewers.   For example, suppose that Interviewer No. 1 obtained, for some reason, too many professional people in his answers to occupation (too many compared with the other interviewers).   His results will raise all 10 subsamples equally.   This does not make right the final result obtained by pooling all the subsamples and all the interviewers, but it does raise or lower all the subsamples equally, and leaves the variance between the subsamples unaffected.   Our estimate of the standard error is then an estimate of the pure error of sampling, not confounded with the differences between the interviewers, which would affect complete counts and samples alike.   Such a plan of allotment is *orthogonal.**

Remark 4.   There will be a small amount of confounding (a) if 1 interviewer at each end has not the same share as the others in all subsamples

* The reader may wish to consult the following references for the measurement of differences between investigators.

1. P. C. Mahalanobis, "Recent experiments in statistical sampling in the Indian Statistical Institute," *J. Roy. Statist. Soc.*, Vol. 109, 1946: pp. 325–370.

2. Morris H. Hansen, William N. Hurwitz, Eli S. Marks, and W. Parker Mauldin, "Response errors in surveys," *J. Amer. Statist. Ass.*, Vol. 37, 1942: pp. 89–94.

3. A. Ross Eckler and Leon Pritzker, "Measuring the accuracy of enumerative surveys," *Bull. Int. Statist. Inst.*, Vol. 33, 1951, Part iv, pp. 7–25.

4. Hansen, Hurwitz, and Madow, *Sample Survey Methods and Theory*, Vol. I (Wiley, 1953), Ch. 2 E.

(as in Appendix B to this chapter); and (*b*) if the number of sampling units in a section is not divisible exactly by the number of interviewers. The statements just made in the text will then be nearly but not exactly true.

Remark 5.   The thick zone for tabulation should be some multiple of a section, in order to balance the interviewers by subsample—i.e., to remove the differences between the interviewers from the differences between the subsamples.

Remark 6.   Even if a full-scale plan of orthogonal allotments is impossible over the whole survey, it is advantageous to use it within areas where it is possible to do so.   The results will still disclose the existence of common causes and of special causes of differences.   They will assist the interpretation of the data from this survey, and they will indicate how to improve the training in the next survey.

Need for speedy correction at the commencement of the survey.   This tool is especially useful at the commencement of a survey.   It indicates whether the interviewers or inspectors have different understandings of the definitions, or whether they go about their work differently.   It is well to discover differences at the start, before it is too late to retrain them.

Supervisors may plot charts at the end of every day or two.   With a little help, a supervisor soon learns to plot his own chart, and to take remedial action at once.   The consulting statistician may require a copy of each chart, and a record of the action taken.

I use this scheme in my own work in any survey that will contribute results as legal evidence, or which will go into a report to the stockholders of a company, or a report that will be signed by an auditor.   This scheme protects the data from misinterpretation from wrong assumptions about the uniformity of performance, or the lack of it.

APPENDIX A.   RANDOM ALLOTMENT OF THE FULL
COMPLEMENT OF INTERVIEWERS

1.   Use this procedure to allot interviewers (or inspectors) in any area in which all the interviewers are free to work in any sampling unit allotted. In the illustration here, there will be 5 interviewers, but adjustment to any other number of interviewers is simple and obvious.   There will be 10 subsamples, but the plan can be adapted to 2 subsamples, or to any number of subsamples.

2. Decide on the starting point in the table of random numbers, and on the system by which to use the digits.

3. Number the interviewers 1, 2, 3, 4, 5, in any order convenient.

4. The sampling units in the 1st zone of Subsamples 1–5 will form the

1st block.   The sampling units in the 1st zone of Subsamples 6–10 will form the 2d block, etc.   (If there were 3 interviewers, the first 3 sub-samples would form the 1st block; the second 3 subsamples would form the 2d block; etc.)

5. Begin in the table of random numbers at the spot appointed in Step 2.   Use the digits in order.   Each digit not a blank will allot an inter-viewer to a sampling unit.   Use the sampling units in order, moving from left to right from one subsample to another, from Subsample 10 to Sub-sample 1, and onward.   For example, if the 1st random number were 2, it would allot Interviewer No. 2 to the 1st sampling unit.

6. Balance the allotments of the interviewers.   To do this, refuse a random number that would allot any interviewer twice within a block. Thus, you will use only 4 random numbers in any block of 5, as the interviewer not allotted to the first 4 sampling units will automatically take the 5th sampling unit.

7. Treat the second block of the 5 sampling units like the first, and continue onward through the area.

8. Add blanks at the end of the area to fill out a complete block.   A blank produces no allotment to that interviewer in the last block.

APPENDIX B. EXAMPLES OF ORTHOGONAL ALLOTMENTS WHERE THE INTERVIEWERS MUST STAY NEAR HOME

1st illustration

1. Use this procedure, or a modification thereof, to allot interviewers when they can not go far from home.   In the 1st illustration here, 2 interviewers are willing to go anywhere within 4 zones.

2. Form sections of 2 zones.   Zones 1 and 2 form Section 1;   Zones 3 and 4 form Section 2; etc.

3. Number the interviewers in any order convenient, but so that 1 and 2 are those who will work in Section 1;   2 and 3 in Section 2; etc.

4. Draw up the following scheme.

Section	Zones	Interviewers
1	1 and 2	1 2
2	3 and 4	2 3
3	5 and 6	3 4
4	7 and 8	4 5
Etc.		

5. Follow Appendix A to allot the interviewers within the sections.

The above scheme permits comparison between each interviewer and only 1 other. It is, however, the least expensive orthogonal scheme. The cost of travel is least with 2 subsamples; greatest with 10 subsamples.

All the interviewers work equally in all the subsamples, with the exceptions stated in the text.

6. Compile the results by interviewer and by subsample, with a sufficient number of thick zones. Estimate the variance between interviewers; also the variance of any estimates (as X or f) desired. Var X and Var f will include accidental errors of performance, but not the variance between interviewers.

2d illustration

In this illustration, 3 interviewers are willing to go anywhere within 9 zones.

1. Form sections of 3 zones each.

2. Number the interviewers in any order convenient, but so that 1, 2, 3 are those who will work in Section 1; 2, 3, 4 in Section 2; etc.

3. Draw up the following scheme.

Section	Zones	Interviewers
1	1, 2, 3	1 2 3
2	4, 5, 6	2 3 4
3	7, 8, 9	3 4 5
4	10, 11, 12	4 5 6
5	13, 14, 15	5 6 7
Etc.		

4. Follow Appendix A to allot the interviewers within the sections.

The above scheme permits comparison between each interviewer and 2 others. It requires a little more travel than the scheme in the first illustration. Again, all the interviewers work equally in all the subsamples, with the exceptions stated in the text.

5. Same as Step 6 above.

CHAPTER 14

Sampling New Material

6. Some man holdeth his tongue because he hath not to answer: and some keepeth silence knowing his time.

7. A wise man will hold his tongue till he see opportunity: but a babbler and a fool will regard no time.

8. He that useth many words shall be abhorred.—Ecclesiasticus 20.

Range and shape of the distribution of the frame. The aim of the survey-methods taught in this book is to estimate some total, mean, or ratio, signified in previous chapters by X, $\bar{x}$, or f, with precision that is about right for the purpose intended, and to calculate afterward what precision was actually reached. The proper size of sample one computes, with new material or with old, by the use of formulas typified by $\sigma_{\bar{x}}^2 = \sigma^2/n$ or σ_w^2/n, where $\sigma_{\bar{x}}$ is the standard error desired (aimed at), and where σ^2 and σ_w^2 are respectively, as before, the total variance and the variance within strata. In Neyman allocation one needs $(\bar{\sigma}_w)^2$. Other variances enter more complex designs. In any case, one may modify the theoretical sample-size to take account of costs and special conditions (p. 303).

If the material is new (and all material is to some extent new or different from any material previously sampled), one is faced with the problem of discovering in advance a satisfactory numerical approximation for σ or for σ_w or for $\bar{\sigma}_w$, in order that the standard error reached in the survey will not differ too far from the value of $\sigma_{\bar{x}}$ that is desired. The problem is not easy, but there are some fairly sure ways to go about it. What one really needs is a divining rod.

It is sometimes possible to dig up some related prior experience, or to make some quick tests or pilot-studies of new material, to gain quantitative information with respect to proportions, variances, and costs, that one needs for efficient design. We saw an example of the use of prior information in Chapter 7 on the sampling of business establishments, and in Chapter 8, Section B on the sampling of shipments of freight: also in

255

Chapter 9 on a sample to estimate the value of an inventory.   Another example will appear in Chapter 15 on page 309 in a problem in accounting.

There is frequently no time for a pilot-study or experimental work, and one must proceed with the plans, thankful for some knowledge of theory and for fragmentary quantitative information from people familiar with the subject-matter.   An example will appear in a psychological study on page 338.   Another example will appear on page 301 in Chapter 15 in which assumptions for extreme values for the ownership of FM radios in the lower levels of rent will form the basis for a logical choice between 2 plans of stratified sampling.   The panels in Fig. 16 (p. 260), along with the accompanying text, have been of great assistance to the author.   Two examples appear in this chapter, and there are other examples here and there in the book.

Huge and unusual units.   The layman's uneasiness about sampling, although he can not formulate his reasons, usually lies, I believe, in the possible or known existence of huge and unusual units.   He can't understand how a random sample can pick up such units in the right proportion. It may be that none of these units will get into the sample.   It is all too easy for too many to get in, or too few.   And on further thought, if the huge units are greatly different from one another, what does one mean by the right proportion?   These fears probably go through his head.

Well actually, he would have grounds for fear if the statistician were not already aware of the problem and had not already done something about it.   Good sample-design admits no chances.   The 1st step is to discover the existence of huge and unusual units, if there are any; and the 2d step is to give them special treatment (perhaps 100%) so that they can not upset the estimate.

We saw an example in Chapter 8, Section B, where every account above $10,000 was in the sample.   Although most of the accounts in the case were below $10,000, a lot of money changed hands in a few accounts of magnitude $50,000, $100,000, $150,000, $200,000, and rarely even $250,000.   When I explained the sampling plan to the auditors of the railways involved, assembled in Chicago, one of them spoke up to say, "That looks like a sure thing."   Of course it is a sure thing, I explained; it wouldn't be a sampling plan if it were not sure.   He then confessed that he had supposed that to draw a random sample from a million accounts, one would put the million serial numbers on tags in a huge hat, shuffle them, and then pull some out. He had been worried, he said, about some huge accounts.   He was naturally afraid to miss all of them, or to draw too many of them.

We saw another example in the inventory of materials in process, where some lots could run to very high values.   The sampling plan called for a 100% sample of all the lots that the foremen declared to exceed $500 (Ch. 9).

Huge and unusual units may exist in the form of housing projects that contain hundreds or even thousands of families on a piece of ground that a few years ago contained only a few dwelling units. Other examples are huge farms, big banks, big companies whose production, income, and number of employees are many times the average. One may procure lists of large housing projects. These lists, though not always complete, are very useful, as one can supplement them by local enquiry. There are lists of large business concerns.

In the sampling of farms for yield of product, for example, it is essential to find or to make a list of the big farms. This is not as easy as it sounds, because a farm that turns out a huge amount of product is not necessarily a lot of acres of land. One of the biggest farms in the country is a hatchery in the Loop in Chicago. A farm may have 5000 or 10,000 or more turkeys on it, but cover only a few acres. Bees store their product in a small space, but roam all over anybody's flowers to get their honey. Moreover, a farm or any other business may be large in one product, or in some characteristic, yet be only average in respect to another. Considerable knowledge and ingenuity are required to perceive in advance all the difficulties that may arise, and to see how to avoid them.

It may help to take a preliminary sample to investigate the material for variances and for special conditions. An example of a pilot study occurs on page 340 of Chapter 15. I often prefer to study at first some preliminary chunks of the material, chosen deliberately in an attempt to discover extreme conditions.

Let the reader be clear that we are not dealing here with mere roughness in classification, or with reasonably moderate changes that take place month to month in almost any material. One must not expect that a cutoff will be completely clean. One must expect to find some large accounts in the stratum of small accounts; a few large blocks in the stratum of small blocks; some lots of high value mixed up in the stratum of small lots. Human beings and machines ought not to fail, but they do. Business establishments and farms and areas change size; they don't hold back because we may wish to draw a sample of them. The last census or the latest available list will always be to some extent out of date. The skilled statistician distinguishes himself by ability to outguess slips and changes, and to prescribe the right additional sample to hold the variance of the final result to about the magnitude intended. (See the next page for a suggestion on what to do if a large account is found misclassified; and page 153, where we put $h = \$700$ for a nominal cutoff of $500.)

What we wish to focus on here is units so extreme in size that a random variation of 1 more or of 1 less of them in the sample will jolt the estimate one way or the other beyond the permissible bounds of sampling error.

One huge unit that turns up in the sample as a surprise may well indicate the presence of more of them in the frame, yet undetected; hence we dare not finalize the estimate until we make a new investigation, and possibly a new selection. The sampling procedure is still unbiased with the inclusion of whatever turns up in the sample, but this fact is little comfort if the variance is tremendous. To aggravate the anxiety, the presence of huge units (because of high β_2, page 439) impairs the estimate of the variance.

The usual story is that the preliminary investigation was not sufficiently thorough, or that enquiries elicited wrong information about the possible sizes of the units. Another story is hard luck or carelessness in screening, as when an account of $200,000 shows up in a stratum below $10,000. (If this were to happen, I should require a complete rescreening of all the accounts.)

Every statistician in practice must expect to face this problem once in a while. His life would be simpler if he could just sweep the problem under the rug, but he can not: he must proceed in accordance with the principles of sampling. It is possible to put down a few rules for guidance. Knowledge of theory and of operations, combined with effort and ingenuity, will usually find an efficient and unbiased solution.*

1. What to do depends on how many huge and unusual units are in the frame. The sample may throw more light on this question if we defer the decision on what to do until the sample is pretty well completed.

2. If the only huge unit in existence turned up in the sample, the solution would be simple—let it form a separate universe, already sampled 100%.

3. But we can not be sure that there are no more such units in the frame unless we re-investigate the frame, and do a better job on it than we did initially. Once we discover all the huge units, we may group them into a separate stratum, and draw a new sample of them, either 100% or some high proportion of them. The sampling plan might (e.g.) give every dollar or every $1000 in these units equal probability with another; or it might give every 20 d. us. the same probability as any other 20.

4. On the other hand, if the huge units are sufficiently numerous, and not too variable, one may form them into a separate stratum after the returns are in, and make proper adjustment without bias by the method of Plan D in Chapter 15 (*quod vide*).

Use of some simple geometric figures. The main theme of this chapter is the use of some geometric figures to estimate σ in advance. I use it almost daily, and it has served well. It is foolproof, provided the sampling

* See also Hansen, Hurwitz, and Madow, *Sample Survey Methods and Theory*, Vol. I (Wiley, 1953): p. 352.

plan takes care of any jack pots beyond the geometric figure, and uses the geometric method to estimate σ only for the remainder where the distribution is well behaved.

The geometric method depends on the fact that the variance σ^2 between the sampling units will depend on 2 pieces of knowledge: (a) the *range* or the spread between the tails of the distribution of the populations of the sampling units; (b) the *shape* of the distribution.

Questions put to experts in the subject-matter will usually lead to some fairly definite limits that will fix the range or the spread of the distribution. However, experience shows that one must be careful to probe thoroughly, and to choose for questioning experts who are likely to be in possession of quantitative information. Crude but cautious conjecture is safer than trust in the wrong expert. A few figures from previous work on related material, even if they come from a judgment-sample, or from a chunk, will often be helpful.

Once we dispose of the range of the distribution, we turn our attention to its shape. Fortunately, σ is not very sensitive to the shape of the distribution, and one may proceed safely with only a rough idea of the shape, provided he has, where necessary, taken the precaution to screen out successfully by a workable cutoff any sampling units that could do damage.

> **Remark 1.** Any experience in sampling new material will provide valuable information for the next sample. When samples are designed properly, the experience gained from them is cumulative. New material becomes old material after a few trials, and one may then begin to work on refinements in design.

The panels in Fig. 16 on page 260 are exceedingly useful, especially the binomial distribution (Panel A), and the rectangular distribution (Panel B), which represents a more conservative assumption than the triangular distributions in Panels C, D, and E.

PANEL A: 2 CATEGORIES. A useful and conservative assumption that one may often make is that some sampling units all take one extreme value a_1, while the other sampling units all take some other extreme value a_2. This proposal is depicted in Panel A, the frame of 2 categories, of proportions p and q, which we shall study in Chapter 17.

The variance of this distribution will be $\sigma^2 = pqh^2$, as shown in the figure, where $h = a_2 - a_1$, the range or the spread. A sample that is big enough under the condition that every sampling unit be solidly one characteristic or the other will be big enough or more than big enough in case the sampling units are not solid.

PANEL B: THE RECTANGULAR DISTRIBUTION. This distribution is also conservative. It assumes that the sampling units vary all the way from 0 to some maximum population, and that there are an equal number in every equal interval.   This is an extremely useful panel.   An example occurs in the Census of Japan further on (p. 267).

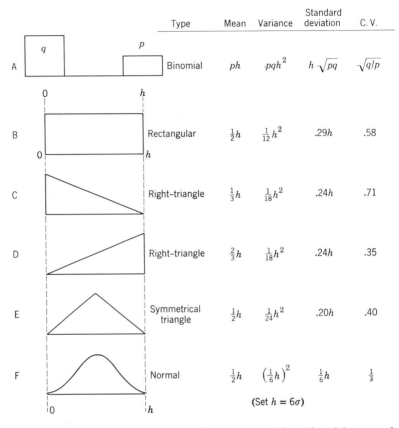

	Type	Mean	Variance	Standard deviation	C. V.
A	Binomial	ph	$\cdot pqh^2$	$h\sqrt{pq}$	$\sqrt{q/p}$
B	Rectangular	$\frac{1}{2}h$	$\frac{1}{12}h^2$	.29h	.58
C	Right-triangle	$\frac{1}{3}h$	$\frac{1}{18}h^2$	.24h	.71
D	Right-triangle	$\frac{2}{3}h$	$\frac{1}{18}h^2$	.24h	.35
E	Symmetrical triangle	$\frac{1}{2}h$	$\frac{1}{24}h^2$	.20h	.40
F	Normal	$\frac{1}{2}h$	$\left(\frac{1}{6}h\right)^2$	$\frac{1}{6}h$	$\frac{1}{3}$

(Set $h = 6\sigma$)

Fig. 16. New material can often be classed in advance roughly as binomial, rectangular, right-triangular, or triangular, and boundaries placed on the extremes.   Quick reference to this figure will give a conservative value of σ on which to plan a sample.   The range of variation is in every panel from 0 to h.

PANEL C: TRIANGLE, BASE AT 0.   This distribution is frequent.   One encounters it in segments of area where most of the segments contain 0 or small amounts, and only a few sampling units run out to the extreme. One encounters this distribution also in samples of accounts, where most accounts are small, but where some may be large.

PANEL D: TRIANGLE, VERTEX AT 0.   This distribution is the opposite of Panel C. Curiously, although its variance is $h^2/18$, as for Panel C, the coefficient of variation is only half that of Panel C. (Why?)

PANEL E: TRIANGLE, CENTERED; ITS PEAK IN THE MIDDLE. There are occasions when there is enough knowledge to indicate that the distribution of the frame is of this type. This distribution is also useful when the material may be nearly normal.

PANEL F: NORMAL. One occasionally meets material that is distributed so close to a normal curve that normal theory will suffice. One circumstance is a quality characteristic in a manufacturing plant where use of the Shewhart charts has brought the process into an excellent "state of statistical control." Another example is the distributions of the estimates that come from the thick zones in replicated sampling, which are usually nearly normal, even though the distributions of the relevant populations in the thin zones be badly skewed. Examples of approximations to the normal distribution occur also in radiation, and in the effects of radiation and nuclear bombardment on germination and other biological growth.

> **Remark 2.** The size of sample calculated in advance with the aid of these panels, modified and tempered by judgment, will lead dependably to about the precision desired. I use these panels almost daily.

> **Remark 3.** It is important to note that although σ is always proportional to h, the range of the distribution, the coefficient of variation is not. Therefore, to design a sample that will yield a coefficient of variation $C_{\bar{x}} = $ (e.g.) 3%, we need not know the range h; we need only learn something in advance about the shape of the distribution.

Some hints on the use of Panel C in Fig. 16. None of the panels, except on rare occasions, fits exactly any distribution encountered in practice. Good fits are not necessary. Moreover, in the usual problem, one sets aside for separate treatment all sampling units that have a value greater than some specified cutoff point, as in the inventory on page 153. In other words, one very often stratifies the material. Beyond the cutoff, if the sample is to be 100%, the shape of the curve is of no consequence. Below the cutoff, the shape may be, in the sampling accounts, or of inventories, or of incomes, sales, employment, and many other characteristics, triangular like Panel C. Fig. 6 on page 132 and Fig. 17 illustrate this point with an actual distribution. Note how, when the tail is cut off, the remainder follows the triangle.

Examination of recorded deeds to estimate the average ratio of assessed value to selling price. A legal problem that recurs from time to time is a claim against the county assessor for overassessment of real estate. The legal basis for overassessment on a parcel of land, in some counties, rests on proof that the ratio of the assessed value to the market-value of this parcel lies "substantially" above the average ratio for property sold in the county during a recent year.   Sales of parcels of land are recorded on deeds in the county clerk's office.   Deeds that merely record the transfer of a parcel of land from one member of the family to another, or easements to the power company, with no transfer of money, are not involved (not members of the universe for this study); they are blanks.

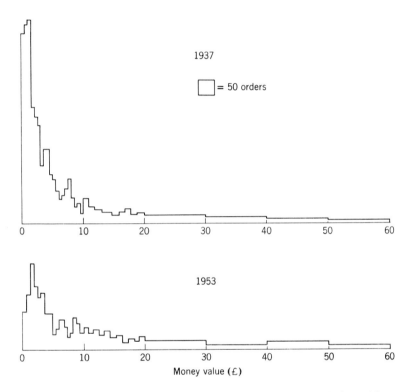

Fig. 17. Examples of money-values of orders in a large industry, on 2 dates, 16 years apart.   From an article by Stanley F. James, "Some sampling problems in connexion with accounting records," *Appl. Statist.*, vol. v, 1956: pp. 86–105.   Reproduced by permission.

A deed covers 2 or 3 pages, sometimes 4 or more pages.   In the illustration for consideration here, photographic copies of the deeds were on file in the County Clerk's office, bound in volumes, with pages running consecutively from 1 to 600, though it was possible for the last page to be 601 or more to finish out the last deed in the volume.   The revenue stamps affixed to a deed gave a satisfactory estimate of the selling-price (decision of the lawyers).

The frame in this case is simple: it is the bound volumes that may contain copies of the recorded deeds dated during the year to be studied. In a recent example,* the year specified was 1 May 1956 to 30 April 1957, and the frame included 4 extra months to allow for most of the stragglers. Any deeds recorded after the cutoff (30 August 1957) were excluded; these deeds constitute the gap in Fig. 2, page 43. The decision on what volumes constituted the frame, the definition of an involved deed, the definition of deeds not involved (blanks), and how to interpret the figures on a deed, were the responsibility of the lawyers.

There was reason to suppose (in the case under discussion) that deeds with a selling price of $75,000 or over could well be out of line with the average ratio; hence the instructions in Step 2 call for segregations of these deeds (called the S-group). These were the "huge and unusual units" (p. 256) that the statistician always enquires about. It was not necessary to search the volumes for them, because they were supposedly all listed in a publication of the real-estate board for the county. There was agreement between the lawyers that this list would be satisfactory. In number, there were about 100 deeds of the S-group. Had any number in the sampling table fallen on an S-deed not on this list, there would have been reason to hold the list under suspicion, but this did not happen. Segregation of the S-group turned out to be a wise precaution, because some of them were of extremely high value, with unusual ratios. In fact, one deed, with selling price $2,750,000, had a ratio of 60%.

The size of sample for the remaining deeds was fairly easy to fix on the basis of prior information. It was known that the actual ratio, though concentrated between 30% and 40%, varied from around 10% up to around 70%, and only in rare instances broke outside these bounds. The S-group ($75,000 and over), however they might vary, could cause no trouble, as they were to be sampled 100%. The nature of the material thus led to the triangle in Panel E of Fig. 16 for an approximation to the required variance. If its range should extend from 10% to 70%, then its variance would be

$$\sigma^2 = \frac{60^2}{24} = 150 \qquad (1)$$

There could be little if any gain from the stratification in zones; hence the size of sample required to deliver a standard error of .5% (client's responsibility) in the estimate of the ratio would be

$$n = \frac{\sigma^2}{.5^2} = 600 \qquad (2)$$

* I am indebted to George Pratt, Esquire, Attorney at Law, of the legal firm Sprague and Stern of Mineola, Long Island, for the privilege of working with him and presenting evidence on a study of this type.

This number does not include the S-group.   The effect of the S-group would be to decrease slightly the variance of the estimate.   On the other hand, one would expect to encounter a few ratios in the sampled group beyond the bounds 10% and 70%, and these would increase the variance. One could say that the 2 opposing forces, both small, would about cancel each other.

The total number of involved deeds in the universe was known to be about 54,000.   Hence the zoning interval for 10 subsamples would be

$$Z = \frac{54,000}{60} = 900 \tag{3}$$

This is the zoning interval in the sampling table (*quod vide*).   The 1st volume in the frame was 6008.   The instructions to the lawyers, for the selection and computation, ran as follows:

1. Define the frame that you would study were the sample 100%. Write instructions that will define an involved deed (symbol D in Step 3), and a blank (symbols B, N, O).   Write instructions on how to read and record the information on an involved deed.

2. Make a list of all the involved deeds with selling price $75,000 or over; show volume and page.   (These deeds take the symbol S in Step 3.)

3. Now use the sampling table (p. 266).   Record the proper symbol, in the column provided, for every page in the sampling table.

D   When the random number falls on the 1st page of an involved deed whose selling price was under $75,000.   The letter D thus indicates a deed that you will study.

B   When the random number falls in an involved deed, but not on the 1st page.

S   When the random number falls in an involved deed whose selling price was $75,000 or over.   Inform the consulting statistician if this deed is not on the list prepared in Step 2.

N   When the random number falls in a deed that would not be involved in this study, even if its date fell within the limits designated.

O   When the random number falls in a deed that satisfies all the criteria for an involved deed, except that the date of the deed is outside the designated limits.

The student should satisfy himself that every deed had a probability of exactly 1 in 90 of coming into the sample, regardless of the number of pages that it covered.

4. Process the deeds with symbol D.   The random number and the subsample must appear as part of the information that you transcribe for every one of these deeds.

5. Compile for each subsample and for all 10 subsamples combined the number of symbols D, S, N, O, B. Submit this summary to the consulting statistician. These numbers will provide tests of compliance with the sampling procedure.

6. Conduct a special investigation on every involved deed for which some of the information is lacking. This effort must be intense.*

7. Find and process all the involved deeds whose market-value is $75,000 or over. This is the S-group. Show the letter S in the space for subsample on the transcription sheet, and leave blank the space for the random number. Compute for this group

A, the sum of all the assessed values

B, the sum of all the selling prices

8. Compile for all the involved deeds in Subsample 1

$X^{(1)}$, the sum of all the assessed values

$Y^{(1)}$, the sum of all the selling prices

$$f^{(1)} = \frac{A + 900X^{(1)}}{B + 900Y^{(1)}} \tag{4}$$

9. Do likewise for Subsamples $2, 3, \ldots, 10$; use superscripts (2), $(3), \ldots, (10)$.

10. Add the assessed values in all 10 subsamples: call this sum X. Add the selling prices in all 10 subsamples: call this sum Y.

11. Compute

$$f = \frac{A + 90X}{B + 90Y} \tag{5}$$

This is the final estimate of the ratio assessed value: selling price. Compute also

$$\hat{\sigma}_f = \frac{f_{max} - f_{min}}{10} \quad \text{[Eq. 25, page 200]} \tag{6}$$

for an estimate of the standard error of f, and take $3\hat{\sigma}_f$ as an estimate of the range of possible variation of the results of repeated samples carried out by the above procedure.

12. Conduct an independent audit of the selection, transcription, and computation of a sample of about 120 deeds, which the consulting statistician will select later on from the main sample.

The results of the survey are in Table 2, whence, by Eq. 6, $\hat{\sigma}_f = .38\%$.

* There was a special investigation of a sample of 1 in 3 of the deeds still not finished by a certain date, as in Chapter 7, but I omit the details, for simplicity. The results were almost 100% successful.

TABLE 1

A PORTION OF THE SAMPLING TABLE FOR A
SAMPLE OF RECORDED DEEDS

Zone	Subsample	Random number	Volume	Page	Symbol
0001–0900	1	555	6008	555	(To be filled
	2	532	6008	532	in as the deeds
	3	028	6008	028	are examined.)
	4	548	6008	548	
	5	883	6009	283	
	6	347	6008	347	
	7	725	6009	125	
	8	782	6009	182	
	9	141	6008	141	
	0	284	6008	284	
0901–1800	1	1218	6010	018	
	2	1408	6010	208	
	3	1773	6010	573	
	etc.				

TABLE 2

RESULTS OF THE STUDY OF RECORDED DEEDS

$A = \$17,921,962$ $\qquad$ $B = \$56,310,207$

Subsample	$X^{(i)}/1000$	$Y^{(i)}/1000$	$f^{(i)}$ (Steps 8 and 9)
1	$198 772	$577 344	.3443
2	168 478	491 898	.3425
3	206 726	649 616	.3182
4	230 475	641 823	.3591
5	183 013	553 441	.3307
6	161 329	462 249	.3490
7	205 194	621 586	.3301
8	164 527	509 143	.3231
9	184 867	550 532	.3358
10	178 144	572 151	.3114
All 10 combined	188 153	562 978	.3342 (Step 11)

Exercise. Show that:

a. $A + 900X^{(i)}$ is the unbiased estimate furnished by Subsample i for the total assessed value in dollars in all the involved deeds in the frame (for practical purposes, all the involved deeds dated within the year specified). $A + 90X$ is the unbiased estimate of this same characteristic furnished by all 10 subsamples combined.

b. $B + 900Y^{(i)}$ is the unbiased estimate furnished by Subsample i for the total selling price in all the involved deeds in the frame. $B + 90Y$ is the unbiased estimate of this same characteristic furnished by all 10 subsamples combined.

The sample of the Census on Japan, 1950. * This will be an example in the sampling of new material. In October 1950 the Japanese took a census of population. It was my privilege to be there during the preceding summer, and to assist in the design of a sample-tabulation of the complete census. The aim was to provide advance tabulations of a number of important characteristics of the people, not for small areas, of course, nor for fine classes, but for Tokyo, possibly Osaka also, and for the country as a whole. The execution and speed of the job, and the results themselves, reflected great credit on the staff of the Japanese Census.

The sampling unit agreed upon was the enumeration district (E. D.). At first thought one might jump to the conclusion that the E. D. would be a crude and inefficient sampling unit. However, there were reasons for the choice:

1. The sampling plan must be simple. There must be no interference with the established procedure for the processing of census returns. These requirements were more important than extreme efficiency.

2. The unit of work in census processing is the E. D., which is never split as it moves from one operation to another; hence the simplest sampling unit would be the whole E. D.

3. This sample could be drawn by 1 man as the returns came in from the field offices. The responsibility for the work was thus not dissipated. Moreover, once the sample E. DD. were put into the stream of processing, they would move onward normally; they require no deviation from the regular census procedures. The coders required no special training in sampling, and the full force might accordingly work on the sample, to speed its completion. In fact, the coders and punchers need not know that they are working on a sample at all.

* I wish to record my indebtedness to Professor Kinichiro Saito of St. Sophia University, Tokyo, and to Professor Akira Asai of Chiba University for the privilege of working with them on this sample.

4. The E. D. in Japan was at that time exceptionally small and uniform, and made a good sampling unit. The Japanese had just completed the enormous task of revising the boundaries of the E. DD. so that they would contain about 50 households, identifiable boundaries being the prime consideration.

5. In Japan, most population characteristics are pretty well scattered. Within an E. D. one may find all sorts and conditions of people—shops and shop keepers, workers at home and workers abroad, small manufacturers, etc. Mining and fishing are an exception; they are concentrated in small areas.

Now let us see what this last paragraph means in terms of the panels on page 260. Mining corresponds to the binomial; an E. D. has either a heavy proportion of miners in it, or it has none. Moreover, unfortunately, the proportion p of E. DD. with miners will be small. For all these reasons the coefficient of variation of the estimated proportion of the population which is miners will be very high. The same thing is true, but to a less extent, with fishermen. There was accordingly no attempt to estimate by a sample the number of men in these 2 occupations.

For the other occupations it seemed reasonable to use the rectangular distribution Panel B on page 260 or possibly the right-triangular distribution Panel C, for which the coefficients of variation are respectively $C = .6$ and $C = .7$.

The sample-size was fixed at 1% replicated in 10 subsamples, each subsample being 1 E. D. drawn at random from every consecutive 1000 E. DD. that contained mostly private households, plus 1 name drawn at random from every consecutive 1000 names in institutional E. DD., such as dormitories. The computations ran along as follows, for the country as a whole.

Total population	80,000,000 (roughly)
Number of households	14,000,000
Number of E. DD. (50 households per E. D.)	$N = 280,000$
Sample of E. DD.	$n = 2800$

$$C_{\bar{x}} = \frac{C}{\sqrt{n}} \quad \text{[Put } C = .7 \text{ and } n = 2800\text{]} \qquad \begin{array}{l}\text{Zoning interval, 1000}\\\text{for 10 subsamples}\end{array}$$

$$= \frac{.7}{\sqrt{2800}} = \frac{.7}{53} = 1.3\% \text{ roughly}$$

This precision was sufficient, and it must be remembered that for many characteristics, especially ratios, the precision would be even better, although for characteristics that were localized, it would not be so good. For Tokyo alone, with a population roughly 1/10th of the whole country, the standard errors would be about treble those calculated above, and for Osaka, about $4\frac{1}{2}$ times those calculated above. Even then, for many ratios, such as proportions in broad age-groups, proportions in the broad occupational groups, proportion male, proportion of in-migrants, the 1% sample would still give useful figures for these cities.

The plan worked well. The number of inhabitants in the sample, when multiplied by 100, gave an estimate of the total number of inhabitants. This estimate agreed with the complete census within 1 part in 800.

The speed of the samples was remarkable. The 1st release of results from the 1% sample for the whole country of Japan was published 7 months following the census. Further releases followed within a few months. The tables from the 1% sample included:

The number of inhabitants by sex and age in 5-year age-groups for urban and rural areas.

Characteristics of the labor force 14 years and over, by 11 broad industry-groups and by 8 broad occupations, by sex, urban and rural.

Characteristics of the labor force, age 10–13 and 14 years and over, by sex, urban and rural.

Marital status, 15 years and over, by sex in 5-year age-groups, urban and rural.

It was easy to calculate the sampling error of any characteristic, as the sample was replicated in 10 subsamples.

The sampling plan was modified for E. DD. that were nearly solid in some characteristic, the chief example being huge dormitories for women. These and certain other institutional E. DD. went into a stream for special processing. The sampling unit for them was the individual person, and the sampling plan was to take 10 persons from every successive 1000 persons. As the probability of selection remained unaltered, the results were additive subsample by subsample to the rest of the sample. The same modification for localized occupations (not used) receives consideration in the exercise below.

Exercise. *a.* Show that the following sampling plan would have given a count of the number of people engaged in fishing and mining.

b. Why not use it?

1. Give a serial number to every household in E. DD. with fishing or mining.

2. Draw 10 random numbers in every 1000 to form 10 subsamples of households. (This would be a 1% sample.)

3. Form for each subsample the total x_i of fisherman.   Form the estimates

$$X_i = 1000x_i \qquad i = 1, 2, \ldots, 10$$

for each of the 10 subsamples.   Do the same for the miners.

Remark 4.   These would be excellent estimates of fishing and of mining, and of any other population.   However, it was necessary in the year 1950 to use a plan that could not possibly clog the regular census procedure.

Incomplete advance knowledge causes no bias.   Inaccurate information impairs the efficiency of a design, and it may lead to too much precision, or to too little precision, but NOT TO BIAS.   One may and should, if the advance estimate of σ is shaky, increase the size of the sample if it is important to meet or to exceed some specified precision.

The advance estimates that the statistician uses represent the statistician's efforts to design an economical or efficient sample.   Whether he is successful or not in designing an efficient sample, the sample is *valid*. The final estimate of the standard error of a result, made from the sample itself, together with tests of the quality of performance (the audit or control, Chapter 5), provide the stamp of quality, good or bad.

Sampling new material sequentially.   A way out of the difficulty of specifying in advance the size of sample required in the sampling of new material is offered by the possibility, when permissible, of laying out the survey, not as a whole, but as 2 interpenetrating portions, the 2d one of which may not be needed.   The procedure is this:

1. Design the sample and specify a size that is almost certainly big enough.

2. Divide the sample by random numbers into 2 portions, 1 and 2.

3. Commence work on Portion 1 only.

4. When Portion 1 is nearly finished, calculate the variance of some of the chief characteristics, and decide whether Portion 1 will deliver the precision desired.   Pay heed that you are not fooled on the variance of Portion 1 (see Remark 6).

5. Proceed into Portion 2 if you need it;   otherwise, finish Portion 1 and close the survey.

The idea of doing a survey in steps is inviting, but the feasibility of doing so is not common.   Such a plan breaks down in crash-surveys that must take place over a given interval of time.   The field-force will thereafter be engaged otherwise, and a 2d portion is impossible.   One may lay out a sample sequentially when the time of completion of the study is to some extent flexible, and when the investigators who will do the work can stay on the job through the 2d portion, if required.   Studies of accounts, and special field-studies in medicine and psychology sometimes offer examples: so do certain studies in hospitals and in clinics.   We saw an example in Remark 3 on page 125.

The fate of Portion 2 may rest on field-costs (or other processing costs), instead of on the need of some specified precision.   One takes a look, as Portion 1 nears completion, at the costs and at the balance in the bank, and decides whether it is possible to carry out Portion 2.   One could, of course, do a random half of Portion 2, if all of Portion 2 would overdraw the account.

Remark 5.   It is important to understand that both Portion 1 and Portion 2 are both valid samples of the whole frame.   Portion 1 is not the first half of the frame, nor Portion 2 the lower half.   Further reading on this subject appears in Stephan and McCarthy's book, *Sampling Opinions* (Wiley, 1958), pages 208 and 398.

Remark 6.   Any estimate of a variance is itself a random variable; consequently, the estimate of the standard error obtained from Portion 1 will occasionally be higher than it really is, and sometimes lower than it really is.   If the estimate of the standard error obtained from Portion 1 is in any instance too low, one will blissfully suppose that what he has is good enough, and that he has no need of Portion 2.   What he accepts, under such circumstances, without knowing it, is a poor estimate of the population of the frame, and a false estimate of the precision of his estimate.

On the other hand, if the estimate of the standard error obtained from Portion 1 is by chance too big, when the precision is actually good enough, one will proceed into Portion 2 without actually needing it.   In this instance he will, after Portion 2 is finished, have a combined estimate of the standard error of the result obtained from both Portions 1 and 2.   He may find out then, too late, that he did not need Portion 2.

It is therefore a good plan to spend a little extra time and money on Portion 1 to examine its variance with care, before one decides that he does not need Portion 2, or that he does.

Hansen, Hurwitz, and Madow (page 79 in Volume I) recommend a coefficient of variation of about 15% or better in the estimate of the standard error from Portion 1, in order not to be misled too often.   This figure is arbitrary, to some extent, but the rule is a good one.   Two thick zones with 10 subsamples will deliver just about the required 15%, provided one forms the thick zones properly (Appendix to Chapter 11, page 223).

Suggested procedure to lay out a sample in 2 subsamples in 2 portions. If the sample is laid out as 2 subsamples, one may use the following procedure to divide the whole sample into 2 interpenetrating halves.

1. Let each successive 4 zones form a block, containing the sampling units A, B, C, . . . , H.

Zone	Subsample 1	Subsample 2
1	A	B
2	C	D
3	E	F
4	G	H

2. Use single digits in a table of random numbers.   Let an odd number designate A, D, F, G for Portion 1; the others for Portion 2.   Conversely, an even number will designate B, C, E, H for Portion 1; the others for Portion 2.   One random number thins both samples over 4 zones.

3. Add blanks at the end to complete the last zone.   A blank if drawn produces no sample.

Remark 7.   The thinning accomplished by the above procedure is balanced over 4 zones.   Every zone will contribute to Portion 1 (hence also to Portion 2).

EXERCISES

1. Establish the variance

$$\sigma^2 = \frac{h^2}{12}$$

for the rectangular distribution in Panel B on page 260.

SOLUTION

The simplest solution is to use calculus.   However, students not familiar with calculus may still obtain the solution, with some interesting by-products.

Start with $2k + 1$ strips, each of width $h/(2k + 1)$.   The variance between the means of the panels is

$$\sigma^2 = \frac{0 + 2(1^2 + 2^2 + \cdots + k^2)}{2k + 1} \left(\frac{h}{2k + 1}\right)^2$$

$$= \frac{2h^2}{(2k + 1)^3} (1^2 + 2^2 + \cdots + k^2)$$

$$= \frac{2h^2}{(2k + 1)^3} \frac{2k^3 + 3k^2 + k}{6} \qquad \text{[See any algebra or the note } infra] \tag{1}$$

$$= \frac{h^2}{12} \left[1 - \frac{1}{(2k + 1)^2}\right] \qquad \text{[By division]} \tag{2}$$

$$\rightarrow \frac{h^2}{12} \text{ as } k \rightarrow 0 \tag{3}$$

Thus, the variance of the original panel of width h must be $h^2/12$.   *Q.E.D.*

The variance $h^2/12$ thus established, it is interesting to calculate σ^2 by another limiting process.   Divide now each of the $2k + 1$ primary strips into $2k + 1$ secondary strips of width $h/(2k + 1)^2$, and divide each secondary strip into $2k + 1$ tertiary strips of width $h/(2k + 1)^3$, etc., ad infinitum.

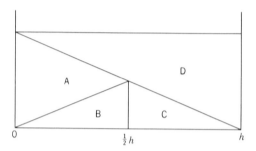

Fig. 18.   Here we break up Panel B of Fig. 16 into triangles, in order to find the variances of the triangles in Fig. 16.

As the width of a secondary strip is $h/(2k + 1)^2$, Eq. 2 tells us that the variance between the secondary strips within a primary strip is

$$\frac{h^2}{12(2k + 1)^2}\left[1 - \frac{1}{(2k + 1)^2}\right]$$

Further, the variance between the tertiary strips within a secondary strip is

$$\frac{h^2}{12(2k + 1)^4}\left[1 - \frac{1}{(2k + 1)^2}\right]$$

And so on.

If we continue through the tth order of strips within strips, the total variance will be

$$\sigma^2 = \frac{h^2}{12}\left[1 - \frac{1}{(k + 12)^{2t}}\right] \rightarrow \frac{h^2}{12} \quad \text{as } t \rightarrow \infty$$

$$Q.E.D. \text{ bis}$$

2. Find the variances of the triangles C and D in Fig. 16.

Hint: One way to do this is to divide the rectangle of Panel B into triangles as shown in Fig. 18.   Let x denote the 2d moment coefficient of the triangle $(A + B + C)$ about the midpoint $\frac{1}{2}h$.   Then x is also the 2d moment of the triangle D about the midpoint $\frac{1}{2}h$.   The distance from the center of gravity of triangle $(A + B + C)$ to the midpoint $\frac{1}{2}h$ is $\frac{1}{6}$th of h; likewise for triangle D.   Therefore

$$x + (\tfrac{1}{6}h)^2 = \tfrac{1}{12}h^2$$

whence

$$x = \tfrac{1}{18}h^2$$

as recorded for Panels C and D in Fig. 16.

3. Find the variance of triangle E in Fig. 16.

Hint: Let x be the variance of triangle E in Fig. 16. This triangle breaks into triangles B and C in Fig. 18, for either of which we now know that the variance is $\frac{1}{18}(\frac{1}{2}h)^2$. It follows that

$$x = \tfrac{1}{18}(\tfrac{1}{2}h)^2 + (\tfrac{1}{6}h)^2 = \tfrac{1}{24}h^2$$

as recorded for Panel E in Fig. 16.

4. Compute a numerical approximation to the variance of the triangle shown as Panel C in Fig. 16 by (*a*) dividing it into 3 strips and finding the

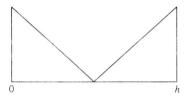

0   h

Fig. 19. The V-shaped distribution of Exercise 5.

variance between the midpoints of the 3 strips. Then (*b*) divide it into 5 strips and find the variance between the midpoints of the 5 strips. (*c*) Plot the 2 variances on the vertical axis, against $1/3^2$ and $1/5^2$ respectively, and then extrapolate by a line to 0—i.e., to strips of infinitesimal width. The result will be an excellent approximation to $h^2/18$. (Use of 3 points will give a curve with small curvature, and presumably a more accurate result.)

Note: The aim here is to teach the student a useful method by which to approximate the variance of any distribution. The main feature of this method is to plot the rough variances against the square of the width of the class interval. This is an application of the Euler-Maclaurin remainder term, known for 200 years. It is far superior and easier than the methods so often taught today in courses and books on mechanics, physics, and college algebra for approximating areas and integrals by summations. For a reference, see Whittaker and Robinson, *Calculus of Observations* (Blackie, 1925): p. 140.

5. The variance of the V-distribution in Fig. 19 is $h^2/8$.

I omitted this distribution from Fig. 16 because it is unusual. One may almost always avoid it by breaking it in the middle to form 2 strata.

6. Compute β_2 for the rectangle in Panel B. *Answer:*

$$\beta_2 = 1.8$$

β_2 is defined on page 440.

SOLUTION

Proceed as in Exercise 1. Let μ_4 be the 4th moment coefficient about the middle. Then

$$\mu_4 = \frac{0 + 2(1^4 + 2^4 + \cdots k^4)}{2k + 1}\left(\frac{h}{2k + 1}\right)^4$$

$$= \frac{2h^4}{(2k + 1)^5}(1^4 + 2^4 + \cdots k^4)$$

$$= \frac{2h^4}{(2k + 1)^5}\left(\frac{k^5}{5} + \frac{k^4}{2} + \frac{k^3}{3} - \frac{k}{30}\right) \qquad \text{[See any algebra or the note \textit{infra}]}$$

$$\to \frac{h^4}{80}$$

whence

$$\beta_2 = \frac{\mu_4}{\sigma^4} = \frac{h^4}{80}\frac{12^2}{h^4} = 1.8$$

Note on the sums of powers. The sum of k natural numbers all raised to the power r is given by the following formula

$$1^r + 2^r + 3^r + \cdots + k^r = \frac{k^{r+1}}{r + 1} + \frac{k^r}{2} + \frac{r}{12}k^{r-1}$$

$$- \frac{r(r - 1)(r - 2)}{720}k^{r-3} + - \cdots$$

which James Bernoulli derived from the binomial coefficients about 1704. His derivation is in his posthumous *Ars Conjectandi* (1713). See Whittaker and Robinson, *Calculus of Observations* (Blackie, 1924): p. 138; or the author's *Some Theory of Sampling* (Wiley, 1950): p. 403.

CHAPTER 15

Stratified Sampling

The practical man is the man who practices the errors of his forefathers.—Thomas Huxley.

A. PURPOSE AND GENERAL REMARKS

The purpose of stratification. In most of the sample-designs that we have studied so far, the material fell into zones just as it came to us in the frame, without rearrangement. There were some painless exceptions, as in Chapter 8, Section B where we set off the big accounts (those of $10,000 and over) into a separate stratum for separate treatment, and in Chapter 11 where we separated a national sample into metropolitan and nonmetropolitan areas, for efficiency and for separate tabulations. We had also faced the fact in Chapter 8, Section A that, had the accounts of Class A shown a strong tendency to occur in bunches, some sort of segregation of pages or of accounts would have been necessary. This observation was in line with Chapter 14, which dealt with the sampling of new material, where one of the main points of emphasis was the necessity to discover in advance the existence of huge sampling units and to segregate them for special treatment.

These suggestions for segregation of sampling units arose mostly from need of protection against huge sampling errors and against failure to get a good measure of the sampling error. We saw also, however, the suggestion in Chapter 11 that the order in which one lists the sampling units in the frame may affect the efficiency of the sampling plan, but we encountered no theory there that would tell us whether this suggestion is important, nor how big the samples should be from the various portions of the frame in order to achieve the best efficiency. The aim of the present chapter is to fill in some of the deficiency. What we need to learn is something about stratified designs. This will be fairly easy with the aid of theory and examples. Although the theory by which to compare the variances of some of the plans may appear complicated, we may take comfort in the fact that any plan, if we carry it out correctly, gives valid results, even though it be not the most efficient one possible under the circumstances.

Stratification means separation of the material into layers or classes of similar characteristics, with a separate sample from each class; although we may sometimes draw the sample first and then form classes in the sample only. Stratification has in fact many meanings and many ways of application.

Stratification is one way to use statistical information in our possession to acquire a better estimate than would be possible otherwise. The use of ratio-estimates, learned in previous chapters and to appear again, is sometimes another way to improve an estimate (p. 424). The statistical information that we use in a stratified design may already exist for all the sampling units (as from the last census), or it may require preliminary tests or interviews of some of the sampling units.

We shall start off with Plan A, which is the plan that we have been using thus far. The only layers or strata in Plan A are the zones themselves. Plan A captures the natural stratification that exists in the frame as it comes. As we shall see, Plan A is, in some circumstances, about the best plan of all. We shall now arm ourselves, however, with theory and examples by which we may be fairly sure when to introduce more powerful methods when the gain in precision would be worth the effort.

See page 506 for a numerical calculation of the natural stratification in a frame.

Another advantage of stratification, often overlooked, is that stratification decreases the 4th moment coefficient as well as the variance of $\bar{x}$, with a resultant improvement in the estimate of Var $\bar{x}$ (see page 441).

Description of plans.

PLAN A. Sample the frame as it is, with no rearrangement. The zones capture the natural stratification in the frame.

THE PROPORTIONS P_i KNOWN

Classify the whole frame

PLAN B. Classify all the sampling units of the frame. Then use proportionate allocation.

PLAN C. Classify all the sampling units of the frame. Then use Neyman allocation.

Classify only the n sampling units of the sample

PLAN D. Draw as in Plan A n sampling units, and classify them into the various strata. Use the entire sample; no thinning.

PLAN E. Draw as in Plan A n' sampling units, and classify them into the various strata. Thin the strata to reach the Neyman ratios $n_i : n_j$.

Choose the number n' so that with as little thinning as possible, the total sample will turn out to be about equal to the desired size n.

Classify only enough sampling units to fill the quotas n_i

PLAN F.   Fix the sample sizes n_i by proportionate allocation, as in Plan B.   Then draw sampling units one by one from the frame and classify them into the various strata until all the quotas n_i are full.   Discard any sampling unit that belongs to a stratum whose quota is full.   Form the estimates as in Plan B.

PLAN G.   Fix the sample-sizes n_i by Neyman allocation, as in Plan C. Proceed otherwise as in Plan F, but form the estimates as in Plan C.

PROPORTIONS P_i NOT KNOWN

*Classify only a preliminary sample of size N' to
estimate the proportions P_i*

PLAN H.   Draw as in Plan A a preliminary sample of fixed size N'. Classify only these sampling units, not the whole frame.   Then thin proportionately the samples from the various strata to reach a total specified size.   Form the estimate X as directed later.

PLAN I.   Draw as in Plan A a preliminary sample of fixed size N', and classify only the sampling units in this preliminary sample, not the whole frame.   Then thin the samples from the various strata by the use of ratios dictated by the Neyman allocation to reach a total specified size.   Form the estimate X as directed later.

Any of these plans may be used to estimate the proportions of individuals in classes finer than the original strata;   and these ratios may often then be used advantageously to form estimates of totals.

> **Remark 1.**   Plan A (no stratification beyond the natural stratification already in the frame) is sometimes the best plan of all, because of its simplicity, and because the gain of any stratified plan may not be worth its cost. One may commence on Plan A as soon as the frame is ready, without waiting to draw up, prepare, and test a stratified sample.

> **Remark 2.**   The theory of variances that we shall study in Chapter 17 is the theory for Plan A.   It is also the theory for the variance that any one stratum contributes to the total variance in any plan of stratified sampling.

> **Remark 3.**   The reader may have been aware that the zones in the replicated method that we used in earlier chapters are strata, and that we drew proportionate samples from all zones.   (This really meant samples of equal size from all zones, because the zones all contained the same number of sampling units.)

Aim of studying the theory of stratified sampling.   The decision on which plan to use will depend on comparisons of costs, on convenience, smoothness, and speed. Theory and experience enable one to make a sensible choice of plan. A little theory will conserve funds and will often save a vast amount of experimentation, because theory points out what it is that we need to know in order to decide which plan to use.

One does not plunge into expensive stratified sampling without knowing why. The gains of stratification are often disappointing; yet sometimes they are so great that it would be a tragedy to fail to stratify. It is important to know when to stratify and when not to. Fortunately, it is possible by the aid of theory and with some ideas about costs, and about the standard deviations σ, σ_w, $\bar{\sigma}_w$, σ_b, to make some quick calculations that will point to the right plan, and will foretell approximately what precision any plan will give, and what it will cost. The equations for the variances, and the table of costs on page 350, will facilitate such calculations.

In my own practice I prefer to use Plan A unless the gain of rearrangement would almost certainly be 10% or more. One must gain enough by rearrangement to offset the extra costs thereof, and to leave a neat profit.

One should be cautious about the Neyman allocation unless it will clearly be better than other plans. Neyman allocation requires some advance approximation to $\bar{\sigma}_w/\sigma_w$ (symbols to be defined later.) The wrong advance information may lead well-intentioned plans to higher variance than proportionate allocation would yield (p. 295). Neyman allocation also requires different sampling ratios in the different strata, and separate estimates for each stratum.

One survey in the sampling of human populations usually provides estimates of many characteristics. There is often 1 chief characteristic whose precision must meet a certain requirement, and such that if we meet this precision, the precision of the remaining characteristics will be sufficient or more than sufficient. This is true, for example, when our sample must estimate the total number of inhabitants with a coefficient of variation of 3%. The proportion male, or the proportion of men 20–39 in the labor force, will automatically have greater precision.*

If, however, the characteristics are competitive, and none should have preference, one will usually find optimum satisfaction by the use of proportionate stratification in some form (Plans B, D, F, H). Attempts to use Neyman allocation in some form (Plans C, E, G, I) may then bring disappointment in some characteristics, and little gain in any of them. Sampling for income, or the sampling of business establishments for sales and employment are something else again: it would be very risky not to

---

* *A Chapter in Population Sampling* (Bureau of the Census, 1947: Superintendent of Documents, Washington 25): pp. 14 and 104.

consider Neyman allocation in such problems.   The reader may profitably turn to Hansen, Hurwitz, and Madow's *Sample Survey Methods and Theory*, Vol. I (Wiley, 1953): Ch. 5, Sec. 13, p. 224 ff.; also p. 302.

Plans D to I (which call for classification of only a sample) are especially suited to studies in which the number of sampling units in the frame is huge, such as 25,000 or 100,000 or more: to classify every unit in the frame is then costly and time-consuming, even when the cost is only a few cents per unit.   However, we must sometimes do it.   We saw an example in Chapter 7 where we classified every interline abstract into 2 groups—(*a*) $10,000 and over; (*b*) under $10,000.   This was necessary because of the appearance here and there of extremely large accounts.   Plan A or Plan D, for example, would have been impossibly inefficient.   Plan C was the right plan.

For frames that contain 3000 sampling units or less, a calculation will often show that it would be better to classify the entire frame, and to use Plan B, C, or A.

Chapter 20 treats the problem of where to demarcate the strata to the best advantage.   Some helpful points on what characteristics may be the important criteria for stratification in the sampling of human populations, and what gains to expect, appear in Hansen, Hurwitz, and Madow's *Sample Survey Methods and Theory*, Vol. I (Wiley, 1953): p. 381 ff. and p. 304 ff.

Ratios of variances and costs sufficient for many purposes.   There are certain important simplifying features in the choice of plan.   For example, to compare one plan with another, one needs not the absolute values of σ, σ_w, $\bar\sigma_w$, but only the ratios of $\bar\sigma_w$ to σ_w or to σ (definition on page 285).   One needs not the costs c_1 and c_2, but only the ratios $c_1 : c_2$.   Moreover, as this ratio only occurs under the root sign, a rough approximation to the ratio will suffice.

One must still face the fact, though, as in unstratified sampling, that he must make advance estimates of the absolute values of the variances and of the costs in order either (*a*) to estimate the precision attainable for a given cost, or (*b*) to meet a prescribed precision.

Some common misconceptions in regard to stratification.   A real universe is dynamic, not static, and any information that one uses for classification of a sampling unit is always to some extent out of date.   Moreover, in any real survey a few blunders may occur; they ought not to, but they do, and a unit will occasionally go into a stratum where no one intended it to go.   A small amount of misclassification is thus to be expected as the natural course of events.   Misclassification does not cause bias: the sample is self-correcting.

There may nevertheless be definite impairment of efficiency if the sample contains areas of heavy growth, or contains other jack pots and surprises, as when some rare but extremely huge account escaped detection and fell into the stratum of small accounts.   Moreover, the high value of β_2 resulting from the tail of the distribution decreases the reliability of our estimate of the precision.   All this we noted in Chapter 14.

A simple example of a random allocation.   If we define strata in any manner, as by geographic location, size of city, proportion colored, etc., but proceed to draw the sample at random from the entire frame, without regard to the strata, the sample-sizes that fall into the various strata will be random variables.

Let us look at a simple illustration in 2 strata.   We take 100 squares, numbered 01 to 00.   Stratum 1 consists of squares 01 to 70, and Stratum 2 consists of the remaining 30 squares numbered 71 to 00.   Let us draw without replacement a sample of 20 squares from the 100 squares (the frame), to see how they distribute themselves between the 2 strata.   If each stratum contributed its proportionate share to the total sample, then:

Stratum 1 would contribute 14 squares
Stratum 2 would contribute 6 squares

Total   20 squares

Now let us see how the samples distribute themselves in one trial.   We open our table of random numbers (Kendall and Smith) to the 23d Thousand, Columns 23 and 24, line 2 (where I had stopped a few days ago on a sampling job), and read out these numbers:

83	39	66*	91	79
88	64	24	57	
93	33	89	31	
66	13	49	48	
74	14	81	84	

These random numbers struck 11 squares in Stratum 1, and 9 squares in Stratum 2.   So we may write, *for this one trial,*

$$n_1 = 11$$
$$n_2 = 9$$

$$n = 20 \quad \text{(fixed)}$$

Remark 4.   The number with the asterisk (*) in the above array of random numbers is a duplicate of some previous number.   As the procedure specified was to sample without replacement, we skip the duplicate (*) and move on to the next number in the table.

Exercise. Show that in repetitions of the foregoing experiment,

$$E\hat{n}_1 = nP_1 = 14$$

$$E\hat{n}_2 = nP_2 = 6$$

where $P_1 = .7$, $P_2 = .3$, and $n = \hat{n}_1 + \hat{n}_2$. Show also that

$$\sigma_{\hat{n}_1} = \frac{N - n}{N - 1} \sqrt{nP_1P_2} = 1.84 \qquad \text{[Eq. 9 in Chapter 17, page 405]}$$

$$\sigma_{\hat{n}_2} = \text{the same} \qquad \text{[Why the same?]}$$

$$\sigma_n = 0 \qquad \text{[Why?]}$$

$$C_{\hat{n}_1} = \frac{\sigma_{\hat{n}_1}}{E\hat{n}_1} = \frac{1.84}{14} = .13$$

$$C_{\hat{n}_2} = \frac{\sigma_{\hat{n}_2}}{E\hat{n}_2} = \frac{1.84}{6} = .31$$

where C denotes coefficient of variation.

The departures of n_1 and n_2 from their expected values are not significant. For n_1,

$$t = \frac{14 - 11}{1.84} = 1.6$$

For n_2,

$$t = -1.6$$

The experiments that we have just performed will serve as an illustration of Plan D.   Let the 100 squares be 100 sampling units, and let the number of the square denote its population.   Thus, 17 shall be the population of sampling unit No. 17, 37 shall be the population of sampling unit No. 37, etc.   The average population per sampling unit in the frame of 100 sampling units is then

$$a = \frac{1}{100} (01 + 02 + \cdots + 99 + 100) = 50.5 \qquad (1)$$

We drew 11 sampling units from Stratum 1.   Their populations are, in order, 66, 39, 64, 33, 13, 14, 24, 49, 57, 31, 48, whose sum is 438; wherefore the average population per sampling unit in Stratum 1

$$\bar{x}_1 = \frac{438}{11} = 39.8 \qquad (2)$$

Likewise, the average population of the 9 sampling units that we drew from Stratum 2 is

$$\bar{x}_2 = \frac{762}{9} = 84.7 \qquad (3)$$

Then our estimate of a is

$$\bar{x} = P_1\bar{x}_1 + P_2\bar{x}_2$$

$$= .7 \times 39.8 + .3 \times 84.7 = 53.3 \qquad \text{[Plan D]} \qquad (4)$$

How does Plan D differ from Plans B and F? If we had used Plan B, we should have classified in advance the 100 sampling units in the frame, and then drawn exactly 14 sampling units from Stratum 1, and 6 sampling units from Stratum 2. To use Plan F, we should compute first of all the sample-sizes 14 and 6, and then continue to draw from the frame until we reach the quota of 14 sampling units from Stratum 1 and 6 sampling units from Stratum 2. The estimates in Plans B, D, and F are all similar: compute $\bar{x}_1$ and $\bar{x}_2$ and use the equation above with the known values of P_1 and P_2.

The result of Plan A (no stratification at all) would be

$$\bar{x} = \frac{438 + 762}{11 + 9} = \frac{1200}{20} = 60 \qquad (5)$$

to compare with $a = 50.5$, and with the estimate in Eq. 4.

Description of Plan A, for comparison with other plans.   Material offered for sampling is never thoroughly mixed; it is always stratified to some extent, just as it comes to us in the frame. This natural stratification is automatic and costs nothing. We can not be sure in the case of new material just how effective the natural stratification is, but Plan A, as we use it, will capture any possible gain from this source. We may think of Plan A as a proportionate stratified sample, where the strata are zones in the frame, just as it comes. Thus, when we speak of stratified sampling in this chapter, we really mean the theory by which to decide whether any rearrangement of the sampling units in the frame would be worth the cost of the rearrangement. We proceed now to describe Plan A, and then to study the plans of stratified sampling.

PLAN A

1. Decide with the help of Eq. 18 ahead the size n of the sample desired. (The form $N/n = N(\sigma_{\bar{x}}/\sigma)^2 + 1$ as given on page 388 may be handier, as it leads directly into step 2.)

The symbol σ^2 in Eq. 18 is supposedly adjusted for the natural stratification already in the frame. In other words, σ^2 in Eq. 18 is the average variance between sampling units *within zones*. In the case of new material (Ch. 14), it is wise to be conservative and to assume that there will be little or no gain from the natural stratification; i.e., to take σ^2 as the total variance of the frame.

2. Compute the zoning interval $Z = 2 N/n$ for 2 subsamples, or $10 N/n$ for 10 subsamples. Form zones of Z consecutive sampling units in the frame. (Illustrations in earlier chapters.)

3. Read out random numbers between 1 and Z to draw 1 sampling unit from every zone.   These drawings form Subsample 1.   Draw another sampling unit from every zone for Subsample 2;   and likewise for the other subsamples if any.   Mark each sampling unit to show which subsample it belongs to, and which zone it came from.

4. Carry out the interviews or the tests on the entire sample.

5. Form the x-population by subsample.

6. Form from Subsample i the estimate

$$X^{(i)} = Zx^{(i)} \tag{6}$$

where $x^{(i)}$ is the x-population in Subsample i.

7. Form the final estimate

$$X = \text{av. } X^{(i)} = Z\bar{x} \tag{7}$$

from all subsamples combined. $\bar{x}$ is the average of the individual $x^{(i)}$, where $x^{(i)}$ is the x-population in Subsample i.   For a ratio f, there will be also individual $y^{(i)}$ and the final estimate Y of the total y-population, and we take $f = X:Y$ or $\bar{x}:\bar{y}$, although this estimate is in many problems not significantly different from the average of the separate estimates $f^{(i)} = x^{(i)}:y^{(i)}$.

8. Estimate the precision attained.   We have seen many examples in previous chapters.   If there are 10 subsamples and 1 thick zone, take

$$\hat{C}_X = \hat{C}_{\bar{x}} = \frac{1}{10\bar{x}} [x^{(i)}_{\max} - x^{(i)}_{\min}]$$

$$= \frac{1}{10X} [X^{(i)}_{\max} - X^{(i)}_{\min}] \qquad \text{[By the range]}$$

Use Eq. 2 in Chapter 11 (p. 197) if you prefer the sum of squares.

Ten thick zones give 9 degrees of freedom.   This is excellent in most work, but if there is need for more degrees of freedom, one may approximately double the degrees of freedom by forming 2 thick zones systematically, as the appendix to Chapter 11 explains (p. 223).

Two subsamples give 1 degree of freedom for each thick zone.   We may build up degrees of freedom here also, following the appendix to Chapter 11.

The ratio f may usually follow the same treatment.   Thus, if there are 10 subsamples, the estimate

$$\hat{C}_f = \frac{1}{10f} [f^{(i)}_{\max} - f^{(i)}_{\min}]$$

is usually excellent.   We have used it many times in earlier chapters.   If you prefer the sum of squares, use Eq. 14 in Chapter 11 for 10 subsamples, or Eq. 9 for 2 subsamples.

B. STRATIFICATION BEFORE SELECTION

Some notation and definitions in the frame. Certain relations between the statistical characteristics of the frame and of the several strata will appear over and over in the theory that is about to follow, and it will save time to introduce first of all some definitions.

$$a = P_1 a_1 + P_2 a_2 + P_3 a_3 = \frac{A}{N} \qquad \begin{array}{l}\text{the average population} \\ \text{per sampling unit}\end{array} \qquad (8)$$

$$\bar{\sigma}_w = P_1 \sigma_1 + P_2 \sigma_2 + P_3 \sigma_3 \qquad \begin{array}{l}\text{the weighted average} \\ \text{standard deviation} \\ \text{within strata}\end{array} \qquad (9)$$

$$\sigma_w^2 = P_1 \sigma_1^2 + P_2 \sigma_2^2 + P_3 \sigma_3^2 \qquad \begin{array}{l}\text{the weighted average} \\ \text{variance within strata}\end{array} \qquad (10)$$

$$\sigma_b^2 = P_1(a_1 - a)^2 + P_2(a_2 - a)^2 + P_3(a_3 - a)^2$$

$$= P_1 a_1^2 + P_2 a_2^2 + P_3 a_3^2 - a^2 \qquad \begin{array}{l}\text{the variance} \\ \text{between strata}\end{array} \qquad (11)$$

$$\sigma^2 = \sigma_b^2 + \sigma_w^2 \qquad \qquad \text{the total variance} \qquad [\text{p. 287}] \quad (12)$$

It is helpful to see these definitions arrayed as operations in Table 1.

Table of estimates and of variances. Table 2 shows the layout of the estimates by strata, and the variances of these estimates. It is important to note that: (1) addition of the separate estimates X_1, X_2, X_3 will give the estimate X for the total population A of the frame; (2) addition of the separate variances will give Var X only if the sample-sizes (n_i) be fixed in advance, as in Plans B, C, F. For Plans D, E, and G, there is an additional term in the variance (*vide infra*), owing to the fact that the sample-sizes (n_i) are random variables, negatively correlated. In Plans H and I, there is still another term for the variance of the estimates of P_1, P_2, P_3.

We proceed on the assumption that we have decided in advance the definitions of the several strata, whether we are going to use stratified sampling or not. Each plan of stratified sampling that we shall propose and study will have its own formula for Var X, which the statistician will use in the planning stages, inserting his advance estimates of σ_w, or of $\bar{\sigma}_w$;

TABLE 1

NOTATION AND DEFINITIONS FOR THE FRAME

Extension to more strata will be obvious. M is the number of strata (here $M = 3$)

Stratum	Number of sampling units		Stratum's proportion of sampling units in the frame	Population		Between the populations of the sampling units within the stratum	
	In the frame	In the sample		Average per sampling unit in the stratum	Total in the stratum	Standard deviation	Variance
1	N_1	n_1	$P_1 = \dfrac{N_1}{N}$	a_1	$A_1 = N_1 a_1$	σ_1	σ_1^2
2	N_2	n_2	$P_2 = \dfrac{N_2}{N}$	a_2	$A_2 = N_2 a_2$	σ_2	σ_2^2
3	N_3	n_3	$P_3 = \dfrac{N_3}{N}$	a_3	$A_3 = N_3 a_3$	σ_3	σ_3^2
Total for the frame	N	n	1	xxx	A	xxx	xxx
Unweighted average per stratum	$\bar{N} = \dfrac{N}{M}$	$\bar{n} = \dfrac{n}{M}$	$\dfrac{1}{M}$	xxx	$\bar{A} = \dfrac{A}{M}$	xxx	xxx
* Weighted average per sampling unit	xxx	xxx	xxx	$a = \dfrac{A}{N}$	xxx	$\bar{\sigma}_w$	σ_w^2

* The weights are P_1, P_2, P_3.

and for some plans, σ_b, and possibly also a rough value of σ_R (p. 323). The unit costs c_1 and c_2 of classification and of interviewing will appear in the cost (Table 15, p. 350), and they will show us in Plans H and I what is the optimum preliminary sample N'.

EXERCISES

1. Show that the 2 forms for σ_b^2 on the right-hand side of Eq. 11 are mere algebraic rearrangements of each other (algebraic identities).

2. Define σ^2 in the usual way as the variance between all N sampling units in the whole frame. Show that it is identically equal to $\sigma_b^2 + \sigma_w^2$ as in Eq. 12.

Proportionate allocation (Plan B). We proceed now to treat Plan B, then Plan C. We assume in these 2 plans that the strata have been formed, and that the sampling units in Stratum 1 have their serial numbers from 1 to N_1, those in Stratum 2 from 1 to N_2, etc.

It is advisable to maintain, in each stratum, the order in which the sampling units appeared in the original frame. By doing so, we capture the benefit of the natural stratification that already existed. This practice is not essential to the theory, as the symbols don't care anything about order. It is essential, however, for efficient design.

TABLE 2

NOTATIONS AND DEFINITIONS FOR THE SAMPLE

Stratum	Population in the sample	Mean population per sampling unit	Estimated total population	Variance of this estimate
1	x_1 x-population in Stratum 1	$\bar{x}_1 = \dfrac{x_1}{n_1}$	$X_1 = N_1 \dfrac{x_1}{n_1}$	Var X_1
2	x_2 x-population in Stratum 2	$\bar{x}_2 = \dfrac{x_2}{n_2}$	$X_2 = N_2 \dfrac{x_2}{n_2}$	Var X_2
3	x_3 x-population in Stratum 3	$\bar{x}_3 = \dfrac{x_3}{n_3}$	$X_3 = N_3 \dfrac{x_3}{n_3}$	Var X_3
Sum	x	xxx	X^*	Var X†

* In proportionate sampling, Plan B, the estimate X reduced to Nx/n; see Eq. 16 on page 289.

† The variances are additive only if the N_i (or P_i) are known and used in the estimate X.

The reader will in practice discover his own short-cuts to the directions that I shall give for the various plans. The directions here show the principles in the steps, not the details of performance, which will vary with local conditions and preferences.

PLAN B

1. Decide with the help of Eq. 21 ahead the size n of the sample desired. The procedure is to make the selections from every stratum as in Plan A.

2. Compute the zoning interval $Z = 2\,N/n$ for 2 subsamples, $10\,N/n$ for 10 subsamples. Form zones of Z sampling units in all strata. (One may prefer to adjust the number of sampling units so that there is no incomplete zone at the end of a stratum; see page 191.)

3. Read out random numbers between 1 and Z to draw 1 sampling unit from every zone, onward through all strata. These drawings form Sub-sample 1. Draw another sampling unit from every zone for Subsample 2, and likewise for the other subsamples if any. Mark each sampling unit to show which subsample it belongs to, and which stratum, and which zone it came from.

4. Carry out the interviews or the tests on the entire sample.

5. Form the x-populations by subsample: designate them $x^{(1)}$, $x^{(2)}$, etc., where $x^{(i)}$ is the x-population in Subsample i through all strata.

6. Form from Subsample i the estimate

$$X^{(i)} = Zx^{(i)} \qquad \text{[As in Plan A]}$$

7. Form

$$X = \text{av.}\ X^{(i)} = Z\ \text{av.}\ x^{(i)} \qquad \text{[As in Plan A]}$$

for the final estimate from all subsamples combined. This will be merely equivalent to Eq. 16 ahead, as $Z = 2\,N/n$ for 2 subsamples, $10\,N/n$ for 10 subsamples.

8. Estimate the precision obtained. For 10 subsamples, there will be the 10 individual x-populations $x^{(i)}$ for $i = 1, 2, \ldots, 10$, and one may proceed to calculate $\hat{C}_X$ or $\hat{C}_{\bar{x}}$ exactly as in Step 8 of Plan A (p. 284). One may even form 2 thick zones systematically to double the degrees of freedom, if he wishes. No weighting is required, because all the sampling units in the frame, regardless of stratum, had in proportionate stratified sampling (Plan B) the same probability of selection. If we wish to form separate estimates by stratum, we may do so; and we may add the estimates of X and of Var X, as in Table 2, page 287. The degrees of freedom are, however, not additive, and the 2 thick zones will produce something less than 18 degrees of freedom, depending on how the variances in the 2 strata compare (see page 223 of the appendix to Ch. 11).

A ratio $f = X:Y$ introduces no new problems (see Step 8 in Plan A, p. 284).

Remark 1.   Two subsamples give 1 degree of freedom for each thick zone, and we may build up degrees of freedom here also, following the appendix to Chapter 11.

Formulation of the gains in proportionate sampling.   We shall now formulate analytically the difference between no stratification and proportionate stratification (Plans A and B).   In the first place, we need a mathematical definition for proportionate sampling, which will be this:

$$n_1:n_2:n_3:\bar{n}:n = N_1:N_2:N_3:\bar{N}:N \quad \text{[Plan B]} \quad (13)$$

or

$$\frac{n_1}{N_1} = \frac{n_2}{N_2} = \frac{n_3}{N_3} = \frac{\bar{n}}{\bar{N}} = \frac{n}{N} \quad (14)$$

In still another form,

$$n_i = n\frac{N_i}{N} = N_i\frac{\bar{n}}{\bar{N}} \quad (15)$$

We recall that in Plan A we formed the estimate X for the total population A of the frame by the equation

$$X = N\bar{x} = N\frac{x}{n} \quad \text{[x is the x-population in the entire sample]} \quad (16)$$

We recall also that if the probabilities for all the sampling units are equal, then X will be an unbiased estimate of A, the total population of the frame (not to be confused with Plan A).   In symbols,

$$EX = A \quad \text{[The total population of the frame]} \quad (17)$$

The variance of X for Plan A is

$$\text{Var } X = \sigma_X{}^2 = N^2\frac{N-n}{N-1}\frac{\sigma^2}{n} = N^2\left(1 - \frac{n}{N}\right)\frac{\sigma^2}{n} \quad \text{[Plan A]} \quad (18)$$

For the total population A of the frame we may form for any type of stratified sample the estimate*

$$X = X_1 + X_2 = \frac{N_1x_1}{n_1} + \frac{N_2x_2}{n_2} \quad \begin{array}{l}\text{[x_1 and x_2 are the x-populations}\\ \text{in the samples from Strata 1 and}\\ \text{2.   Extension to more strata is}\\ \text{obvious]}\end{array} \quad (19)$$

* The symbols x_1 and x_2 have had different meanings in some previous chapters where the subscript i referred to Subsample i.   We shall need in this chapter a notation to indicate the stratum as well as the subsample.   From now on, the subscript will denote the stratum, and a superscript in parenthesis will denote the subsample.   For example, $\bar{x}_1^{(2)}$ will denote the average x-population in Subsample 2 taken from Stratum 1. A total over all strata (e.g., $X^{(i)}$) will have no subscript, and a total over all subsamples (e.g., X) will have no superscript.

Now if the numbers N_1 and N_2 are known, and if the sample-sizes n_1 and n_2 are independent of each other, the variances in the 2 strata will be additive, and

$$\text{Var } X = \text{Var } X_1 + \text{Var } X_2 \quad [\text{Each term will follow Eq. 18}]$$

$$= N_1^2\left(1 - \frac{n_1}{N_1}\right)\frac{\sigma_1^2}{n_1} + N_2^2\left(1 - \frac{n_2}{N_2}\right)\frac{\sigma_2^2}{n_2} \tag{20}$$

Further, if the sampling is proportionate, we replace n_1 by nP_1 and n_2 by nP_2 in Eq. 19, which thereupon reduces to Eq. 16, so that the same formula for X serves both Plan A and Plan B.   With the same replacement for n_1 and for n_2, Eq. 20 reduces to

$$\text{Var } \bar{x} = \left(1 - \frac{n}{N}\right)\frac{P_1\sigma_1^2 + P_2\sigma_2^2}{n}$$

$$= \left(1 - \frac{n}{N}\right)\frac{\sigma_w^2}{n} \quad [\text{Proportionate sampling, Plan B}] \tag{21}$$

Remark 2.   The computation for X is the same for Plan B as for Plan A, as already observed.   We merely pool the x-populations from the several strata to form x; then multiply x by N/n to form the estimate $X = xN/n$, which is Eq. 16.   A proportionate sample is what we call a self-weighted sample.

Remark 3.   The student will perceive, when we come to it, that Neyman sampling is not self-weighting; the computer can not pool the samples from the several strata: he must first form the estimates X_1 and X_2 separately; then add them.

Remark 4.   Eq. 21 for the variance of a proportionate sample is identical in form with Eq. 18 for the variance of Plan A, except that Eq. 21 contains σ_w^2 where Eq. 18 contains σ^2.

Remark 5.   The difference between σ^2/n for Plan A (no additional stratification) and σ_w^2/n for Plan B (proportionate allocation) is $(\sigma^2 - \sigma_w^2)/n = \sigma_b^2/n$.   As σ^2 is independent of the arrangement of the sampling units in the zones, the differences between the means of the strata must be wholly responsible for the difference between the variances of Plan A and of Plan B.   See page 298 for further comparisons.

Exercise.   Prove that in proportionate sampling (Plan B), X is an unbiased estimate of A, the total x-population in the frame.   In symbols,

$$EX = A$$

Neyman allocation to strata (Plan C).   One may be able in some problems to improve on proportionate sampling by altering n_1 and n_2 in proportion to σ_1 and σ_2.   This is so when it is possible to form strata so that their variabilities (as measured by σ_1 and σ_2) are distinctly different. Such a plan was first put into practice by Neyman.* Plan C will denote Neyman allocation where the sample-sizes are fixed in advance by the equations

$$
\left.
\begin{aligned}
n_1 &= n\,\frac{P_1\sigma_1}{\bar{\sigma}_w} \\[2ex]
n_2 &= n\,\frac{P_2\sigma_2}{\bar{\sigma}_w}
\end{aligned}
\right\}
\quad
\begin{array}{l}
\text{[Extension to more} \\
\text{strata is obvious]}
\end{array}
\qquad (22)
$$

The steps in Plan C follow. (See Eq. 31 ahead for an alternate form of this equation.)

PLAN C

1. Decide with the aid of Eq. 30 ahead the size n of the sample desired. Compute for Stratum i the sample-size $n_i = nP_i\sigma_i/\bar{\sigma}_w$ (Eq. 22).

2. Compute the zoning intervals $Z_i = 2\,N_i/n_i$ for 2 subsamples, $10\,N_i/n_i$ for 10 subsamples (illustration page 311). See Step 2 under Plan B. Form the zones in the various strata with the zoning intervals just calculated; see Step 2 under Plan B. The procedure is to make the selections from every stratum as in Plan A.

3. Draw with random numbers between 1 and Z_1, 1 sampling unit from every zone of Stratum 1; with random numbers between 1 and Z_2, 1 sampling unit from each zone of Stratum 2; etc. The sample so drawn is Subsample 1. Draw another sampling unit from every zone for Subsample 2, and likewise for the other subsamples if any. Mark each sampling unit to show which subsample it belongs to, and which stratum, and which zone it came from.

4. Carry out the interviews or tests on the entire sample.

5. Form from Subsample i the x-populations $x_1^{(i)}$, $x_2^{(i)}$, $x_3^{(i)}$ stratum by stratum. Do this for every subsample.

6. Compute for Subsample i

$$
\begin{aligned}
X^{(i)} &= Z_1 x_1^{(i)} + Z_2 x_2^{(i)} + Z_3 x_3^{(i)} \\
&= X_1^{(i)} + X_2^{(i)} + X_3^{(i)}
\end{aligned}
\qquad (23)
$$

This is the estimate that Subsample i furnishes for the total x-population in the frame. The 3 terms, one by one, are estimates from Subsample i of the x-population stratum by stratum.

---

* J. Neyman, "On the two different aspects of the representative method," *J. Roy. Statist. Soc.*, vol. xcvii, 1934: pp. 558–606. The mathematical equations for Neyman allocation were nevertheless published earlier by A. Tschuprow, *Metron*, No. 3, 1923: p. 672, but so far as I know, Tschuprow made no use of his formulas.

Note that if the zoning intervals were equal in all strata, then Eq. 23 would collapse to the simple equation $X^{(i)} = Zx^{(i)}$, as we had it in Plan B.

7. Calculate $\hat{C}_X = \hat{C}_{\bar{x}}$, following the advice in Plan A (p. 284). For example,

$$\hat{C}_X = \hat{C}_{\bar{x}} = \frac{1}{10X} [X^{(i)}_{\max} - X^{(i)}_{\min}]$$

for 10 subsamples. Note, however, that in the formation of thick zones by combinations of thin zones, the thin zones will require weights proportionate to the zoning intervals (note the coefficients Z_1, Z_2, Z_3 in Eq. 23).

One may require separate estimates by stratum, in which case the procedure is to compute

$$X_1 = Z_1 \bar{x}_1 \quad \text{for Stratum 1}$$

$$X_2 = Z_2 \bar{x}_2 \quad \text{for Stratum 2}$$

Etc.

and

$$\hat{\text{Var}}\ X_1 = Z_1^2 \frac{1}{k(k-1)} \Sigma\ (x_1^{(i)} - \bar{x}_1)^2 \quad \text{for Stratum 1}$$

$$\hat{\text{Var}}\ X_2 = Z_2^2 \frac{1}{k(k-1)} \Sigma\ (x_2^{(i)} - \bar{x}_2)^2 \quad \text{for Stratum 2}$$

Etc.

for k subsamples. $\bar{x}_1$ is the average of the k individual x-populations $x_1^{(i)}$ in Stratum 1; $\bar{x}_2$ has a similar definition in Stratum 2. In the case of 2 subsamples, $\Sigma(x_1^{(i)} - \bar{x}_1)^2$ is merely $\frac{1}{2}w_1^2$, where w_1 is the range, the difference between $x_1^{(1)}$ and $x_1^{(2)}$. The estimates and their variances are additive, as in Step 8 of Plan B (p. 288).

If the variance of one stratum dominates the other variances, break this stratum into substrata (see the example in the appendix to Ch. 11, page 224).

Formulation of the gains in Neyman sampling (Plan C). Two strata will be sufficient for illustration of the theory. We start with

$$\left.\begin{aligned} n_1 &= n\frac{P_1\sigma_1}{\bar{\sigma}_w} + h \\[2mm] n_2 &= n\frac{P_2\sigma_2}{\bar{\sigma}_w} - h \end{aligned}\right\} \tag{24}$$

h is the amount by which the allocation fails to satisfy Eq. 22. We wish to see what happens to Var $\bar{x}$ for small values of h. The solution of this problem will not only discover the optimum allocation (Eq. 22); it will also tell us how much precision we lose by making an approximate Neyman allocation (as we can only do in practice) instead of an exact one. So now let us substitute the above values of n_1 and n_2 into Eq. 20. Here is what we get:

$$\text{Var } \bar{x} = P_1^2\left(1 - \frac{n_1}{N_1}\right)\frac{\sigma_1^2}{n_1} + P_2^2\left(1 - \frac{n_2}{N_2}\right)\frac{\sigma_2^2}{n_2} \quad \text{[From Eq. 20, page 290]}$$

$$= \frac{(P_1\sigma_1)^2}{n_1} + \frac{(P_2\sigma_2)^2}{n_2} - \frac{1}{N}(P_1\sigma_1^2 + P_2\sigma_2^2) \quad \text{[By rearrangement]}$$

$$= \frac{(P_1\sigma_1)^2}{nP_1\sigma_1/\bar{\sigma}_w + h} + \frac{(P_2\sigma_2)^2}{nP_2\sigma_2/\bar{\sigma}_w - h} - \frac{\sigma_w^2}{N}$$

$$= \frac{\bar{\sigma}_w}{n}\frac{P_1\sigma_1}{1 + h\bar{\sigma}_w/nP_1\sigma_1} + \frac{\bar{\sigma}_w}{n}\frac{P_2\sigma_2}{1 - h\bar{\sigma}_w/nP_2\sigma_2} - \frac{\sigma_w^2}{N}$$

$$= \frac{\bar{\sigma}_w}{n}P_1\sigma_1\left\{1 - \frac{h\bar{\sigma}_w}{nP_1\sigma_1} + \left(\frac{h\bar{\sigma}_w}{nP_1\sigma_1}\right)^2 + \cdots\right\}$$

$$+ \frac{\bar{\sigma}_w}{n}P_2\sigma_2\left\{1 + \frac{h\bar{\sigma}_w}{nP_2\sigma_2} + \left(\frac{h\bar{\sigma}_w}{nP_2\sigma_2}\right)^2 + \cdots\right\} - \frac{\sigma_w^2}{N}$$

[By long division, through 2 terms only, as h is small]

$$= \frac{\bar{\sigma}_w}{n}\left\{P_1\sigma_1 + P_2\sigma_2 + 0 + \frac{h^2\bar{\sigma}_w}{nn_1} + \frac{h^2\bar{\sigma}_w}{nn_2}\right\} - \frac{\sigma_w^2}{N}$$

$$= \frac{(\bar{\sigma}_w)^2}{n}\left\{1 + \frac{h}{n_1}\frac{h}{n_2}\right\} - \frac{\sigma_w^2}{N} \tag{25}$$

This is a very important result.

1. The term h^2/n_1n_2 is positive whether h be positive or negative: it is 0 only if $h = 0$. Therefore, Var $\bar{x}$ is at its minimum if $h = 0$.
2. If we put $h = 0$ in Eq. 25 we have left

$$\text{Var } \bar{x} = \frac{(\bar{\sigma}_w)^2}{n} - \frac{\sigma_w^2}{N} \quad \text{[Minimum variance, Plan C]} \tag{26}$$

The 2d term arises from sampling without replacement. We may usually neglect it.

3. The variance just written is the minimum variance obtainable under

the assumption that the total sample-size n is fixed. It arises from the Neyman allocation shown by Eq. 22, which Eq. 25 reduces to if $h = 0$. We note here in passing the following alternative form of Eq. 22:

$$n_i : n_j = P_i\sigma_i : P_j\sigma_j = N_i\sigma_i : N_j\sigma_j \qquad \begin{matrix} \text{[Alternative form of Eq. 22} \\ \text{for Neyman allocation]} \end{matrix} \qquad (27)$$

i and j are any 2 strata.

4. The term $h^2/n_1 n_2$ in Eq. 25 is the approximate relative increase in Var $\bar{x}$ that arises from failure to make an exact Neyman allocation. Or, it is the relative increase in the sample-size n that is necessary to restore

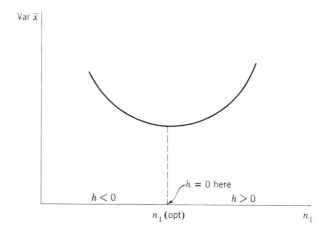

Fig. 20. This is a graph of Var $\bar{x}$ according to Eq. 25. Var $\bar{x}$ is at its minimum where $h = 0$ and where n_1 has its optimum value $N_1\sigma_1 n/\bar{\sigma}_w$, the Neyman allocation. As n_1 varies either way from this point by an amount h, Var $\bar{x}$ increases only very slow at first where h is small, but it increases rapidly upward when h becomes large.

the Var $\bar{x}$ to what it would have been for exact Neyman allocation. We shall soon see how helpful this term is. Exercise 2 on the next page will extend Eq. 25 to any number of strata.

5. We may now plot Var $\bar{x}$ against n_1 for the case of 2 strata (Fig. 20). The curve is a parabola, vertex down; it is flat in the neighborhood of the optimum value of n_1 in Eq. 22. Hence any reasonable approximation to the Neyman allocation will give excellent results. This is not to say that any allocation whatever will be good, but that an honest sincere attempt to learn something about the material may enable us to achieve efficiency almost as good as the exact values of σ_1 and σ_2 would yield.

EXERCISES

1. Show that in the Neyman allocation (Plan C), X is an unbiased estimate of A, the total x-population in the frame.   In symbols,

$$EX = A$$

exactly as with proportionate sampling (Plan B), and with unstratified sampling (Plan A).

2. *a.* Generalize Eq. 25 to any number of strata by setting

$$n_i = \frac{P_i \sigma_i}{\bar{\sigma}_w} + h_i \qquad \text{with } \Sigma h_i = 0 \tag{28}$$

h_i is the failure to make an exact Neyman allocation in Stratum i.   Prove that

$$\text{Var } \bar{x} = \frac{(\bar{\sigma}_w)^2}{n} \left\{ 1 + \Sigma \frac{h_i}{n} \frac{h_i}{n_i} \right\} - \frac{\sigma_w{}^2}{N} \tag{29}$$

b. Show that if the failures h_i are all small, Var $\bar{x}$ will be but little bigger than if the allocation were Neyman exactly.

Note that for 2 strata, $h_1 = -h_2$, and that the summation above reduces to $h^2/nn_1 + h^2/nn_2 = h^2/n_1 n_2$, as written in Eq. 25.

3. Show from Eq. 20 that an unfortunate choice of sample-sizes n_1 and n_2 in Plan C (through misinformation about the variances in the strata) may lead to a higher variance than no stratification at all (Plan A).

Remark 6. Although small departures from the Neyman allocation cause only insignificant losses, big departures, this exercise shows, may wipe out the gains.   An interesting illustration of this fact was kindly published as an aid to statistical practice by Austin A. Hasel in his chapter "Problems in forestry-inventory" in the book edited by Kempthorne, Bancroft, et al., *Statistics and Mathematics in Biology* (Iowa State College Press, Ames, 1954).   The problem was to measure both the volume and the growth in a forest since the last inventory.   The plan that he used was an intended Neyman allocation (Plan C).   In the sampling of new material (Ch. 14), one can only use his best judgment and latest estimates on similar material for the standard deviations σ_i in Eq. 22.   Table 3 (from page 268 in the book cited) shows what happened.   Note that the actual coefficient of variation that the sample yielded for both volume and growth was higher than if he had used proportionate allocation.

4. In a sample of equipment inspected for the purpose of estimating the physical condition of all the equipment owned by a telephone company, the total variance was .0376, of which the underground plant contributed .0236.   Show that slightly less variance would have resulted from doubling

TABLE 3

COEFFICIENT OF VARIATION OBTAINED IN
AN INVENTORY OF A FOREST

Type of allocation	Volume	Growth
Plan B: proportionate sampling	3.10%	2.87%
Intended Plan C, Neyman: actual survey	3.27	3.77
Plan C: Neyman sampling, had he known in advance:		
The σ_i for volume	2.85	2.95
The σ_i for growth	3.27	2.60

the size of the sample from the underground plant, and taking a sample only 2/3 as big in the remainder.   (The proposed sample would have been cheaper.)

SOLUTION

As carried out: .0376   total
.0236   underground
——
.0140   all but underground

Proposed: .0210   all but underground (.0140 × 3/2)
.0118   underground (.0236 × 1/2)
——
.0328   total

Q.E.D.

What do we lose by small departures from Neyman allocation? The answer is that we lose very little by small departures, as we saw in Fig. 20 on page 294.   In an actual example, records and previous experience indicated that the ratio $P_1\sigma_1:P_2\sigma_2$ was about 46:54, wherefore by Eq. 22 $n_1:n_2$ should also be 46:54.   Because of the numbers N_1 and N_2, there would have been a difficult thinning ratio to work with, and I hesitated to prescribe it: I decided to set $n_1 = n_2$, which made the work much easier.   How much precision did this simplification cost?   It may be obvious that this decision was equivalent to setting

$$\frac{h}{n_1} = \frac{4}{46}, \qquad \frac{h}{n_2} = \frac{4}{54} \tag{30}$$

whence

$$\frac{h}{n_1}\frac{h}{n_2} = \frac{4}{46}\frac{4}{54} \doteq .08^2 \tag{31}$$

Then the bracketed term in Eq. 29 will be

$$1 + .08^2 = 1.0064 \tag{32}$$

which indicates a loss of only 6 interviews in 1000—far too trivial to mention, and far within the limits of the uncertainty in the advance knowledge of $P_1\sigma_1:P_2\sigma_2$. The cost of the simplification was thus trifling, but I did not feel safe in prescribing it until I saw this delightful numerical result.   Generalization to more strata appeared in Exercise 2 on page 295.

Comparison of variances to date.   We may now collect the formulas for the variances of Plans A, B, C.   It is useful to start with Eq. 20, in the following form:

$$\operatorname{Var} \bar{x} = P_1{}^2\left(1 - \frac{n_1}{N_1}\right)\frac{\sigma_1{}^2}{n_1}$$

$$+ P_2{}^2\left(1 - \frac{n_2}{N_2}\right)\frac{\sigma_2{}^2}{n_2} \quad \begin{array}{l}\text{[For any allocation if}\\ P_1 \text{ and } P_2 \text{ are known]}\end{array} \tag{33}$$

A: $\quad \operatorname{Var} \bar{x} = \left(1 - \frac{n}{N}\right)\frac{\sigma^2}{n}$ [No additional stratification] (34)

B: $\quad \operatorname{Var} \bar{x} = \left(1 - \frac{n}{N}\right)\frac{\sigma_w{}^2}{n}$ [Proportionate allocation] (35)

C: $\quad \operatorname{Var} \bar{x} = \frac{(\bar{\sigma}_w)^2}{n} - \frac{\sigma_w{}^2}{N}$ [Neyman allocation] (36)

Comparison is only a matter of algebra, and as an exercise the student may show that if the total sample (n) is the same in all 3 plans, then

$$\frac{A - B}{A} = 1 - \left(\frac{\sigma_w}{\sigma}\right)^2 = \left(\frac{\sigma_b}{\sigma}\right)^2$$

or

$$B = A\left(\frac{\sigma_w}{\sigma}\right)^2$$

[The relative gain of proportionate allocation over no stratification] (37)

$$\frac{B - C}{B} = 1 - \left(\frac{\bar{\sigma}_w}{\sigma_w}\right)^2 = \frac{C_{\sigma_i}{}^2}{1 + C_{\sigma_i}{}^2}$$

or

$$C = B\left(\frac{\bar{\sigma}_w}{\sigma_w}\right)^2 = B\frac{1}{1 + C_{\sigma_i}{}^2}$$

$$= A\left(\frac{\bar{\sigma}_w}{\sigma}\right)^2$$

[The relative gain of Neyman allocation over proportionate allocation and over no stratification] (38) wherein $C_{\sigma_i}^2 = (1/\bar{\sigma}_w)^2 \Sigma P_i (\sigma_i - \bar{\sigma}_w)^2$ is the rel-variance of the σ_i (see Exercise 4 ahead). These are very important equations.* They tell us that the gain of Neyman sampling (Plan C) over proportionate sampling (Plan B) arises from the rel-variance of the σ_i. We see now why Neyman sampling often shows huge gains in the sampling of highly skewed distributions, as in sampling accounts, areas with highly variable production, income, number of employees, or industrial production, and why it will not show similar gains in opinion-studies, where each person counts 0 or 1.

In my own practice, I usually make calculations based on 2 strata. If 2 strata show no appreciable gain, then there is no use to try 3. But if 2 strata show some appreciable gain, then 3 or more strata, carefully defined, may show a further gain. Some theory for the formation of strata is in Chapter 20.

The table of relative costs of the various plans on page 350 is supplementary to the above equations, and more comprehensive. The student may wish to turn to it now and derive the relative costs of Plans A, B, and C.

EXERCISES

1. Show that if there are only 2 strata, Eqs. 37 and 38 may take the forms

$$\frac{A-B}{A} = P_1 P_2 \left[\frac{a_2 - a_1}{\sigma}\right]^2 \tag{39}$$

$$\frac{B-C}{B} = P_1 P_2 \left[\frac{\sigma_2 - \sigma_1}{\sigma_w}\right]^2 \tag{40}$$

2. Rewrite Eqs. 37 and 38 in the following form:

$$\frac{A-B}{A} = \sum_{j>i} P_i P_j \left[\frac{a_j - a_i}{\sigma}\right]^2 \tag{41}$$

$$\frac{B-C}{B} = \sum_{j>i} P_i P_j \left[\frac{\sigma_j - \sigma_i}{\sigma_w}\right]^2 \tag{42}$$

If there were 4 strata, the indexes i and j would take the 6 values

$$
\begin{array}{lll}
1,2 & & \\
1,3 & 2,3 & \\
1,4 & 2,4 & 3,4
\end{array}
$$

3. Comparison of Eqs. 35 and 36 shows that the difference between Var $\bar{x}$ for the Neyman allocation and for the proportionate allocation lies

---

* The form in terms of $C_{\sigma_i}^2$ comes from Hansen, Hurwitz, and Madow, *Sample Survey Methods and Theory*, Vol. I (Wiley, 1953): p. 212.

in the difference between $\bar{\sigma}_w$ and σ_w.   The student should turn to the table of notation on page 285 and illustrate for himself the difference between $\bar{\sigma}_w$ and σ_w.   For example, let

$$\sigma_1 = 1 \qquad P_1 = \tfrac{1}{3}$$

$$\sigma_2 = 2 \qquad P_2 = \tfrac{1}{3}$$

$$\sigma_3 = 3 \qquad P_3 = \tfrac{1}{3}$$

Then (as the student should prove)

$$\bar{\sigma}_w = 2 \qquad (\bar{\sigma}_w)^2 = 4$$

whereas

$$\sigma_w = 2.16 \qquad \sigma_w{}^2 = 4.67$$

4. Prove that $\sigma_w{}^2 \geq (\bar{\sigma}_w)^2$.   Hence, Neyman allocation can not give less precision than proportionate allocation PROVIDED we really know the $P_i \sigma_i$ for the Neyman   allocation (on page 291).   (*Hint:* the mean square of 2 numbers σ_1 and σ_2 with weights P_1 and P_2 is $P_1\sigma_1{}^2 + P_2\sigma_2{}^2 - (P_1\sigma_1 + P_2\sigma_2)^2 = \sigma_w{}^2 - (\bar{\sigma}_w)^2 = \sigma_{\sigma_i}{}^2$, which must be ≥ 0.)   But note that if we are unfortunate in our predictions of the σ_i, an attempted Neyman allocation can turn out worse than proportionate sampling, and worse even than no stratification at all (Exercise 3, page 295).

Summary of comparisons between Plans A, B, C.   The following remarks summarize the conclusions that one may draw from Exercises 1, 2, and 3.

>   **Remark 7.**   The gain of proportionate sampling over no stratification will be prominent when the variance $\sigma_b{}^2$ between the means of the strata makes up a good share of the total variance σ^2.   In particular, if the means of the prominent strata are nearly equal, proportionate sampling will show little gain over unstratified sampling.

>   **Remark 8.**   The gain of Neyman sampling over proportionate sampling will be prominent if there are large differences in the standard deviations between the sampling units within prominent strata.

Some simple numerical comparisons.   In order to see some numerical results, suppose that we are going to take a sample over a region to discover the total number of readers of a certain newspaper.   The publisher's mailing list, and the number of copies sold by dealers, enables us to divide the area into 2 parts such that

$p_1 = .10$, the average proportion of people in Stratum 1 that read the paper
$p_2 = .03$, the average proportion of people in Stratum 2 that read the paper
$P_1 = .5$,   the proportion of sampling units in Stratum 1
$P_2 = .5$,   the proportion of sampling units in Stratum 2

A sampling unit is, in both strata, a segment of area of about 10 d. us., or with about 30 people over 10 years of age.   Then there are about 3 readers per segment in Stratum 1, and about 1 reader per segment in Stratum 2.

If the distribution of the number of readers per segment were Poisson (Ch. 18), then σ_1^2 would be 3, and σ_2^2 would be 1.   Variation in the size of the segment will increase these variances.   For good measure, we may assume

$$\sigma_1^2 = 5 \qquad \sigma_2^2 = 2$$

Then

$$\bar{\sigma}_w = P_1\sigma_1 + P_2\sigma_2 = \tfrac{1}{2}(\sqrt{5} + \sqrt{2}) = 1.83$$

$$(\bar{\sigma}_w)^2 = 3.35$$

$$\sigma_w^2 = P_1\sigma_1^2 + P_2\sigma_2^2 = \tfrac{1}{2}(5 + 2) = 3.5$$

$$p = P_1p_1 + P_2p_2 = .065, \text{ the overall proportion of readers}$$

$$\sigma_b^2 = P_1(p_1 - p)^2 + P_2(p_2 - p)^2 = .0012 \qquad \text{[Eq. 11, page 285]}$$

$$\sigma^2 = \sigma_w^2 + \sigma_b^2 = 3.501 \qquad \text{[Eq. 12, page 285]}$$

Let us compare the 3 variances A, B, C in this problem.   First, to compare Plans A and B, we go back to Eq. 37 and see that

$$B = A\left(\frac{\sigma_w}{\sigma}\right)^2 \doteq A\frac{3.500}{3.501} \doteq A \qquad (43)$$

Thus, no stratification (Plan A) in this problem would give us about the same precision as proportionate sampling (Plan B).

Now let us see if Neyman allocation will be any better.   We use Eq. 38 and see that

$$C = B\left(\frac{\bar{\sigma}_w}{\sigma_w}\right)^2 = B\frac{3.35}{3.50} = .96B \qquad (44)$$

Thus, 96 interviews by Neyman sampling will give us about the same precision as 100 interviews allocated proportionately, or without stratification.   One would probably use proportionate allocation because it is easier than to draw a sample with no stratification at all.

Remark 9.   If we were wrong in our assumed variances, the expected gain of Neyman sampling would be less than our calculations showed.

Remark 10.   One must be careful not to generalize from 1 illustration. Sometimes the difference $B - C$ between proportionate and Neyman sampling will be negligible; sometimes large.   One must depend on theory, not hunches.   Fortunately, the calculations require only a few minutes.

Exercise.   Take

$$\sigma_1^2 = 6 \qquad \sigma_2^2 = 1$$

$$P_1 = P_2 = .5$$

in the above examples and show that

$$\bar{\sigma}_w = 1.72 \qquad (\bar{\sigma}_w)^2 = 2.95$$

$$\sigma_w^2 = 3.5$$

in which case

$$C = B\left(\frac{\bar{\sigma}_w}{\sigma_w}\right)^2 = B\frac{2.95}{3.5} = .85B$$

The gain of Neyman sampling might here repay its cost. Eighty-five interviews with Neyman allocation would be the equivalent of 100 by proportionate sampling.

Use of plausible extreme values in new material for decision on type of stratified design.   It is sometimes possible and entirely adequate to rely on plausible assumptions about extreme values in order to choose between 2 plans of stratified sampling. What constitutes a plausible assumption is of course the responsibility of the expert in the subject-matter (in this case, consumer-studies).

A survey was to cover Seattle, Portland, San Francisco, Los Angeles, and San Diego, to estimate the number of homes with FM radios, and to learn the characteristics of homes with FM radios, with separate estimates for the 2 groups: (a) Seattle and Portland combined; (b) San Francisco, Los Angeles, and San Diego combined.* This was an example in the sampling of new material, as little was known about the characteristics of home with FM radios, nor how the proportion of homes varied from one level of income to another. There was in fact only a fanciful figure for the overall proportion, which we shall assume in the calculations to be 30%. The frame was Census block-statistics, and maps.

To start with, an FM radio is relatively expensive. It was therefore strongly suspected that FM ownership is more frequent in areas of high income. The block statistics provided data on average rent for every block within the 5 cities. Rent and income are highly correlated. Would it pay to stratify blocks, or at least tracts, and to concentrate the sample in the areas of high rent, by use of the Neyman allocation?

Let P_1 and P_2 be the proportions of households (and hence of segments) in the strata of low rent and of the high rent, respectively. Let p_1 and p_2

* I thank Alfred Politz Research, Inc., New York, for the privilege of working on this survey.

be the proportions of homes in these strata with FM radios.   Then, if the overall average ownership of FM radios be 30%,

$$P_1 p_1 + P_2 p_2 = .3$$

whence one may solve for p_2, once he inserts numerical values of P_1, P_2, and p_1.

Assumptions I and III in the table below represent extremes for p_1, on the basis of knowledge of the subject-matter.   Assumption II is intermediate.   We can see from Eqs. 39–42 above that it will not pay in this problem to create a stratum that contains only a small portion of the sampling units: hence we adopt $P_1 = .25$ for the proportion of blocks with low rent in Assumption I.

We shall assume binomial variances in the 2 strata.   If people try to keep up with the Joneses next door, then door-to-door correlation will partially invalidate this assumption.   The relative gain of Neyman sampling over proportionate sampling will be only slightly affected by distortion of the binomial variances; hence I calculate the variances between households within segments in the 2 strata with the simple formulas $\sigma_1^2 = p_1 q_1$ and $\sigma_2^2 = p_2 q_2$.   The relative gain of Neyman allocation (Plan C) over proportionate allocation (Plan B) is given by Eq. 40 on page 298.   The calculations are in the accompanying table.

TABLE OF THE RELATIVE GAIN OF NEYMAN ALLOCATION (PLAN C) OVER PROPORTIONATE ALLOCATION (PLAN B) FOR AN ESTIMATE OF THE OVERALL PROPORTION OF HOMES WITH FM RADIOS

$$\sigma_w^2 = P_1 \sigma_1^2 + P_2 \sigma_2^2 \qquad \text{Relative gain} = P_1 P_2 (\sigma_2 - \sigma_1)^2 / \sigma_w^2 \text{ (Eq. 42)}$$

Assumption	Low income			High income			$P_1 P_2$	$(\sigma_2-\sigma_1)^2$	σ_w^2	Relative gain
	P_1	p_1	$\sigma_1^2=p_1 q_1$	P_2	p_2	$\sigma_2^2=p_2 q_2$				
I	.25	.05	.05	.75	.38	.23	.19	.067	.18	.070
II	.33	.08	.07	.67	.41	.24	.22	.044	.19	.051
III	.50	.15	.13	.50	.45	.25	.25	.012	.19	.016

The relatively small gain (7% at the most) shown in the table provides ground for immediate decision: the gain of Neyman sampling will not pay its cost in this problem; in fact, the expense of classifying blocks or even tracts into strata would incur a serious deficit.   Moreover, strata formed on the basis of rent in 1950 (the latest figures available) would give even less gain than the figures indicate, not only because of changes since 1950, but because of variability between households within the block.   No reasonable set of assumptions as a basis for the calculations could lead to a different decision.

The conclusion arrived at here was by no means obvious prior to calculation of the table. This was an instance where esteemed colleagues had supposed, in advance of calculation, that Neyman allocation might well pay a handsome dividend.

Modification when the costs vary greatly from stratum to stratum. The cost of an interview (or of a test) in the foregoing pages was the same in one stratum as in another. Sometimes, however, the costs will be greatly different in the various strata. It will then be advisable, if administratively feasible, to decrease the size of the sample in any stratum where the costs are relatively excessive, and to increase the sample where the costs are relatively cheap, keeping the total cost the same as it would have been otherwise. Stated otherwise, it may be wise to decrease the probability of selection for sampling units from which the cost per unit of information will be excessive in comparison with the returns from other units.

Specifically, under such circumstances, we replace proportionate allocation (Plan B) by the modification

$$n_i : n_j = \frac{P_i}{\sqrt{c_i}} : \frac{P_j}{\sqrt{c_j}} \qquad \text{[Plan B}'] \qquad (45)$$

and we replace Neyman allocation (Plan C) by the modification

$$n_i : n_j = \frac{P_i \sigma_i}{\sqrt{c_i}} : \frac{P_j \sigma_j}{\sqrt{c_j}} \qquad \text{[Plan C}'] \qquad (46)$$

wherein c_i and c_j are the costs to investigate 1 sampling unit in Strata i and j respectively.* The costs enter these equations only under the root sign; hence the modifications will be important only when the costs c_i and c_j are greatly different. Plan B$'$ reduces to Plan B if $c_i = c_j$, and Plan C$'$ reduces to Plan C. The proof of Eq. 46 is in the exercise on page 307, and Eq. 45 follows at once if $\sigma_i = \sigma_j$.

A recent example in my own experience encountered costs that varied considerably with the size of the unit. The aim of the study was to estimate the cost of psychiatric treatment in addition to other medical service, and to discover which educational and income classes would make most use of the psychiatric service, and which ones would not use it. The sampling units were entire groups of employees. It was not allowable

---

* Reduction in the size of the sample in any stratum by the square root of the cost was first published by P. C. Mahalanobis, "On large-scale sample-surveys," *Phil. Trans. Roy. Soc.*, vol. 231B, 1944: pp. 329–451.

See also Hansen, Hurwitz, and Madow, *Sample Survey Methods and Theory*, Vol. I (Wiley, 1953), Ch. 6, for further examples.

to draw some employees of a group into the sample and to omit others in the same group, as the ones omitted would enter complaints of discrimination.   The groups varied in size from 1 employee on up to 870. Theory showed that any group of 400 or more should be in the sample with probability 1.   The problem arose, though, concerning the selection of groups of size below 400.   There were several thousand small groups containing 1, 2, 3, 4, 5 employees; some, 9; some, 13; and so on.   The cost of administering the questionnaires and records would be disproportionately high in the small groups, as the big groups had full-time leaders who would take initiative in collecting the necessary information. The cost per employee might easily run 10 times as high or more for a group of 5 or 10 employees as for a group of 300.   The decision was to vary the probabilities as shown in the accompanying table.

Stratum	Number of employees in the group	Probability of selection in the whole sample	Zoning interval for 10 sub-samples	Weight in the tabulation of a subsample
1	400 and over	1	xxx	1
2a	200–399	3	30	30
2	100–199	3	30	30
3a	50– 99	10	100	100
3	10– 49	10	100	100
4	3– 9	10	100	100
5	1 and 2	10	100	100

A numerical comparison of plans may help.   Suppose that a survey is to cover a region that consists of an urban area (Stratum 1), where the cost of an interview averages $5; also the surrounding rural area (Stratum 2), where the cost is $10.   Suppose that the proportions of sampling units in the 2 areas are 60% and 40% of the total, and that the variances between sampling units within Stratum 1 and within Stratum 2 (for some one characteristic) are in the ratio 2:1.   Then

$$c_1:c_2 = 1:2 \tag{47}$$

$$P_1:P_2 = .6:.4 \tag{48}$$

$$\sigma_1:\sigma_2 = \sqrt{2}:1 \tag{49}$$

Suppose that the allowable cost of the field-work is $2000.   We here compute the expected Var $\bar{x}$ under the 4 allocations, Plans B, B', C, C', for comparison.

Plan B

$$\frac{n_1}{n_2} = \frac{P_1}{P_2} = \frac{6}{4}$$

$$n_2 = \frac{2n_1}{3}$$

$$n_1 c_1 + n_2 c_2 = n_1\left(5 + \frac{20}{3}\right) = 2000$$

$$n_1 = 171$$

$$n_2 = \frac{2n_1}{3} = 114$$

$$n = 285$$

$$\text{Var } \bar{x} = \frac{\sigma_w^2}{n} = \frac{P_1 \sigma_1^2 + P_2 \sigma_2^2}{n} \qquad \text{[P. 297]}$$

$$= \frac{.6 \times 2 + .4 \times 1}{285} \sigma_2^2 = .00563\sigma_2^2 \qquad (50)$$

Plan B′

$$\frac{n_1}{n_2} = \frac{6}{4}\sqrt{\frac{c_2}{c_1}} = \frac{6}{4}\sqrt{2} \doteq 2.12$$

$$n_1 c_1 + n_2 c_2 = (10.60 + 10)n_2 = 2000$$

$$n_1 = 206$$

$$n_2 = \frac{3n_1}{14} = 97$$

$$n = 303$$

$$\text{Var } \bar{x} = P_1^2 \frac{\sigma_1^2}{n_1} + P_2^2 \frac{\sigma_2^2}{n_2} \qquad \text{[Eq. 33, page 297]}$$

$$= \left(.6^2 \frac{2}{206} + .4^2 \frac{1}{97}\right)\sigma_2^2$$

$$= .00515\sigma_2^2 \qquad (51)$$

Plan C $$\frac{n_1}{n_2} = \frac{P_1\sigma_1}{P_2\sigma_2} = \frac{6\sqrt{2}}{4} \doteq 2.12$$

The sample-sizes will be exactly those of Plan B′, namely,

$$n_1 = 206$$

$$n_2 = \ \ 97$$

$$\overline{n = 303}$$

Var $\bar{x}$ will also be the same as for Plan B′, but for variety we use Eq. 26 on page 293 to find

$$\text{Var }\bar{x} = \frac{(\bar{\sigma}_w)^2}{n} = \frac{(P_1\sigma_1 + P_2\sigma_2)^2}{n}$$

$$= \frac{(.85 + .4)^2 \, \sigma_2^2}{n} = .00515\sigma_2^2 \qquad \text{[As in Eq. 51]} \qquad (52)$$

Plan C′ $$\frac{n_1}{n_2} = \frac{P_1\sigma_1}{P_2\sigma_2}\sqrt{\frac{c_2}{c_1}} = \frac{6}{4}\frac{\sqrt{2}}{1}\sqrt{2} = \frac{3}{1}$$

$$n_2 = \frac{n_1}{3}$$

$$n_1 c_1 + n_2 c_2 = n_1\left(5 + \frac{10n_1}{3}\right) = 2000$$

$$n_1 = \ \ 240$$

$$n_2 = \frac{n_1}{3} = \ \ \ 80$$

$$\overline{n = \ \ \ 320}$$

$$\text{Var }\bar{x} = P_1^2\frac{\sigma_1^2}{n_1} + P_2^2\frac{\sigma_2^2}{n_2} \qquad \text{[Eq. 33, page 297]}$$

$$= \left(.6^2\frac{2}{240} + .4^2\frac{1}{80}\right)\sigma_2^2$$

$$= .00500\sigma_2^2 \qquad\qquad (53)$$

The reader should now recalculate the variances of the 4 plans, taking $\sigma_1 : \sigma_2$ and $P_1 : P_2$ as before, but with $c_1 : c_2 = 9 : 1$. The 4 variances of $\bar{x}$ will then be:

For Plan B, $.0168\sigma_2{}^2$

For Plan B', $.0107\sigma_2{}^2$

For Plan C, $.0138\sigma_2{}^2$

For Plan C', $.0105\sigma_2{}^2$

Here, the comparisons are more striking than in the example in the text. There are important points to recommend consideration of Plan B'. Costs are often easier to predict than variances, especially if one contracts at a fixed price for each interview or test completed. We may thus be better able to predict the gain of Plan B' over B, and of Plan C' over C, than to predict the gain of Plan C over Plan B. Moreover, Plan B' shows the same gain over Plan B for all characteristics estimated, which is not true in a comparison of Plan C with Plan B.

We have seen another way to cut costs where travel is expensive, namely, to use bigger sampling units in such areas, to decrease the probability of selection (p. 193).

Exercise. Establish Eq. 46.

SOLUTION

Start with Eq. 36, page 297.

$$\text{Var } \bar{x} = \Sigma \frac{P_i \sigma_i{}^2}{n_i}$$

Now set

$$\Sigma n_i c_i = K$$

This equation fixes the total cost at K. We require the n_i that will make Var $\bar{x}$ a minimum. Let ∂ denote variation away from the optimum values of n_i that we seek. Then

$$\partial \text{ Var } \bar{x} = -\Sigma \frac{(P_i \sigma_i)^2}{n_i{}^2} \partial n_i = 0$$

and

$$\Sigma c_i \, \partial n_i = 0$$

Multiply the last equation through by $-\lambda^2$ and add it to the one before. The result is

$$\Sigma \left\{ \frac{(P_i \sigma_i)^2}{n_i{}^2} - \lambda^2 c_i \right\} \partial n_i = 0$$

This equation must hold for all permissible values of ∂n_i, all but 1 of which are arbitrary. Set the coefficient of ∂n_1 equal to 0. The 1st term in the sum then vanishes no matter what be ∂n_1. Then all the other coefficients in the sum must also be 0, because ∂n_2, ∂n_3, etc. are all arbitrary.

Thus, the coefficient of every ∂n_i vanishes, which is to say,

$$\frac{(P_i \sigma_i)^2}{n_i^2} = \lambda^2 c_i \qquad \text{for every } i$$

which gives

$$n_i = \frac{P_i \sigma_i}{\lambda \sqrt{c_i}}$$

$$n_j = \frac{P_j \sigma_j}{\lambda \sqrt{c_j}}$$

whence Eq. 46 follows at once.

Q.E.D.

The student may wish to show further that

$$\lambda^2 = \frac{1}{K} \Sigma \frac{(P_i \sigma_i)^2}{n_i}$$

Optimum allocation when sampling for a ratio. The reader will see from Eq. 7 on page 421 in Chapter 17 and in Table 9 on page 422 that the variance of a ratio is expressible in exactly the same form as the variance of a mean. The problem of optimum allocation for a ratio is thus in principle no different from that of the optimum allocation for a total. The 2 allocations will often be very different numerically, however, because the variance between the x-populations of the sampling units may be large, likewise the variance between the y-populations, yet the variance in the ratio $f = x/y$ may be small. A simple example is the weights of bales of wool in kilograms, and the weights of the same bales in pounds. The bales might vary from 150 to 400 kilograms, yet the ratio of pounds to kilograms might be nearly constant, depending on how much time has elapsed between weighings, and whether the bales had equal opportunity meanwhile to absorb or give up moisture, or to suffer losses through abrasion. Another example would be the yield of wheat in bushels per acre: the yield might be about the same for big fields as for small ones. Wage-rates likewise: they might be about the same for big companies as for little ones, for comparable work.

For further reading, the reader may now turn to William G. Cochran's treatise *Sampling Techniques* (Wiley, 1953): p. 131 for the combined ratio-estimate; p. 132 for the variance of f; p. 135 for the optimum allocation for a combined ratio-estimate.

Another excellent reference is P. V. Sukhatme, *Sampling Theory of Surveys with Applications* (Iowa State College Press, Ames; and the Indian Society of Agricultural Statistics, New Delhi, 1953); Ch. IV.

An example of Neyman allocation in a marketing problem. A manufacturer of soap wishes to discover his share of the market in various areas of the country. The figures in Table 4 show some historical information from a judgment sample for the City of Milwaukee averaged

TABLE 4

AVERAGE WEEKLY SALES OF A CERTAIN MANUFACTURER'S BRAND OF SOAP,
AND AVERAGE SALES OF ALL SOAP, FOR THE CITY OF MILWAUKEE
These figures were assembled as an aid in the design of a
new probability sample

Type of store	Average weekly sales in dollars, all soaps (y_i)	Average weekly sales in dollars, this soap (x_i)	Range	Proposed sample
$i = 1$ Chains	$934	$44	2.3–10.1	57%
2 Very large	$500	$31	1.6–10.6	31%
3 Large	$142	$13	3.6–14.6	9%
4 Medium	$38	$2	2.5– 9.3	2%
5 Small	$15	$1	3.3–18.0	1%
All types	$1629	xxx	xxx	100%

over a week. There is similar information for most other cities, or at least for wide areas, on which to base the design of the sample. The allocation of the total sample to the various types of stores in the City of Milwaukee should be approximately in the proportion shown in the right-hand column of Table 4. The reasoning follows.

For an estimate of this manufacturer's share of the market, we may write

$$f = w_1 f_1 + w_2 f_2 + \cdots \tag{54}$$

where each term refers to a type of store. The w_i are weights in proportion to the sales of all soaps in all the stores in Class i, and f_i is the estimate, made from the sample of stores in Class i, of this manufacturer's share of the market in this class.

Remark 1. The equation just written calculates the overall ratio f by weighting the separate ratios (f_i) found in the individual strata. We could form also here, as an alternative, the overall ratio

$$f = \frac{\Sigma(N_i/n_i)w_i x_i}{\Sigma(N_i/n_i)w_i y_i} \tag{55}$$

which corresponds to the ratio that we have used heretofore. The variances of the 2 ratios do not in practice differ much, and as the latter ratio behaves better in the presence of a small value of y_i in the denominator, I use it almost exclusively. Eq. 25 on page 200 will in practice give a good estimate of the coefficient of variation of either ratio, after the results are in.

The usual approximation for the variance of either ratio by the sum of squares is not difficult. Each term of Eq. 54 will contribute variance w_i^2

Var f_i, and Var f_i is calculable by Eq. 18 on page 199, with a separate f_i for each term. For the variance of the overall ratio in Eq. 55, we adapt the same equation, in which f remains fixed, but in which we replace x_i by the numerator of the right-hand member of Eq. 55 above, and y_i by the denominator. The variances of the 2 ratios will be about the same unless the ratio f_i varies greatly from stratum to stratum; see Cochran's *Sampling Techniques* (Wiley, 1953), page 131; Hansen, Hurwitz, and Madow's *Sample Survey Methods and Theory;* Vol. I (Wiley, 1953), page 190. See also page 448.

As the ranges in f_i are about the same for all types of stores (except for the small stores, which are not important enough for separate consideration), we may assume that the standard deviations are also about the same, in which case the allocation of the sample should be in proportion to the average sales of all soaps by class: i.e., $n_i = nw_i$.

An example of Neyman allocation in a problem in accounting. A manufacturing firm has in one plant an inventory of materials in process valued at about $21,000,000. The firm may decrease taxes and obtain more meaningful information for management, by estimating the change in value of this inventory over the course of the year. This change in value is the so-called LIFO adjustment.* The sampling problem is to design a sample of material that will provide the required information with the precision desired.

The total number of lots of material is near 20,000 in the one plant that we shall deal with here (Table 5). Some preliminary investigations on a sample of lots of material furnished information by which to estimate in advance the standard deviation of the changes in value (the LIFO adjustment) in the 4 classes into which the inventory will be divided. The definitions of the 4 classes and the results of the preliminary investigation on variances and sizes are shown in the 1st 3 columns of Table 5. The counts (N_i) of the lots in the 4 classes were made by rough but quick separations. More reliable preliminary information would have contributed but little to the precision of the final results (p. 293). The remaining columns show the basic derived features of the sample design, which was to be as follows:

a. 100% sample in Class 1.
b. Neyman allocation in the remaining 3 classes.
c. 10 systematic subsamples.
d. Standard error aimed at, $50,000 (decision of management; Ch. 1).

To be more precise, the decision of management was that they could tolerate sampling error to the extent of $150,000. This amount is then the allowable 3-sigma variation, whereupon 1 standard error is $50,000.

---

* For the meaning of the LIFO system, see any book on accounting.

The formula for the variance of an estimate of a total in Neyman allocation comes from Eq. 24, page 292.   Here we write it

$$\text{Var}\,(X - Y) = \frac{1}{n}\,N^2(\bar{\sigma}_w)^2 - N\sigma_w{}^2$$

$$= \frac{1}{n}\left(\sum_2^4 N_i\sigma_i\right)^2 - \sum_2^4 N_i\sigma_i{}^2 \qquad (56)$$

$X - Y$ is the LIFO adjustment that originates in Classes 2, 3, 4. X is the estimate of the total value of the inventory today in these classes, and Y is the estimate for a year ago.   The contribution to the LIFO adjustment from Class 1 is substantial, but Class 1 contributes 0 variance; hence the symbols n and N here refer only to Classes 2, 3, 4.

To find n, we put $\text{Var}\,(X - Y) = 50{,}000^2$, and look in Table 5 for the numerical values of the sums.   Then

$$n = \frac{(\Sigma N_i\sigma_i)^2}{\text{Var}\,(X - Y) + \Sigma N_i\sigma_i{}^2}$$

$$= \frac{2.45 \times 10^{12}}{(25 + 9.8)10^8} = \frac{6 \times 10^4}{34.8} = 1720 \qquad (57)$$

The sample-sizes in Classes 2, 3, and 4 will be proportional to $N_i\sigma_i$ in Table 5.   That is,

$$n_2 = \frac{1.92n}{2.45}, \qquad n_3 = \frac{.37n}{2.45}, \qquad n_4 = \frac{.16n}{2.45} \qquad (58)$$

wherein $n = 1720$.

TABLE 5

BASIC DATA AND BASIC CALCULATIONS FOR THE SAMPLE DESIGN

Class	N_i	σ_i	$N_i\sigma_i$	$N_i\sigma_i{}^2$	n_i (calcu-lated)	Zoning interval, Z
1 $10,000 and over	395	2750	xxx	xxx	395	xxx
2 $1000–$9999	3837	500	1.92×10^6	9.60×10^8	1350	30
3 $100– $999	7467	50	.37	.19	260	280
4 0– $99	8022	20	.16	.03	110	700
Sum	19,721	xxx	$N\bar{\sigma}_w = 2.45$	$N\sigma_w{}^2 = 9.82$	2115	xxx

The zoning intervals will be $N_i/.1n_i$, rounded to some convenient figure (Table 5).   The next step is to make sure that every lot in every class has a serial number 1, 2, 3, etc.; then to read out from a table of random numbers 10 unduplicated random starts in each class between 1 and the

312 PART II. REPLICATED SAMPLING DESIGNS

zoning interval for that class (Table 6). We may then build up the sampling tables by adding the proper zoning interval to the random starts, as we have done before. The sampling tables so produced will draw 10 systematic subsamples in each class.

TABLE 6

THE RANDOM STARTS IN EACH CLASS

Kendall and Smith, *Random Sampling Numbers* (Cambridge, 1954), 64th Thousand, Columns 6, 7, 8, line 1

Class	Zoning interval	1	2	3	4	5	6	7	8	9	10
2	30	15	30	25	16	04	06	22	24	03	23
3	280	119	108	040	098	168	106	058	070	076	224
4	700	124	486	661	056	229	181	575	689	473	630

To process the sample, determine the number of items in each lot in the sample, the standard cost per item both now and a year ago, and by multiplication (extension), the dollar value of each lot now (x) and a year ago (y). Sums of the dollar values by subsample in each class will furnish the figures needed in the equations for the estimates.

The estimate from Subsample i for the total inventory at present is calculable by the formula

$$X^{(i)} = A_1 + 30x_2^{(i)} + 280x_3^{(i)} + 700x_4^{(i)} \tag{59}$$

where A_1 is the present value of Class 1, $x_2^{(i)}$ is the present value of the inventory in Subsample i in Class 2, with similar definitions in Classes 3 and 4. The estimate X from the total sample will be the average of the individual 10 estimates. Or, we may write directly

$$X = A_1 + 3x_2 + 28x_3 + 70x_4 \tag{60}$$

where x_2, x_3, and x_4 are the present values of the inventories in Classes 2, 3, and 4 in all 10 subsamples combined. There will be similar equations for the estimates of the inventory a year ago, with Y in place of X, and with B_1 in place of A_1.

The results are in Table 7. The averages at the bottom give

$X = \$21,062,000$, the estimate of the inventory now
$Y = \$19,977,000$, the estimate of the inventory a year ago
$X - Y = \$1,085,000$, the estimate of the LIFO adjustment
$f = \dfrac{X}{Y} = 1.0544$, the estimate of the relative LIFO adjustment, or the LIFO index

For the standard error of $X - Y$ we compute from Table 7 the 10 values of $X^{(i)} - Y^{(i)}$ in the accompanying array. The maximum is 1288; the minimum is 850:

1.	1215	6.	1142
2.	1129	7.	1288
3.	862	8.	1210
4.	1005	9.	850
5.	1100	10.	1058

hence an estimate of the standard error of $X - Y$ is

$$\hat{\sigma}_{X-Y} = \tfrac{1}{10}(1288 - 850)1000 = \$43{,}800 \tag{61}$$

TABLE 7

RESULTS OF THE INVENTORY NOW AND A YEAR AGO

$A_1 = \$7,714,000 \quad B_1 = \$7,226,000$

Sub-sample	$30x_2$	$280x_3$	$700x_4$	$X^{(i)}$	$30y_2$	$280y_3$	$700y_4$	$Y^{(i)}$	$f^{(i)}$
$i = 1$	10 068	3 118	231	21 131	9 364	3 020	306	19 916	1.0610
2	9 515	3 419	440	21 088	9 141	3 191	402	19 959	1.0566
3	10 044	2 823	168	20 746	9 650	2 796	212	19 884	1.0434
4	10 859	2 614	228	21 417	10 365	2 597	224	20 412	1.0492
5	9 769	2 938	319	20 741	9 206	2 894	315	19 641	1.0560
6	9 763	2 959	549	20 985	9 405	2 667	545	19 843	1.0576
7	10 317	3 200	197	21 428	9 617	3 103	194	20 140	1.0639
8	9 703	2 808	326	20 551	9 045	2 756	314	19 341	1.0626
9	10 150	3 048	275	21 186	9 833	3 014	264	20 336	1.0418
10	9 803	3 566	270	21 353	9 384	3 415	270	20 295	1.0521
Average	9 999	3 049	300	21 062	9 501	2 945	305	19 977	1.0544

Add 000 to every figure.

For the standard error of f we note from Table 7 that the maximum estimate of f is 1.0639 and that the minimum is 1.0418, whence

$$\hat{\sigma}_f = \tfrac{1}{10}(1.0639 - 1.0418) = .0022 \tag{62}$$

The standard errors calculated above do not take advantage of the finite multiplier in Class 2. As Class 2 dominates the variance, and as n_i/N_i in Class 2 is $\tfrac{1}{3}$, a new calculation with retention of the finite correction would reduce the standard error by about 11 %. Our corrected estimate of the standard error of $X - Y$ is thus about \$40,000, which is safely below the standard error aimed at, and in any event is good enough. (See further discussion in Exercise 3, page 317.)

It is important to compare the estimate X with the book-value of the inventory, in order to discover gross blunders in carrying out the sampling procedure, or whether the frame omitted part of the inventory, or to discover differences in methods of accounting.

Remark 2. The prices collected in the sample made possible the calculation of the standard deviations of the LIFO changes in the 4 classes shown in Table 8, and we may now compare these standard deviations with those assumed in advance in Table 5 as the basis for the sample design. The agreement is good in Classes 2, 3, and 4. The agreement is bad in Class 1, but fortunately the sample in Class 1 was 100%, so the precision of the sample was unaffected by the difference. As a matter of fact, the preliminary work on Class 1 was curtailed because of an early decision to sample it 100%.

TABLE 8

THE STANDARD DEVIATIONS IN THE 4 CLASSES

Class	Assumed	Calculated from the entire sample in each class
1	2750	$4700
2	500	528
3	50	77
4	20	17

Remark 3. Note that one may compute the LIFO adjustment for any stratum, along with its standard error.

The sample design for the audit. The sample just described was audited by finding anew and independently the quantities and the prices of each item in a small sample of the main sample, and by computing with the new prices the LIFO adjustment (also the unrealized earnings and certain other characteristics of the inventory not mentioned in the preceding section). The steps below describe the sampling procedure for the audit, which covered not only the plant that formed the subject of the preceding section, but also several other plants owned by the same corporation.

Incidentally, as a result of comparing the original quantities and prices in the main sample with those in the audit, and by comparing estimates made from the main sample and from the audit, it was deemed necessary to refine the procedures for determining prices and quantities, especially for items that had come in during the year to replace items discontinued, and to rework part of the main sample. Without the audit, there would have been no quantitative measure of the reliability of the equal complete coverage that the main sample was drawn from.

Sampling Procedure for the Audit

1. The frame for the audit will consist of the items in the main sample. The arrangement and the sampling ratios are in Table 9.

TABLE 9

SCHEME FOR THE SAMPLING RATIOS IN THE AUDIT

Class	Size	Sampling ratio
Stratum 00	$100,000 and over	1 : 1
Stratum 0	50,000–$99,999	1 : 2
Stratum 1	10,000– 49,999	1 : 9
Stratum 2	1,000– 9,999	1 : 20
Stratum 3	100– 999	1 : 20
Stratum 4	0– 99	1 : 20

2. Give a local serial number to every item in the main sample. Subsample 1 of the audit (designated hereafter as Subsample 1A) will consist of all of Stratum 00, plus a sample of every other stratum, according to the scheme shown in the table. The procedure may be clear from the following steps.

3. To draw Subsample 1A:

Include all of Stratum 00.
Draw 1 item at random from every consecutive 20 in Stratum 0.
Draw 1 item at random from every consecutive 90 in Stratum 1.
Draw 1 item at random from every consecutive 20 in Subsample 1 of the main sample from Strata 2, 3, 4.

4. To draw Subsample 2A:

Include all of Stratum 00, as this stratum will belong to every subsample of the audit.
Draw 1 item at random from the 19 that remain in every consecutive 20 in Stratum 0.
Draw 1 item at random from the 89 that remain in every consecutive 60 in Stratum 1.
Draw 1 item at random from every consecutive 20 in Subsample 2 of the main sample from Strata 2, 3, 4.

5. To draw Subsamples 3A, 4A, . . . , 10A, proceed likewise with appropriate changes. Subsample 3A comes from Subsample 3 of the main sample in Strata 2, 3, 4; Subsample 4A from Subsample 4; etc.

6. Punch into every card in the audit the quantity and prices shown in the main sample, and the quantity and prices obtained independently in the audit.

7. To expand Subsample 1A to the original frame whence came the main sample; proceed as shown below. The characteristic to be estimated might be any extensive characteristic such as the book-inventory with the original prices from the main sample; the book-inventory with prices obtained in the audit; the LIFO adjustment obtained in the audit.

a. Compute the contribution from Stratum 00.   This stratum belongs to every subsample.

b. Multiply by 20 the result in Subsample 1A in Stratum 0.

c. Multiply by 90 the result in Subsample 1A in Stratum 1.

d. Multiply by $20Z_2$ the result in Subsample 1A in Stratum 2, where Z_2 is the zoning interval in the main sample in Stratum 2.

e. Multiply by $20Z_3$ the result in Subsample 1A in Stratum 3, where Z_3 is the zoning interval in the main sample in Stratum 3.

f. Multiply by $20Z_4$ the result in Subsample 1 of the audit in Stratum 4, where Z_4 is the zoning interval in the main sample in Stratum 4.

g. Add parts *a, b, c, d, e, f.*

8. Do likewise for Subsamples 2A, 3A, . . . , 10A.   Part *a* from Stratum 00 belongs to every subsample.

9. The final estimate derived from the audit will be the average of the results so obtained for the 5 subsamples.

10. Compare, for every item in the audit, the price in the audit with the corresponding price in the main sample; also the quantities and any other information.   Make a tally of the number of equalities, and the number of prices higher in the audit, and the number of prices lower; the number of quantities higher in the audit, and the number of quantities lower.   Look for persistent differences, and for their causes.

11. Form here any ratio desired, such as the LIFO index.   Each subsample of the audit will furnish a ratio; the final estimate will come from all 10 subsamples combined.

12. Estimate the standard error of any estimate as $R/10$, where R is the range between the 10 results of the 10 subsamples in the audit.

13. Compare (*a*) each estimate obtained from the audit with (*b*) the corresponding estimate obtained from the main sample.   Look further for indications of persistent differences, and for their causes.

14. Decide whether the main sample is acceptable.

Results of a replicated design of an audit.   I present here a set of 10 ratios calculated for one of the characteristics studied in the audit described

in the preceding paragraph.   These ratios are not typical: no illustration
is.   They are interesting because it so happened here that all 10 ratios
were greater than 1, which fact pointed definitely to the existence of
persistent errors in the main sample that caused it to produce an underestimate (p. 412).   In fact, the lower limit of underestimate is 1.0019.   An

RATIO OF AUDIT TO MAIN SAMPLE, BY SUBSAMPLE

1. 1.1045	6. 1.0144
2. 1.0066	7. 1.0974
3. 1.0019	8. 1.0167
4. 1.0122	9. 1.0250
5. 1.2739	10. 1.0150

underestimate of 1% meant loss of $100,000 in taxes.   Study of the reason
for the high ratio in Subsample 5 disclosed the existence of a special type of
error.   Investigation and study of the procedures used in the main sample
led to the discovery of various other errors and their causes, and led to a
decision of management to reprice the main sample with improved
definitions and with special training and care.

EXERCISES

1. Compute the gain in Neyman sampling over proportionate sampling
in Classes 2, 3, 4, for the sample described on page 310.

SOLUTION

Eq. 38 on page 297 gives $B/C = \sigma_w^2/(\bar{\sigma}_w)^2$, whose numerical value in
this example is $(9.82 \times 10^8/19{,}721)(19{,}721/2.45 \times 10^6)^2 = 3.2$.   Thus, a
proportionate sample, to give equivalent precision, would have been 3.2
times as big in Classes 2, 3, 4.   This example shows the importance of using
the right plan.   The student should guard himself from expecting similar
gains in all problems: the difference between proportionate and Neyman
allocation is often small.

2. Compute the estimate of the standard deviation of $X - Y$ by the sum
of the 10 squares.   (*Answer:* $46,500 with no allowance for nonreplace-
ment, to compare with $43,800 by the range.)

3. *a.* Compute with the new standard deviations (σ_i) in Table 8 the
standard deviation of $X - Y$ by substitution into Eq. 20 to take
account of nonreplacement. (*Answer:* $\operatorname{Var}(X - Y) = \sum \left(1 - \dfrac{n_i}{N_i}\right)\dfrac{\sigma_i^2}{n_i} =$
35.376×10^8, the square root of which gives $59,500 for the standard
error.)

b. How do you reconcile this calculation of the standard error with the lower values calculated from the range and from the sum of squares ($43,800 and $46,500)?

There are 2 reasons possible. (1) One may expect some gain from the fact that the use of zones introduces proportionate stratified sampling within each class, although it is difficult without a test to say how much this gain may be. (See pages following for an example of a test.) (2) The previous estimates were subject to a standard error of about 23%, because they had only 9 degrees of freedom (p. 441), and could by accident have been low as well as high. My own interpretation is that the stratification in zones yielded some gain, and that our estimate thereof is the difference between $59,500 and $40,000.

Simultaneous Neyman allocation in several locations. It sometimes happens that the aim of a sampling plan is to deliver an estimate X for the consolidated inventory at several locations. It has been determined, let us suppose, that the Neyman allocation, like the one in the preceding section, is the best plan in most of the locations. The size of sample (n_j) in Location j will then be proportional to $\Sigma N_i \sigma_i$ for that location. Formally:

1. Define the classes in each location. (It is not necessary to define the classes uniformly in all locations.)
2. Compute $N_i \sigma_i$ for Class i separately for every class in every location.
3. Compute the sum $\sum_i N_i \sigma_i$ for every location. Use the subscript j for Location j. Thus, $(N_i \sigma_i)_j$ will refer to Class i in Location j, and $(\sum_i N_i \sigma_i)_j$ will be the sum over all classes in Location j.

Omit from the summation any class that will be sampled 100%; see the preceding example.

4. Compute the grand sum $\sum_j (\sum_i N_i \sigma_i)_j$ over all locations. Then the size of the sample required at Location j will be

$$n_j = n \frac{(\sum_i N_i \sigma_i)_j}{\sum_j (\sum_i N_i \sigma_i)_j} \tag{63}$$

where n is the required total sample in all plants combined.

5. The ratio of the samples in 2 plants, j and k, will be

$$\frac{n_j}{n_k} = \frac{(\sum_i N_i \sigma_i)_j}{(\sum_i N_i \sigma_i)_k} \tag{64}$$

6. The sample in Class i in Location j will be

$$n_{ji} = n_j \frac{(N_i \sigma_i)_j}{(\sum\limits_i N_i \sigma_i)_j}$$

$$= n \frac{(N_i \sigma_i)_j}{\sum\limits_j (\sum\limits_i N_i \sigma_i)_j} \tag{65}$$

7. The ratio of the sizes of any 2 classes i and i' in Locations j and k will be

$$\frac{n_{ji}}{n_{ki'}} = \frac{(N_i \sigma_i)_j}{(N_{i'} \sigma_{i'})_k} \quad \begin{array}{l} \text{[Set } j = k \text{ if the 2 classes} \\ \text{are both in Location } j] \end{array} \tag{66}$$

Note that this plan is not aimed at a reliable separate figure for any one location, but only for the consolidated inventory in all locations. If one wishes a separate figure for any one location, he should design a separate sample therefor.

EXERCISES

These 5 exercises deal with biases and variances that arise from non-standard but possible modes of allocation in stratified sampling.* The statistician sometimes meets such allocations, especially in the analysis of samples already selected and tested, and needs to have methods by which to compute the bias and variance. Methods that one may use for solution are in Chapter 16 on pages 369 ff.: simple theorems on variance are in Chapter 17.

1. Write the estimate

$$\hat{a}_{(1)} = \frac{1}{n} \Sigma n_i x_i$$

The summation here and hereafter will run over all M strata.

Show that:

a. This estimate is a biased estimate of a unless the sampling is proportionate. The bias is

$$E\hat{a}_{(1)} - a = \frac{1}{n} E\Sigma n_i(\bar{x}_i - a) = \Sigma \frac{n_i}{n}(a_i - a)$$

b. The estimate is consistent; i.e., the bias decreases to 0 as all the sample-sizes n_i approach N_i.

c. $$\text{Var } \hat{a}_{(1)} = \frac{1}{n^2} \Sigma \left(1 - \frac{n_i}{N_i}\right) n_i \sigma_i^2$$

* These exercises were inspired by Section 15 in P. Thionet's *Méthodes Statistiques Modernes* (Hermann, Paris, 1946).

2. Write the estimate

$$\hat{a}_{(2)} = \frac{1}{M} \Sigma \bar{x}_i = \text{av. } \bar{x}_i$$

Show that

a. The bias of this estimate is

$$E\hat{a}_{(2)} - a = \Sigma \left\{ \frac{1}{M} a_i - \frac{N_i}{N} a \right\} = \frac{1}{M} \Sigma \left(a_i - \frac{N_i}{\bar{N}} a \right)$$

The bias is therefore 0 only if $N_1 = N_2 = N_3 = \cdots = \bar{N}$, where $\bar{N} = N/M$.

b. The bias of this estimate is inconsistent; i.e., the bias if any does not diminish to 0 as n_i approaches N_i.

3. Write the estimate

$$\hat{a}_{(3)} = \Sigma c_i \bar{x}_i \qquad [\text{where } \Sigma c_i = 1]$$

Show that

a. The bias of this estimate is

$$E\hat{a}_{(3)} - a = \Sigma c_i (a_i - a)$$

which is 0 if $c_i = N_i/N$.

b. This estimate is inconsistent if $c_i \neq N_i/N$; the bias does not then diminish to 0 as n_i approaches N_i.

c.

$$\text{Var } \hat{a}_{(3)} = \Sigma c_i^2 \left(1 - \frac{n_i}{N_i} \right) \frac{\sigma_1^2}{n_i}$$

4. If the sampling is proportionate, and if $c_i = N_i/N$ in $\hat{a}_{(3)}$, then $\hat{a}_{(1)}$ and $\hat{a}_{(3)}$ are identical and are unbiased.

5. The variance of $\hat{a}_{(3)}$ is a minimum for fixed sample-sizes n_i, if

$$c_i = \frac{1}{\text{Var } \bar{x}_i} \div \Sigma \frac{1}{\text{Var } \bar{x}_i}$$

C. STRATIFICATION AFTER SELECTION

Why stratify the entire frame before we draw the sample? If the cost of classifying a sampling unit were zero, one could always safely recommend fantastic plans of stratified sampling, with no worry about costs. The fact is, though, that there is always a price to pay: the cost to classify a sampling unit and to rearrange the frame may be small, or it may be large, but it is never zero. There is a price to pay for rearrangement even when the information for classification already exists, as in Census tables. The total cost of a survey includes the cost of stratification. When the frame contains thousands of sampling units, the cost of classifying every sampling unit within the frame may outrun any possible savings that could accrue from further rearrangement. Cost and time will run especially high if the information by which to classify a sampling unit requires a test or an interview or a search to discover which class a sampling unit belongs to. It is fortunately possible to draw up plans of stratified sampling that require only the classification of the sampling units in the sample. This idea, when we can use it, will shrink the cost of classification, and it will also speed up the work. However, we must not jump too rapidly. We must look at costs and variances. We shall learn first something about variances.

We start off by drawing a sample with only the natural stratification that exists in the frame, just as if we were going to use Plan A. We have to decide first how many sampling units to draw. Anyway, we classify only these sampling units, and pay no further attention to the sampling units that remain in the frame, except that we shall certainly use, in our estimates, some of the statistical information that we have concerning the frame. For example, if we know the proportions P_1, P_2, P_3, we shall certainly use them, as we shall do in Plans D, E, F, G. If we know not the proportions P_1, P_2, P_3 in the frame, we must estimate them from the sample. We may even classify a fairly large preliminary sample of size N', just to get good estimates of P_1, P_2, P_3; then this preliminary sample will serve as a miniature frame whence to draw the final sample (proportionate, Plan H; or Neyman, Plan I).

Plans D, E, F, G. These plans (like Plans B and C) all require advance knowledge of P_1, P_2, P_3.

Plan D

1. Compute by Eq. 68 ahead the size n of the sample desired. Proceed to draw this sample exactly as in Plan A.

2. Compute the required zoning interval $Z = 2 N/n$ for 2 subsamples, $10 N/n$ for 10 subsamples. Form zones of Z sampling units in the frame.

3. Read out random numbers between 1 and Z to draw 1 sampling unit from every zone. These drawings form Subsample 1.

4. Classify each sampling unit into its proper stratum. Mark each sampling unit to show that it belongs to Subsample 1, and to show which stratum and which zone it came from.

5. Repeat Steps 3 and 4 for Subsample 2, and for the other subsamples if any. The sample-sizes $n_1^{(i)}$, $n_2^{(i)}$, $n_3^{(i)}$ thus drawn will be random

variables. (The subscripts refer to the strata; the superscript is the subsample.)

6. Carry out the interviews or the tests on the entire sample.

7. Let $x_1^{(i)}$ denote as before the x-population in Subsample i from Stratum 1, with similar definitions for $x_2^{(i)}$, $x_3^{(i)}$. Form for Subsample i the estimate

$$X^{(i)} = N_1 \frac{x_1^{(i)}}{n_1} + N_2 \frac{x_2^{(i)}}{n_2} + N_3 \frac{x_3^{(i)}}{n_3}$$
$$= X_1^{(i)} + X_2^{(i)} + X_3^{(i)} \tag{67}$$

of the total x-population. Do this for every sample. The 3 terms, one by one, are the estimates that Subsample i furnishes for the x-population in the 3 strata.

8. Form the final estimate

$$X = \text{av. } X^{(i)}$$

and if desired

$$\bar{x} = \frac{X}{N}$$

for the average x-population per sampling unit over the whole frame.

9. Calculate $\hat{C}_X = \hat{C}_{\bar{x}}$, following the advice in Plan A (p. 283). For example,

$$\hat{C}_X = \hat{C}_{\bar{x}} = \frac{1}{10X} [X_{\max}^{(i)} - X_{\min}^{(i)}]$$

for 10 subsamples by use of the range.

Remark 1. Convenient modifications will occur to the user in practice. For example, it may be necessary to defer the classification of a sampling unit until after the final test or interview. This state of affairs introduces no complication. One simply proceeds into Step 7 after the final tests are complete. The reader may note that under these circumstances, Plan E is impossible, as the special feature of Plan E is the reduction of some of the strata *before* the final test.

Another point is that one may calculate the estimates X_1, X_2, X_3 for the strata separately, if there be need of them.

Then also, the user will in practice probably prefer, as I do, to make all his drawings from Zone 1 before he proceeds into Zone 2. The 1st sampling unit drawn from any zone belongs to Subsample 1; the 2d sampling unit belongs to Subsample 2; etc. The above description of the steps, as written, has the advantage, I hope, that it leaves no doubt about the independence of the subsamples, except for the fact that one will usually draw them without replacement.

The formula for Var $\bar{x}$, which one will use in the planning stages is this:

$$\text{Var } \bar{x} = \left(\frac{1}{n} - \frac{1}{N}\right)\left(\sigma_w{}^2 + \frac{1}{n}\sigma_R{}^2\right)$$

$$\doteq \frac{1}{n}\left(\sigma_w{}^2 + \frac{\sigma_R{}^2}{n}\right) \qquad \text{[Plan D]} \qquad (68)$$

where

$$\sigma_w{}^2 = P_1\sigma_1{}^2 + P_2\sigma_2{}^2 + \text{etc.} \qquad \begin{array}{l}\text{[The variance within} \\ \text{strata, as before]}\end{array} \qquad (69)$$

$$\sigma_R{}^2 = Q_1\sigma_1{}^2 + Q_2\sigma_2{}^2 + \text{etc.} \qquad \begin{array}{l}\text{[The reverse internal} \\ \text{variance]}\end{array} \qquad (70)$$

Remark 2. The derivation of the variance for Plan D will come in the exercises, page 331. The variance of Plan D exceeds the variance in Plan B only by the addition of the term $\sigma_R{}^2/n$ to $\sigma_w{}^2$. This additional term will in practice often be negligible, but one must be careful.

As an exercise, the student may prove that although $n_1^{(i)}$, $n_2^{(i)}$, etc., are random variables, $\bar{x}_1^{(i)}$, $\bar{x}_2^{(i)}$, etc. are unbiased estimates of the stratum means a_1, a_2, etc.; and that $X^{(i)}$ is an unbiased estimate of A. (The superscript denotes Subsample i.) The same is true of the remaining plans.

PLAN E

1. Decide with the help of Eq. 72 the size n of the final sample desired. Decide also with the help of Eq. 76 the size n' for the preliminary sample. Proceed to draw the sample of n' exactly as in Plan A.

2. Compute the zoning interval $Z = 2 N/n'$ for 2 subsamples, 10 N/n' for 10 subsamples. Form the zones in the frame; see Step 2 under Plan B.

3, 4, 5. Proceed as in Steps 3–5 of Plan D. The sizes of the preliminary subsamples so drawn will be random variables in every stratum and in every subsample.

6. Decide on the most likely ratios $\sigma_1:\sigma_2:\sigma_3$ for the chief characteristic that the sample is to measure (hereafter, the x-population).

7. Fix for the next step the thinning ratios by the Neyman allocation

$$\frac{n_1}{n_1'} : \frac{n_2}{n_2'} : \frac{n_3}{n_3'} = \sigma_1:\sigma_2:\sigma_3 \qquad (71)$$

in which n_1', n_2', etc., are the sizes of the preliminary sample in the several strata.

8. Thin the strata that require thinning. (See the appendix on thinning, page 337.)

The thinning ratios are relative, and the size n' of the preliminary sample was supposedly chosen so that the class with the biggest σ_i will not take any thinning at all.   Thus, if $\sigma_i : \sigma_2 : \sigma_3$ were $1 : 2 : 4$, we should leave Stratum 1 as it is;   retain 1 sampling unit at random from every 2 of Stratum 2;   retain 1 sampling unit from every 4 of Stratum 3.   A convenient arrangement is to tie the beginning of Subsample 2 to the end of Subsample 1, to leave no gap.   The final sample-sizes $n_1^{(i)}$, $n_2^{(i)}$, etc., in Subsample i will all be random variables.

9, 10, 11.   Proceed as in Steps 7, 8, 9 of Plan D.   The formula for Var $\bar{x}$, which one will need in the planning stages, is

Plan C

$$\text{Var } \bar{x} = \overbrace{\frac{(\bar{\sigma}_w)^2}{n} - \frac{\sigma_w{}^2}{N}}^{} + \frac{1}{n}\left(\frac{1}{n'} - \frac{1}{N}\right)\bar{\sigma}_w\bar{\sigma}_R \qquad \text{[Plan E]}$$

$$\doteq \frac{1}{n}\left\{(\bar{\sigma}_w)^2 + \frac{1}{n'}\,\bar{\sigma}_w\bar{\sigma}_R\right\} \qquad \begin{array}{l}\text{[If } N \text{ is big compared} \\ \text{with } n'\text{]}\end{array} \qquad (72)$$

where

$$\bar{\sigma}_R = Q_1\sigma_1 + Q_2\sigma_2 + \text{etc.} \qquad \begin{array}{l}\text{[The reverse internal standard} \\ \text{deviation]}\end{array} \qquad (73)$$

We may replace n' by $n/\bar{T}$, where $\bar{T}$ is the average thinning ratio, defined by

$$\bar{T} = \Sigma P_i T_i \qquad (74)$$

where T_i is the thinning ratio in class i.   Thus, if we were to retain all the sampling units of Class 1 that appear in the preliminary sample n', and half the sampling units of Class 2, then would

$$\bar{T} = P_1 + \tfrac{1}{2}P_2 \qquad (75)$$

and

$$n = n'\bar{T} \qquad (76)$$

The reader may satisfy himself that $\bar{T} \leq 1$.

Remark 3.   The derivation of the variance for Plan E will come in the exercises, page 333.   The variance of Plan E exceeds the variance of Plan C only by the addition of the term in $\bar{\sigma}_R$.   This additional term will in practice often be negligible.

Exercise.   I am designing at this moment a sample of retail stores for a study of sales and employment.   There will be 10 subsamples.   Calculations show that for the most efficient design there should be in each subsample about the same number of stores of Class 1 (big stores) as there are of Class 2 (medium and small stores).   The 2 classes are mixed in the frame, although it is possible, at some small expense, to procure information concerning any store to discover whether it belonged in Class 1 or in Class 2 a few months ago, and this information will be recent enough. We may assume from other sources of information that there are about 3 times as many stores of Class 2 now as there are of Class 1.

The procedure will be to draw 10 preliminary subsamples of stores; then to thin out the stores of Class 2 with random numbers by retaining 1 in 3. I wish to be 2-sigma sure that there are at least 30 stores of Class 1 in each subsample.   This is Plan E.   (It would be Plan I if we knew not in advance sufficiently good approximations to P_1 and to P_2.)   Show that the preliminary subsamples should each contain about 168 stores of both classes combined.

Hint: Let $p = \frac{1}{4}$, the proportion of Class 1, and let n be the size of preliminary sample.   The square-root transformation (p. 461) gives

$$(\sqrt{np} - 1)^2 = 30$$

$$\sqrt{np} = \sqrt{30} + 1 = 6.5$$

$$np = 42$$

$$n = 168 \qquad \text{if } p = \tfrac{1}{4}$$

PLAN F

1. Decide the desired sample-size n, as in Plan B.   Compute also the sample-sizes $n_i = nP_i$ for Stratum i as in Plan B.

2. Compute the zoning intervals $Z_i = 2\,N_i/n_i$ for 2 subsamples, $10\,N_i/n_i$ for 10 subsamples.

3. Divide the frame into zones of Z_1 sampling units.   Draw a sampling unit from Zone 1.   If it belongs to Stratum 1, mark it so.   If it belongs to some other stratum, ignore it and draw another and another until you get one for Stratum 1.   Move into Zone 2 and repeat the procedure. Continue thus through all the zones.   The units so obtained belong to Subsample 1 of Stratum 1.

4. Repeat the same procedure for Subsample 2 and for the other subsamples, if any, to complete the subsamples for Stratum 1.

5. Divide the frame into zones of Z_2 sampling units and repeat Steps 3 and 4 to obtain the sample for Stratum 2.

6. Use the same procedure for Stratum 3, and for the other strata if any.

7–11. Proceed as in Steps 4–8 of Plan B.

Remark 4.   The user may often in practice simplify the above procedure by using a zoning interval that is some common denominator of the zones calculated in Step 2, adjusted slightly as necessary.   One draws sampling units from this common denominator until he has obtained the required multiples of the sample for the several strata. For example, if there were 2 strata, and if $Z_1 = 2Z_2$ after adjustment, Z_1 would be the common denominator. The procedure would be to draw sampling units from each zone of size Z_1 until one obtains 1 sampling unit for Stratum 1 and 2 for Stratum 2.

Caution: do not alter too far the zoning intervals calculated in Step 1 (why?).

PLAN G

1. Decide the desired sample-size n, as in Plan C. Compute also the sample-sizes $n_i = n P_i \sigma_i / \bar{\sigma}_w$ for Stratum i as in Plan C.

2. Compute the zoning intervals $Z_i = 2 N_i / n_i$ for 2 subsamples, 10 N_i / n_i for 10 subsamples.

3–6. Proceed as in Steps 3–6 in Plan F.

7–11. Proceed as in Steps 4–8 of Plan C. See Remark 4 under Plan F.

Remark 5. In Plans F and G, the sample-sizes are fixed in advance: in Plans D and E they are not; they are random variables. That is why, as the student may have observed, in Plans F and G we may form the estimate X directly, whereas in Plans D and E we must form separate estimates by stratum and then add them. The variances of the plans with fixed sample-sizes are slightly smaller than the variances of the plans with variable sample-sizes for the same total number (n) of sampling units. The difference lies in the term σ_R^2 / n. However, in Plans F and G with fixed sizes there is the additional cost of classifying some units, only to find that we can not use them in the final sample because the quotas (n_i) are already filled. This additional cost appears in the table of costs on page 350.

Choice between Plans D and F. It was presupposed in the treatment of both these plans that information exists in records that are already on hand or obtainable (as by purchase of a directory or of Census tables) by which to classify a sampling unit into one stratum or another. In respect to costs, Table 15 on page 350 shows that they are about equal for a prescribed precision. However, one must remember that this table shows costs only up to the stage where tabulation commences. If the tabulations are simple, and if there is little extra weighting* to do in the formation of the estimates, then there will be little difference between the 2 plans.

But this is not the whole story. Both Plans D and F will often be used to obtain estimates of proportions in fine classes, such as the proportion of males that are of age 20–29 and employed in a particular occupation; also for ratio-estimates of the total population in such

---

* By extra weighting I mean multiplications or other weighting that is sometimes required in the tabulation. Examples: (*a*) multiplication by 3 of every interview obtained in a family of 3 persons when the rules called for a random selection within such a family; (*b*) use of the Politz plan in the formation of the estimates, to correct for people not at home (Ch. 5); (*c*) weighting the results for people not interviewed.

classes. In a heavy tabulation program, the weighting feature of Plan F may thus be a distinct advantage.

Bias from the use of erroneous proportions. It may be that the proportions P_i that we should like to use in Plans D, E, F, or G are not known exactly, being derived from data now to some extent outmoded. Wrong proportions P_i, if we use them, will introduce a certain amount of bias. One way out is to take a preliminary sample that is big enough to get respectable estimates of the proportion P_i: this is exactly what Plans H and I will do. As we shall see, though, we do not get something for nothing: the estimates of the P_i will introduce no bias, but they will add a term to the variance (pp. 329 and 331). This problem will not arise in Plans B and C because there we ordinarily count the sampling units as we classify them into the various strata.

It often turns out that the effect of wrong proportions is very small, as will be the case when the means of the strata are not far different. On the other hand, the effect may be uncomfortable if the means of the strata are wide apart.

Sukhatme* makes the point that the bias from incorrect proportions does not diminish as the size of the sample increases; hence, if we contemplate taking a big sample in the hope of achieving good precision, we may deceive ourselves in the use of incorrect proportions: it may be preferable to estimate them from the sample. We fortunately have the proper theory, or shall soon have it, by which we can make a choice, based on comparison between: (a) the possible bias from incorrect weights, and (b) the expected increase in the standard error that will result from estimating the proportions from the sample itself.

We divert our attention now to an arithmetic calculation of the bias to expect under assumed numerical values of the correct proportions, and of the stratum means. I have borrowed the illustration from Sukhatme. This is a very real and important type of calculation, and easy to make. I could draw from my files a dozen like it. Suppose that (with Sukhatme)

$$P_1 \text{ (correct value)} = .85; \quad \text{(incorrect)} = .80$$
$$P_2 \text{ (correct value)} = .15; \quad \text{(incorrect)} = .20$$
$$\text{Bias in } \bar{x} = (.80 - .85)a_1 + (.20 - .15)a_2 = .05(a_1 - a_2)$$

The statements made earlier follow at once. If a_1 and a_2 are not far apart, say $.20a_1$, then the bias in $\bar{x}$ is only $.05 \times .20a_1 = .01a_1$. But if $a_1 = 5a_2$, then the bias in $\bar{x}$ will be $.05 \times 4a_1 = .2a_1$, and it might be

* P. V. Sukhatme, *Sampling Theory of Surveys with Applications* (Iowa State College, Ames; and the Indian Society of Agricultural Statistics, New Delhi, 1953): p. 109. See also William G. Cochran's *Sampling Techniques* (Wiley, 1953): p. 102.

preferable to use Plan H or Plan I.   Each problem requires consideration
on its own merits, with its own arithmetic calculations.

Plans H and I.   In the previous plans, the proportions P_i were known,
and we made use of them in fixing the sizes of the samples in the various
strata, and in forming the estimates.*   We now encounter the problem
where the proportions P_i are not known.   What we do is to estimate the
proportions P_i from classification of a preliminary sample of size N',
which, the reader will observe, serves as a new frame.   Once we have
this new frame, Plan H will resemble Plan B; Plan I will resemble Plan C.

PLAN H

Proportions P_i Not Known

1. Decide with the help of Eq. 79 the size n of the final sample desired.

2. Compute the optimum size N' of the preliminary sample.   The
equation for this is

$$\frac{n}{N'} = \frac{\sigma_w}{\sigma_b} \sqrt{c_1/c_2} \qquad \text{[P. 336]} \qquad (77)$$

c_1 is the cost of classifying a sampling unit in the preliminary sample, and
c_2 is the cost of interviewing or testing a sampling unit in the final sample.

3. Compute the zoning interval $Z = 2 \, N/N'$ for 2 subsamples, $10 \, N/N'$
for 10 subsamples.   Form zones of Z consecutive sampling units in the
frame.   (See Remark 4 on page 325.)

4. Draw from the frame, just as you would in Plan A, a preliminary
sample of size N'.   Mark each sampling unit to show which subsample
it belongs to, and which zone it came from.

5. Carry out the preliminary tests or interviews on the entire preliminary
sample to acquire information by which to classify every sampling unit
therein.   Mark each sampling unit to show which stratum it belongs to.
Maintain in each stratum the order drawn.

Remark 6.   We meet here for the first time the possibility that the
sampling units in the various strata have never been classified nor counted.

6. Record for Subsample i the number of sampling units in each
stratum, $N_1^{(i)}$, $N_2^{(i)}$, etc.   These counts give the estimates

$$\hat{P}_1^{(i)} = (N_1' : N')^{(i)}, \qquad \hat{P}_2^{(i)} = (N_2' : N')^{(i)}, \qquad \text{etc.} \qquad (78)$$

of the proportions P_1, P_2, etc.   Form these estimates for every subsample,
and for all subsamples combined.

* An excellent discussion of Plans H and I occurs in P. V. Sukhatme, *Sampling Theory
of Surveys with Applications* (Iowa State College Press, Ames; and the Indian Society
of Agricultural Statisticians, New Delhi, 1953).

7. Reduce the number of sampling units proportionately in all strata and in all subsamples, to reach the required total sample of size n, decided in advance.   For example, to retain 1/3d of the preliminary sample, select by random numbers, for the final sample, 1 out of every 3 consecutive sampling units in the preliminary sample (p. 337).

8. Carry out on the final sample the main interviews or the main tests (as distinct from any preliminary tests required in Step 5).

9. Calculate for Subsample i

$$\bar{x}^{(i)} = \Sigma P_1^{(i)} \frac{x_1^{(i)}}{n_1^{(i)}} \qquad \text{[Sum over all strata]}$$

$$\bar{x} = \text{av. } \bar{x}^{(i)}$$

$x_1^{(i)}$ is the population in Subsample i from Stratum 1.

10. Calculate

$$\hat{C}_{\bar{x}} = \frac{1}{10\bar{x}} [\bar{x}_{\max}^{(i)} - \bar{x}_{\min}^{(i)}]$$

for 10 subsamples, or use the sum of squares if you prefer.   For 2 subsamples, form enough thick zones, as in Plan A, page 284.

The formula for Var $\bar{x}$, which one will need in the planning stages, is

$$\text{Var } \bar{x} = \left(\frac{1}{n} - \frac{1}{N}\right)\sigma_w^2 + \frac{1}{n}\left(\frac{1}{N'} - \frac{1}{N}\right)\sigma_R^2 + \left(\frac{1}{N'} - \frac{1}{N}\right)\sigma_b^2 \qquad \text{[Plan H]}$$

$$\doteq \frac{\sigma_w^2}{n} + \frac{\sigma_b^2}{N'} \qquad \text{[N large compared with N']} \tag{79}$$

The derivation of this variance will come in the exercises, page 334.

PLAN I

1. Decide with the help of Eq. 82 the size n of the final sample desired.

2. Compute the optimum size N' of the preliminary sample.   The equation for this is

$$\frac{n}{N'} = \frac{\bar{\sigma}_w}{\sigma} \sqrt{c_1/c_2} \qquad \begin{array}{l}\text{[Due to Neyman, 1938.}\\ \text{Reference on page 336]}\end{array} \tag{80}$$

c_1 is the cost of classifying a sampling unit in the preliminary sample, and c_2 is the cost of interviewing or testing a sampling unit in the final sample.

3 and 4.   Draw the preliminary sample N', as in Plan H.

5. Carry out the preliminary tests or interviews on the entire preliminary

sample to acquire information by which to classify every sampling unit therein.   Mark each sampling unit to show which stratum it belongs to. Maintain in each stratum the order drawn.

6. Decide on the most likely ratios $\sigma_1 : \sigma_2 : \sigma_3$ for the chief characteristic that the sample is expected to measure (hereafter, the x-population).   Fix the thinning ratios $n_i : N_i'$ by the Neyman relations

$$\frac{n_1}{N_1'} : \frac{n_2}{N_2'} : \frac{n_3}{N_3'} = \sigma_1 : \sigma_2 : \sigma_3 \tag{81}$$

in which N_1', N_2', etc., are the sizes in the various classes of the preliminary sample of total size N' (next step).

7. Reduce the number of units in each stratum by the thinning ratios decided in Step 6, to reach the final total sample-size n.   (See the appendix on thinning, page 337.)

For example, suppose that $\sigma_1 : \sigma_2 = 5 : 1$.   Then Step 6 gives

$$\frac{n_1}{N_1} : \frac{n_2}{N_2}$$

to be $5 : 1$. $n_1 : N_1$ is the thinning ratio in Stratum 1 ; $n_2 : N_2$ is the thinning ratio in Step 2.   More specifically, suppose that

$$n = 500, \text{ fixed in advance}$$
$$n : N' = 0.208, \text{ calculated by Eq. 80}$$

whence

$$N' = 2400, \text{ the preliminary sample}$$

Suppose that

$$N_1' = 900 \text{ in Stratum 1 after Step 3}$$
$$N_2' = 1500 \text{ in Stratum 2 after Step 3}$$

Let x be the thinning ratio in Stratum 1.   Then the thinning ratio in Stratum 2 will be $x/5$.   Moreover

$$900x + 1500x/5 = 500$$

whence $x = 5/12$, which leads to the following rules for thinning the 2 strata in this particular example:

1.   Out of every consecutive 12 sampling units in Stratum 1, retain 5 for the final sample.

2.   Out of every consecutive 12 sampling units in Stratum 2, retain 1 for the final sample.

The final sample-sizes will be $n_1 = 375$, $n_2 = 125$, with a total of $n = 500$, as required.

8–10. Same as Steps 8–10 under Plan H.

The formula for Var $\bar{x}$, which one will need in the planning stages, is

$$\text{Var }\bar{x} = \frac{(\bar{\sigma}_w)^2}{n} - \frac{\sigma_w^2}{N} + \frac{1}{n}\left(\frac{1}{N'} - \frac{1}{N}\right)\bar{\sigma}_w\bar{\sigma}_R + \left(\frac{1}{N'} - \frac{1}{N}\right)\sigma_b^2 \qquad \text{[Plan I]}$$

$$\doteq \frac{(\bar{\sigma}_w)^2}{n} + \frac{\sigma_b^2}{N'} \qquad [N \text{ large compared with } N' \text{ and with } n] \qquad (82)$$

The derivation of this variance will come in the exercises, page 335. Examples of Plan I occur in Part D.

Remark 7. Note that the variance σ_b^2 between strata is eliminated from the Var $\bar{x}$ in Plans B, C, D, E, F, and G, but not entirely in Plans H and I. In Plans H and I the variance σ_b^2 appears in the term σ_b^2/N', which diminishes with increase in N'.

Remark 8. The revised accounting system designed by Mr. John Perrin of the American Telephone and Telegraph Company for the inventory of telephone apparatus used by the companies of the great Bell System and by other telephone companies in America makes use of Plan I. Each sampling unit is some possible telephone number in a central office. Information ascertained for each telephone number in the preliminary sample determines whether it is a blank or is in service, and if so, which class of service. The samples of some of the classes are then thinned to economical sizes by Plan I, the thinning ratios being determined by the average value of the apparatus attached to a telephone number in its class of service.*

EXERCISES ON THE VARIOUS PLANS

1. Derive Eq. 68 for the Var $\bar{x}$ for Plan D.

SOLUTION

We know that

$$\text{Var }\bar{x} = P_1^2\sigma_1^2\left(\frac{1}{n_1} - \frac{1}{N_1}\right) + P_2^2\sigma_2^2\left(\frac{1}{n_2} - \frac{1}{N_2}\right) \qquad (83)$$

This holds for any fixed sample-sizes n_1 and n_2. It will also hold if we pick out these special sample-sizes from a long series of experiments in which we have permitted n_1 and n_2 to be random variables: if for these particular sample-sizes, we plot the distribution of X, we shall find that its theoretical variance is exactly what is written in the equation above.

What we need now is the theoretical average of Var $\bar{x}$ for the random sample-sizes in the long series of experiments. Let the symbol n_i denote the random sample-size in Stratum i. As the n_i sampling units are drawn 1 at a time at random, and as each sampling unit either does or does not belong to Stratum i, the variances of the n_i will be hypergeometric as shown in Table 10.

* For other examples in telephone accounting and management see Elbert T. Magruder, *Some Sampling Applications* (Chesapeake and Potomac Telephone Companies, Washington 5, 1955).

TABLE 10

EXPECTED SIZES AND VARIANCES OF THE SAMPLES
IN THE STRATA FOR PLAN D

Stratum	Random size	Expected size	Rel-variance
1	$\hat{n}_1$	$n_1 = nP_1$	$\dfrac{Q_1}{P_1}\left(\dfrac{1}{n} - \dfrac{1}{N}\right)$
2	$\hat{n}_2$	$n_2 = nP_2$	$\dfrac{Q_2}{P_2}\left(\dfrac{1}{n} - \dfrac{1}{N}\right)$
3	$\hat{n}_3$	$n_3 = nP_3$	$\dfrac{Q_3}{P_3}\left(\dfrac{1}{n} - \dfrac{1}{N}\right)$
All strata	n	n	xxx

We borrow a theorem from page 372 in Chapter 16, which states that

$$E\frac{1}{x} \doteq \frac{1}{Ex}(1 + C_x^2)$$

We are now ready to calculate $E \text{ Var } \bar{x}$.   To do this, we take the expected value of each term in the equation for $\text{Var } \bar{x}$.   The result is

$$E \text{ Var } \bar{x} = E \Sigma P_i^2 \sigma_i^2 \left\{\frac{1}{\hat{n}_i} - \frac{1}{N_i}\right\} \qquad \text{[The summation runs over all strata]}$$

$$= \Sigma P_i^2 \sigma_i^2 \left\{\frac{1}{nP_i}\left[1 + \frac{Q_i}{P_i}\left(\frac{1}{n} - \frac{1}{N}\right)\right] - \frac{1}{N_i}\right\}$$

$$= \frac{1}{n}\left\{\Sigma P_i \sigma_i^2 + \left(\frac{1}{n} - \frac{1}{N}\right)\Sigma Q_i \sigma_i^2\right\} - \frac{1}{N}\Sigma P_i \sigma_i^2$$

$$\overset{\text{Plan B}}{= \left(\frac{1}{n} - \frac{1}{N}\right)\left(\sigma_w^2 + \frac{1}{n}\sigma_R^2\right)} \qquad \text{[Plan D]}$$

$$= \frac{1}{n}\left(\sigma_w^2 + \frac{1}{n}\sigma_R^2\right) \qquad \text{[If } N \text{ is big compared with } n\text{]}$$

$$Q.E.D.$$

2. Derive Eq. 72 for the variance of Plan E.

SOLUTION

We may follow the solution for Plan D except that here $E\hat{n}_i = nP_i\sigma_i/\bar{\sigma}_w$ and the rel-variance of n_i is $Q_i/n'P_i$. Then

$$E \text{ Var } \bar{x} = E \Sigma P_i^2\sigma_i^2\left(\frac{1}{\hat{n}_i} - \frac{1}{N_i}\right)$$

$$= \Sigma P_i^2\sigma_i^2\left\{\frac{\bar{\sigma}_w}{nP_i\sigma_i}\left[1 + \frac{Q_i}{P_i}\left(\frac{1}{n'} - \frac{1}{N}\right)\right] - \frac{1}{N_i}\right\}$$

$$= \frac{\bar{\sigma}_w}{n}\left\{\Sigma P_i\sigma_i + \left(\frac{1}{n'} - \frac{1}{N}\right)\Sigma Q_i\sigma_i\right\} - \frac{1}{N}\Sigma P_i\sigma_i^2$$

Plan C

$$= \overbrace{\frac{(\bar{\sigma}_w)^2}{n} - \frac{\sigma_w^2}{N}} + \frac{1}{n}\left(\frac{1}{n'} - \frac{1}{N}\right)\bar{\sigma}_w\bar{\sigma}_R \qquad \text{[Plan E]}$$

3. Show that

$$\text{Cov } \hat{P}_i\hat{P}_j = -P_iP_j\left(\frac{1}{N'} - \frac{1}{N}\right) \doteq -\frac{P_iP_j}{N'} \qquad (84)$$

(We need this result in preparation for the variance of Plans H and I.)

SOLUTION

$$\hat{P}_i = \frac{\hat{N}_i}{N} \quad \text{and} \quad \hat{P}_j = \frac{\hat{N}_j}{N} \qquad (85)$$

in which I have for convenience omitted the stroke on N'. The probability of the simultaneous occurrence of any special values of $\hat{N}_1, \hat{N}_2, \hat{N}_3$ is

$$\frac{N!}{\hat{N}_1!\hat{N}_2!\hat{N}_3!}P_1^{\hat{N}_1}P_2^{\hat{N}_2}P_3^{\hat{N}_3} \qquad \text{[The multinomial probability, page 418]}$$

Hence

$$\text{Cov } \hat{N}_i\hat{N}_j = E(\hat{N}_i - N_i)(\hat{N}_j - N_j) \qquad \text{[Definition]}$$

$$= \Sigma\hat{N}_i\hat{N}_j \frac{N!}{\hat{N}_1!\hat{N}_2!\hat{N}_3!}P_1^{\hat{N}_1}P_2^{\hat{N}_2}P_3^{\hat{N}_3} - N_iN_j$$

$$= N(N - 1)P_iP_j - N_iN_j \qquad (86)$$

Here N_i stands for $E\hat{N}_i$. It follows that

$$\text{Cov } \hat{P}_i \hat{P}_j = \frac{1}{N^2} \text{Cov } \hat{N}_i \hat{N}_j$$

$$= -\frac{P_i P_j}{N} \tag{87}$$

This is the result required; we only need restore the stroke on N to get $-P_i P_j / N'$. If we use the more exact hypergeometric probability, we get $-P_i P_j (1/N' - 1/N)$.

Q.E.D.

4. Derive Eq. 79 for the variance of Plan H.

The sample gives

$$\bar{x} = \hat{P}_1 \bar{x}_1 + \hat{P}_2 \bar{x}_2 + \hat{P}_3 \bar{x}_3$$

$$= \Sigma \hat{P}_i \bar{x}_i \tag{88}$$

where the sum runs over all strata. $\hat{P}_i$ is the estimate N_i'/N' for the proportion of sampling units in Stratum i. I shall now for convenience omit the bar over x. Then

$$\text{Var } x = \Sigma P_i^2 \text{ Var } x_i + \Sigma a_i^2 \text{ Var } \hat{P}_i + 2 \sum_{j>i} a_i a_j \text{ Cov } \hat{P}_i \hat{P}_j \tag{89}$$

We note now that

$$Ex_i = a_i$$

$$E\hat{P}_i = P_i$$

$$\text{Var } \hat{P}_i = \left(\frac{1}{N'} - \frac{1}{N}\right) P_i Q_i$$

$$\text{Cov } \hat{P}_i \hat{P}_j = \left(\frac{1}{N'} - \frac{1}{N}\right) P_i P_j \qquad \text{[Preceding exercise]}$$

Now for any fixed n_i, regardless of the size of N_i',

$$\text{Var } x_i = \left(1 - \frac{n_i}{N_i}\right) \frac{\sigma_i^2}{n_i} = \left(\frac{1}{n_i} - \frac{1}{N_i}\right) \sigma_i^2 \tag{90}$$

But n_i is a random variable, with $En_i = nP_i$, and with rel-variance equal to $Q_i/N'P_i$, or, more precisely, $(1/N' - 1/N)Q_i/P_i$. Then for random repetitions of the sampling plan,

$$E \text{ Var } x_i = \sigma_i^2 E \left\{\frac{1}{n_i} - \frac{1}{N_i}\right\}$$

$$\doteq \sigma_i^2 \left\{\frac{1}{nP_i}\left[1 + \frac{Q_i}{P_i}\left(\frac{1}{N'} - \frac{1}{N}\right)\right] - \frac{1}{N_i}\right\} \tag{91}$$

We may now observe from Eq. 89 that

$$E \operatorname{Var} x = \Sigma P_i^2 E \operatorname{Var} x_i + \left(\frac{1}{N'} - \frac{1}{N}\right)\{\Sigma a_i^2 P_i Q_i \qquad \text{[The sum runs over all strata, but with } j > i]$$
$$- 2\Sigma a_i a_j P_i P_j\}$$

$$= \Sigma P_i^2 \sigma_i^2 \left\{\frac{1}{nP_i}\left[1 + \frac{Q_i}{P_i}\left(\frac{1}{N'} - \frac{1}{N}\right)\right] - \frac{1}{N_i}\right\} \qquad [N_i = NP_i]$$

$$+ \left(\frac{1}{N'} - \frac{1}{N}\right)\{\Sigma P_i a_i^2 - (\Sigma P_i a_i)^2\}$$

Plan D if $N' = n$

$$= \overbrace{\left(\frac{1}{n} - \frac{1}{N}\right)\sigma_w^2 + \frac{1}{n}\left(\frac{1}{N'} - \frac{1}{N}\right)\sigma_R^2} + \left(\frac{1}{N'} - \frac{1}{N}\right)\sigma_b^2 \qquad \text{[Plan H]}$$

Plan B

$$= \frac{\sigma_w^2}{n} + \frac{\sigma_b^2}{N'} \qquad \text{[If } N \text{ is big compared with } n \text{ and with } N', \text{ and if we may neglect the term in } 1/nN']$$

$$Q.E.D.$$

5. Derive Eq. 82 for the variance of Plan I.

SOLUTION

Follow the solution for Plan H in the preceding exercise, but put $En_i = nP_i\sigma_i/\bar{\sigma}_w$. Then

$$E \operatorname{Var} x_i = \sigma_i^2 E\left\{\frac{1}{\hat{n}_i} - \frac{1}{N_i}\right\}$$

$$= \sigma_i^2 \left\{\frac{\bar{\sigma}_w}{nP_i\sigma_i}\left[1 + \frac{Q_i}{P_i}\left(\frac{1}{N'} - \frac{1}{N}\right)\right] - \frac{1}{N_i}\right\} \qquad (92)$$

As in the preceding exercise,

$$E \operatorname{Var} x = \Sigma P_i^2 E \operatorname{Var} x_i + \left(\frac{1}{N'} - \frac{1}{N}\right)\{\Sigma a_i^2 P_i Q_i - 2\Sigma a_i a_j P_i P_j\}$$

$$= \Sigma P_i^2 \sigma_i^2 \left\{\frac{\bar{\sigma}_w}{nP_i\sigma_i}\left[1 + \frac{Q_i}{P_i}\left(\frac{1}{N'} - \frac{1}{N}\right)\right] - \frac{1}{N_i}\right\} + \left(\frac{1}{N'} - \frac{1}{N}\right)\sigma_b^2$$

Plan C if $N' = n$

$$= \overbrace{\frac{(\bar{\sigma}_w)^2}{n} - \frac{\sigma_w^2}{N}} + \frac{1}{n}\left(\frac{1}{N'} - \frac{1}{N}\right)\bar{\sigma}_w\bar{\sigma}_R + \left(\frac{1}{N'} - \frac{1}{N}\right)\sigma_b^2 \qquad \text{[Plan I]}$$

Plan C

$$= \frac{(\bar{\sigma}_w)^2}{n} + \frac{\sigma_b^2}{N'} \qquad \text{[If } N \text{ is big compared with } n \text{ and with } N', \text{ and if we may neglect the terms in } 1/nN']$$

6. Derive the optimum ratio

$$\frac{n}{N'} = \frac{\sigma_w}{\sigma_b} \sqrt{\frac{c_1}{c_2}} \qquad \text{[Eq. 77]}$$

for Plan H.   Show that for Plan I we merely replace σ_w by $\bar{\sigma}_w$, to get Eq. 80 on page 329.

SOLUTION

The approximate variance of Plan H is

$$\sigma_{\bar{x}}^2 = \frac{\sigma_w^2}{n} + \frac{\sigma_b^2}{N'} \qquad \text{[P. 329]}$$

The total cost will be

$$K = N'c_1 + nc_2 \qquad \text{[P. 351]}$$

whatever be N' and n.   Now if we desire to keep the total cost constant while we vary N' and n, we perceive that if we increase n by h units, we must decrease N' by hc_2/c_1 units.   What happens to Var $\bar{x}$?   The term σ_w^2/n will decrease, and the other term will increase.   Their sum will decrease, pass through a minimum, the n increase, as $n : N'$ passes through its optimum ratio.

Let us assume that this minimum in Var $\bar{x}$ does exist.   Set $N' = N_0$ at this point, and $R = n : N_0$.   Let us see what will happen if we increase n by h units and decrease N' by hc_2/c_1 units from N_0.   Then

$$\left. \begin{aligned} n &= RN_0 + h \\ N' &= N_0 - \frac{hc_2}{c_1} \end{aligned} \right\} \qquad (93)$$

The total cost remains K whether h be 0 or any number positive or negative. Then

$$\begin{aligned} \text{Var } \bar{x} &= \frac{\sigma_w^2}{RN_0 + h} + \frac{\sigma_b^2}{N_0 - hc_2/c_1} \\ &= \frac{\sigma_w^2}{RN_0(1 + h/RN_0)} + \frac{\sigma_b^2}{N_0(1 - hc_2/N_0c_1)} \\ &= \frac{\sigma_w^2}{RN_0}\left[1 - \frac{h}{RN_0} + \frac{h^2}{R^2N_0^2} \right] \\ &\quad + \frac{\sigma_b^2}{N_0}\left[1 + \frac{hc_2}{N_0c_1} + \left(\frac{hc_2}{N_0c_1}\right)^2 \right] \qquad \text{[By division, curtailed at } h^2] \\ &= \frac{\sigma_w^2}{RN_0} + \frac{\sigma_b^2}{N_0} + h\left[-\frac{\sigma_w^2}{(RN_0)^2} + \frac{\sigma_b^2 c_2}{N_0^2 c_1} \right] + h^2\left[\frac{\sigma_w^2}{(RN_0)^3} + \frac{\sigma_b^2 c_2^2}{N_0^3 c_1^2} \right] \quad (94) \end{aligned}$$

The coefficient of h will be 0 if

$$R = \frac{\sigma_w}{\sigma_b} \sqrt{\frac{c_1}{c_2}}$$

and the term in h^2 will then be positive, whether h be positive or negative.

The value of R just written is then the optimum ratio for $n : N'$. The total cost will be

$$K = N'c_1 + RN'c_2 \qquad \text{[for fixed } N'] \qquad (95)$$

or

$$K = \frac{nc_1}{R} + nc_2 \qquad \text{[for fixed } n] \qquad (96)$$

Eq. 80 for the optimum value of n/N' in Plan I is due to J. Neyman, "Contribution to the theory of sampling human populations," *J. Amer. Statist. Ass.*, vol. 33, 1938: pages 101–116; Eq. 49 on page 110.

7. Show that $EX = A$ in all the plans that we have studied. That is, the estimate X is, in each case, an unbiased estimate of the total x-population A in the frame. It follows that $\bar{x}$ is an unbiased estimate of a, the average x-population per sampling unit in the frame.

APPENDIX: INSTRUCTIONS FOR THINNING A SAMPLE

1. The directions here will retain 1 out of 3. Modifications required for any other thinning ratio will be obvious.

2. Three successive sampling numbers in Subsample 1 will form a block, and they will have the ordinal numbers 1, 2, 3.

3. Decide on a starting point in the table of random numbers, and on the system by which to read the numbers.

For example, to retain 1 in 3, use the system shown below.

1	2	3
1	2	3
4	5	6
7	8	9

0	blank

The digits 1, 4, 7 select the 1st member of a block of 3; the digits 2, 5, 8 select the 2d member; the digits 3, 6, 9 select the 3d member; 0 is a blank.

To retain 1 out of 4, the system would be this:

1	2	3	4
1	2	3	4
6	7	8	9

0 and 5 blanks

The digits 1 and 6 select the 1st member of a block of 4; 2 and 7 select the 2d member; etc. 5 and 0 are blanks.

We continue now.

4. The 1st random number (unless it is a blank) that you encounter in the table draws an item from the 1st block in Subsample 1.   If the 1st random number is a blank, skip it and go on to the next number.

5. The 2d random number that you encounter (unless it is a blank) draws an item from the 2d block in Subsample 1, but you are to refuse a number that would draw the same ordinal number 1, 2, or 3 that you drew from the 1st block.

6. Proceed to the 3d block, in which the drawing is automatic, as you will retain the remaining ordinal number of the triplet 1, 2, 3 not retained in the first 2 blocks.

7. Treat the 4th block like the 1st block.   Treat the 5th block like the 2d, etc.

8. At the end of Subsample 1 there may be an incomplete block. Complete it by tying it to the beginning of Subsample 2.   Continue onward in Subsample 2 in the manner already specified.   Tie the end of Subsample 2 to the beginning of Subsample 3, etc.   The block at the end of Subsample 10 may be incomplete.   Complete it by adding blanks, if necessary. A blank drawn from an incomplete block discards the whole block.

9. If 2 people work on the job, each one will have his own starting point in the table of random numbers.

> **Remark 9.**   Note that the above selections are balanced.   Every consecutive 3 blocks contribute a 1st member, a 2d, and a 3d.   This is very important where the order of listing is systematic, as by the order of position in the family, or by age.   Such blocks are badly loaded, but the sample therefrom is balanced perfectly.   I found balancing to be very important in the selection of a sample of 1 telephone in 4 from the wire-chiefs' line-cards. The 1st space is nearly always filled: it is loaded with 1-party lines.   The 1st 2 spaces are loaded with 2-party lines; etc.

> **Remark 10.**   The above procedure is excellent for the selection of 1 nonrespondent out of 3 for recalls on nonresponse (p. 106).

D. EXAMPLES OF PLAN I

A sample for a study of mental retardation.* This was a study of a sample of families in the State of Delaware to determine *inter alia* the prevalence of mental retardation.   It turned out to be possible, by

---

* It is a pleasure to express my appreciation to Professor Joseph Jastak of the University of Delaware, for the privilege of working with him on the survey described here. Professor Jastak and Dr. Martin Whiteman supplied the basic knowledge of psychology required for the design of the survey.

stratifying a preliminary sample on the basis of the relatively cheap Wechsler-Bellevue test, to cut the cost of the field-work to about 46% of what its cost would have been without benefit of statistical theory (see Table 12, page 343). The calculations for the design of the sample follow. From a statistical standpoint there were 2 main aims:

a. To make an estimate of the prevalence of mental retardation (the average number of retarded persons per family).

b. To carry out certain psychological tests and measurements in both retarded and nonretarded families (a retarded family being one in which 1 member or more over 10 years of age is retarded), to compare the 2 groups.

Certain facts presented themselves: (1) the interviews and tests in aim *b* are long and expensive (hereafter this will be known as the L-test); (2) the L-test should therefore concentrate on the retarded families, and not dissipate itself on the much larger stratum of nonretarded families; (3) in aim *b* it was best, from the standpoint of statistical efficiency, that the number of tests should not be greatly unequal in the 2 groups.

Enquiry disclosed the fact that there is a simple, brief, and inexpensive test, called the Wechsler-Bellevue test (hereafter the W-B test), which will classify a person above or below any designated point on the psychological scale, in almost exact agreement with the L-test (which, however, served many other purposes).   Use of the W-B test would thus permit quick and inexpensive classification of the families in a preliminary sample into 2 classes:

Class 1. Families that are retarded, according to the W-B test.
Class 2. All other families.

Because of a fortunate relation between proportions and costs, this sample turned out to be efficient for both aims *a* and *b*.

The plan of sampling.   Once the usefulness of the W-B test became clear, Plan I emerged as the best plan to use.   Here is a brief description of the procedure:

1. Draw a master sample of about 2600 d. us. in 10 subsamples, about 260 d. us. in each subsample.   The size of the master sample was chosen so that it would be big enough to furnish the preliminary sample N'.

2. Conduct a pilot study on 1 subsample (Subsample 10) to learn something about the proportions P_1 and P_2, and to get some experience and some figures on costs; to classify the families in the pilot study into Classes 1 and 2, on the basis of the W-B test.

3. Study by the L-test the families in Class 1, to estimate the distribution of retarded persons in Class 1.

4. Draw up the sampling plan and procedure.   Decide on the thinning ratios and on the size of the final sample.   Proceed with the field-work and tabulations.

The pilot study would cost very little extra, as it was a portion of the main study, carried out in advance up to the point of thinning Class 2.

The results of the pilot study are in Table 11.   The statistical characteristics of the 2 classes and for both classes combined appear at the

TABLE 11

DISTRIBUTION OF RETARDATION

Class 1 (From the pilot L-test in the study of 57 families)	Class 2 (By expert knowledge)
1 have 0 persons retarded 39 ,, 1 ,, ,, 11 ,, 2 ,, ,, 6 ,, 3 ,, ,, 0 ,, 4 ,, ,,	196 have 0 persons retarded 4 ,, 1 ,, ,, 0 ,, 2 or more persons re- tarded

$a_1 = 1.39$	Mean	$a_2 = \dfrac{4}{200} = .02$	(binomial, p. 404)
$\sigma_1^2 = 0.48$	Variance	$\sigma_2^2 = \dfrac{4}{200}\,\dfrac{196}{200} = .02$	
$\sigma_1 = 0.69$	Standard deviation	$\sigma_2 = \sqrt{\dfrac{4}{200}\,\dfrac{196}{200}} = .14$	

Both classes combined

$$P_1 = \frac{57}{257} = .22$$

$$P_2 = \frac{200}{257} = .78$$

$$a = P_1 a_1 + P_2 a_2 = .32$$

$$\sigma_w^2 = P_1 \sigma_1^2 + P_2 \sigma_2^2 = .12$$

$$\sigma_b^2 = P_1(a_1 - a)^2 + P_2(a_2 - a)^2$$

$$= P_1 P_2 (a_2 - a_1)^2 = .32 \qquad \sigma_b = .57$$

$$\sigma^2 = \sigma_w^2 + \sigma_b^2 = .44$$

$$\bar{\sigma}_w = P_1 \sigma_1 + P_2 \sigma_2 = .26$$

bottom of the table. The L-test conducted in Class 1 disclosed the fact that 1 family classed as retarded by the W-B test was actually not retarded. Thus, the W-B test made an excellent separation, though not perfect.

Of the 257 families in the pilot study, 57 went into Class 1 and 200 into Class 2. These figures gave the preliminary estimates of P_1 and P_2 that appear in the table. These estimates were good enough for planning, but not good enough to permit use of Plan E.

It was too expensive to conduct the L-test on the 200 families of Class 2, but it was possible by the aid of expert knowledge (not mine) to predict close enough the number of retarded families amongst the 200 families of Class 2, and thus to fix the sample-sizes and the thinning ratios. (The exercise on page 343 may be of interest here.)

This example provides an illustration of the sampling of new material. The pilot study gave the information that we needed about Class 1, and expert knowledge furnished information we needed about Class 2.

Optimum sizes of the samples. For the Neyman allocation of the final sample to the 2 strata we take

$$n_2 = n_1 \frac{N_2 \sigma_2}{N_1 \sigma_1} \qquad \text{[Eq. 27, p. 294]}$$

$$= \frac{200 \times .14}{57 \times .69} n_1 = .71 n_1 \tag{97}$$

whence

$$n = n_1 + n_2 = 1.7 n_1 \tag{98}$$

Considerations of cost and of the expected precision shown by Eq. 106 *infra* limited the final sample to about 400 families, which should be distributed as follows:

$$n_1 = \frac{400}{1.7} = 235 \text{ in Class 1}$$
$$n_2 = 165 \text{ in Class 2} \tag{99}$$

For the size N' of the preliminary sample we make use of Eq. 91 on page 334, in which c_1 is the average cost of using the W-B test in 1 family of the preliminary sample, and c_2 is the average cost of the L-test in 1 family of the final sample. We learned from the pilot study that

$$c_1 = \$6$$
$$c_2 = \$50 \tag{100}$$

whereupon Eq. 90 gives

$$\frac{n}{N'} = \frac{\bar{\sigma}_w}{\sigma_b} \sqrt{\frac{c_1}{c_2}} = \frac{26}{57} \sqrt{\frac{6}{50}} = .16 \tag{101}$$

If $n = 400$, then the optimum preliminary sample should be

$$N' = \frac{400}{.16} = 2500$$

which will divide itself approximately into

$$\left. \begin{array}{l} N_1' = N'P_1 = 550 \text{ families in Class 1} \\ N_2' = N'P_2 = 1950 \text{ families in Class 2} \end{array} \right\}$$

On this basis we calculate the thinning ratios

$$\frac{n_1}{N_1'} = \frac{235}{550}$$

$$\frac{n_2}{N_2'} = \frac{165}{1950}$$

practical approximations being

$$\frac{n_1}{N_1'} = \frac{1}{2} \tag{102}$$

$$\frac{n_2}{N_2'} = \frac{1}{12} \tag{103}$$

That is, we select 1 family at random from each successive 2 families in the preliminary sample of Class 1, and 1 family at random from each successive 12 in Class 2. We know from Eq. 25 that these convenient ratios will give almost the same precision as the exact ratios. With these numbers the proposed sample should give

$$\text{Var } \bar{x} = \frac{(\bar{\sigma}_w)^2}{n} + \frac{\sigma_b^2}{N'} \quad \text{[Eq. 82, page 331]}$$

$$= \frac{.26^2}{400} + \frac{.32}{2500}$$

$$= .00017 + .00013 = .00030 \tag{104}$$

$$\sigma_{\bar{x}} = \sqrt{.00030} = .017 \tag{105}$$

$$C_{\bar{x}} = \frac{\sigma_{\bar{x}}}{a} = \frac{.017}{.32} = \text{about } 5\% \tag{106}$$

The precision is sufficient, in view of the uncertainties of the tests, and of difficulties of definition of residence, nonresponse, and the like.

For reduction of the master sample from 2600 to the preliminary sample of size $N' = 2500$, it was convenient to delete 1 sampling unit in every successive 26 through all 10 subsamples.

Calculation of the expected saving over Plan A.   Now let us see what size of sample this same precision would require if there were no stratification.   Then would

$$\text{Var } \bar{x} = \frac{\sigma^2}{n} \quad \text{[Plan A, no additional stratification; page 283]}$$

With Var $\bar{x} = .00030$, $\sigma^2 = .44$, we find that $n = 1500$.

Now how about the relative costs of Plans A and I?   The computations are in Table 12, whence we see that Plan I will cost only about 46% as much as Plan A (no stratification).

TABLE 12

COMPARISON OF THE COSTS OF PLANS A AND I IN THE
SURVEY OF MENTAL RETARDATION

| Plan | Preliminary sample | | Final sample | | Total cost |
	Number of families	Cost per family	Number of families	Cost per family	
A	xxx	xxx	1500	$50	$75,000
I	2400	$6	400	50	34,400
Difference					40,600
Saving effected by Plan I, 40,600/75,000					54%

The moral of this calculation lies in the fact that (as I learned later) some members of the advisory committee for this study expressed their "considered opinion" that there would be no point in engaging a statistician, as this was a psychological study, and the survey-methods that it would require were well known.   It was the statistician who enquired whether there might possibly be some sort of inexpensive preliminary test that would classify families, roughly at least, into 2 classes, retarded and nonretarded, so as to concentrate the long and expensive interviews amongst the retarded families, where these interviews could most effectively decrease the variance of the estimates desired.

Exercise.   Suppose that only 3 families of the 200 in Class 2 in Table 11 had a retarded member, instead of 4.   Show that the optimum sample-sizes would have been $n_1 = 250$ and $n_2 = 150$, but that by Eq. 25 the loss wrought by the allocation actually made ($n_1 = 235$ and $n_2 = 165$ in Eq. 99) would have been only $(15/250)(15/150) = .006$, or 6 final interviews in 1000.

Thus, expert knowledge, if it was able to fix the number of retarded families as 3 or 4 or 5 in the 200 families of Class 2, furnished all that we needed to know about Class 2.

Application of Plan I to determine the physical condition of property. Stratification is sometimes very useful in a survey where the aim is to determine the value or the per cent condition of physical property, when some of the items are worth (in dollars) much more than the others. The best allocation of the sample to any class is in proportion to the total value (in dollars) of that class, regardless of the number of items in the class (Eq. 121 ahead). The allocation would be improved further by making it proportional also to the standard deviation of the class, were the standard deviations known, and if they were very different.

A specific example is to determine the per cent condition of the aerial plant of a telephone company. There are many kinds of items in the aerial plant, but they are all on or connected to a pole. Hence, the company's property records which show the locations of all poles will be an excellent frame. Every pole has or can have a serial number. Expert inspectors will examine all the items on a pole that comes into the sample, and will report the physical condition (good as new, slightly used, etc.) of each item (pole, cable, wire, crossarm, strand, terminal, etc.).

It often happens that the aerial cable is by far the most important part of the aerial plant, and that it is concentrated on a small portion of the poles. The maps that show the locations of the poles do not show which ones carry cable; but it is possible, by reference to another set of records (called the cable-records), to ascertain pretty definitely whether any pole carries cable. It is thus possible to draw a preliminary sample of poles, and to thin out a portion of those that do not carry cable, so as to concentrate the efforts of the inspectors on the poles that carry cable, where the bulk of the dollars are located.

It is not necessary to separate the 2 kinds of pole (with cable and without), but it is sometimes important to do so, as calculations ahead will show. In any case, separation or no separation, the tabulation of the results will resemble the example shown in Table 10. The weights of the various kinds of items are supposed known from the accounting records; the weights are thus not random variables.

Remark. One meets other conditions. Sometimes the engineers, instead of getting the weights from the accounting records, will prefer to estimate the number of each kind of item, and hence the weight of that item, from the sample itself, or from a larger preliminary sample. The weights will then vary from subsample to subsample: moreover, they may be correlated with the per cent conditions.   The easiest way, under these conditions, to compute the variance of the final estimate of the overall per cent condition (which comes from all 10 subsamples combined, both weights and per cent conditions) is to compute the overall per cent condition of each subsample with its own proper set of weights.   The 10 results for the overall per cent condition will then give an unbiased estimate of the total variance, properly adjusted for the varying weights and for any correlation that exists.

One may use the same system of computation even though the weights came from 10 subsamples that are completely independent of the subsamples used for the per cent conditions.

If the weights came from an independent sample that was not replicated in 10 subsamples, the weight of each kind of item will be the same for all 10 subsamples that show the per cent conditions (as in Table 13), but the variances of the weights will contribute an additional term to the variance of the 10 results for per cent condition.

TABLE 13

EXAMPLE OF THE RESULTS OF INSPECTION OF THE
AERIAL PLANT OF A TELEPHONE COMPANY

Kind of item	w_i	Subsample			
		1	2 $\cdots$	10	All 10
Poles	.245	85.1	79.3	81.2	78.8
Cable	.342	89.8	89.5	85.8	87.3
Copper wire	.152	94.7	94.3	93.8	94.3
Iron wire	.068	76.9	81.0	74.9	76.4
Crossarms	.059	85.4	80.9	80.9	80.8
Terminals	.064	90.5	86.9	87.4	87.5
Strand	.070	88.8	89.6	85.6	86.8
All the above items	1.000	88.2	86.5 $\cdots$	84.9	85.1

It turns out that the preliminary sample by which to estimate the proportions P_1 and P_2 need be only big enough to furnish the required number of poles with cable (vide Eq. 110); hence Plan H will be indistinguishable from Plan D or A, and Plan I will be indistinguishable from Plan E.   However, we shall retain the symbols H and I.   We now compare the 2 plans.   The symbols w_1 and w_2 will denote weights.

For Plan I we desire that

$$\left. \begin{array}{l} n_1 = nw_1 \\ n_2 = nw_2 \end{array} \right\} \tag{107}$$

for which the accounting records show that

$$w_1 : w_2 = 7 : 3$$

Suppose further that the engineering department has in advance an approximate figure of 25% for the proportion P_1 of the poles that carry aerial cable, wherefore, for purposes of planning, we take

$$P_1 : P_2 = 25 : 75$$

Previous experience shows that one may expect σ_1, σ_2, and $\bar{\sigma}_w$ to be about 12%, and that σ_b may be anywhere from 1 to 2%: as an approximation we set $\sigma_b = 1.5\%$. As for the costs c_1 and c_2, I had learned that a girl that earns \$20 per day can classify about 100 poles per day, wherefore c_1 is about 20¢ per pole. A pair of inspectors, with a truck and tools, can inspect about 10 poles per day, wherefore c_2 is about \$10 per pole.

We next observe that we may ignore the term σ_b^2/N' in either Eq. 79 or 82 for the variances: this is so because σ_b is small and because we shall soon see that the preliminary sample for either plan will be large. The proportionate efficiency of Plan I over Plan H now appears in the simple equation

$$\frac{H}{I} = \frac{w_1^2}{P_1} + \frac{w_2^2}{P_2} \qquad \text{[Derivation on page 349]} \qquad (108)$$

Here

$$\frac{H}{I} = \frac{.7^2}{.25} + \frac{.3^2}{.75}$$

$$= 1.96 + .12$$

$$= 208 : 100 \qquad (109)$$

Seeing this numerical result, we adopt Plan I, because 100 inspections carried out by Plan I will be equivalent to 208 by Plan H.

Now comes the question of the sizes of the samples. First, the preliminary sample N'. If $\bar{\sigma}_w = 12$, the optimum ratio $n : N'$ turns out to be

$$\frac{n}{N'} = \frac{\bar{\sigma}_w}{\sigma_b} \sqrt{\frac{c_1}{c_2}} \qquad \text{[Eq. 80, page 329]}$$

$$= \frac{12}{1.5} \sqrt{\frac{20}{1000}} > 1 \qquad (110)$$

As this ratio is greater than 1, our preliminary sample need only be big enough to supply the required number of poles in Class 1.

Suppose that we desire $\sigma_{\bar{x}}$ to be about .4%.   Then, if $\sigma = 12\%$, Plan I would require the inspection of equipment on approximately

$$n = \left(\frac{\sigma}{\sigma_{\bar{x}}}\right)^2$$

$$= \left(\frac{12}{.4}\right)^2 = 900 \text{ poles} \qquad (111)$$

The procedure of selection will be this (after each pole has a serial number):

1. Draw a preliminary sample of poles in 10 subsamples.   (Size $N' = 2520$, as shown later.)
2. Classify every pole in the preliminary sample—
    Class 1. Poles that according to the cable-record carry cable.
    Class 2. All the other poles.
3. Thin Class 2 (thinning ratio $T = 7$ to come later).
4. Inspect all the property on the poles of Class 1, and on the poles that remain in Class 2.
5. Compile for each kind of item the average per cent condition.   Do this by subsample and for all 10 subsamples combined.   Multiply by T the result of any item in Class 2 before you combine it with Class 1.   In symbols, the computation for any kind of item (wire, or cable, etc.) will be

$$\bar{x} = \frac{\sum_1^{n_1} x_i + T \sum_1^{n_2} x_i}{n_1 + Tn_2} \qquad \begin{array}{l}\text{[This } T \text{ is not} \\ \text{the symbol } T \\ \text{on page 199.]}\end{array} \qquad (112)$$

The 2 terms in the numerator refer to Class 1 and to Class 2.
Each x_i is the per cent condition of some item.   This calculation is the origin of the figures in Table 10.
6. Compute the standard error of the final estimate (the result of all 10 subsamples combined).   This step is easy because the sample is replicated in 10 subsamples.

The thinning ratio T for Class 2 must produce the optimum allocation, which means that

$$\frac{N'P_1}{N'P_2/T} = \frac{w_1}{w_2} \qquad (113)$$

where N' is the size of the preliminary sample.   This leads to

$$T = \frac{w_1}{w_2}\frac{P_2}{P_1}$$

$$= \frac{7}{3}\frac{75}{25} = 7$$

The preliminary sample must produce the right number of poles for Class 1.   Hence N' must satisfy the equation

$$N'P_1 + N'P_2/T = 900 \qquad (114)$$

which with $T = 7$ gives $N' = 2520$.   Table 14 shows the predicted distribution of the sample before and after thinning.   The reader will note that the relative number of poles in the 2 classes in the final sample is $7:3$, which is precisely the ratio of the weights.

TABLE 14

THE PRELIMINARY SAMPLE AND THE FINAL
SAMPLE OF POLES

Class	Preliminary sample		Final sample	
Both classes	2520	100	900	100%
With cable	630	25	630	70%
Without cable	1890	75	270	30%

The gain came in this instance from the fact that w_1 was large and P_1 was small.   If P_1 were larger, the gain would be less.   Thus, if 33% of the poles carried aerial cable, the relative efficiency of Plan I over Plan H would be, by Eq. 108,

$$\frac{H}{I} = \frac{.7^2}{.33} + \frac{.3^2}{.67} = 160:100$$

This is still a sizable gain, but not as much as before.

As the gain obviously falls off sharply with an increase in P_1, one must prepare himself to accept some loss in precision from the fact that the prior information on the number of poles that carry aerial cable may be in error, and that the thinning ratio prescribed on the basis of this information may consequently not produce the required precision.   In anticipation, it is wise, in the absence of firm information, to specify samples a bit bigger than theory indicates.

Exercise 1.   Use

$$\bar{x} = \Sigma w_i \bar{x}_i \qquad (115)$$

for the estimate of the overall per cent condition of all classes combined. w_i is the weight of Class i, and $\bar{x}_i$ is the average per cent condition of the items in the sample of Class i.   Weight is relative dollar-value, and $\Sigma w_i = 1$.

a. Show that the optimum allocation will be

$$n_i = nw_i \frac{\sigma_i}{\bar{\sigma}_w \sqrt{c_i}} \qquad (116)$$

where

$$\bar{\sigma}_w = \Sigma w_i \sigma_i$$

and c_i is the cost of inspecting an item in Class *i*.

b. If the costs are all equal, and the σ_i all equal, then the optimum allocation is

$$n_i = nw_i \qquad \text{[As in Eq. 107]}$$

Thus, the ratio of n_1 to n_2 should be $7 : 3$ in the problem just treated.

Exercise 2. Derive Eq. 108 for the relative variances of Plan H and I, under the condition that we may neglect the term $\sigma_b{}^2/N'$ in Eqs. 79 and 82, and that $\sigma_1 = \sigma_2 = \bar{\sigma}_w = \sigma$.

SOLUTION

$$\bar{x} = w_1 \bar{x}_1 + w_2 \bar{x}_2 \qquad (w_1 \text{ and } w_2 \text{ known})$$

$$\text{Var } \bar{x} = w_1{}^2 \text{ Var } \bar{x}_1 + w_2{}^2 \text{ Var } \bar{x}_2$$

$$= w_1{}^2 \frac{\sigma_1{}^2}{n_1} + w_2{}^2 \frac{\sigma_2{}^2}{n_2}$$

$$= \sigma^2 \left(\frac{w_1{}^2}{n_1} + \frac{w_2{}^2}{n_2} \right) \qquad (117)$$

In Plan H, put $n_1 = nP_1$ and $n_2 = nP_2$. Then

$$\text{Var } \bar{x} = \frac{\sigma^2}{n} \left(\frac{w_1{}^2}{P_1} + \frac{w_2{}^2}{P_2} \right) \qquad (118)$$

In Plan I, put $n_1 = nw_1$ and $n_2 = nw_2$. Then

$$\text{Var } \bar{x} = \frac{\sigma^2}{n} \qquad (119)$$

The ratio of the 2 variances for the same sample *n* is then

$$\frac{H}{I} = \frac{w_1{}^2}{P_1} + \frac{w_2{}^2}{P_2} \qquad (120)$$

Q.E.D.

E. COMPARISON OF COSTS

Table of comparative costs. It is possible now to construct the table of comparative costs (Table 15, p. 350), which shows how the average cost of any plan will compare with the average cost of Plan A, when both

TABLE 15

COMPARATIVE COSTS OF THE VARIOUS PLANS OF STRATIFIED
SAMPLING THAT WILL DELIVER THE SAME PRECISION AS
PLAN A WILL DELIVER WITH SIZE OF SAMPLE EQUAL TO ν.

Assumptions: (1) the frame is not already stratified; (2) the unit costs of classification, and of interviewing or of testing, are the same in all strata. c_1 is the cost to classify one unit. In Plans H and I the cost c_1 may include the cost of a preliminary test or short interview. c_2 is the cost to interview or to test 1 unit in the final sample for the main study.

	Plan	Average cost	Remarks
A	No stratification	$\nu c_2 = K$	This cost K and the sample-size ν furnish the bases for reference. $\sigma_{\bar{x}} = \dfrac{\sigma}{\sqrt{\nu}}$

The proportions P_i known

Classify all N sampling units in the frame

	Plan	Average cost	Remarks
B	Proportionate allocation; sample-sizes n_i fixed in advance	$Nc_1 + \nu c_2\left(\dfrac{\sigma_w}{\sigma}\right)^2$ $= Nc_1 + K\left(\dfrac{\sigma_w}{\sigma}\right)^2$	Size of sample, $\nu\left(\dfrac{\sigma_w}{\sigma}\right)^2$
C	Neyman allocation; sample-sizes n_i fixed in advance	$Nc_1 + K\left(\dfrac{\bar{\sigma}_w}{\sigma}\right)^2$	Size of sample, $\nu\dfrac{(\bar{\sigma}_w)^2}{\sigma^2}$

The proportions P_i known

Classify only the n sampling units of the sample

	Plan	Average cost	Remarks
D	Draw and classify a sample of specified size n, which shall be also the final sample. The individual sample-sizes n_i are random variables.	$n(c_1 + c_2) \doteq K\left(\dfrac{\sigma_w}{\sigma}\right)^2$ $+ c_1\dfrac{\nu\sigma_w^2}{\sigma^2} + (c_1 + c_2)$ $\times \left(\dfrac{\sigma_R}{\sigma_w}\right)^2$	Size of sample, $n \doteq \nu\left(\dfrac{\sigma_w}{\sigma}\right)^2$ $+ \left(\dfrac{\sigma_R}{\sigma_w}\right)^2$

TABLE 15—*continued*

Plan	Average cost	Remarks
E Draw and classify a specified number n' of sampling units. Thin the strata by use of the Neyman ratios. The total sample n and the individual sample-sizes n_i are all random variables	$n'c_1 + nc_2 = n'c_1$ $+ \dfrac{Kn}{v} = n'c_1$ $+ K\left(\dfrac{\bar\sigma_w}{\sigma}\right)^2\left\{1 + \dfrac{\bar T\bar\sigma_R}{\bar\sigma_w}\right\}$	Average size of the final sample, $n = v\left(\dfrac{\bar\sigma_w}{\sigma}\right)^2 + \dfrac{\bar T\bar\sigma_R}{\bar\sigma_w}$ $n' = \dfrac{n}{\bar T}$

The proportions P_i known

Classify only enough sampling units to fill the quotas n_i

Plan	Average cost	Remarks
F Fix the sample-sizes n_i in advance by proportionate allocation. Draw sampling units and classify them until all the quotas n_i are filled. n is fixed; also the n_i.	$n'c_1 + nc_2 = n'c_1$ $+ K\left(\dfrac{\sigma_w}{\sigma}\right)^2$ Note that n' in this plan is not equal numerically to the n' in Plan E, nor to the n' in Plan G.	Total sample n and the quotas n_i as in Plan B. Variance, the same as the variance of Plan B. The number n' of sampling units that require classification will be, on the average, a bit bigger than n.
G Fix the sample-sizes n_i in advance by the Neyman allocation. Draw sampling units and classify them until all the quotas n_i are filled. n is fixed.	$n'c_1 + nc_2 = n'c_1$ $+ v\left(\dfrac{\bar\sigma_w}{\sigma}\right)^2 c_2 = n'c_1$ $+ K\left(\dfrac{\bar\sigma_w}{\sigma}\right)^2$	Total sample n and the quotas n_i as in Plan C. Variance, the same as the variance of Plan C. The number n' of sampling units that require classification will be, on the average, a bit bigger than $n/\bar T$.

The proportions P_i not known in advance

Classify a preliminary sample of size N' to estimate the proportions P_i

Plan	Average cost	Remarks
H Classify the preliminary sample, and thin all classes proportionately to reach a specified final size n. The individual sample-sizes n_i are random variables.	$N'c_1 + nc_2 \doteq N'c_1$ $+ K\left(\dfrac{\sigma_w}{\sigma}\right)^2$	Opt $N' = \dfrac{\sigma_w}{\sigma_b}\sqrt{\dfrac{c_1}{c_2}}$ Size of final sample, $n = v\left(\dfrac{\sigma_w}{\sigma}\right)^2\left\{1 + \dfrac{1}{N'}\right.$ $\times \left[v\dfrac{\sigma_b^2}{\sigma^2} + \left(\dfrac{\sigma_R}{\sigma_w}\right)^2\right]\right\}$

TABLE 15—*continued*

Plan	Average cost	Remarks
I Classify the preliminary sample, and thin the classes by the Neyman ratios to reach a specified final size n. The individual sample-sizes are random variables.	$N'c_1 + nc_2 = N'c_1$ $+ K\left(\dfrac{\bar{\sigma}_w}{\sigma}\right)^2$	Opt $N' = \dfrac{\bar{\sigma}_w}{\sigma_b}\sqrt{\dfrac{c_1}{c_2}}$ Size of final sample, $n \doteq v\left(\dfrac{\bar{\sigma}_w}{\sigma}\right)^2\left\{1 + \dfrac{1}{N'}\right.$ $\times \left[v\dfrac{\sigma_b{}^2}{\sigma^2} + \left(\dfrac{\bar{\sigma}_R}{\bar{\sigma}_w}\right)\right]\right\}$

plans yield the same variance, viz., $(1 - v/N)\sigma^2/v$. This table is helpful in the choice of plan. The assumptions are in the heading of the table. There is the further assumption that the cost (c_2) or test of an interview is the same in all strata.

There are occasional small intangible costs, besides those in the table, not easy to evaluate: these are mentioned in the notes at the end of the table: though small, they may serve to tip the balance one way or another in case the costs in the table for 2 plans turn out to be about the same.

One may do well to reduce the size of the sample in a stratum where the cost of interviewing or of testing is excessive, and to build it up in other strata, according to the suggestion on page 303. The table of costs will then not be exactly applicable.

The exercises further on will give the student ideas by which he may compute the sample-sizes and the costs in Table 15.

As an illustration of the use of Table 15, we may compare the cost of Plan B with that of Plan A. Suppose that

$$N = 10,000$$
$$v = 500$$
$$c_1 = 20¢$$
$$c_2 = \$5.00$$
$$\sigma_w : \sigma = .8$$
$$K = vc_2 = \$2500, \text{ the cost of Plan A}$$

Then the cost of Plan B will be

$$Nc_1 + K\left(\frac{\sigma_w}{\sigma}\right)^2 = 10,000 \times .20 + 2500 \times .64$$
$$= \$2000 + \$1600 = \$3600 \qquad (121)$$

which is more than the cost $K = \$2500$ for Plan A.

Suppose that $\sigma_R : \sigma_w$ were 3.   Then Plan D would cost

$$K\left(\frac{\sigma_w}{\sigma}\right)^2 + c_1 \nu \left(\frac{\sigma_w}{\sigma}\right)^2 + (c_1 + c_2)\left(\frac{\sigma_R}{\sigma_w}\right)^2 = \$1600 + .20 \times 500 \times .64$$

$$+ \$5.20 \times 3^2 = \$1600 + \$64 + \$46.80 = \$1710.80 \quad (122)$$

which is less than the cost of Plan A or of Plan B.

EXERCISES ON THE TABLE OF COSTS

1. Derive the size of sample for Plan C that appears in the table of costs.   (Note that ν is the size of sample required for Plan A, to which all the other costs and variances in the table are related.)

2. Derive the size of sample required in Plan D to match the variance of Plan A (no additional stratification) with size of sample ν.

$$n = \nu \left(\frac{\sigma_w}{\sigma}\right)^2 + \left(\frac{\sigma_R}{\sigma_w}\right)^2$$

as shown in the table on page 350.

SOLUTION

The variance of Plan D appears on page 323.   Omit the finite multiplier and equate the variances of Plans A and D.   This equation of variance gives

$$\frac{{\sigma_w}^2}{n} + \frac{{\sigma_R}^2}{n^2} = \frac{\sigma^2}{\nu}$$

whence

$$n = \frac{\nu}{\sigma^2}\left[{\sigma_w}^2 + \frac{{\sigma_R}^2}{n}\right]$$

$$\doteq \nu \left(\frac{\sigma_w}{\sigma}\right)^2 + \left(\frac{\sigma_R}{\sigma_w}\right)^2$$

Q.E.D.

3. Derive the size of sample required for Plan E in terms of ν, to match the variance of Plan A (no additional stratification) with size of sample ν.

SOLUTION

The variance of Plan E is on page 324.   Equation of the approximate variances of Plans A and E gives

$$\frac{(\bar{\sigma}_w)^2}{n} + \frac{\bar{T}\bar{\sigma}_w\bar{\sigma}_R}{n^2} = \frac{\sigma^2}{\nu}$$

whence

$$n = \frac{\nu}{\sigma^2}\left\{(\bar{\sigma}_w)^2 + \frac{\bar{T}\bar{\sigma}_w\bar{\sigma}_R}{n}\right\}$$

$$= \nu \left(\frac{\bar{\sigma}_w}{\sigma}\right)^2 + \frac{\bar{T}\bar{\sigma}_R}{\bar{\sigma}_w}$$

as shown in the table on page 351.

4. Derive the size of sample required in Plan H to match the variance of Plan A (no additional stratification) with size of sample v.

SOLUTION

Equation of the approximate variances for Plans A and H gives

$$\frac{\sigma_w^2}{n} + \frac{\sigma_R^2}{nN'} + \frac{\sigma_b^2}{N'} = \frac{\sigma^2}{v}$$

whence

$$n = \frac{v}{\sigma^2}\left\{\sigma_w^2 + \frac{1}{N'}\left[n\sigma_b^2 + \sigma_R^2\right]\right\}$$

$$\doteq v\left(\frac{\sigma_w}{\sigma}\right)^2\left\{1 + \frac{1}{N'}\left[v\frac{\sigma_b^2}{\sigma^2} + \left(\frac{\sigma_R}{\sigma_w}\right)^2\right]\right\}$$

as shown in the table on page 351.

5. Derive the size of sample required for Plan I to match the variance of Plan A (no additional stratification) with size of sample v.

SOLUTION

Equation of the approximate variances for Plans A and I gives

$$\frac{(\bar{\sigma}_w)^2}{n} + \frac{\bar{\sigma}_w\bar{\sigma}_R}{nN'} + \frac{\sigma_b^2}{N'} = \frac{\sigma^2}{v}$$

$$n = \frac{v}{\sigma^2}\left\{(\bar{\sigma}_w)^2 + \frac{\bar{\sigma}_w\bar{\sigma}_R}{N'} + n\sigma_b^2\right\}$$

$$\doteq v\left(\frac{\bar{\sigma}_w}{\sigma}\right)^2\left\{1 + \frac{1}{N'}\left[v\frac{\sigma_b^2}{\sigma^2} + \frac{\bar{\sigma}_R}{\bar{\sigma}_w}\right]\right\}$$

as shown in the table on page 351.

Conditions under which Plan H may be cheaper than Plan A. *

PLAN A: 1. Draw v sampling units from the frame.
2. Carry out the survey and form the estimate

$$\bar{x} = \frac{N}{v}x \tag{123}$$

PLAN H: 1. Draw a random sample of N' sampling units from the frame.
2. Classify these units (perhaps by the aid of a simplified survey). We now have the estimates $\hat{P}_1, \hat{P}_2$.
3. Draw from these classes a proportionate sample of size n.
4. Carry out the survey and form the estimate

$$\bar{x} = \hat{P}_1\bar{x}_1 + \hat{P}_2\bar{x}_2 \tag{124}$$

* Contributed by Professor Akira Asai of Chiba University in Japan.

Under what conditions will Plan H be cheaper for the same Var $\bar{x}$? The following analysis (kindly supplied by Professor Asai) is applicable to comparisons of other plans as well. For example, to compare Plan I with Plan A, we need only replace σ_w by $\bar{\sigma}_w$ in these results.

Plan A: Cost, $K = \nu c_2$

$$\text{Size, } \nu = \frac{K}{c_2}$$

$$\text{Var } \bar{x} = \frac{\sigma^2}{\nu} = c_2 \frac{\sigma^2}{K} \quad \text{[Call this } A\text{]}$$

Plan H: Opt. $N' = n \dfrac{\sigma_b}{\sigma_n} \sqrt{c_2/c_1}$ [P. 328]

$$\text{Cost } K = N'c_1 + nc_2$$

$$= n\left[\frac{\sigma_b}{\sigma_w} \sqrt{c_2/c_1}\, c_1 + c_2\right] \tag{125}$$

$$\text{Sizes, } n = \frac{\sigma_w}{\sqrt{c_2}} \frac{K}{\sigma_b\sqrt{c_1} + \sigma_w\sqrt{c_2}} \tag{126}$$

$$N' = \frac{\sigma_b}{\sqrt{c_1}} \frac{K}{\sigma_b\sqrt{c_1} + \sigma_w\sqrt{c_2}} \tag{127}$$

$$\text{Var } \bar{x} = \frac{\sigma_w{}^2}{n} + \frac{\sigma_b{}^2}{N'} \quad \text{[P. 329]}$$

$$= \frac{1}{K}(\sigma_w\sqrt{c_2} + \sigma_b\sqrt{c_1})^2 \quad \text{[Call this } H\text{]} \tag{128}$$

We may now form the ratio of the 2 variances A and H.

$$\frac{H}{A} = \frac{(\sigma_w\sqrt{c_2} + \sigma_b\sqrt{c_1})^2}{c_2\sigma^2} \quad \text{[As } K = c_2\sigma^2/A\text{]}$$

whence

$$\sqrt{\frac{H}{A}} = \frac{\sigma_w}{\sigma} + \frac{\sigma_b}{\sigma}\sqrt{\frac{c_1}{c_2}} \tag{129}$$

This relation is useful when we wish to choose between Plan A and Plan H. A condition favorable to plan H is obviously that σ_w be small compared with σ, and c_1 very small compared with c_2. This favorable ratio of costs is sometimes achieved in a series of surveys, where the cost $N'c_1$ appears only once, and not in the successive subsequent surveys of small samples drawn from the original preliminary sample. A condition unfavorable to H is a fairly high value of σ_w combined with a high ratio of c_1 to c_2.

A precisely similar equation for the ratio $I:A$, for the variance of Plan I to the variance of Plan A, would hold, and the arguments likewise, with $\bar{\sigma}_w$ in place of σ_w.

F. STRATIFICATION FOR ANALYSIS

Enumerative and analytic uses of data. * Thus far in this chapter the problem has been to allocate the sample to get the best precision for a total (X) or for an average ($\bar{x}$ or f). An equally important problem is to estimate the magnitude of a difference between 2 means or between 2 percentages. This problem occurs frequently in marketing research. For example, one might need to estimate the number of people that use soluble coffee, and information as well on the reasons why they do, and on the difference between the average per capita consumption in 2 areas. The estimate of the number of users is an enumerative problem, and this is what we have been dealing with so far. The study of the reasons that people give, and the estimate of the per capita differences, is analytic, and requires another kind of sample-design.

In agriculture, in industry, and in science, one meets the analytic problem even oftener than he meets the enumerative one. Wherever there is a study of the causes of variation in dimensions, hardness, sales, production rates, error rates, the problem is analytic. We shall learn now something about the best allocation of a sample when the purpose is analytic.

Formula for allocation for purpose of analysis. Suppose that there are 2 strata, or 2 areas, or 2 methods, x and y, and that we wish to estimate the difference $a_x - a_y$ between the 2 means. We wish to learn how big the samples should be from the 2 strata. The variance of the difference $\bar{x} - \bar{y}$ will be

$$\text{Var}(\bar{x} - \bar{y}) = \frac{\sigma_x^2}{n_x} + \frac{\sigma_y^2}{n_y} \tag{130}$$

We note: (1) that there is no finite multiplier on the right for analysis: the absence of the finite multiplier is not an approximation nor a

---

* These terms originated in Chapter 7 of my book *Some Theory of Sampling* (Wiley, 1950).

See also William G. Cochran, *Sampling Techniques* (Wiley, 1953): p. 106 ff.

convenience: it is correctly omitted.*   (2) The samples from the 2 strata are independent, so there will be no correlation term.

Let us suppose first that the costs are the same in the 2 strata. Then, as the reader may wish to prove to himself, by methods similar to those on page 292, or otherwise, sample-sizes proportional to σ_x and to σ_y will yield the greatest precision for a given cost. Formally, we put

$$
\left.
\begin{aligned}
n_x &= \tfrac{1}{2}n\,\frac{\sigma_x}{\bar{\sigma}} \\[2ex]
n_y &= \tfrac{1}{2}n\,\frac{\sigma_y}{\bar{\sigma}}
\end{aligned}
\right\}
\tag{131}
$$

where $n = n_x + n_y$ and $\bar{\sigma} = \tfrac{1}{2}(\sigma_x + \sigma_y)$ for the greatest precision. Then

$$
\operatorname{Var}(\bar{x} - \bar{y}) = \frac{4(\bar{\sigma})^2}{n} = \frac{1}{n}(\sigma_x + \sigma_y)^2
\tag{132}
$$

If $\sigma_x = \sigma_y = \sigma$, these equations reduce to

$$
n_x = n_y
\tag{133}
$$

$$
\operatorname{Var}(\bar{x} - \bar{y}) = \frac{4\sigma^2}{n}
\tag{134}
$$

Thus, suppose that $\sigma_x = \sigma_y = 15$, and that we wish the standard error of $\bar{x} - \bar{y}$ to be 3. Then would

$$
n = \frac{4\sigma^2}{\operatorname{Var}(\bar{x} - \bar{y})} = \frac{4 \times 15^2}{3^2} = 100
\tag{135}
$$

wherefore $n_x = n_y = 50$.

If the problem is to measure the difference between 2 proportions p_x and p_y, and if the sampling is binomial in both strata, we may usefully put $\sigma_x{}^2 = \sigma_y{}^2 = pq$ where $p = \tfrac{1}{2}(p_x + p_y)$. The justification for this equality is that in most surveys or experiments where this type of problem comes up, the difference $p_x - p_y$ is small—so small that it takes a survey or an experiment to be sure that it exists at all. Eq. 132 then reduces to

$$
\operatorname{Var}(\hat{p}_1 - \hat{p}_2) = \frac{4}{n}pq
\tag{136}
$$

---

* For a proof, see my book, *Some Theory of Sampling* (Wiley, 1950): Ch. 7; also *J. Amer. Statist. Ass.*, vol. 48, 1953: pp. 224–255.

If p_x and p_y were both near to $\frac{1}{4}$, and if we wished the standard error of $\hat{p}_1 - \hat{p}_2$ to be $\frac{1}{12}$, then would

$$n = \frac{4pq}{\text{Var}\,(\hat{p}_1 - \hat{p}_2)}$$

$$= 4pq \times 12^2 = 36 \tag{137}$$

wherefore $n_x = n_y = 18$.

If the costs c_x and c_y per unit test are different in the 2 strata, the optimum allocation will be

$$\left.\begin{aligned} n_x &= \tfrac{1}{2}n\,\frac{\sigma_x}{\bar{\sigma}\sqrt{c_x}} \\[2mm] n_y &= \tfrac{1}{2}n\,\frac{\sigma_x}{\bar{\sigma}\sqrt{c_y}} \end{aligned}\right\} \tag{138}$$

where $\bar{\sigma} = \tfrac{1}{2}(\sigma_x/\sqrt{c_x} + \sigma_y/\sqrt{c_y})$.

Further examples appear in William G. Cochran's *Sampling Techniques* (Wiley, 1953): p. 106 ff.

PART III

Some Theory Useful in Sampling

CHAPTER 16

Evaluation of Expected Value
and of Bias in Sampling Procedures

And I have heard of thee, that thou canst make interpretations, and dissolve doubts: now if thou canst read the writing, and make known to me the interpretation thereof, thou shalt be clothed with scarlet, and have a chain of gold about thy neck, and shalt be the third ruler in the kingdom.—Daniel 5:16.

Purpose and content of Part III. We have thus far dealt with a number of examples and with various types of sample-designs and special tools. We have used theory, but we have not paused to derive the fundamental theorems of bias and of variance on which our theory and practice depend so heavily. We proceed here to fill in some of the gaps, so that we may have more confidence in our ability to handle new problems. We shall examine in this chapter the mathematical biases that may exist in certain estimates; and we shall learn why it is that this kind of bias, when it exists at all, disappears rapidly as n increases, and may be of no consequence under conditions of practice. Chapter 17 will derive some of the variances of random sampling, and the variance of the estimate of a variance, with adaptation to replicated sampling.

Chapters 16 and 17 together will provide a rational basis for decision between a biased and an unbiased estimate, with a view to keeping the total error a minimum (p. 62). Chapter 18 will help us when the characteristic that we wish to estimate is scattered in the sampling units like a Poisson variate, and it will show some simple ways to compute the margin of sampling error under these conditions. Chapter 19 will help us to decide how many segments there should be in 1 sampling unit, and what we need to know in advance about the material to fix this number. It will show that we need not worry about moderate roughness in our advance information, nor about moderate or even large inequalities in the sizes of the segments. Chapter 20 will aid us in the formation of strata, and Chapter 21 will aid our choice of the number of subsamples.

Rule for the calculation of an expected value. We learned in Chapter 4 that the expected value of any sampling procedure is the mean of the distribution of the estimates X, each X being calculated by the rules contained in the sampling procedure for all the possible samples that one can draw by applying the procedure to a given frame. Fortunately, a simple rule will find an expected value, without the construction of the distribution of X for all possible samples, nor even for a sample of all the possible samples. The rule is this:

1. Write down all the possible values a_i that the random variable may take.

2. Multiply each of these possible values a_i by P_i, where P_i designates the probability that the random variable will take the value a_i.

Formally, the E-operation for the expected value of the random variable x is this:

$$Ex = P_1 a_1 + P_2 a_2 + P_3 a_3 \tag{1}$$

where a_1, a_2, a_3 are the possible values that the random variable x may take, with the respective probabilities P_1, P_2, P_3.

Some simple applications of the expected value. Suppose that the plan of selection is to draw n random numbers between 1 and N, where N is the number of sampling units in the frame. The populations of the N sampling units are a_1, a_2, a_3, ... , a_N. A random number may draw any one of the N sampling units. Use the symbol x_1 to denote the population in the sampling unit drawn by the 1st random number. Denote by Ex_1 the expected value of x_1 under the rules stated above. All N sampling units have the same probability, $1/N$. Hence

$$Ex_1 = \frac{1}{N}(a_1 + a_2 + \cdots + a_N) = a \tag{2}$$

where a is, as before, the mean population per sampling unit in the frame. Hence, for the given plan of selection, x_1 is an unbiased estimate of a.

But now, curiously, if I let x_2 denote the 2d member of the sample, I find also that

$$Ex_2 = a \tag{3}$$

As a matter of fact, in a sample of n, all n members have the same expected value. That is,

$$Ex_1 = Ex_2 = \cdots = Ex_n = a \qquad \text{[True whether the drawings are made with or without replacement]} \tag{4}$$

Thus, every drawing produces an unbiased estimate of a. The proof is left as an exercise, which the student should now undertake.

Exercise 1.   The sampling units in a frame have serial numbers 1 to N. A sample is drawn by reading out from a table n random numbers between 1 and N. Find the probability that any specific sampling unit in the frame will be the mth member of a sample of size n.

SOLUTION

With replacement. The probabilities remain unchanged and equal to $1/N$ in all drawings, because the unit drawn is restored after each drawing.

Without replacement.

Probability to fail at the 1st drawing $\qquad 1 - \dfrac{1}{N} = \dfrac{N-1}{N}$

Probability to fail also at the 2d drawing $\dfrac{N-1}{N} \dfrac{N-2}{N-1} = \dfrac{N-2}{N}$

Probability to fail also at the 3d drawing $\qquad\qquad \dfrac{N-3}{N}$

Etc.

To be the mth member of the sample, the unit must fail on the 1st, 2d, 3d, $\ldots$, $(m-1)$th drawings, but succeed on the mth, the probability thereof being

$$\frac{N-(m-1)}{N} \frac{1}{N-(m-1)} = \frac{1}{N}$$

exactly as in sampling with replacement.

Thus, $Ex_m = a$ for every m, whether we draw with or without replacement, so we have now established Eq. 4. It will follow, after the theorem in the next section, that $E\bar{x} = a$ (Exercise 3, page 365).

Exercise 2.   Find the probability that any specific sampling unit of the frame in the preceding exercise will be in the sample.

SOLUTION

With replacement. The probability to appear in any drawing is $1/N$, and the probability to fail is $1 - 1/N$. The probability to fail n times running is $(1 - 1/N)^n$. The probability that it will be in a sample of size n is 1 less the probability that it will fail to appear in the 1st n drawings. This probability is obviously

$$1 - \left(1 - \frac{1}{N}\right)^n$$

which probability clearly approaches 1 as n increases, but is not yet 1 if $n = N$.

Note that this probability approaches n/N if N is very large compared with n. (Compare with the result n/N for sampling without replacement.)

Without replacement. The probability to be any member of the sample is $1/N$, already established in the preceding exercise. The probability that any specific sampling unit will be in a sample of size n is therefore n/N, being merely $1/N$ multiplied by n. This probability differs from the result obtained for sampling with replacement, although the difference will be small if N be very large compared with n. The probability without replacement is 1 if $n = N$, as is obviously correct.

Interchange of E and Σ.   We digress now for an important theorem. The expected value of the sum of 2 random variables x and y is the sum of their expected values.   Formally,

$$E(x + y) = Ex + Ey \tag{5}$$

Extension to any finite number of random variables follows at once; i.e.

$$E\Sigma x_i = \Sigma E x_i \tag{6}$$

A constant factor may go either side of the symbol E, thus

$$E(ax + by + cz) = aEx + bEy + cEz \qquad \text{[Exercise 2, page 371]} \tag{7}$$

PROOF

Let there be 2 frames, A and B, containing sampling units as shown. M and N may or may not be equal.   Let P_{ij} be the probability that a_i and b_j will be the results of the ith draw from A and the jth draw from B.   The

Frame	Populations of the sampling units
A	$a_1, a_2, a_3, \ldots, a_N$ for the x-population
B	$b_1, b_2, b_3, \ldots, b_M$ for the y-population

student may wish to write out a 2-way table of all the possible values of $a_i + b_j$, with $N = 5$ and $M = 3$, and observe that

$$\begin{aligned} E(x + y) &= \sum_i \sum_j P_{ij}(a_i + b_j) \\ &= \sum_i P_i a_i + \sum_j P_j b_j \\ &= Ex + Ey \end{aligned} \qquad \left[\begin{aligned} P_i &= \sum_j P_{ij} \\ P_j &= \sum_i P_{ij} \end{aligned} \right]$$

This result establishes Eq. 8.   In other words, E and Σ are interchangeable.

Remark 1.   This result holds under any probabilities whatever.   It matters not whether there is or is not correlation between a_i and b_j.   If there is no correlation between the drawings from A and from B, then $P_{ij} = P_i P_j$.   If there is only 1 frame, and the 2 drawings both come out of it without replacement, we merely put $P_{11} = P_{22} = P_{33} = \cdots = P_{NN} - 0$. With replacement, $P_{ii} = P_i^2$.

Remark 2.   Although the operations E and Σ are always interchangeable, E and other operations are not generally so.   Thus, $Ex^2 \neq (Ex)^2$ (p. 368); $E(x/y)$ may not always equal (Ex/Ey), as we shall observe for the ratio-estimate (pp. 394, 425); $Exy \neq ExEy$.   As a matter of fact, the student may wish to prove that

$$Exy = ExEy + \rho \sigma_x \sigma_y \tag{8}$$

where ρ is the correlation between the x- and y-populations of the sampling units in the 2 frames.   Thus, the 2 successive operations, E and multiplication, are interchangeable only if there is no correlation ($\rho = 0$).

Exercise 3.   Draw a sample of size n from a frame of N sampling units, by reading out n random numbers between 1 and N from a table of random numbers.   We learned in the exercise on page 362 that every sampling unit in the frame has the same probability $1/N$ of being the 1st, 2d, 3d, . . . , nth member of the sample.   Show that whether we draw with or without replacement,

$$E\bar{x} = a \tag{9}$$

where $\bar{x}$ is the mean of the sample, and a is the mean population of the sampling units in the frame.   In other words, $\bar{x}$ is an unbiased estimate of a.

SOLUTION

$$\bar{x} = \frac{1}{n}(x_1 + x_2 + \cdots + x_n) \quad \text{[Definition of } \bar{x}]$$

$$E\bar{x} = E\frac{1}{n}(x_1 + x_2 + \cdots + x_n)$$

$$= \frac{1}{n}(Ex_1 + Ex_2 + \cdots + Ex_n) \quad \text{[Interchange } E \text{ and } \Sigma]$$

$$= \frac{1}{n}(a + a + \cdots + a) \quad [n \text{ terms.   Recall } Ex_1 = Ex_2 = Ex_3$$

$$= \cdots = Ex_n = a; \text{ page 362]}$$

$$= a \qquad\qquad\qquad Q.E.D.$$

Definition of a function of a random variable.   If x is a random variable, then any function of x is also a random variable.   Thus, $cx, x + c, ax + c,$ $x^2, \sqrt{x}, x - 2$, and $\sin x^2$ are also random variables.   The combination of 2 or more random variables creates a random variable.   The mean, the variance, and any estimate made from a sample are all random variables, because they will vary with definite probabilities from one sample to another as the sampling procedure is repeated.

The expected value of a function of a random variable.   Let $f(x)$ be some specific function of the random variable x.   Then $f(x)$ is also a random variable.   Formally, it is a simple matter to write down the expected value $E f(x)$, although for any function other than a linear function of x or for a polynomial in x, we shall require a satisfactory approximation. To get started, we only need to keep in mind this fundamental principle: if x may take on the value $a_1, a_2, \ldots$ with probabilities $P_1, P_2, \ldots$, then $f(x)$ may take on the values $f(a_1), f(a_2), \ldots$ with the same probabilities $P_1, P_2, \ldots$.   For the expected value of $f(x)$ we return to the fundamental E-operation defined in Eq. 1, which tells us to write

$$E f(x) = P_1 f(a_1) + P_2 f(a_2) + \cdots + P_N f(a_N) \tag{10}$$

We saw already in Chapter 4 the formal definition of the bias of a sampling procedure. If $B(X)$ denotes the bias in a sampling procedure that yields the estimate X of the total population A in the frame, then

$$B(X) = EX - A \qquad (11)$$

where EX is the expected value of the random variable X. There is a similar definition for the bias of the sampling procedure that yields the esitmate $\bar{x}$ of the mean population a per sampling unit in the frame. This bias would be

$$B(\bar{x}) = E\bar{x} - a \qquad (12)$$

General formula for the bias in an estimate. There are many occasions when we wish to know $f(Ex)$, but in which we know not Ex. We may be fortunate, though, to have a random numerical value of the random variable x. Just as we sometimes use x as an estimate of Ex, so we sometimes use $f(x)$ as an estimate of $f(Ex)$. The question then arises, what will be the average error in $f(x)$ if we replace Ex by x in $f(Ex)$?   Let

$$\Delta x = x - Ex \qquad (13)$$

be the sampling error in x.   Then

$$\Delta f = f(x) - f(Ex) \qquad (14)$$

will be the resulting error in $f(x)$.   We note that x, Δx, $f(x)$, and Δf are all random variables, and that

$$E\,\Delta x = 0 \qquad \text{[By definition of } \Delta x] \qquad (15)$$

$$E(\Delta x)^2 = \sigma^2 \qquad \text{[By definition of } \sigma^2] \qquad (16)$$

We now define $E\,\Delta f$ as the bias or average error in the repeated use of $f(x)$ in place of $f(Ex)$.   In symbols,

$$B f(x) = E\,\Delta f \qquad \text{[Definition of the bias of } f(x)] \qquad (17)$$

The *relative* bias in $f(x)$ is by definition

$$\frac{B f(x)}{f(Ex)}$$

which is the bias $B f(x)$ measured in units of $f(Ex)$.

If $E\,\Delta f$ is 0 for some function $f(x)$, then the average error in the repeated use of $f(x)$ in place of $f(Ex)$ will be 0.   That is, $f(x)$ is then an unbiased estimate of $f(Ex)$.   If $E\,\Delta f \neq 0$, there will be a bias in the repeated use of $f(x)$.

Our problem is to evaluate the bias $Bf(x)$; and to show, if there is a bias, how rapidly this bias will diminish with an increase in the sample whence comes the estimate x.   Two examples may help, before we write down the general formula for the bias in a function, which we shall see in Eqs. 31 and 32 (p. 369).

There remains also the problem to evaluate $\text{Var} f(x)$: this we do in Section B of the next chapter.

EXAMPLE 1. A LINEAR FUNCTION OF x.   Let

$$f(x) = h + kx \tag{18}$$

Then

$$f(Ex) = h + kEx \tag{19}$$

$$\Delta f = f(x) - f(Ex) = k(x - Ex) = k\,\Delta x \tag{20}$$

whence by the definition of bias given in Eq. 17,

$$Bf(x) = E\,\Delta f = 0 \qquad [\text{Because } E\,\Delta x = 0]$$

wherefore in this case

$$Ef(x) = f(Ex) \tag{21}$$

and the linear function $f(x) = h + kx$ is an unbiased estimator of this function $f(x)$.

EXAMPLE 2. A QUADRATIC FUNCTION OF x.   Let $f(x) = \pi x^2$.   Then $f(Ex) = \pi(Ex)^2$, which is the area of a circle whose radius is Ex. πx^2 is an estimate of this area, x being a measurement of the radius.

$$f(Ex) = \pi(Ex)^2 \tag{22}$$

$$f(x) = f(Ex + \Delta x) = \pi(Ex + \Delta x)^2$$
$$= \pi\{(Ex)^2 + 2Ex\,\Delta x + (\Delta x)^2\} \tag{23}$$

By subtraction, the error in the area is

$$\Delta f = f(x) - f(Ex) = \pi\{2Ex\,\Delta x + (\Delta x)^2\}$$

Now take the expected value of both sides, and remember Eqs. 17 and 18. Then

$$Bf(x) = E\,\Delta f = \pi\{0 + \sigma^2\} \tag{24}$$

The *relative* bias [the bias measured in units of $f(Ex)$] is then

$$\frac{Bf(x)}{f(Ex)} = E\frac{\Delta f}{f(Ex)} = C^2 \tag{25}$$

where C is the coefficient of variation of x.   Thus, πx^2 is a biased estimate of $\pi(Ex)^2$, and the *relative* bias $E\,\Delta f / f(Ex)$ is for this function just the square of the coefficient of variation C of x.

Remark 3. We may easily illustrate numerically the 2 above examples. Let x be a random variable with the distribution shown in the accompanying table. Use the notation f_1 for $h + kx$ and f_2 for πx^2. The rel-variance of the random variable x is

$$C^2 = \frac{\sigma^2}{(Ex)^2} = \frac{.5}{2^2} = \frac{1}{8} \tag{26}$$

The table shows that the mean of the distribution of $h + kx$ is precisely $h + k\ Ex$. That is, the function $h + kx$ is an unbiased estimate of

DISTRIBUTIONS OF x AND OF THE FUNCTIONS $h + kx$ AND πx^2

Probability	x	$(x - Ex)^2$	$f_1 = h + kx$	$f_2 = \pi x^2$
.25	1	1	$h + k$	π
.50	2	0	$h + 2k$	4π
.25	3	1	$h + 3k$	9π
Mean	$Ex = 2$	$\sigma^2 = .5$	$Ef_1 = h + 2k$	$Ef_2 = 4.5\pi$

$h + k\ Ex$. But the mean of the distribution of πx^2 is 4.5π, whereas $\pi(Ex)^2 = 4\pi$. In other words, the function πx^2 is a biased estimate of $\pi(Ex)^2$, and the amount of the bias is

$$Bf(x) = E\Delta f = 4.5\pi - 4\pi = .5\pi \tag{27}$$

which is just $4\pi C_x^2$, as predicted by Eq. 25.

Remark 4. There is a lesson here in the science of physical measurement. Suppose that we have a series of n measurements of the radius of a circle; denote them by $x_1, x_2, \ldots, x_n$, and suppose that they behave as if they were random drawings from a frame whose mean is $E\bar{x} = 2$ and whose variance is $\sigma_x^2 = .50$. If for every x_i we compute the area πx_i^2; then take the average of the n areas so computed, we should find that the average area is too high by the absolute amount $\sigma_x^2 = .5$, and by the relative amount of $\frac{1}{8}$th, as predicted by Eq. 25. The more measurements we take, the closer the average of the n calculated areas settles onto $4.5\pi^2$, which is $\frac{1}{8}$th too high, regardless of how many measurements we take.

On the other hand, if we compute $\bar{x}$ from the n measurements, and then calculate once for all the area πx^2, this area is subject only to the bias $\pi\sigma_{\bar{x}}^2$ or $\pi\sigma_x^2/n$, which diminishes toward 0 as n increases.

We have therefore this rule: it is better to compute $\bar{x}$ first and use it in the function, than to compute the function n times and then to take the average of all the n functions $f(x_1), f(x_2)$, etc: the easiest way is therefore the best. This rule appears in some books on least squares. It originated with Gauss about 1813, who derived it for use in astronomy, where originated much modern statistical theory. A procedure to eliminate the bias in the estimate $\pi\bar{x}^2$ appears on page 428.

We may now write down the following general formula for the bias in $f(x)$. We start with Taylor's series

$$f(x) = f(Ex) + (x - Ex)f_x + \frac{1}{2!}(x - Ex)^2 f_{xx}$$

$$+ \frac{1}{3!}(x - Ex)^3 f_{xxx} + \frac{1}{4!}(x - Ex)^4 f_{xxxx} + \cdots \quad (28)$$

Then

$$\Delta f = f(x) - f(Ex) = \frac{1}{2!}f_{xx}(\Delta x)^2 + \frac{1}{3!}f_{xxx}(\Delta x)^3 + \frac{1}{4!}f_{xxxx}(\Delta x)^4 + \cdots \quad (29)$$

The expected value term by term gives

$$Bf(x) = E\,\Delta f = \frac{1}{2!}f_{xx}E(\Delta x)^2 + \frac{1}{3!}f_{xxx}E(\Delta x)^3 + \frac{1}{4!}f_{xxxx}E(\Delta x)^4 + R \quad (30)$$

where R is the remainder of the series. f_x, f_{xx}, etc., are derivatives of f, evaluated at $x = Ex$.

Our usual problem is to find the bias in $f(\bar{x})$, where $\bar{x}$ is the mean of a sample drawn by reading out n random numbers between 1 and N. The equation just derived then gives

$$Bf(\bar{x}) = E\{f(\bar{x}) - f(E\bar{x})\}$$

$$= \frac{1}{2!}f_{xx}E(\Delta\bar{x})^2 + \frac{1}{3!}f_{xxx}E(\Delta\bar{x})^3 + \frac{1}{4!}f_{xxxx}E(\Delta\bar{x})^4 + R_{\bar{x}}$$

$$= \frac{1}{2!}f_{xx}\frac{\sigma^2}{n} + \frac{1}{3!}f_{xxx}\frac{\mu_3}{n^2} + \frac{1}{4!}f_{xxxx}\left\{\frac{3}{n^2}\sigma^4 + \frac{1}{n^3}(\mu_4 - 3\sigma^4)\right\} + R_{\bar{x}}$$

$$(31)$$

where $R_{\bar{x}}$ is the remainder of the series. σ^2 is the variance between the N sampling units in the frame, as heretofore. The evaluations of $E(\Delta\bar{x})^3$ and $E(\Delta\bar{x})^4$ which appear here are in the exercises on page 443. We omit for the moment the finite multipliers on the right, as they are difficult and we shall later neglect all the terms but the 1st one anyhow.

The expected values of the next 2 higher powers of $\Delta\bar{x}$ will contain $1/n^3$ as a factor.* The moments of any real frame are finite (i.e., no moment is infinite); hence there will always be a sample-size n so big that we may neglect all the terms on the right beyond the 1st term. That is,

$$Bf(\bar{x}) \doteq \frac{1}{2!}f_{xx}\frac{\sigma^2}{n} \quad (32)$$

---

* Harald Cramér, *Mathematical Methods of Statistics* (Princeton, 1946); p. 345 ff. Oskar Anderson, *Probleme der Statistischen Methodenlehre* (Physica Verlag, Wurzburg, 1957): Ch. VI.

This is the equation that we shall use for a practical evaluation of the bias in a function of $\bar{x}$. It is called the 1st approximation because it neglects all the other terms. It will serve us well if we do not misuse it. Retention of the terms in $1/n^2$ would give the 2d approximation. The finite multiplier $(N - n)/(N - 1)$ is easy to insert when required. We may in fact neglect the bias altogether in many estimates.

Remark 5. An unbiased estimate is one for which all the terms on the right of Eq. 31 are 0 absolutely, regardless of the distribution of x in the frame. An example is $f(x) = X = N\bar{x}$, for which the 2d derivative and all others beyond it are 0 absolutely, wherefore there is no bias in X, even if $n = 1$. We have learned this already, of course, on page 365. We investigated the quadratic πx^2 a few pages back. The 2d derivative therefor is 2π. The 3d derivative and all higher derivatives are 0; hence the bias in $\pi \bar{x}^2$ is exactly $\pi \sigma^2/n$, regardless of the distribution of x^2 in the frame.

Remark 6. What one must do when the derivatives beyond a certain point are not all 0 absolutely, if he wishes to use Eq. 32, is to evaluate the remainder after the 1st term on the right of Eq. 31. This is a task that differs from one function to another, and from one frame to another. Actually, with experience, knowledge of theory, the problem of evaluating the bias is usually not difficult. The exceptions are freakish, and are recognizable.

Remark 7. The actual bias in a function, when there is any bias, is simple to evaluate numerically, by use of the replicated method, and with the help of Eq. 32, after the results are in (p. 425).

Remark 8. The student should bear in mind that n is not necessarily the number of people in the sample, nor the number of dwelling units or of other items; n denotes (as hitherto) the number of sampling units in the sample. A random number draws 1 sampling unit, which may contain 1 or a number of people, dwelling units, or other items.

Functions of 2 or more random variables. We sometimes deal with functions of 2 or more random variables. An important example is the ratio $f = x/y = \bar{x}/\bar{y}$, which we have calculated a number of times. Some theory for the bias and for the variance of this ratio appears in the next chapter (see Table 1 on page 394; also page 422). The general formula for the bias in 2 random variables $\bar{x}$ and $\bar{y}$ is

$$B f(\bar{x}, \bar{y}) \doteq \frac{1}{2n} \{f_{xx}\sigma_x^2 + 2f_{xy}\,\rho\sigma_x\sigma_y + f_{yy}\sigma_y^2\}$$

$$= \frac{1}{2n} \{b^2 f_{xx}C_x^2 + 2ab f_{xy}C_{xy} + a^2 f_{yy}C_y^2\} \qquad (33)$$

σ_x and σ_y are the standard deviations between the x- and the y-populations of the sampling units in the frame; $a = Ex_i$; $b = Ey_i$,

$$C_x^2 = \frac{E(x_i - a)^2}{a^2} \qquad (34)$$

is the rel-variance* between the x-populations of the sampling units in the frame. C_y^2 has a similar definition, and

$$\rho \sigma_x \sigma_y = \text{Cov } x, y = E(x_i - a)(y_i - b) \tag{35}$$

is the covariance between these populations. ρ is the correlation between them, and

$$C_{xy} = \frac{\rho \sigma_x \sigma_y}{ab} = \frac{E(x_i - a)(y_i - b)}{ab} \tag{36}$$

is their rel-covariance. The finite multiplier $(N - n)/(N - 1)$ is easy to insert as a factor on the right when required.

EXERCISES ON THE EXPECTED VALUES OF FUNCTIONS OF RANDOM VARIABLES

1. $f(x) = c$ [c is a constant]

$Ec = c$

The meaning of $f(x) = c$ is this: write down the constant number c no matter what be the random variable x. The result of the operation f is merely the sequence $c, c, c, \ldots$, whose mean is c, and whose standard deviation is 0.

2. $f(x) = cx$

$Ecx = cEx$

Here, every random drawing gives $f(x) = cx$ where x is the result of the drawing. Suppose that we plot the distribution of cx, after many drawings. The mean of the distribution will be Ecx, which is equal to cEx, or c times the mean of the distribution of x. Derived formally,

$$Ecx = P_1 ca_1 + P_2 ca_2 + \cdots P_M ca_M \quad \text{[From Eq. 1]}$$
$$= c[P_1 a_1 + P_2 a_2 + \cdots P_M a_M]$$
$$= cEx \quad \text{[By definition of } Ex\text{]}$$

3. Let $X = N\bar{x}$, where N, as usual, is the number of sampling units in the frame. Show that

$$EX = A$$

Thus, when the probabilities are all equal, X is an unbiased estimate of A, the total population in the frame.

* Rel-variance was defined on page 124 as $C_x^2 = \sigma_x^2/(Ex)^2$. The term rel-covariance was also invented by Hansen, Hurwitz, and Madow to signify $C_{xy} = \rho_{xy}\sigma_x\sigma_y/ExEy$.

For the variance of $\bar{x}$ and of X see Eqs. 7 and 20 of Chapter 17 (pages 381 and 383).

4. Show that if $f(x) = x^2$, then $Ef(x) \neq f(Ex)$. That is, $f(Ex)$ is a biased estimate of $f(x)$. (See the illustration on page 368.)

5. Let $f(x) = x^2$, and suppose that x may take the values $a_1, a_2, \ldots, a_N$ with probabilities $P_1, P_2, \ldots, P_N$. Show that

$$Ex^2 = P_1 a_1{}^2 + P_2 a_2{}^2 + \cdots$$

$$= a^2 + \sigma^2 \qquad [\text{Compare with Eq. 14, p. 382.}] \qquad (37)$$

where

$$a = P_1 a_1 + P_2 a_2 + \cdots + P_N a_N$$

and

$$\sigma^2 = P_1 (a_1 - a)^2 + P_2 (a_2 - a)^2 + \ldots + P_N (a_N - a)^2$$

6. Show that

$$E \frac{1}{x} \doteq \frac{1}{Ex} (1 + C_x{}^2) \qquad (38)$$

Illustrate this result with the distribution on page 395.

This and further examples occur on page 394.

Some theory for an inconsistent biased estimate. The bias that we dealt with in earlier pages of this chapter is known as a consistent bias*—that is, it disappears as the size of the sample increases. We now encounter a bias (Plan 2 ahead) that is inconsistent; it does not disappear as the number m of primary sampling units in the sample increases. The theory and illustrations here will refer to samples of names drawn from samples of cards in a file, though the applications are much broader and are very common.

The cards in a file are numbered 1 to M (Fig. 21). There are 5 lines on the front of a card, and 5 lines on the back. Some lines show names, and some lines are blank. For example, there might be a card at each address where a company has customers. At some addresses there might be 1 name; at others, 2 or 3 or more names. There will be more than 1 card at any address where there are more than 10 names. If a person moves away, the name goes off, and the line becomes a blank. A new

---

* The terms consistent and inconsistent originated with Sir Ronald Fisher in his monumental work "On the mathematical foundations of theoretical statistics," *Phil. Trans. Royal Soc.*, vol. A222, 1922: pp. 309–368.

name goes into a blank at the end of the card, or into a space made blank by removal.   Each name shows a numerical entry, which may be 0. The numerical entry might be number of purchases, or amount purchased. In one example that I have in mind, there was a card for each motor vehicle in a fleet, and the entries showed repairs, with amounts in dollars.

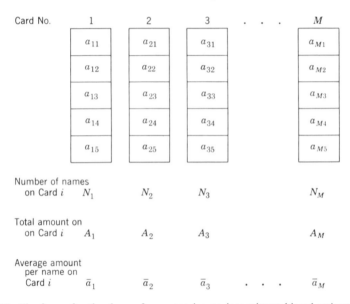

Card No.	1	2	3	. . .	M
	a_{11}	a_{21}	a_{31}		a_{M1}
	a_{12}	a_{22}	a_{32}		a_{M2}
	a_{13}	a_{23}	a_{33}		a_{M3}
	a_{14}	a_{24}	a_{34}		a_{M4}
	a_{15}	a_{25}	a_{35}		a_{M5}

Number of names on Card i	N_1	N_2	N_3		N_M
Total amount on on Card i	A_1	A_2	A_3		A_M
Average amount per name on Card i	$\bar{a}_1$	$\bar{a}_2$	$\bar{a}_3$	. . .	$\bar{a}_M$

Fig. 21. The frame for the theory for comparing an inconsistent biased estimate with unbiased estimates. a_{ij} is the amount shown on Line j of Card i.   Some of the spaces may be blank, with no name, in which case a_{ij} on that line is treated as 0.

Other examples:

1. The families in a sample came in with equal probabilities.   Random numbers selected on the spot 1 adult from each family for interview, following Appendix C to Chapter 12, page 240.   In Chapter 17, where we shall examine the variances of 2 possible estimates (Plans 2 and 3 ahead), each household under consideration contains either 1 reader or 2 readers of a certain newspaper.   If there is 1 reader in a household, we interview him (or her).   If there are 2 readers in a household, we draw 1 of them and interview only him (or her).   A household thus corresponds to a card with 2 lines, either or both of which may show a name.   The probability of selection thus depends on whether there are 1 or 2 readers in the household.

2. In another example, where a sample of families came in with equal probabilities, the procedure called for the selection on the spot, from each

family in the sample, 1 female of age 16 or over for interview.   The probability of selection varies here also, depending on whether there are 1, 2, or 3 females in the family.   (This was a survey on the use of lipstick and other aids to beauty.   The procedure of selection within the family again followed Appendix C to Chapter 12, page 240.)

3. Draw 1 segment from the list of segments in a block.   Canvass this segment.   The probability of selection will vary from block to block depending (a) upon the number N_i of segments in the block, and (b) on the number S_i of segments accorded to the block in the list of blocks. Weighting factors N_i/S_i then compensate the unequal probabilities. (Weighting factors may or may not be wise, as we shall see.   The reader will sense the fact that this procedure differs from the procedure that he has learned here in Chapters 10 and 11, where no weighting was necessary, as the rules called for the selection of a constant proportion of segments from any area, and not a constant number of segments from all areas.)

4. Draw a sample of ore from each load of ore in a shipment that consists of M loads in total.   Send each sample to the laboratory for assay.   The samples that go to the laboratory come in with probability dependent on the size of the load.

Heretofore, sampling units came in with equal probabilities, and the straight average ($\bar{x}$) obtained from the sample was an unbiased estimate of the average population per sampling unit in the frame.   Will unequal probabilities lead to bias?   Maybe yes, maybe no, as we shall see.   The answer depends on the formula of estimation (the 3d part of a sampling plan; see page 38).   We may apply weights and remove whatever bias there be (Plan 3).   But even without weights, there may be no bias, depending on whether certain characteristics of the sampling units are correlated.

Moreover, even if an estimating procedure is subject to bias, should we discard it?   Our decision should rest on the total error, which arises from bias and sampling variation together.   A biased procedure may in fact be preferable to an unbiased estimate, as it may be subject to smaller standard deviation and to less total error than a biased estimate.*

* The usefulness of biased estimates was first recognized, I believe, by my colleagues Morris H. Hansen and William N. Hurwitz, about 1941 or 1942.   Till then most writers had supposed that an estimate, to be useful, must be unbiased.   This assumption is common even now.   One sees instances where the sampling procedure calls for the application of weights (as in Plan 3 ahead) to eliminate the bias, where tests and comparisons might easily show the unbiased estimate (Plan 2) to be preferable and simpler.

Meanwhile, we remark here that unless one possesses in advance sufficient information by which to make a rational choice in the formula of estimation, he would be wise to defer his decision until an examination of the returns may provide quantitative information for the choice.   An example appears in Chapter 17, page 398.

For notation, let

M	be the total number of cards
N_i	be the number of names on Card i
$N = \sum_1^M N_i = M\bar{N}$	be the total number of names on all M cards
$\bar{N} = \dfrac{N}{M}$	be the average number of names per card
a_{ij}	be the amount shown for Name j on Card i
$A_i = \sum_{j=1}^{N_i} a_{ij}$	be the total amount for all N_i names on Card i
$A = \sum_1^M A_i$	be the total amount for all N names on all M cards
$\bar{a}_i = \dfrac{A_i}{N_i}$	be the average amount per name on Card i
$a = \dfrac{A}{N}$	be the average amount per name for all N names on all M cards
k	be the number of lines per card (10 in the above description)
$c = \dfrac{A}{Mk}$	be the average amount per line on all Mk lines on all M cards

We shall describe several sampling plans by which to estimate A, a, and a'. We investigate in this chapter the biases in certain estimating procedures that we shall propose; and in the next chapter, the variances of 2 of the procedures (pp. 398 ff.).

Plan 1.   Draw m cards without replacement by reading out m unduplicated random numbers between 1 and M. For any card in the sample, add up the amount for all the names on that card, and designate this amount by X_i. If there is no name on a card, X_i will be 0. The reader may show that

$$X_{(1)} = \frac{M}{m} \sum_1^m X_i \quad \text{is an unbiased estimate of } A \quad (39)$$

The subscript (1) will denote that the estimate comes from Plan 1. There will be subscripts in like manner for Plans 2, 3, 4.

SOLUTION

There is nothing new here. $\sum_1^m X_i$ is the total amount in the sample, and m is the number of sampling units. The sampling unit is a card, and all

M cards in the frame have the same probability to come into the sample.
$\frac{1}{m} \sum_1^m X_i$ is the average amount per sampling unit in the sample, and is an
unbiased estimate of the average amount per card in the frame (like $\bar{x}$ on
page 365).   As there are M cards in total, $\frac{M}{m} \sum_1^m X_i$ must be an unbiased
estimate of A.

Plan 2.   Draw a sample of m cards, as before;   then read out a random
number between 1 and N_i to draw 1 name from Card i $(i = 1, 2, \ldots, m)$.
Let x_i denote the amount shown for the name so drawn.   If there is no
name on a card, let $x_i = 0$ for that card.   The reader may, as an exercise:

a. Show that

$$X_{(2)} = \frac{M\bar{N}}{m} \sum_1^m x_i \quad \text{is a biased estimate of } A \qquad (40)$$

b. Find the magnitude of the bias in this estimate.

SOLUTION

Here x_i is the amount that falls into a sampling unit, and $\sum_1^m x_i$ is the total
amount in the m sampling units. $\frac{1}{m} \sum_1^m x_i$ is the average amount per
sampling unit, but the sampling units (names) will have different probabilities
of selection, as N_i may vary from card to card.

$$EX_{(2)} = \frac{M\bar{N}}{m} E \sum_1^m x_i = \frac{M\bar{N}}{m} \sum_1^m Ex_i \qquad \text{[P. 364]}$$

$$= \frac{M\bar{N}}{m} m \left\{ \frac{1}{MN_1} (a_{11} + a_{12} + \cdots + a_{1_{N_1}}) \right.$$

$$+ \frac{1}{MN_2} (a_{21} + a_{22} + \cdots + a_{2_{N_2}})$$

$$\vdots$$

$$\left. + \frac{1}{MN_M} (a_{M1} + a_{M2} + \cdots + a_{MN_M}) \right\}$$

$$= \frac{M\bar{N}}{M} \{ \bar{a}_1 + \bar{a}_2 + \cdots + \bar{a}_M \} = \sum_1^M \bar{N} \bar{a}_i \qquad (41)$$

wherein $\bar{a}_i$ is the average consumption per family on Card i.   Now

$$A = \sum_1^M A_i = \sum_1^M N_i \bar{a}_i \qquad \text{[By definition of } \bar{a}_i]$$

Hence the bias in Plan 2 will be

$$BX_{(2)} = EX_{(2)} - A \quad \text{[By definition of bias]}$$

$$= \sum_{1}^{M} (\bar{N} - N_i)\bar{a}_i$$

$$\left. \begin{aligned} &= 0 \quad \text{if } N_i \text{ and } \bar{a}_i \text{ are uncorrelated} \\ &\neq 0 \quad \text{otherwise} \end{aligned} \right\} \tag{42}$$

Remark 9. Unequal probabilities of selection thus lead to bias only if N_i and $\bar{a}_i$ are uncorrelated. One may estimate this bias from a subsample of the returns by plotting (say) 100 points on a scatter-diagram, N_i on one axis, x_i on the other.

If the scatter-diagram shows no correlation, Plan 2 would be preferable over Plan 3 (below). If the diagram shows possible correlation, the simple calculation of

$$\frac{M}{100} \sum_{1}^{100} (\bar{N} - N_i)\bar{x}_i \tag{43}$$

for a trial sample of 100 cards would give a useful estimate of the bias. One usually needs to know the relative bias: this would be $\left\{ \sum_{1}^{100}(1 - N_i / \bar{N})\bar{x}_i \right\} \Big/ \sum_{1}^{100} \bar{x}_i$, in which one might have to estimate $\bar{N}$ from the sample. Even if there is obvious bias, Plan 2 may still be preferable (p. 398).

Remark 10. A replicated sample furnishes an alternative to the scatter-diagram mentioned above. One may estimate the bias of Plan 2 by the procedure to appear on page 425. The replication also furnishes easy estimates of the variances of this and any other plan, enabling one to make a rational choice, on the basis of the total error. It is important to use enough thick zones to furnish sufficient degrees of freedom, as I have remarked elsewhere (page 189).

Remark 1. Note that the bias in Plan 2 is independent of m. That is, if we take a fixed number of names (1 in this illustration) from each card, the bias will remain the same whether we take a sample of 1 card, 100 cards, or all M cards. In other words, the bias is inconsistent (see the footnote on page 372).

Plan 3. The selection here is the same as in Plan 2, but the formula for the estimate is different. The reader may show that

$$Plan\ 3 \qquad X_{(3)} = \frac{M}{m} \sum_{1}^{m} N_i x_i \quad \text{is an unbiased estimate of } A \tag{44}$$

Remark 2. This estimate is unbiased, but the weighting factors N_i, if they vary greatly from one to another, may add so much variance that we might well prefer Plan 2. Some theory by which to make a decision between Plans 2 and 3 begins on page 398.

Plan 4.   Draw a sample of m cards as before, and draw 1 of the k lines at random from each card.   Record the amount x_i for the name (if any) on the line so drawn.   A blank line draws 0.   The reader may show that

Plan 4 $$X_{(4)} = \frac{Mk}{m} \sum_1^m x_i \qquad \text{is an unbiased estimate of } A \qquad (45)$$

Remark 3.   If some cards bear no names, the sample of m cards in Plans 1, 2, and 3 will not yield m names.   Likewise, if some lines are blank, Plan 4 will not yield m names.

If we desire a sample of m names, we may simply draw more cards. Thus, if 40% of the lines are vacant, we should take in Plan 4 a sample of 100 cards for an expected 60 names.

Remark 4.   Plan 4 has much to recommend it, but the cost will be high if the proportion of blanks is high, and if the cost of finding a card is high.

Remark 5.   We may fix in advance the number of names in the sample for Plans 2, 3, 4, by specifying that the procedure of selection will continue until we reach the required number of names.   The cost will not be exactly predictable, however, unless we know in advance the proportion of blanks.

Remark 6.   All families in Plan 4 have equal probabilities, regardless of blanks.

Remark 7.   We may derive from any of the foregoing estimates an estimate of the average amount per name.   Let X be any estimate of A. Then X/N or $X/M\bar{N}$ will be an estimate of the amount per name.   If we know not N, we may easily estimate $\bar{N}$ from the sample, and take $N = M\bar{N}$.

Remark 8.   If one draws a constant proportion of names or of lines from each card, all names have equal probabilities of selection, and the unweighted arithmetic average per name will be an unbiased estimate of a.   Plan 1 was an example; we drew there 100% of the names from every card in the sample.   Plan 4 is another example, where we drew the constant proportion 1 line in k; also on the average, 1 name in k.

Remark 9.   To take care of the case where a name covers 2 lines, we may specify in the sampling procedure that the 2d line shall be a blank. See page 137, where the instructions give a rule for the selection of interline abstracts that cover 2 or more pages.

EXERCISES

1. Make proper modifications of the estimates for Plans 2 and 3 in case you wish to draw 2 names from every card in the sample.   (Merely replace x_i by $\frac{1}{2}(x_{i1} + x_{i2})$, where the subscripts $i1$ and $i2$ refer to the 2 selections from Card i.)

2. Show that the bias in Plan 2 takes the form

$$BX_{(2)} = -M\sigma_N\sigma_b\rho = -M\bar{N}\sigma_b C_N\rho$$

where

$$\rho = \frac{E(N_i - \bar{N})(\bar{a}_i - a)}{\sigma_N\sigma_b}$$

$$\sigma_N{}^2 = \frac{1}{M}\sum_1^M (N_i - \bar{N})^2$$

$$C_N = \frac{\sigma_N}{\bar{N}}$$

$$\sigma_b{}^2 = \frac{1}{M}\sum_1^M [\bar{a}_i - E\bar{x}_{(2)}]^2$$

ρ is the correlation between the M values of N_i and $\bar{a}_i$—i.e., the correlation between the number of names on Card i and the average amount thereon. $\sigma_b{}^2$ is the unweighted variance between the means $\bar{a}_i$ of the M cards. $\sigma_N{}^2$ is the variance between the numbers N_i, and $C_N = \sigma_N/\bar{N}$ is the coefficient of variation between the N_i. $E\bar{x}_{(2)} = (1/N)EX_{(2)}$.

CHAPTER 17

Theory of Variances

How poor are they that have not patience!—Iago to Roderigo in Shakespeare's *Othello*, Act III, Scene iii.

A. VARIANCES OF ESTIMATES FORMED FROM A RATIO WITH DENOMINATOR FIXED

The variance of a function of a random variable. The variance σ_x^2 of the theoretical distribution of the random variable x (i.e., for short, the variance of x) is,

$$\sigma_x^2 = E[x - Ex]^2 \quad \text{[Definition]} \tag{1}$$

$$= Ex^2 - (Ex)^2 \quad \text{[Identical with the definition]} \tag{2}$$

Or, in reverse,

$$Ex^2 = \sigma_x^2 + (Ex)^2 \quad \text{[A very important theorem]} \tag{3}$$

These last 2 equations will be useful continually in our study of theory. The former evaluates the variance of x in terms of Ex^2 and $(Ex)^2$; the latter evaluates the Ex^2 in terms of the variance of x. The symbol x here is perfectly general: the above relations hold for any random variable x, and hence for any function of x. Thus, we may write, at any time,

$$\text{Var} f(x) = E[f(x)]^2 - [E f(x)]^2 \tag{4}$$

as another form of Eq. 2; and likewise the transposition,

$$E[f(x)]^2 = \text{Var} f(x) + [E f(x)]^2 \tag{5}$$

as another form of Eq. 3.

We have heretofore borrowed formulas from theory and used them in our examples, without much talk about the conditions under which the formulas would have the interpretation that we gave to them. We proceed in this chapter to derive some of the formulas, and to try to understand how far we may strain the ideal conditions without distorting too badly our interpretations. This chapter teaches only some rudiments of theory,

380

but the careful student will pursue also works in mathematical statistics, of which fortunately there is now a good supply in almost every language.

EXERCISES ON VARIANCE

1. *a.* Prove that

$$\sigma_{cx} = c\sigma_x \qquad (6)$$

SOLUTION

$$\sigma_{cx}^2 = \text{Var } cx = E[cx - Ecx]^2$$
$$= E[c(x - Ex)]^2 = c^2 E[x - Ex]^2 \qquad [\text{P. 371}]$$
$$= c^2 \sigma_x^2$$
$$\sigma_{cx} = c\sigma_x$$

Q.E.D.

In particular, if $\bar{x}$ denotes the mean of a sample of size n, and if

$$X = N\bar{x} \qquad (7)$$

then

$$\sigma_X = \sigma_{N\bar{x}} = N\sigma_{\bar{x}} \qquad (8)$$

That is, the standard error of $X = N\bar{x}$ is just N times the standard error of $\bar{x}$.

b. Prove that

$$\text{Var } (x + c) = \text{Var } x \qquad (9)$$

or

$$\sigma_{x+c} = \sigma_x \qquad (10)$$

SOLUTION

$$\text{Var } (x + c) = E[(x + c) - E(x + c)]^2$$
$$= E[x - Ex]^2$$
$$= \sigma^2 \qquad [\text{By definition}] \qquad (11)$$

For complicated functions it will be convenient to write Var for variance, instead of the symbol σ^2 with a subscript. We could write either σ_{x+c}^2 or Var $(x + c)$; the meaning is the same, and a writer may take his choice.

2. A sample of n sampling units is to be drawn at random from a frame of N sampling units, by reading out n random numbers between 1 and N. Let

$$a = \frac{1}{N} \sum_1^N a_i, \qquad \begin{array}{l} \text{the mean population} \\ \text{per sampling unit in} \\ \text{the frame} \end{array} \qquad (12)$$

$$\sigma^2 = \frac{1}{N} \sum_1^N (a_i - a)^2, \qquad \begin{array}{l} \text{the variance between} \\ \text{the sampling units in} \\ \text{the frame (definition)} \end{array} \qquad (13)$$

Let x_i be the population of the sampling unit drawn by the ith random number. Recall that $Ex_i = a$ (p. 362).

a. Show that

$$Ex_i^2 = \frac{1}{N}\sum_1^N a_i^2$$

$$= \sigma^2 + a^2 \qquad \text{[For any member of the sample,} \qquad (14)$$
$$\text{with or without replacement]}$$

b. Show that

$$Ex_i x_j = a^2 \qquad\qquad\qquad \text{with replacement}$$
$$\scriptstyle j \neq i$$

$$= a^2 - \frac{\sigma^2}{N-1} \qquad \text{without replacement} \qquad\qquad (15)$$

$j \neq i$ means here that x_i and x_j refer to 2 different members of the sample.

c. Let ρ be the correlation between the x-populations of 2 different members of the sample.   Show that

$$\rho = 0 \qquad\qquad\qquad \text{with replacement}$$

$$= -\frac{1}{N-1} \qquad \text{without replacement} \qquad\qquad (16)$$

The results hold regardless of the size of the sample (just so *n* is 2 or greater) and regardless of the shape of the distribution of the x-populations in the frame.

This negative correlation between the members of a sample is the reason why the Var $\bar{x}$ is diminished by nonreplacement: see the next exercise.

SOLUTION

$$\sigma^2 = \frac{1}{N}\sum_1^N (a_i - a)^2 \qquad \text{[Definition, Eq. 13]}$$

$$= \frac{1}{N}\sum_1^N a_i^2 - a^2 \qquad \text{[By algebra]}$$

Therefore

$$\frac{1}{N}\sum_1^N a_i^2 = a^2 + \sigma^2 \qquad\qquad (17)$$

or

$$Ex_i^2 = a^2 + \sigma^2 \qquad \text{[As in Eq. 14, and as seen already in Exercise 5, page 372]}$$

Note now that

$$A^2 = (a_1 + a_2 + \cdots + a_N)^2 \qquad \text{[\textit{A} is, as before, the total } x\text{-population in the frame]}$$

$$=
\begin{matrix}
a_1^2 & a_1 a_2 & a_1 a_3 & a_1 a_4 \\
a_2 a_1 & a_2^2 & a_2 a_3 & a_2 a_4 \\
a_3 a_1 & a_3 a_2 & a_3^2 & a_3 a_4 \\
a_4 a_1 & a_4 a_2 & a_4 a_3 & a_4^2
\end{matrix}$$

it being understood that all the terms in the array are to be connected with plus signs (+). This array is written for $N = 4$, but it can be extended easily to a frame of any size.

There are N^2 terms altogether in the array, N on the diagonal, and $N^2 - N$ off the diagonal. The sum of all the terms is A^2 or $N^2 a^2$. The sum of the N terms on the diagonal is by Eq. 17 equal to $N(a^2 + \sigma^2)$. Hence the sum of the terms off the diagonal is $N^2 a^2 - N(a^2 + \sigma^2)$.

Now if the sample is drawn with replacement, $x_i x_j$ ($j \neq i$) may be any of the N^2 terms in the array. If the sample is drawn without replacement, $x_i x_j$ ($j \neq i$) is a term off the diagonal, as terms on the diagonal are forbidden. So, it follows immediately,

$$\begin{aligned}
\text{with replacement} \quad & E x_i x_j \atop (j \neq i) = \frac{N^2 a^2}{N^2} = a^2 \\
\text{without replacement} \quad & E x_i x_j \atop (j \neq i) = \frac{N^2 a^2 - N(a^2 + \sigma^2)}{N^2 - N} \\
& = a^2 - \frac{\sigma^2}{N-1}
\end{aligned} \right\} \quad (18)$$

Then

$$\begin{aligned}
\rho &= \frac{E(x_i - a)(x_j - a)}{\sigma^2} \, (j \neq i) \quad \text{[Definition]} \\
&= \frac{E x_i x_j - a^2}{\sigma^2} \\
&= 0 \qquad \text{with replacement} \\
&= -\frac{1}{N-1} \quad \text{without replacement}
\end{aligned} \right\} \quad (19)$$

$$Q.E.D.$$

The fundamental forms. We are now ready to derive the variance of the mean $\bar{x}$ of a sample of predetermined size, drawn by reading out n random numbers between 1 and N. The fundamental forms which we shall derive are these:

$$\begin{aligned}
\sigma_{\bar{x}}^2 &= \frac{\sigma^2}{n} \qquad \text{with replacement} \\
&= \frac{N-n}{N-1} \frac{\sigma^2}{n} \quad \text{without replacement}
\end{aligned} \right\} \quad (20)$$

The factor $(N - n)/(N - 1)$ is the *finite multiplier* for nonreplacement.

$\sigma^2 =$ the variance between the populations of the N sampling units of the frame, whatever be their size or shape

$N =$ the number of sampling units in the frame

$n =$ the number of sampling units in the sample, drawn by reading out random numbers between 1 and N

$\bar{x} =$ the mean population per sampling unit in the sample

$\sigma_{\bar{x}}^2 = $ the variance of the sampling procedure

$\sigma_{\bar{x}} = $ the standard error of the sampling procedure

SOLUTION*

By definition,

$$\sigma_{\bar{x}}^2 = E(\bar{x} - E\bar{x})^2$$
$$= E\bar{x}^2 - (E\bar{x})^2$$
$$= E\bar{x}^2 - a^2 \qquad [\text{P. 365}]$$

Written out more fully,

$$\sigma_{\bar{x}}^2 = E\left[\frac{1}{n}(x_1 + x_2 + \cdots + x_n)\right]^2 - a^2$$

$$= E\left[\frac{1}{n^2}(x_1^2 + x_2^2 + \cdots + x_n^2)\right.$$

$$\left. + \frac{1}{n^2}(x_1 x_2 + x_2 x_1 + \text{all other cross-products of the sample})\right] - a^2$$

$$= \frac{1}{n} Ex_1^2 \qquad [\text{Because all the } n \text{ values of } Ex_i^2 \text{ are equal}]$$

$$+ \frac{n(n-1)}{n^2} E_{\substack{j \neq i}} x_i x_j - a^2 \qquad \begin{array}{l}[\text{As there are } n(n-1) \text{ cross-products, all} \\ \text{having the same expected value, which} \\ \text{appears in Exercise 2c, page 382}]\end{array}$$

$$= \frac{1}{n}(a^2 + \sigma^2) + \frac{n(n-1)}{n^2} a^2 - a^2 \qquad \text{with replacement}$$

$$= \frac{1}{n}(a^2 + \sigma^2) + \frac{n(n-1)}{n^2}\left[a^2 - \frac{\sigma^2}{N-1}\right] - a^2 \quad \text{without replacement}$$

The rest is simple algebraic reduction, not statistical theory.

Remark 1. To draw a sample without replacement, we reject any random number that appears a second time within the same zone. To draw with replacement, we take every random number within the zone, just as it comes, even though it appear twice or even thrice in the same zone.

Remark 2. It is very convenient at times to draw up a sampling table with replacement, to avoid the problem of eliminating duplicates. For example, if n/N be 5%, then about 5% of the random numbers, as one reads them out of the table, will be duplicates within the same zone. It is a convenience to leave the duplicates in the table, just as they appear, and to make no effort to eliminate them. One may use for the sample-size the number of distinct sampling units actually hit.† The unequal numbers of

---

* A much simpler derivation appears on page 392, but we must wait for theorems on the propagation of variance. See also Frank Yates, *Sampling Methods for Censuses and Surveys* (Griffin, 1949): p. 183.

† Des Raj and Salem B. Khamis, "Some remarks on sampling with replacement," *Ann. Math. Statist.*, vol. 29, 1958: pp. 550–557.

CH. 17. THEORY OF VARIANCES

sampling units in the subsamples create no bias and introduce very little additional variance in an estimate of a ratio, $f = x/y$.

Alternative forms for the variance of the mean. This section continues use of the same sampling plan, by which we read out n random numbers between 1 and N. We then form $\bar{x}$; also $X = N\bar{x}$. Eqs. 9 and 37 of Chapter 16 (pp. 365 and 372) then tell us that $E\bar{x} = a$ and that $EX = A = Na$, where a is the mean population per sampling unit in the frame, and A or Na is the total for the whole frame. Then

$$\sigma_X = N\sigma_{\bar{x}} \qquad \text{[Eq. 8, page 381]}$$

where $\sigma_{\bar{x}}$ has the value shown in Eq. 20. Now let

$$C = \frac{\sigma}{a},$$
the coefficient of variation of the x-populations of the N sampling units of the frame (21)

$$C_{\bar{x}} = \frac{\sigma_{\bar{x}}}{E\bar{x}} = \frac{\sigma_{\bar{x}}}{a},$$
the coefficient of variation of the distribution of the estimate $\bar{x}$ in repeated samples (22)

$$C_X = \frac{\sigma_X}{EX} = \frac{N\sigma_{\bar{x}}}{Na} = C_{\bar{x}},$$
the coefficient of variation of the distribution of the estimate X in repeated samples (23)

Then divide Eq. 20 through by a^2 to find that

$$C_X^2 = C_{\bar{x}}^2 = \frac{N-n}{N-1}\frac{C^2}{n} \qquad \text{without replacement} \qquad (24)$$

$$C_X = C_{\bar{x}} = \sqrt{\frac{N-n}{N-1}}\frac{C}{\sqrt{n}} \qquad \text{without replacement} \qquad (25)$$

Set the factor $(N-n)/(N-1)$ equal to 1 for sampling with replacement.

Remark 3. For simplicity we often replace

$$\frac{N-n}{N-1} \quad \text{by} \quad 1 - \frac{n}{N}$$

$$\sqrt{\frac{N-n}{N-1}} \quad \text{by} \quad 1 - \frac{n}{2N}$$

if N is 10 or 15 times as big as n, or still bigger

Remark 4. n/N is the fraction of sampling units removed by the sample. $1 - n/N$ is the fraction of sampling units still in the frame after the sample is drawn. Hence, we may say that nonreplacement (a) reduces the variance by about the fraction of sampling units removed by the sample; (b) reduces the standard error by about half the fraction of the sampling units removed. Thus, a sample that decimates the frame will produce a variance about 10% less (and a standard error about 5% less) than the same size sample would produce if drawn from a much bigger frame.

Exercise.   Show that under the same plan of sampling that we used in the derivation of Eq. 20,

$$E(\bar{x} - a)(\bar{y} - b) = \frac{N - n}{N - 1}\frac{1}{n} E(x_i - a)(y_i - b) \qquad (26)$$

Remember that $E\bar{x} = Ex_i = a$, $E\bar{y} = Ey_i = b$.

SOLUTION

$$E(\bar{x} - a)(\bar{y} - b) = \frac{1}{n^2} \sum_1^n (x_i - a) \sum_1^n (y_i - b)$$

$$= \frac{1}{n^2} E\{ \quad (x_1 - a)(y_1 - b) + (x_1 - a)(y_1 - b) + (x_1 - a)(y_1 - b)$$
$$+ (x_2 - a)(y_2 - b) + (x_2 - a)(y_2 - b) + (x_2 - a)(y_2 - b)$$
$$+ (x_3 - a)(y_3 - b) + (x_3 - a)(y_3 - b) + (x_3 - a)(y_3 - b)\}$$

I have here written out the terms for the sample-size $n = 3$. Extension to a bigger sample is obvious. Observe that the n terms on the diagonal all have the same expected value, namely, $E(x_i - a)(y_i - b)$ or $\frac{1}{N} \sum_1^N (a_i - a)$ $(b_i - b)$; next, that the $n(n - 1)$ terms off the diagonal all have the same expected value, namely, $\frac{1}{N(N - 1)} \sum_{i=1}^N \sum_{j \neq i}^N (a_i - a)(b_j - b)$. It thereupon follows that

$$E(\bar{x} - a)(\bar{y} - b) = \frac{1}{n^2} \left\{ \frac{n}{N} \sum_1^N (a_i - a)(b_i - b) \right. \quad \text{[For terms on the diagonal]}$$

$$+ \frac{n(n - 1)}{N(N - 1)} \sum_{i=1}^N \sum_{j \neq i}^N (a_i - a)(b_i - b) \right\} \quad \begin{array}{l}\text{[For terms off}\\\text{the diagonal]}\end{array}$$

$$= \frac{1}{n^2} \left\{ \frac{n}{N} \sum_1^N (a_i - a)(b_i - b) + \frac{n(n - 1)}{N(N - 1)} \left[\sum_1^N (a_i - a) \right.\right.$$

$$\left.\left. \times \sum_1^N (b_j - a) - \sum_1^N (a_i - a)(b_i - b) \right] \right\} \quad [j = i]$$

$$= \frac{1}{n^2} \left\{ \frac{n}{N} \sum_1^N (a_i - a)(b_i - b) + \frac{n(n - 1)}{N(N - 1)} \right.$$

$$\left. \times \left[0 - \sum_1^N (a_i - a)(b_i - b) \right] \right\}$$

$$= \frac{1}{n^2} \frac{n}{N} \left\{ 1 - \frac{n - 1}{N - 1} \right\} \sum_1^N (a_i - a)(b_i - b)$$

$$= \frac{N - n}{N - 1}\frac{1}{n} E(x_i - a)(y_i - b)$$

Q.E.D.

We shall need this result in the construction of an estimate of the variance of a ratio $f = x/y$ (p. 438).

Remark 5.   Eq. 26 may take the following alternate forms:

$$\text{Cov } \bar{x}, \bar{y} = \frac{N-n}{N-1} \frac{1}{n} \text{ Cov } x, y \qquad (27)$$

or

$$C_{\bar{x}\bar{y}} = \frac{N-n}{N-1} \frac{1}{n} C_{xy} \qquad (28)$$

where Cov $x, y = E(x - a)(y - b)$ is the covariance between the x- and y-populations of the sampling units in the frame, and $C_{xy} = (\text{Cov } x, y)/ExEy$ is the rel-covariance between the x- and y-populations of the sampling units in the frame.

Remark 6.   The students should note that Cov $\bar{x}, \bar{y}$ and $C_{\bar{x}\bar{y}}$ take the same multiplier on the right, as we saw in Eq. 25.

Remark 7.   We now have the important fact that

$$\rho_{\bar{x}, \bar{y}} = \rho_{x_i, y_i} \qquad (29)$$

where $\rho_{x_i, y_i} = E(x_i - Ex)(y_i - Ey)/\sigma_x \sigma_y$. That is, under the sampling plan specified, the correlation between the means $\bar{x}$ and $\bar{y}$ of the x-populations in successive samples of any size, with or without replacement, is the same as the correlation between the x- and y-populations of the sampling units in the original frame.

Chart to show the combined effect of the size of the sample and the size of the frame (Fig. 22).   Fig. 22 shows the standard error of the mean of a sample as a function of the size (n) of the sample, for frame-sizes $N = 500, 1000, 10,000$, and ∞. The horizontal scale is laid off so that distances toward the left from the point $n = \infty$ are proportional to $1/\sqrt{n}$. This scale produces a straight line for $N = \infty$, for which $\sigma_{\bar{x}}:\sigma = 1/\sqrt{n}$. The student should note that—

i.   The standard error $\sigma_{\bar{x}}$ is practically independent of N, the number of sampling units in the frame. This statement follows from the fact that the curves are all practically straight and almost coincident until the sample-size n reaches a value of about 20% of N. The size N of the frame plays little part in the precision of a sample unless the sample is 20% or more of the total.

ii.   The curves all bend downward to 0 at the right, which shows that the standard error $\sigma_{\bar{x}}$ diminishes rapidly to 0 as n approaches a complete census.

iii.   It follows that the size of sample required to deliver a specified precision for a study that is to cover the whole country is about the same as the sample required for a city of 100,000 inhabitants. Thus, the sample for the Monthly Report on the Labor Force for the United States is of the same size, design, and precision as the sample used in Canada, which has 1/10th as many inhabitants as the United States.

iv.   The reader will enjoy continuation of this point in Wallis and Roberts' book, *Statistics: A New Approach* (The Free Press, 1956); pp. 371 ff.

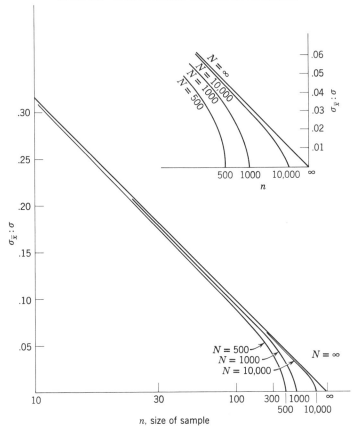

Fig. 22. Curves that show how $\sigma_{\bar{x}}$ varies with the size n of the sample, for selected sizes of frame.   There are n sampling units in the sample, N in the frame.   Note that $\sigma_{\bar{x}}$ is practically independent of N, except when n/N is .1 or bigger.

EXERCISES

1. Start with Eq. 20 for sampling without replacement and show by mere algebraic rearrangement that

$$\frac{N}{n} = (N - 1)\left(\frac{\sigma_{\bar{x}}}{\sigma}\right)^2 + 1 \doteq N\left(\frac{\sigma_{\bar{x}}}{\sigma}\right)^2 + 1$$

$$= (N - 1)\left(\frac{C_{\bar{x}}}{C}\right)^2 + 1 \doteq N\left(\frac{C_{\bar{x}}}{C}\right)^2 + 1 \qquad \text{[Without replacement]}\quad (30)$$

The numeral 1 on the right arises from the finite multiplier.   These forms are convenient because they give the zoning interval very conveniently, corrected for the finite multiplier. For 2 samples, $Z = 2\,N/n$; for 10 samples, $Z = 10\,N/n$.

2. A sample of 1000 turns out to be about right for a certain questionnaire in a city of 200,000 inhabitants. The same sample-design and the same questionnaire are to be used next month in a city of 1,000,000 inhabitants. Assume that the variances between segments are the same in the 2 cities. Show that the same size of sample in the 2d city would yield the same precision as it did in the 1st city. (Fig. 22.)

3. A sample of 1100 telephone poles and the aerial equipment carried thereon is sufficient to determine with a standard error of 1% the per cent condition of the aerial portion of a telephone company that owns 1,000,000 poles. Show that the same sample would be required for a smaller company that owns 50,000 poles, if the smaller company must have the same standard error. (Assume that the variances are equal in the 2 companies. Use Fig. 22 again.)

4. There are 2 areas, 1 and 2, containing N_1 and N_2 households respectively. At first, only Area 1 was to be sampled. The aim of the sampling plan that had been drawn up for Area 1 was to deliver a desired coefficient of variation $C = 2.5\%$ for a special characteristic of the population, viz., the proportion of men of age 20–29 engaged in nonagricultural employment. After the plans for the sample of Area 1 had been frozen (or, in the actual incident, after the sample had been taken), came a decision to cover Area 2 as well, and to aim at the same coefficient of variation C for the 2 areas combined. The problem arose to determine the required size of sample in Area 2.

SOLUTION

For Area 1,

$$C^2 = \frac{\sigma_1{}^2}{a_1{}^2 n_1} \tag{31}$$

Herein a_1 is the average number of men per sampling unit in Area 1, aged 20–29, and engaged in nonagricultural employment. $\sigma_1{}^2$ is the variance between the sampling units in Area 1 for the characteristic mentioned, and n_1 is the number of sampling units in the sample. The same symbols with the subscript 2 will refer to Area 2. Formally, the problem is to determine n_2 so that

$$\frac{N_1{}^2\,\sigma_1{}^2/n_1 + N_2{}^2\,\sigma_2{}^2/n_2}{(N_1 a_1 + N_2 a_2)^2} = C^2 \tag{32}$$

The numerator on the left is the variance of the sample in the 2 areas combined, and the denominator is the square of the number of men in the 2 areas that have the characteristic mentioned. By equating the 2 values of C in Eqs. 31 and 32, and by performing some algebraic reduction, the answer is found to be

$$\frac{N_2}{n_2} = \frac{N_1}{n_1}\left(\frac{\sigma_1\,a_2}{a_1\,\sigma_2}\right)^2\left(2\frac{a_1}{a_2} + \frac{N_2}{N_1}\right) \tag{33}$$

If there were to be 2 samples per zone, then the zoning interval Z in Area 2 would be $2\,N_2/n_2$. If there were to be 10 samples per zone, Z would be $10\,N_2/n_2$.

If we require the finite multipliers, Eq. 33 will change to

$$\frac{N_2}{n_2} - 1 .= \left(\frac{\sigma_1\,a_2}{a_1\,\sigma_2}\right)^2\left(2\frac{a_1}{a_2} + \frac{N_2}{N_1}\right)\left(\frac{N_1}{n_1} - 1\right) \tag{34}$$

5. Continue with the preceding exercise and show that if $a_1 = a_2$, and if $\sigma_1 = \sigma_2$, then Eq. 33 would give

$$\frac{N_2}{n_2} = \frac{N_1}{n_1}\left(2 + \frac{N_2}{N_1}\right) \tag{35}$$

or

$$\frac{n_1}{n_2} = 2\frac{N_1}{N_2} + 1 \tag{36}$$

Thus, suppose further that $N_2 = \frac{1}{4}N_1$; then would $n_2 = n_1/9$. Or, handier,

$$Z_2 = 2.25Z_1 \tag{37}$$

That is, an additional sample of only $\frac{1}{9}$th the size of n_1 would build up the precision for both areas combined equal to the precision intended for Area 1 alone, even though Area 2 be $\frac{1}{4}$th the size of Area 1.

Remark 8. It is important not to confuse the above problem with allocation in stratified sampling. The solution of the above problem tells us how to obtain a specified precision for Area 1, and the same precision for both areas combined. It pulls us out of the difficulty that we find ourselves in when the survey has gone so far in Area 1 that we can not stop it, and the administration or client suddenly decides to add Area 2 to the survey.

The solution to the above problem does not give us the optimum allocation that will deliver a specified precision for both areas combined. This would be a problem in stratified sampling, which we learned about in Chapter 15.

B. THE PROPAGATION OF VARIANCE

An example. As x varies from sample to sample, so will vary also the calculated value of a function $f(x)$. We saw in Chapter 16 that the average value of $f(x)$ may be above or below $f(Ex)$ by the amount $E\,\Delta f$

or $Bf(x)$, called the *bias* in $f(x)$. We now calculate the variance of $f(x)$ that arises from the variance in x.

$$\sigma_f^2 = E\{f - Ef\}^2 \quad \text{[By definition]}$$

$$\doteq E\{f(Ex) + f_x\,\Delta x + \cdots - E[f(Ex) + f_x\,\Delta x + \cdots]\}^2 \quad \begin{array}{l}\text{[By Taylor's}\\ \text{series as in}\\ \text{Chapter 16]}\end{array}$$

$$= E\{f_x\,\Delta x + \cdots\}^2$$

$$= f_x^2\,E(\Delta x)^2 = f_x^2\sigma_x^2 \tag{1}$$

Or,

$$\sigma_f = \left[\frac{df}{dx}\right]_{x=Ex}\sigma \tag{2}$$

For a function of 2 variables x and y, one may write for large samples,

$$\sigma_f^2 \doteq \frac{1}{n}[f_x^2\sigma_x^2 + f_y^2\sigma_y^2 + f_xf_y\rho\sigma_x\sigma_y]$$

$$= \frac{1}{n}a^2b^2\{f_x^2C_x^2 + f_y^2C_y^2 + 2f_xf_yC_{xy}\}^2 \tag{3}$$

where ρ, C_x^2, C_y^2, and C_{xy} have the definitions given on pages 385–387. f_x is the derivative of f with respect to x, evaluated at $x = Ex$. f_y is defined in a similar way. The finite multiplier $(N - n)/(N - 1)$ is easy to insert as a factor on the right when required.

Variances of some useful functions. The student may now use Eq. 3 in the following exercises, to derive the variances in Table 1 (p. 394), which contains a number of functions that one may encounter in practice.

EXERCISES

1. Variance of a sum.   Variance of a difference.   Prove:

$$\left.\begin{array}{ll}a. & \text{Var}\,(x + y) = \text{Var}\,x + \text{Var}\,y + 2\,\text{Cov}\,x, y\\ b. & \text{Var}\,(x - y) = \text{Var}\,x + \text{Var}\,y - 2\,\text{Cov}\,x, y\end{array}\right\} \tag{4}$$

Here x and y are random variables, the results (such as the sums or the means) of 2 samples. These results are exact; no approximation. $\text{Cov}\,x, y = E(x - Ex)(y - Ey) = \rho\sigma_x\sigma_y$.

$$\begin{aligned}\text{Var}\,(x + y) &= E[(x + y) - E(x - y)]^2 \quad \text{[By definition]}\\ &= E[(x - Ex) + (y - Ey)]^2 \quad \text{[By rearrangement]}\\ &= E(x - Ex)^2 + E(y - Ey)^2 + 2E[(x - Ex)(y - Ey)]\\ &= \quad,, \quad + \quad,, \quad + 2\,\text{Cov}\,x, y\end{aligned}$$

In other symbols,

$$\sigma^2_{x+y} = \sigma_x{}^2 + \sigma_y{}^2 + 2\rho\sigma_x\sigma_y$$

Drop the correlation term if Cov $x, y = 0$.

$$Q.E.D.$$

The solution to part b follows similar steps with a minus sign in its proper place.

Remark 1.   Corollaries.   If $\bar{x}_1$ and $\bar{x}_2$ are the means of 2 samples of sizes n_1 and n_2, drawn with random numbers with replacement from 2 frames with variances $\sigma_1{}^2$ and $\sigma_2{}^2$, then $\rho = 0$ and

$$\left.\begin{array}{ll} \text{Var}\ (\bar{x}_1 + \bar{x}_2) = \dfrac{\sigma_1{}^2}{n_1} + \dfrac{\sigma_2{}^2}{n_2} & \text{[No correlation]} \\[3mm] \text{Var}\ (\bar{x}_2 - \bar{x}_1) = \text{the same} & \text{[No correlation]} \end{array}\right\} \quad (5)$$

These theorems form the basis for much theory of sampling and statistical design of experiment.   Thus, if $\sigma_1 = \sigma_2 = \sigma$,

$$\text{Var}\ (\bar{x}_2 - \bar{x}_1) = \sigma^2\left(\frac{1}{n_1} + \frac{1}{n_2}\right) \quad (6)$$

If p_1 and p_2 are estimates of 2 proportions in binomial sampling from 2 frames that have the same proportions p and q of red and white respectively, then

$$\text{Var}\ (\hat{p}_2 - \hat{p}_1) = pq\left(\frac{1}{n_1} + \frac{1}{n_2}\right) \quad (7)$$

2. Variance of a linear function.   Show that

$$\text{Var}\ (k + ax + by + cz) = a^2\sigma_x{}^2 + b^2\sigma_y{}^2 + c^2\sigma_z{}^2 + 2ab\ \text{Cov}\ x, y$$
$$+ 2ac\ \text{Cov}\ x, z + 2bc\ \text{Cov}\ y, z \quad (8)$$

where x, y, z are random variables, and k, a, b, c are constants.   Also

$$\text{Var}\ (k + ax + by + cz) = a^2\sigma_x{}^2 + b^2\sigma_y{}^2 + c^2\sigma_z{}^2 \quad \begin{array}{l} \text{[If there is no}\quad(9) \\ \text{correlation} \\ \text{between } x, y, z] \end{array}$$

3. Classical derivation of the variance of the mean.*   Draw n sampling units from the frame by reading out n random numbers between 1 and N. Compute $\bar{x}$.   Derive Eq. 20 for the variance of $\bar{x}$ by use of Eq. 4.

SOLUTION (GAUSS)

Let

$$x = n\bar{x} = x_1 + x_2 + \cdots + x_n$$

* Gauss, *Theoria Motus Corporum Coelestium* (Goettingen, 1809): Sec. 181.

All the x_i have the same variance σ^2.   Then

$$\sigma_x^2 = n\sigma^2 + n(n-1)\rho\sigma^2 \qquad \text{[By Eq. 8 above, as there are}$$
$$n(n-1) \text{ cross-products]}$$

where ρ is the correlation between any 2 members of the sample.
With replacement. $\rho = 0$; hence this equation gives $\sigma_x^2 = n\sigma^2$, wherefore $\sigma_{\bar{x}}^2 = \sigma_x^2/n^2 = \sigma^2/n$.   *Q.E.D.*

Without replacement. $\rho \neq 0$ here.   Let $n = N$; then $x = A$, the total
population in the frame, and

$$\sigma_x^2 = 0 = N\sigma^2 + N(N-1)\rho\sigma^2$$

whence

$$\rho = -\frac{1}{N-1} \qquad \text{[In agreement with Eq. 16, p. 382]}$$

and for any n,

$$\sigma_x^2 = n\sigma^2 - \frac{n(n-1)}{N-1}\sigma^2$$

which leads at once to

$$\sigma_{\bar{x}}^2 = \frac{\sigma_x^2}{n^2} = \frac{N-n}{N-1}\frac{\sigma^2}{n}$$

                                                                          Q.E.D.

4. Variance of a product.   Variance of a quotient.   Let

$$v = \frac{axy}{u}$$

Then

$$C_v^2 = C_x^2 + C_y^2 + C_u^2 + 2C_{xy} - 2C_{xu} - 2C_{yu} \qquad (10)$$

where

$$C_{xy} = \frac{E(x-Ex)(y-Ey)}{ExEy} = \frac{\text{Cov } x, y}{ExEy} \qquad (11)$$

with similar definitions for the other covariances.   Omit the covariance
for any 2 variables that are not correlated.
As a special important case, let

$$f = \frac{x}{y} = \frac{\bar{x}}{\bar{y}} \qquad (12)$$

as we have used it in Chapter 6 and elsewhere.   Then

$$C_f^2 \doteq C_{\bar{x}}^2 + C_{\bar{y}}^2 - 2C_{\bar{x}\bar{y}} \qquad (13)$$

For samples of size n, drawn by reading out n unduplicated random
numbers between 1 and N, we turn back to Eqs. 24 and 26 on pages
385 and 386 to see that

$$C_f^2 = \frac{N-n}{N-1}\frac{1}{n}(C_x^2 + C_y^2 - 2C_{xy}) \qquad \text{[Without replacement]} \qquad (14)$$

where C_x^2, C_y^2, and C_{xy} refer to the x- and y-populations of the N sampling
units of the frame.   This result is now ready to go into the table that the

TABLE 1

BIASES AND VARIANCE OF SOME USEFUL FUNCTIONS

Function	Bias, $Ef(\bar{x}) - f(E\bar{x})$ (Eq. 32 in Chapter 16)	Variance (Eq. 1 or 2, this chapter)
$f = h + k\bar{x}$	none	$\sigma_f^2 = \dfrac{1}{n} k^2 \sigma_x^2$
$f = h + k\bar{x} + g\bar{y}$	none	$\sigma_f^2 = \dfrac{1}{n}(k^2\sigma_x^2 + g^2\sigma_y^2 + 2gk\rho\sigma_x\sigma_y)$
		$= \dfrac{1}{n} a^2 b^2 \{k^2 C_x^2 + g^2 C_y^2 + 2gk C_{xy}\}$
$f = k\bar{x}^2$	$\dfrac{1}{n} k\sigma_x^2 = \dfrac{1}{n} f_0 C_x^2$	$C_f^2 = \dfrac{4}{n} C_x^2$
$f = k\bar{x}^a$	$\dfrac{1}{2!} a(a-1) f_0 C_x^2$	$C_f^2 = a^2 C_x^2$
$f = \dfrac{k}{\bar{x}}$	$\dfrac{1}{n} f_0 C_x^2$	$C_f^2 = \dfrac{1}{n} C_x^2$
$f = k\bar{x}\bar{y}$	$\dfrac{1}{n} f_0 \rho C_x C_y$ ($=0$ if x and y are uncorrelated)	$C_f^2 = \dfrac{1}{n}(C_x^2 + C_y^2 + 2C_{xy})$
$f = \dfrac{k\bar{x}\bar{y}}{\bar{z}}$	$\dfrac{1}{n} f_0 \{C_{xy} - C_{xz} - C_{yz} + C_z^2\}$	$C_f^2 = \dfrac{1}{n}(C_x^2 + C_y^2 + C_z^2 + 2C_{xy}$ $- 2C_{xz} - 2C_{yz})$
$f = \ln \bar{x}$	$-\dfrac{1}{2n} C_x^2$	$\sigma_f^2 = \dfrac{1}{n} C_x^2$
$f = \dfrac{\bar{x}}{\bar{y}} = \dfrac{\Sigma x_i}{\Sigma y_i}$	$\dfrac{f_0}{n}(-C_{xy} + C_y^2)$ (See Section D of this chapter for more details on the bias of the ratio $\bar{x}/\bar{y}$.)*	$C_f^2 = \dfrac{1}{n}(C_x^2 + C_y^2 - 2C_{xy})$ (See page 439 for another form of this equation, and for an estimate of this variance.)
$f = \sqrt{\bar{x}}$	$-\dfrac{1}{8n} f_0 C_x^2$	$C_{\sqrt{\bar{x}}}^2 = \dfrac{1}{4} C_{\bar{x}}^2 = \dfrac{1}{4n} C_x^2$

* Sukhatme gives the 2d approximation to the bias as $B_1(1 + 3C_y^2/n)$, where B_1 is the 1st approximation $(-C_{xy} + C_y^2)/n$. This is on page 146 in his book *Sampling Theory of Surveys with Applications* (Iowa State College Press, Ames; and the Indian Society of Agricultural Statistics, New Delhi, 1953). He also gives on page 153 the 2d approximation to the variance.

next exercise calls for. We shall see in Eq. 16 of Section F how to estimate this variance from the sample itself (p. 439).

5. Derive the bias and the variance of each function in Table 1. The sampling plan is to draw n random numbers between 1 and N; form $\bar{x}$ (and $\bar{y}$ if required). σ_x and σ_y are the standard deviations between the x- and y-populations of the sampling units in the frame; C_x, C_y, and C_{xy} refer likewise to the sampling units in the frame. $f_0 = f(Ex, Ey)$. The symbol ρ stands for the correlation between x and y unless the subscripts indicate otherwise. $E\bar{x} = a$, $E\bar{y} = b$.

Note that the biases and the variances all contain $1/n$ as a factor; hence they decrease with the size of the sample. Most of the biases are negligible for the sizes of sample that are used in practice.   Divide each bias by f_0 to get the relative bias.   Insert the finite multiplier $1 - n/N$ for non-replacement as a factor for both bias and variance, when required.

Remark 2.   The student may easily teach himself the meaning of the formulas for the bias and the variance.   For example, (a) let $f = 1/\bar{x} = n/\Sigma x_i$.   (b) Take a frame of 3 numbers, say $a_1 = 1$, $a_2 = 2$, $a_3 = 3$. $E\bar{x} = 2$ and $f(E\bar{x}) = 1/E\bar{x} = \frac{1}{2}$.   (c) Write down all the possible samples of 2, drawn with equal probabilities, and without replacement.   (There are only 3 possible samples.)   (d) Find the average value of $f = 1/\bar{x}$ over all possible samples; this is Ef.   (e) Find the variance of f.   (f) Compare the actual bias and the actual variance with the bias and the variance calculated by the formulas in the table.

Table 2 shows the results of the samples.   The bias in f is $Ef - f(E\bar{x}) = 47/90 - \frac{1}{2} = 1/45$, and the variance of f is $98/90^2 = .0121$.   Table 3 shows the results of applying the formulas in Table 1.   The formula for the bias gives $1/48$, and the formula for the variance gives $1/96 = .0104$.   The small discrepancies between the experimental results and the formulas arise from terms in $1/n^2$, which the formulas neglect.

TABLE 2

THE RESULTS OF SAMPLES OF 2 DRAWN FROM THE
FRAME $a_1 = 1$, $a_2 = 2$, $a_3 = 3$

Sample	Σx_i	$\bar{x}$	$f = 1/\bar{x}$	$f - Ef$	$(f - Ef)^2$
1 and 2	3	$\frac{3}{2}$	$\frac{2}{3}$	$\frac{13}{90}$	$\frac{169}{90^2}$
1 and 3	4	$\frac{4}{2}$	$\frac{2}{4}$	$-\frac{2}{90}$	$\frac{4}{90^2}$
2 and 3	5	$\frac{5}{2}$	$\frac{2}{5}$	$-\frac{11}{90}$	$\frac{121}{90^2}$
Sum	xxx	$\frac{12}{2}$	$\frac{47}{30}$	0	$\frac{294}{90^2}$
Average		$E\bar{x} = 2$ $Ef = \frac{47}{90}$		0	$\sigma_f^2 = \frac{98}{90^2}$
		$Ef - 1/E\bar{x} = \frac{1}{45}$			$= .0121$

TABLE 3

THE BIAS AND THE VARIANCE OF $1/\bar{x}$ CALCULATED
FROM THE FORMULAS IN TABLE 1

Note: The formulas require the finite multiplier in this example

Function	Bias	Variance
$f = \dfrac{1}{\bar{x}} = \dfrac{n}{\Sigma x_i}$	$\dfrac{N-n}{N-1}\dfrac{1}{n}f_0 C_x^2$	$\dfrac{N-n}{n-1}\dfrac{1}{n}f_0^4 \sigma_x^2$
	$= \dfrac{1}{2}\dfrac{1}{2}\dfrac{1}{2}\dfrac{1}{6} = \dfrac{1}{48}$	$= \dfrac{1}{2}\dfrac{1}{2}\dfrac{1}{16}\dfrac{2}{3} = \dfrac{1}{96}$
		$= .0104$

6. In manufacturing, where n piece-parts go in an assembly side by side, the allowable tolerance for the group is $\sqrt{t_1{}^2 + t_2{}^2 + \cdots + t_n{}^2}$, where t_1, t_2, etc., are the tolerances for the individual parts. Justify this rule by the theory of this chapter.

7. A rule in surveying is that the allowable error in measuring any distance of n miles is proportional to $\sqrt{n}$. (This has in fact been a rule for 150 years.) Justify it on the basis of the theory of this chapter.

Theory for the choice between the weighted and unweighted estimates. We encountered in Chapter 16 an important type of problem in which the sampling units came in with unequal probabilities. One estimate that we proposed was subject to possible inconsistent bias (Plan 2, page 376); another was unbiased (Plan 3, page 377). We dare not assume without investigation that the unbiased estimate is always preferable: the biased estimate sometimes wins when we look at the total error, bias plus sampling error combined. Some theory that we now derive will tell us under what conditions one estimate is better than the other. The *mean square error* of a sampling procedure is the square of the bias plus the variance. In symbols,

$$(MX)^2 = (BX)^2 + \sigma_X{}^2 \tag{15}$$

MX stands for the root-mean-square error (or the total error) in X; BX stands for the bias in X. One may depict the relationship so written by a right triangle in which the root-mean-square error (or the total error) is the hypotenuse, the legs being the bias and the standard error.

The notation will be the same as it was on page 375. We may form by Plans 2 and 3 the following 2 estimates of a, the average amount per name:

Plan 2 $\quad \bar{x}_{(2)} = \dfrac{1}{m} \sum_1^m x_i \qquad$ [The unweighted average amount per name, page 376] $\tag{16}$

Plan 3 $\quad \bar{x}_{(3)} = \dfrac{1}{m\bar{N}} \Sigma N_i x_i \qquad$ [The weighted average amount per name, page 377] $\tag{17}$

$\bar{N}$ is the average of N_i in the entire frame of M primary units (cards). $M\bar{N} = N$ is the total number of secondary units (names) in the frame. Then the bias in the estimate that comes from Plan 2 takes the form

$$B\bar{x}_{(2)} = -\sigma_b C_N \rho \tag{18}$$

This equation comes from page 379, and the symbols are the same as they were there.

We shall for simplicity draw the sample of cards with replacement, wherefore there will be no finite multiplier.   Then

$$\text{Var } \bar{x}_{(2)} = \frac{1}{m} E(x_i - E\bar{x}_{(2)})^2 \qquad \begin{array}{l}\text{[By the classical derivation}\\ \text{of variance, page 392]}\end{array}$$

$$= \frac{\sigma'^2}{m} \qquad (19)$$

where

$$\sigma'^2 = \frac{1}{M} \sum_{i=1}^{M} \frac{1}{N_i} \sum_{j=1}^{N_i} \{a_{ij} - E\bar{x}_{(2)}\}^2 \qquad (20)$$

is the total unweighted variance between all N names on all M cards in the frame.

To find Var $\bar{x}_{(3)}$, we recall from page 393 the variance of a product:

$$\sigma_{xy}^2 = (Ey)^2\sigma_x^2 + (Ex)^2\sigma_y^2 + 2ExEy\rho_{xy}\sigma_x\sigma_y \qquad (21)$$

Now N_i and x_i in the estimate $\bar{x}_{(3)}$ are random variables; hence

$$\text{Var } N_i x_i = [E\bar{x}_{(2)}]^2\sigma_N^2 + (\bar{N})^2\sigma'^2 + 2E\bar{x}_{(2)}\bar{N}\rho\sigma_N\sigma_\mu \qquad (22)$$

Then

$$\text{Var } \bar{x}_{(3)} = \frac{1}{(m\bar{N})^2} \text{Var } \sum_1^m N_i x_i$$

$$= \frac{1}{(m\bar{N})^2} m\{[E\bar{x}_{(2)}]^2\sigma_N^2 + (\bar{N})^2\sigma'^2 + 2E\bar{x}_{(2)}\bar{N}\rho\sigma_N\sigma'\}$$

$$= \frac{\sigma'^2}{m} + \frac{[E\bar{x}_{(2)}]^2}{m}\{C_N^2 + 2\rho C'C_N\}$$

$$= \text{Var } \bar{x}_{(2)} + \frac{[Ex_{(2)}]^2}{m}\{C_N^2 + 2\rho C'C_N\} \qquad (23)$$

wherein the notation again follows page 379 in Chapter 16, and wherein $C' = \sigma'/E\bar{x}_{(2)}$.

We are now ready to compare the mean square errors of the 2 estimates. We need only write

$$\{M\bar{x}_{(2)}\}^2 = \text{Var } \bar{x}_{(2)} + \{Bx_{(2)}\}^2$$

$$= \text{Var } \bar{x}_{(2)} + \{\sigma_b C_N\rho\}^2 \qquad \text{[P. 379]} \qquad (24)$$

$$\{M\bar{x}_{(3)}\}^2 = \text{Var } \bar{x}_{(3)} + 0$$

$$= \text{Var } \bar{x}_{(2)} + \frac{\{Ex_{(2)}\}^2}{m}\{C_N^2 + 2C'C_N\rho\} \qquad (25)$$

and then take the difference

$$\{M\bar{x}_{(3)}\}^2 - \{M\bar{x}_{(2)}\}^2 = \frac{\{Ex_{(2)}\}^2}{m}\{C_N{}^2 + 2C'C_N\rho\} - (\sigma_b C_N\rho)^2 \quad (26)$$

We see now what happens for various sizes of sample. For m small, the positive term on the right may be bigger than the negative term (the bias in $\bar{x}_{(2)}$), depending on a complex set of relations between m, a, C_N, C', σ_b, and ρ. Thus, for small samples, the unweighted average of Plan 2 will usually be preferable. But the positive term will for any frame diminish as m increases, and there will be some sample-size m at which the 2 terms become equal, at which point the 2 estimates have equal mean errors. For still bigger samples, the unbiased estimate $\bar{x}_{(3)}$ will have the smaller mean error and will be preferable.

Exercise. Show that if $\rho = 0$ (no correlation between N_i and $\bar{a}_i$), then

$$B\bar{x}_{(2)} = 0$$

and

$$\frac{\{M\bar{x}_{(3)}\}^2}{\{M\bar{x}_{(2)}\}^2} = \frac{\text{Var }\bar{x}_{(3)}}{\text{Var }\bar{x}_{(2)}} = 1 + \frac{C_N{}^2}{C'^2} \quad \begin{array}{l} [\text{Only if } \rho = 0, \ C_N = \sigma_N/\bar{N}, \\ \quad C' = \sigma'/E\bar{x}_{(2)}] \end{array} \quad (27)$$

Thus, if there is no correlation between N_i and $\bar{a}_i$, we should most certainly use the unweighted result $\bar{x}_{(2)}$, because $\bar{x}_{(2)}$ is then not only unbiased, but has the smaller variance, and hence (under this condition) the smaller mean error as well. The estimate $\bar{x}_{(3)}$ loses because of the term in $C_N{}^2$, which arises from the variable weighting factors N_i.

The 2 estimates are of course identical if $\rho = 0$ and if in addition the weighting factors N_i are all equal.

Comparison of actual results, weighted and unweighted. This is the example that Chapter 16 referred to on page 376. Tables 4 and 5 present 2 sets of results for comparison; one unweighted and subject to possible bias, the other weighted and unbiased.* The survey covered 2 counties. The questions elicited certain information about the families in which the head of a family or his wife read a certain journal. (Other readers in the family and families with no readers were not part of the universe.) The procedure of interviewing was first to enquire of the man and of his wife which one of them read the journal. If the male head was the only reader, or if there was no wife, he was automatically in the sample. If he did not read the journal but his wife did, the wife was automatically in the sample. If both the man and his wife read the journal, random

---

* I am indebted to the firm O'Brien-Sherwood of New York for Tables 4 and 5 and for the privilege of working with them on the survey.

numbers drew one of them for interview. Obviously, the man had only half a chance to answer if both he and his wife read the journal.
There are 2 groups of male readers: m_1 of them are in households where the only reader was male; $2m_2$ of them are in households where the man and his wife were both readers. Let

$x_i = 1$ if the male reader owns stocks and bonds

$x_i = 0$ if he does not

We now compare the following 2 estimates of the proportion of male readers in the 2 counties that own stocks and bonds:

$$\text{Unweighted, Plan 2 (p. 376)} \quad \bar{x}_{(2)} = \frac{\sum_1 x_i + \sum_2 x_i}{m_1 + m_2} \tag{28}$$

$$\text{Weighted, Plan 3 (p. 377)} \quad \bar{x}_{(3)} = \frac{\sum_1 x_i + \sum_2 2x_2}{m_1 + 2m_2} \tag{29}$$

Both estimates come from the same interviews. $\sum_1 x_i$ is the number of males that own stocks and bonds in the m_1 households where the only reader was male. $\sum_2 x_i$ is the number of males that own stocks and bonds in the m_2 households where there were 2 readers, and from which the random numbers selected the male for interview.
The sample was replicated in 10 subsamples. Tables 4 and 5 show the 2 estimates for ownership of stocks and bonds; also for 3 other characteristics. The estimates under the heading "All 10 subsamples" are the final results of the survey. The purpose of the 10 subsamples was, as usual, to furnish easy estimates of the standard errors. Table 6 compares the 2 estimates for the 4 characteristics. I have estimated the bias of Plan 2 as the difference $\bar{x}_{(2)} - \bar{x}_{(3)}$. The 10 separate differences from the 10 subsamples show that the estimates of the biases are not statistically significant: we shall use them nevertheless to get estimates of the mean square errors of Plan 2.
The variance of f for any characteristic in Table 6 comes from the formula

$$\hat{\sigma}_f^2 = \frac{1}{90\bar{y}^2} \sum_1^{10} (x_i - fy_i)^2 \quad \text{[P. 439]} \tag{30}$$

The 10 values of x_i for any specific characteristic are shown under the 10 subsamples in Tables 5 and 6. The 10 values of y_i are common to all 4

TABLE 4

PLAN 2, EQ. 28

Unweighted results of the answers that come from male readers. Some of the male readers had only half a chance to come into the sample, but there is no adjustment in this table

Characteristic	All 10 sub-samples	Subsample									
		1	2	3	4	5	6	7	8	9	10
1. Number of men interviewed	518	55	53	36	54	46	69	56	62	42	45
Per cent	100.0	100.0	100.0	100.0	100.0	100.0	100.0	100.0	100.0	100.0	100.0
2. Number that own stocks and bonds	215	20	24	17	20	19	27	26	23	21	18
Per cent	41.5	36.4	45.3	47.2	37.0	41.3	39.1	46.4	37.1	50.0	40.0
3. Number that have a safe deposit box	178	23	18	12	13	12	28	20	15	20	17
Per cent	34.4	41.8	34.0	33.3	24.0	26.1	40.6	35.7	24.2	47.6	37.8
4. Number that have one or more checking accounts	474	48	47	30	51	39	63	54	61	39	42
Per cent	91.5	87.3	88.7	83.3	94.4	84.8	91.3	96.4	98.4	92.9	93.3
5. Number that have one or more insurance policies	496	54	52	33	50	42	67	54	59	41	44
Per cent	95.8	98.2	98.1	91.7	92.6	91.3	97.1	96.4	95.2	97.6	97.8

TABLE 5

PLAN 3, EQ. 29

Here, the answers from the male head were multiplied by 2 (weighted) whenever the man and his wife both read the journal and the random numbers selected the man

Characteristic	All 10 sub-samples	Subsample									
		1	2	3	4	5	6	7	8	9	10
1. Number of households in which there was a male reader	839	83	92	59	85	78	116	92	97	71	66
Per cent	100.0	100.0	100.0	100.0	100.0	100.0	100.0	100.0	100.0	100.0	100.0
2. Weighted number of male readers that own stocks and bonds	356	34	44	26	30	34	45	44	35	38	26
Per cent	42.4	40.1	47.8	44.1	35.3	43.6	38.8	47.8	36.1	53.5	39.4
3. Weighted number that have a safe-deposit box	295	37	34	22	23	20	47	34	22	34	22
Per cent	35.2	44.6	37.0	37.3	27.1	25.6	40.5	37.0	22.7	47.9	33.3
4. Weighted number that have one or more checking accounts	764	73	81	48	81	66	105	89	95	66	60
Per cent	91.1	88.0	88.0	81.4	95.3	84.6	90.5	96.7	97.9	93.0	90.9
5. Weighted number that have one or more insurance policies	801	82	90	53	80	70	113	88	92	69	64
Per cent	95.5	98.8	97.8	89.8	94.1	89.7	97.4	95.7	94.8	97.2	97.0

TABLE 6

COMPARISON OF THE VARIANCES OF THE RESULTS IN THE
2 TABLES PRECEDING

B denotes the bias of the unweighted result $\bar{x}_{(2)}$, estimated here as $\bar{x}_{(2)} - \bar{x}_{(3)}$.
The weighted result $\bar{x}_{(3)}$ has, by definition, no bias

Characteristic and method of estimate	All 10 subsamples Per cent	Bias B Per cent	Variance $\hat{\sigma}_f{}^2$	Mean square error $\hat{\sigma}_f{}^2 + B^2$ (Eq. 15)
2. Unweighted	42.4	−.9	2.20	3.01
Weighted	41.5	0	3.33	3.33
3. Unweighted	35.2	−.8	6.39	7.03
Weighted	34.4	0	6.93	6.93
4. Unweighted	91.1	.4	2.29	2.45
Weighted	91.5	0	2.52	2.52
5. Unweighted	95.5	.3	.63	.72
Weighted	95.8	0	.83	.83
2–5. Unweighted	xxx	xxx	11.50	13.21
Weighted	xxx	xxx	13.61	13.61

characteristics: they appear as characteristic No. 1. $f = \Sigma x_i / \Sigma y_i$ for any characteristic.

The variance of Plan 2 (unweighted) is, for every characteristic, the lower of the 2 variances—so much lower than even when we handicap it by adding the square of the bias to get the mean square error, it still wins 3 out of 4.

My decision was to use the unweighted result. The theory in the preceding section leads us to expect that had the sample been smaller, there might have been a more decided difference in favor of the unweighted (biased) procedure. Had the sample been bigger, there might have been a difference definitely in favor of the weighted (unbiased) procedure.

C. THEORY FOR 2 CLASSES: DENOMINATOR FIXED

Population of 2 classes. There is a special class of problem of frequent occurrence and great importance in which the aim is to determine the proportion of sampling units that belong in one or the other of 2 classes.

The 2 classes are very often attributes, the presence or absence of some characteristic. Table 7 contains some examples.

In the counting of attributes, a sampling unit takes the value 0 or 1. Thus, if one of us were to count the number of females in a room or in any area, he would assign 1 to a female and 0 to a male, and add up the

TABLE 7

EXAMPLES OF QUESTIONS AND ANSWERS THAT WILL PLACE
A SAMPLING UNIT IN CLASS, WHITE OR RED, 0 OR 1

Kind of survey or test	Sampling unit	Possible answers (only 1 of 2 permitted)	
		(White beads, 0)	(Red beads, 1)
1. Test of manufactured article	Article	Passed	Failed
2. Sex	Person	Male	Female
3. Did you catch the train?	Person	Yes	No
4. Are you 21 or over?	Person	Yes	No
5. Is your income $3000 or more?	Person (or household)	$3000 or more	Below $3000
6. Did you vote in the election last November?	Person	Yes	No
7. Are you seeking work?	Person	Yes	No
8. Grade	Student	Passed	Failed
9. Are you satisfied with the vacuum cleaner that you now own?	Housewife	Yes	No
10. Did you buy any bread yesterday?	Housewife	Yes	No
11. Do you own or rent this dwelling unit?	Dwelling unit	Own it	Rent it
12. If you own it, is there a mortgage on it?	Owned dwelling unit	Yes	No
13. Does this farm contain 10 acres or more?	Farm	Yes	No
14. Did you sell any of your cows last year?	Farmer	Yes	No
15. Is this area urban or rural?	Area	Urban	Rural

result of his count, and call it N_{fem} for the total female population. If three were N people of both sexes in the room, the proportion female would be N_{fem}/N. For the male population he would merely take $N_{male} = N - N_{fem}$. Or, he could recount, assigning 0 to every female and 1 to every male.

Sometimes the 2 classes have numerical values. An example would be an apartment house where the flats all rent for either $40 or $60. If a sampling unit is a flat, then a sampling unit must have one or the other of these 2 possible numerical values.

We shall use the letter p to denote the proportion of sampling units in the frame that possess a specified characteristic, like one of those in the list above. Thus, if in a room there are 50 people, 20 male and 30 female, the proportion female is $p = 30/50 = 3/5$. On the average, 3 out of 5 people in the room are females. In another example, 1 out of 10 pieces of a particular manufacturer's product is on the average defective; 1 out of 50 people in the labor force is on the average unemployed. These are common ways of expressing proportions.

Results proper to 2 classes. The theory that we need for sampling a frame of 2 classes comes immediately from Section B. Let each sampling unit be a poker chip. Paint a chip red and give it the numerical value 1 if the sampling unit possesses a certain characteristic; let it be white and 0 otherwise. Sort the N sampling units of the frame into a distribution whose classes are 0 and 1. Let p be the proportion of chips that are red. Then

The mean of this distribution is at p.
Its variance is pq.

PROOF

The general formula for the mean a of any discrete distribution is

$$a = P_1 a_1 + P_2 a_2 + P_3 a_3 + \text{etc.} \tag{1}$$

where a_1, a_2, a_3, etc., are the means of the classes, and P_1, P_2, P_3, etc., are the proportions of the sampling units in the classes. Now here we have only 2 classes: $a_1 = 0$ and $a_2 = 1$; $P_1 = q = 1 - p$ and $P_2 = p$. The general formula reduces instantly here to

$$a = p \tag{2}$$

The general formula for the variance of any discrete distribution is

$$\sigma^2 = P_1(a_1 - a)^2 + P_2(a_2 - a)^2 + P_3(a_3 - a)^2 + \text{etc.}$$

In our problem here, the mean $a = p$, and there are only 2 cells. Hence

$$\sigma^2 = q(0 - p)^2 + p(1 - p)^2$$
$$= qp^2 + pq^2 = pq(q + p)$$
$$= pq \tag{3}$$

The coefficient of variation of this frame will be

$$C = \sqrt{q/p} \tag{4}$$

so we may use any equation in Section A by replacing—

The mean of the frame	a	by p, the proportion red chips
The variance of the frame	σ^2	by pq
The mean of the sample	$\bar{x}$	by $\hat{p} = r/n$, the proportion red in the sample; i.e., the number of red chips in the sample divided by the total red and white in the sample. $\hat{p}$ is an estimate of p.
An estimate of a total	$X = N\bar{x}$	by $X = N\hat{p}$, an estimate of the total number of red chips in the frame

Now number the N chips serially from 1 to N, and read out n random numbers to draw a sample. Denote by r the number of red chips in the sample; by $n - r$ the number white. Let $\hat{p}$ be the proportion red in the sample; that is, write

$$\hat{p} = \frac{r}{n} \qquad \text{[A random variable]} \qquad (5)$$

It is important to note that p is a constant, a property of the frame, whereas $\hat{p}$ is a random variable. The sampling procedure, repeated over and over, generates a random sequence of values of $\hat{p}$. As we know already that $E\bar{x} = a$ (p. 365), we know now that

$$E\hat{p} = p \qquad \text{[With or without replacement]} \qquad (6)$$

That is, $\hat{p}$ or r/n, the proportion red in the sample, is an unbiased estimate of p, the proportion red in the frame. The estimate of the total number of red chips in the frame will be

$$X = N\hat{p} = N\frac{r}{n} \qquad \text{[A random variable]} \qquad (7)$$

The estimate of the total number of white chips in the frame will be

$$Y = N - X \qquad \text{[A random variable]} \qquad (8)$$

Both X and Y are unbiased estimates, of the total red and of the total white chips in the bowl.

The variance of $\hat{p}$ follows readily from Eq. 20 (p. 383): replace σ^2 by pq and $\bar{x}$ by $\hat{p}$, whereupon

$$\left. \begin{array}{ll} \sigma_{\hat{p}}^2 = & \dfrac{pq}{n} \qquad \text{with} \quad \text{replacement} \\[3mm] = & \dfrac{N - n}{N - 1}\dfrac{pq}{n} \qquad \text{without replacement} \end{array} \right\} \qquad (9)$$

Also, if $X = N\hat{p}$, as before, for an estimate of the number of red chips in the frame, it is convenient to remember that,

$$C_{\hat{p}} = C_X = \left.\begin{array}{ll} \dfrac{C}{\sqrt{n}} & \text{with} \quad \text{replacement} \\[18pt] = \sqrt{\dfrac{N-n}{N-1}} \cdot \dfrac{C}{\sqrt{n}} & \text{without replacement} \end{array}\right\} \tag{10}$$

where $C = \sqrt{q/p}$, the coefficient of variation of the frame.

The binomial and the hypergeometric series of probabilities. We continue with the same sampling plan; viz., read out n random numbers between 1 and N to draw a sample of n chips from a list of N chips, Np being red, Nq white. A chip counts 1 if red, 0 if white. The probability that the sample will contain r red chips and $n - r$ white is

$$P(r) = \binom{n}{r} q^{n-r} p^r \qquad \begin{array}{l}\text{[With replacement;}\\ \text{binomial probability]}\end{array} \tag{11}$$

$$P(r) = \frac{\binom{Nq}{n-r}\binom{Np}{r}}{\binom{N}{r}} \qquad \begin{array}{l}\text{[Without replacement;}\\ \text{hypergeometric probability]}\end{array} \tag{12}$$

for $r = 0, 1, 2, \ldots, n$. The time-honored symbol

$$\binom{n}{r} = \frac{n!}{(n-r)!r!} = \binom{n}{n-r} \tag{13}$$

in which $\binom{n}{0} = \binom{n}{n} = 1$. Thus,

$$(a + b)^3 = a^3 + 3a^2b + 3ab^2 + b^3$$
$$= \binom{3}{0}a^3b^0 + \binom{3}{1}a^2b + \binom{3}{2}ab^2 + \binom{3}{3}b^3 \tag{14}$$

or a simple illustration, suppose that we have a list of 12 farms total: 7 with electricity; 5 without What is the probability that if we draw 3 farms by reading out 3 random numbers between 1 and 12, 2 of the 3 farms will have electricity? The answer is

$$P(2) = \binom{3}{2}\left(\frac{5}{12}\right)\left(\frac{7}{12}\right)^2 = .426 \qquad \text{[with \quad replacement]}$$

$$P(2) = \frac{\binom{7}{2}\binom{5}{1}}{\binom{12}{3}} = \frac{\left(\frac{7!}{5!2!}\right)\left(\frac{5!}{4!1!}\right)}{\left(\frac{12!}{9!3!}\right)} = \frac{21}{44} = 477 \qquad \text{[without replacement]}$$

The student may wish to finish Table 8.

TABLE 8

Constitution of the sample	Probability	
	With replacement	Without replacement
0 with electricity, 3 without	$\dfrac{2}{44} = .0454$	$\dfrac{125}{1728} = .0723$
1 with electricity, 2 without	$\dfrac{14}{44} = .3182$	$\dfrac{175}{576} = .3038$
2 with electricity, 1 without	$\dfrac{21}{44} = .4773$	$\dfrac{245}{576} = .4254$
3 with electricity, 0 without	$\dfrac{7}{44} = .1591$	$\dfrac{343}{1728} = .1985$
Any combination	$\dfrac{44}{44} = 1$	$\dfrac{1728}{1728} = 1$

Both series of probabilities are skewed if p is small and n not large. The skewness disappears rapidly, however, as the size n of the sample increases,* even for small values of p. The Poisson series of probabilities will under certain conditions give excellent approximation to binomial probabilities (Ch. 18, p. 457). The square-root-transformation for sums of Poisson terms, and hence also for sums of binomial terms under the proper conditions, is easy and rapid (same chapter). A still simpler approximation to the 2-sigma or 3-sigma limits, or to other multiples of $\sigma_{\hat{p}}$ for Poisson and binomial probabilities is provided by the Mosteller-Tukey paper,† which we saw in Chapter 13.

The hypergeometric series is sometimes approximated well enough by the binomial series, or by the normal integral with mean at p and with

* See Exercise 4 on page 443, which shows that $E(\Delta\bar{x})^3 = \mu_3/n^2$, which means that the skewness in $\hat{p}$ disappears rapidly with increasing n. Some notes and history on approximations of the binomial series appear in the author's book *Some Theory of Sampling* (Wiley, 1950), p. 407. The incomplete Beta function for the exact evaluation of binomial probabilities is on page 481.

A splendid summary of various approximations to the binomial probabilities, with theory, tests, and bibliography, appeared in a paper by Morton S. Raff, "On approximating the point binomial," *J. Amer. Statist. Ass.*, vol. 51, 1956: pp. 293–303.

† For a description of the Mosteller-Tukey paper and how to use it, see Wallis and Roberts's *Statistics: A New Approach* (The Free Press, 1956): pp. 604 ff. Also, Mosteller and Tukey's original article, *J. Amer. Statist. Ass.*, vol. 44, 1949: pp. 174–212.

standard deviation $\sqrt{\{(N - n)/(N - 1)\}}\sqrt{pq}/n$. There are conditions under which the corresponding Poisson series, or the square-root-transformation, or the Mosteller-Tukey paper gives good results (Chapter 18).

Size of sample required to estimate a proportion with a desired degree of precision. The chief of the credit department of a large department store wishes to know what proportion of his customers return 20% of their purchases for credit over the course of a year. He suspects that the proportion is low, perhaps 5%, and he would be satisfied to know that it is anywhere below $7\frac{1}{2}$%. If it is above $7\frac{1}{2}$%, he would like to know about it. Must he be absolutely sure of the proportion? No. He may at first say yes, but when confronted with the cost of a big sample or of a complete coverage of all the accounts, he backs down, especially when he sees the scientific nature of a sample enquiry, which gears the risk of an error to the permissible cost, and even permits a sliding approach, by which the cost is kept very low unless the first sample indicates an alarming possibility of a high proportion.

If the actual but unknown proportion is around 5%, then the examination of a sample of $n = 300$ accounts will give 2-sigma limits of $2\frac{1}{2}$% and this should provide him with the required information. This number is arrived at by solution of the equation

$$\sigma_{\hat{p}} = \sqrt{\frac{pq}{n}} \qquad [\text{P. 383}]$$

in which we put $2\sigma_{\hat{p}} = .025, p = .05, q = .95$. n turns out to be 304, but I should prefer to take a round number like 300, especially as inclusion of the finite multiplier would reduce the computed size of n (see Remark 1).

If the actual proportion p is less than 5%, this sample will give a smaller standard error; $2\sigma_{\hat{p}}$ will then be less than $2\frac{1}{2}$%. If this sample shows $p = 6$%, it may be advisable to go to a bigger sample.

Remark 1.   An alternative computation is to use the equation

$$\frac{N}{n} = N\left(\frac{\sigma_{\hat{p}}}{\sigma}\right)^2 + 1 \qquad [\text{Page 388.   The 1 arises from the finite multiplier}]$$

Suppose that the total number of accounts is $N = 30{,}000$. Then if $p = .05$, σ^2 will be $.05 \times .95$; the calculated zoning interval will then be

$$\frac{N}{n} = 30{,}000\,\frac{.0125^2}{.05 \times .95} + 1 = 99 + 1 = 100$$

To lay out the sample in 10 subsamples, one would draw 10 accounts at random from every consecutive 1000 accounts.

Chart showing the standard deviation and the coefficient of variation of a binomial frame.   We have learned that if 0 and 1 are the only possible values that a sampling unit can take, the frame has the following properties:

$$\text{Mean} = p \tag{15}$$

$$\sigma = \sqrt{pq} \tag{16}$$

$$C = \sqrt{\frac{q}{p}} \tag{17}$$

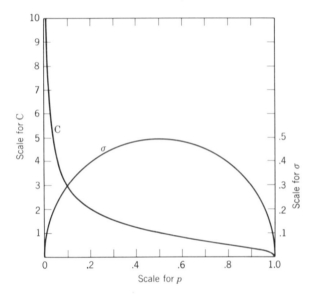

Fig. 23. σ and C for the binomial, plotted against p. $\sigma = \sqrt{pq}$; $C = \sqrt{q/p}$.

It is helpful to look at the chart in Fig. 23 to see how σ and C vary with p and q, and to make the following observations:

1. The standard error $\sigma = \sqrt{pq}$ is 0 at each end and high in the middle.

2. The coefficient of variation $C = \sqrt{q/p}$ is lowest at the right, where $p = 1$; and it increases as p decreases.   It finally rises precipitously at the lowest values of p.   This precipitous rise displays the difficulty of sampling to estimate a small proportion.

Special difficulties in sampling for a small proportion.   Fig. 23 shows that C increases as p decreases toward 0.   Hence, a sample is under a handicap if its aim is to estimate some rare proportion.   For example, we might wish to estimate the proportion of parts defective in a shipment, in which the proportion, though not known exactly, lies somewhere around 1% or 2%.   In a study of consumers, the aim might be to talk to people that own a certain type of vacuum cleaner *and* a certain type of rug.   The handicap comes from the fact that we must carry out 99 tests to find 1 defective, or 99 interviews to discover the 1 home that qualifies as a member of the universe, and in which the rest of the questionnaire applies.   An operational difficulty can easily make matters even worse; the monotony of finding no defect deadens the inspector's chances of finding even the few that do exist, and there is a similar monotony in the consumer study.

Such samples are uneconomical, and require special administrative art to avoid error from monotony.   There is sometimes no escape from the expense.   In some problems, however, it is possible (a) to isolate certain lots or areas most of which contain a relatively high proportion of the rare units, and (b) to allocate most of the tests or interviews to these lots or areas, with a small sample from the remaining lots or areas.   The most economical allocation of the sample was the subject of Chapter 15.

Fortunately, one seldom needs high precision in the estimate of a rare proportion.   Thus, if the proportion of dwelling units vacant in a city is around 2%, or below, it will suffice to estimate this proportion with a coefficient of variation of 30% or even of 50%, as there is no need for greater precision; and moreover, the difficulties of defining a vacancy, or of defining a habitable dwelling, cause even bigger uncertainties, and too much refinement in the sampling error only creates an illusion of accuracy that can not exist at any price.

EXERCISES

1. Assume the sampling plan described on page 405.   Define $\hat{p}$ as r/n. Show by direct calculation (i.e., not by use of the general equations derived in Section A of this chapter) that:

a. $\qquad E\hat{p} = p \qquad$ for both series

b. $\qquad \sigma_{\hat{p}}^2 = \dfrac{pq}{n} \qquad$ if we draw with replacement (binomial series)

c. $\qquad \sigma_{\hat{p}}^2 = \dfrac{N-n}{N-1}\dfrac{pq}{n} \qquad$ if we draw without replacement (hypergeometric series)

2. State the conditions under which the formula

$$C_{\hat{p}} = \sqrt{\frac{q}{np}} \qquad \text{[From Eq. 10]}$$

is strictly valid.

Answer: There are 2 conditions: 1.   Any sampling unit may have but 1 and only 1 of 2 values, such as 0 and 1, or a_1 and a_2; nothing else.   2.   The sample must be drawn by reading out *n* random numbers between 1 and *N*, just as they come, duplicates accepted. (If we reject the duplicates, the only change is to require the finite multiplier.)

Remark 2. We could only disregard the 2d stipulation if the material were thoroughly mixed, in which case we could draw our sample off the top, or from any convenient layer. We have already learned, however, to be skeptical about thorough mixing (Ch. 2).

3. Criticize the following remark, which I copied from a book on marketing research:

For reasons that can be found in any text on the mathematical aspects of statistics, the standard error of a frequency is $(pq/n)^{\frac{1}{2}}$ where p is the number of successes, q the number of failures, and n the sum of the two, or the number of answers.

Note: the passage contained no description of the sampling plan, and no warning about the conditions under which the formula applies or does not apply. Such passages occur commonly in books that deal with subject-matter, and which mention statistical techniques as a side line.

4. A sample of 100 areas was drawn from a frame of 10,000 areas, and the housewives in the sample of areas were interviewed. A proportion $\hat{p}$ of the housewives in the sample have a specified characteristic. The number of housewives is 753. Is $\hat{p}\hat{q}/753$ an estimate of the variance of $\hat{p}$? (The answer is No.) State your reasons.

5. Suppose that a frame contains a list of dwelling units, and that the rent of a dwelling unit is either $50 per month or $75 per month; nothing else. Their proportions are q and p respectively. Derive these statistical properties of the frame:

a. The mean rent $= a = 50q + 75p$

$$= 50 + 25p$$

b. The variance of the rent between dwelling units $= \sigma^2 = 25^2 pq$

6. The cash receipts taken in by tellers, banks, salesgirls, and other sources, are counted and compared with the record at a central clearing office, day after day. Disagreements are to be expected as normal once in a while: sometimes the cash will be long, sometimes short. Assume that positive and negative errors have equal expected frequencies ($p = q = \frac{1}{2}$) in what follows.

a. Show that if over an interval of time some salesgirl is short 6 times and long not at all, the probability is only $(\frac{1}{2})^5$ or 1 in 32 that the 6 discrepancies have arisen by chance.

b. If an account were short 10 times and long not at all, the odds are overwhelming that some mischief is taking place, as the probability is then only $(\frac{1}{2})^9$ or 1 in 512.

c. Show that these conclusions hold even if the shortage is very small in any or all instances.   They hold whether the shortages were \$1000 or 10¢, or a mixture of various amounts.

d. If occasional mistakes are normal, then if the cash turned in by some salesgirl never shows disagreement with the total shown by the cash register, one may conclude that she must be replenishing an occasional shortage.

7. A file contains the test-scores of several hundred students.   The scores on one particular test may run from perhaps 30 to 95 or 100.   It is necessary to gain some idea quick about:

i. The 25-percentile.
ii. The 50-percentile (median).
iii. The 75-percentile.

A sample of 10 cards is drawn with random numbers: they show 70, 65, 90, 70, 80, 75, 58, 82, 62, 92.   Show that

a. There is about 1 chance in 18 that these 10 scores are all below the 25-percentile.   In other words, if we adopt the rule that the lowest of 10 scores sets an upper limit to the 25-percentile, we shall be wrong in only 1 statement in 18.   (*Hint:* $(\frac{3}{4})^{10}$ = about $1/18$.)

b. There is less than 1 chance in 1000 that these 10 scores are all above the median.   (*Hint:* $(\frac{1}{2})^{10} = 1/1024$.)

8. A man leaves his home every morning between 8 and 8:10 to take the tramway to his work.   He walks to the tramway in exactly 2 minutes.

a. Suppose that he leaves at a random second between 8 and 8:10, all seconds having equal probability.   The tram passes regularly at 8:07 and at 8:17.   Show that the distribution of the time when he arrives at the tramway is rectangular like Panel B on page 260, lying between 8:02 and 8:12.

b. Show that his expected use of the tramway is half the time at 8:07, and half at 8:17.

c. Show that if *p* represents the proportion of days on which he takes the tram at 8:07, and if *q* represents the proportion of days on which he takes it at 8:17, the expected frequencies follow the binomial probabilities with $p = q = \frac{1}{2}$.

d. The probability is only $(\frac{1}{2})^{10}$ or $1/1024$ that he will take the tram at 8:07 10 times running: the same for the tram at 8:17.

e. Suppose that he leaves home always at exactly 8:05 but that the tram that he wishes to take is randomized between 8 and 8:10.   Show that his expectation is to miss it 70% of the time.

f. Continue on the supposition that the tram is randomized.   Let the man sprint 1 morning in 4, on the average, depending on the random toss of 2 coins; and suppose that he sprints the distance to the tram in 1 minute instead of the 2 minutes that he requires when he walks. Show that his expected arrival at his office is 15 seconds earlier than if he always walked, never sprinted; and that the distribution of his time of arrival for the mornings that he sprints is exactly the same but displaced 1 minute earlier than the distribution of arrival for the mornings when he walks.

9. (On feeding a number of people.) Why is it that when a cook prepares food for a number of people, he can judge more accurately the relative amount required as the number of people increases? That is, the relative amount left over, or the relative amount short, decreases with the number of people to be fed.*

10. Matching 2 lists.† The publisher of a magazine wishes to discover how many of his subscribers are contained in a list of executives. Comparison of names in a sample from each list will provide the information. The accompanying table shows the scheme of notation. $a_1, a_2, \ldots, a_M$, $b_1, b_2, \ldots, b_N$ are names on the lists. D names are common to both lists. This number D is important: it is the number that the publisher wishes to know. Let

$$p = \frac{D}{M} \tag{18}$$

$$P = \frac{D}{N} \tag{19}$$

It will suffice to estimate either p or P.

	List 1	List 2
	a_1	b_1
	a_2	b_2
	a_3	b_3
	.	.
	.	.
	.	.
	a_M	b_N
Number on the list	M	N
Number common to both lists	D	D
Proportion common to both lists	p	P

* Edward U. Condon, "Food and the theory of probability," *Proc. U.S. Nav. Inst.*, vol. 60, 1934: pp. 74–78.

† W. Edwards Deming and Gerald J. Glasser, *J. Amer. Stat. Assoc.*, vol. 54, 1959: pp. 403–415.

SAMPLING PROCEDURE

1. Draw by random numbers and without replacement m names from List 1.

2. Draw by random numbers and without replacement n names from List 2.

3. Compare every name in the sample from List 1 with every name in the sample from List 2 to discover how many names are common to both samples. Let d be this number. Symbolically, in the notation that we shall adopt later, $d = \Sigma x_i y_j$. The summation runs over all combinations of i and j.

4. Let

$$\hat{p} = \frac{N}{n}\frac{d}{m} = \frac{N}{mn}\Sigma x_i y_j \tag{20}$$

$$\hat{P} = \frac{M}{m}\frac{d}{n} = \frac{M}{mn}\Sigma x_i y_j \tag{21}$$

Then $\hat{p}$ is an unbiased estimate of p, and $\hat{P}$ is an unbiased estimate of P. The student may prove that these estimates are unbiased and that

$$\text{Var } \hat{p} = \frac{Np}{mn}\left\{1 + \frac{m-1}{M-1}\frac{n-1}{N-1}(D-1)\right\} - p^2 \tag{22}$$

$$\doteq \frac{Np}{mn} \qquad \text{[If } m \text{ and } n \text{ are small compared with } M \text{ and } N] \tag{23}$$

$$\text{Var } \hat{P} = \frac{MP}{mn}\left\{1 + \frac{m-1}{M-1}\frac{n-1}{N-1}(D-1)\right\} - P^2 \tag{24}$$

$$\doteq \frac{MP}{mn} \qquad \text{[If } m \text{ and } n \text{ are small compared with } M \text{ and } N] \tag{25}$$

If $C_{\hat{p}}$ denotes the coefficient of variation of $\hat{p}$, the approximate forms may be written

$$C_{\hat{p}}^2 \doteq N/mnp \tag{26}$$

$$C_{\hat{P}}^2 \doteq M/mnP \tag{27}$$

The optimum sizes of the samples occur in the next exercise.

SOLUTION

Compare the 2 names a_i and b_j. Then we shall have a convenient notation if we let

$$a_i b_j = 1 \text{ if the 2 names are identical} \atop = 0 \text{ otherwise} \Bigg\} \tag{28}$$

Then the number D of names common to both lists is

$$\Sigma a_i b_j = D \tag{29}$$

where i in the summation runs through List 1, and j runs through List 2. Let the samples be

$$x_1, x_2, \ldots, x_m \qquad \text{from List 1}$$
$$y_1, y_2, \ldots, y_n \qquad \text{from List 2}$$

$$\left.\begin{aligned} x_i y_j &= 1 \text{ if the 2 names } x_i \text{ and } y_j \text{ are identical} \\ &= 0 \text{ otherwise} \end{aligned}\right\} \tag{30}$$

$$\hat{p} = \frac{N}{mn} \Sigma x_i y_j \tag{31}$$

where i and j in the summation run over both samples.   We need first these expected values:

$$E\Sigma x_i y_j = \frac{mn}{MN} \Sigma a_i b_j = \frac{mn}{MN} D$$

$$= \frac{mn}{N} p = \frac{mn}{M} P \tag{32}$$

$$E[\Sigma x_i y_j]^2 = E[\Sigma x_i y_j x_i y_j + \Sigma x_i y_j x_{i'} y_{j'}] \qquad i' \neq i \quad j' \neq j$$

$$= E\Sigma x_i y_j + E\Sigma x_i y_j x_{i'} y_{j'} \qquad i' \neq i \quad j' \neq j$$

$$= \frac{mn}{MN} D + \frac{m}{M} \frac{m-1}{M-1} \frac{n}{N} \frac{n-1}{N-1} (\Sigma a_i a_{i'} b_j b_j - \Sigma a_i b_j a_{i'} b_{j'})$$

$$= \frac{mn}{MN} D + \frac{m}{M} \frac{m-1}{M-1} \frac{n}{N} \frac{n-1}{N-1} (D^2 - D)$$

$$= \frac{mn}{N} p \left[1 + \frac{m-1}{M-1} \frac{n-1}{N-1} (D-1) \right] \tag{33}$$

It follows that

$$E\hat{p} = \frac{N}{mn} E\Sigma x_i y_j$$

$$= \frac{N}{mn} \frac{mn}{N} p = p \tag{34}$$

Likewise or by symmetry

$$E\hat{P} = P \tag{35}$$

$$\operatorname{Var} \hat{p} = E(\hat{p} - p)^2 = E\hat{p}^2 - p^2$$

$$= \left(\frac{N}{mn}\right)^2 E[\Sigma x_i y_j]^2 - p^2$$

$$= \frac{Np}{mn} \left[1 + \frac{m-1}{M-1} \frac{n-1}{N-1} (D-1) \right] - p^2 \tag{36}$$

$\operatorname{Var} \hat{P}$ follows by symmetry.

Q. E. D.

Remark 3.   If the sample from List 2 is complete, then $n = N$, and the above formula for Var $\hat{p}$ reduces to

$$\text{Var } \hat{p} = \frac{M - m}{M - 1} \frac{pq}{m} \tag{37}$$

If the sample from List 1 is complete, then $m = M$, and the above formula for Var $\hat{P}$ reduces to

$$\text{Var } \hat{P} = \frac{N - n}{N - 1} \frac{PQ}{n} \tag{38}$$

11. Derive the following approximations for the variances in the last exercise.   (*Hint:* $1/(1 - x) = 1 + x + x^2 + \cdots$ if x is small.)

$$\text{Var } \hat{p} = \frac{Np}{mn} - \frac{p}{M} + \frac{p^2}{N} \tag{39}$$

$$\text{Var } \hat{P} = \frac{MP}{mn} - \frac{P}{N} + \frac{P^2}{M} \tag{40}$$

Remark 4.   Only the first term on the right is important.   The other terms are finite corrections for nonreplacement, to ignore in most practice.

12. Optimum sample-sizes for matching 2 lists.   Suppose that these are the costs:

> c_1 to draw a name from List 1, and to write it down or to prepare a card therefor, in preparation to compare it with the sample from List 2. c_1 includes also a proper share of the cost of sorting the cards of the sample to put them in alphabetic order.
> c_2 the same for List 2.
> c_3 to compare a name in one sample with a name in the other sample, and to record the comparison as 1 or 0.

Then the total cost of the job will be

$$K = mc_1 + nc_2 + mnc_3 \tag{41}$$

Show that the optimum sizes are such that

$$mc_1 = nc_2 \tag{42}$$

which means that to get the most precision for our money, we equate the costs of drawing the 2 samples, and choose m and n big enough to yield the required precision in $\hat{p}$ or in $\hat{P}$.

SOLUTION

Let us first express the precision of $\hat{p}$ as its coefficient of variation, $C_{\hat{p}}$. Then the main term is

$$C_{\hat{p}}^2 = N/mnp \quad \text{[Ignore the finite corrections; see Eq. 26]} \quad (43)$$

This equation determines mn.   Suppose that we write it in the form

$$mc_1 \cdot nc_2 = Nc_1c_2/pC_{\hat{p}}^2 \quad (44)$$

The right-hand side is a number, once we fix N, c_1, c_2, $C_{\hat{p}}^2$ and insert some plausible value of p.

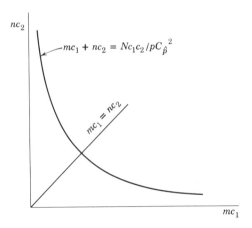

Fig. 24. The value of mn, and hence the cost mnc_3 of matching the 2 samples, is the same at all points of the hyperbola, which is fixed to deliver the precision required.   The sum of the coordinates of any point on the hyperbola is the cost of drawing the 2 samples. The costs of drawing the 2 samples, and hence also the total cost K, are at their minima where the hyperbola meets the 45°-line $mc_1 = nc_2$.

Suppose that we plot Eq. 44 with mc_1 on one axis and nc_2 on the other. The graph is the hyperbola in Fig. 24.   The value of mn, and hence also the cost mnc_3 of matching the 2 samples, is the same for all points of the hyperbola.   Now we observe that the coordinates mc_1 and nc_2 of any point are the costs of drawing the 2 samples.   The sum of these 2 costs, and hence also the total cost $K = mc_1 + nc_2 + mnc_3$, have their minima where the hyperbola is closest to the origin, which is at the point where $mc_1 = nc_2$, as recorded in Eq. 42.

Remark 5.   The procedure to find the optimum sizes of the samples could then be this.

1.   Choose a plausible value of p.

2.   Choose the desired coefficient of variation, $C_{\hat{p}}$ (45)

3.   Find $mn = N/pC_{\hat{p}}^2$.   (46)

4.   Find $m = \sqrt{mnc_2/c_1}$ (47)

$$n = \quad mc_1/c_2 \quad (48)$$

Thus, suppose that N is 20,000, and that the publisher thinks that p is about 5%. He says that $C_{\hat{p}} = 50\%$ will be sufficient for his purpose. The costs are about like this:

$$c_1 = 50¢, \quad c_2 = 25¢, \quad c_3 = .001 \tag{49}$$

Then

$$mn = N/pC_{\hat{p}}^2 = 20,000/.05 \times .25 = 1,600,000$$

$$m = \sqrt{mnc_2/c_1} = \sqrt{1,600,000 \times \tfrac{1}{2}} = \sqrt{800,000} \doteq 900 \tag{50}$$

$$n = mc_1/c_2 = 1800 \tag{51}$$

The total cost of the job would be

$$K = mc_1 + nc_2 + mnc_3$$

$$= .50 \times 900 + .25 \times 1800 + 1,600,000 \times .001 = \$2500 \tag{52}$$

To compare this cost with proportionate allocation, we keep $mn = 1,600,000$, but we must assume some value for M: let $M = 2N$, and $m = 2n$, $mn = 2n^2$. Then

$$C_{\hat{p}}^2 = \frac{N}{mnp} = \frac{N}{2n^2 p} \tag{53}$$

$$n^2 = \tfrac{1}{2}N/pC_{\hat{p}}^2$$

$$= 10,000/.05 \times .5^2 = 800,000$$

$$n = 895$$

$$m = 1790$$

$$mn = 1,600,000 \qquad \text{as before}$$

The total cost would be

$$K = mc_1 + nc_2 + mnc_3 = \$895 + \$447.50 + \$1600$$

$$= \$2942.50 \tag{54}$$

to compare with \$2500 by the optimum allocation.

The multinomial. Suppose that the frame contains Np_1 chips colored red, Np_1 colored yellow, Np_3 green. Draw a sample of n chips by reading out n random numbers between 1 and N. Count the number of red, yellow, and green. The probability that the counts of red, yellow, and green will be n_1, n_2, n_3 is

$$
\left.
\begin{aligned}
P(n_1, n_2, n_3) &= \frac{n!}{n_1!\,n_2!\,n_3!}\, p_1{}^{n_1} p_2{}^{n_2} p_3{}^{n_3} \quad \text{with} \quad \text{replacement} \\[2mm]
&= \frac{\dbinom{Np_1}{n_1}\dbinom{Np_2}{n_2}\dbinom{Np_3}{n_3}}{\dbinom{N}{n}} \quad \text{without replacement}
\end{aligned}
\right\} \tag{55}
$$

These terms are known as multinomial probabilities. n_1, n_2, and n_3 are random variables subject to the condition

$$n_1 + n_2 + n_3 = n \tag{56}$$

p_1, p_2, and p_3 are constants satisfying the condition

$$p_1 + p_2 + p_3 = 1 \tag{57}$$

EXERCISES ON THE MULTINOMIAL

The sampling plan for these exercises is the one described in the text.

1. *a.* Show that

$$En_1 = np_1$$

whether the drawing is made with or without replacement.

b. Define $\hat{p}_1 = n_1/n$. Show that

$$E\hat{p}_1 = p_1$$

with or without replacement.

2. Show that

$$\frac{n}{n-1} E\hat{p}_1\hat{p}_2 = \frac{N}{n-1} p_1 p_2$$

3. Let each red chip in the frame count a_1, each yellow chip count a_2, each green chip count a_3. Draw without replacement a sample of size by the sampling plan described in the text. Define

$$\bar{x} = \hat{p}_1 a_1 + \hat{p}_2 a_2 + \hat{p}_3 a_3$$

as the mean of the sample, and

$$a = p_1 a_1 + p_2 a_2 + p_3 a_3$$

as the mean of the frame. Prove that

$$E\bar{x} = a$$

$$\mathrm{Var}\, \bar{x} = \frac{N-1}{N-1} \frac{\sigma^2}{n}$$

where

$$\sigma^2 = p_1(a_1 - a)^2 + p_2(a_2 - a)^2 + p_3(a_3 - a)^2$$

D. SAMPLING VARIANCES FOR ESTIMATES FORMED FROM A RATIO WITH DENOMINATOR RANDOM (RATIO ESTIMATE)

A sample estimates a ratio. Suppose that we read out n random numbers between 1 and N to draw n sampling units from a frame of N units.

Compute

$$\bar{x} = \frac{\sum\limits_1^n x_i}{n} = \frac{x_1 + x_2 + x_3 + \cdots + x_n}{1 + 1 + 1 + \cdots + 1} \tag{1}$$

as usual.   Clearly, $\bar{x}$ is the ratio of the x-population in the sample to the number of sampling units in the sample.   Or, $\bar{x}$ is the average x-population per sampling unit in the sample.   The ith random number adds x_i to the numerator of $\bar{x}$, and 1 to the denominator, and we may fix the denominator n in advance.

The sampling units in the frame are very often small pieces of area (segments or a scattered group of segments).   We may govern n, the number of sampling units in the sample, but we can not govern the total dwelling units that will come into the sample, nor the number of people, nor the number of people that have any various characteristics.   Hence a ratio like the proportion of people unemployed, or the proportion that prefer a certain type of music with their meals, or the proportion of homes owned, will have not only a variable numerator but a variable denominator as well.   In fact, this is true of most of the ratios that we deal with in practice.

Fortunately, the required theory for the variance of such a ratio is not difficult to remember if we lay out the formulas alongside those for $\bar{x}$ and X.   As in previous chapters, we let

$$f = \frac{x\text{-population in the sample}}{y\text{-population in the sample}} = \frac{x_1 + x_2 + x_3 + \cdots + x_n}{y_1 + y_2 + y_3 + \cdots + y_n} \tag{2}$$

We may say that f is the ratio of the x-population in the sample to the y-population in the sample; or, that f is the average x-population per unit of y-population in the sample.   The ith random number adds x_i to the numerator, and y_i to the denominator.   Both x_i and y_i are random variables, even though we fix n in advance.

The symbol a has heretofore denoted the average x-population per sampling unit in the frame.   Likewise, we may (as in Chapter 11) denote by $\phi = A/B$ the average x-population per unit of y-population in the frame. A is the total x-population in the frame, and B is the total y-population in the frame. $\bar{x}$ is an estimate of a; f is an estimate of ϕ.

The ratio f as an estimate of ϕ is itself very often the aim of the survey. There are times, however, when there is need of an estimate X of a total population A.   We may estimate A in 2 ways:

$$X = N\bar{x} \qquad \text{[Formula X, unbiased]} \tag{3}$$

$$X' = Bf \qquad \text{[Formula X', a ratio-estimate]} \tag{4}$$

wherein we recall that N is the number of sampling units in the frame, and that B is the total y-population in the frame.   The constant B comes from outside sources (as a previous census, or a count of cards).   It is not one of the random variables of this survey; hence

$$C_{X'} = C_f \tag{5}$$

Formula X' will under certain conditions show a gain in precision over Formula X, as we shall learn soon (p. 424).

Remark 1.   The effect of possible error in B.   Any error or obsolescence in B carries over into the estimate X'.   The standard errors in f and X' make no allowance for error in B.

Comparison of formulas for the variances of $\bar{x}$, X, f, and X'.   We saw in Table 1 (p. 394) a formula for C_f^2, which is a good approximation for the sampling plan specified there, provided n is big enough.   The formula involves $C_x^2 - 2C_{xy} + C_y^2$, which is the algebraic equivalent of $\frac{1}{N}\sum_1^N \left(\frac{a_i - \phi b_i}{a}\right)^2$, as the student may wish to prove as an exercise (p. 423). ϕ stands for a/b, as in Chapter 11.

The purpose of the present section is to observe that the rel-variances of $\bar{x}$, X, f, and X' all take the same form, as we see them here:

$$C_{\bar{x}}^2 = C_X^2 = \frac{N-n}{N-1}\frac{C^2}{n} \tag{6}$$

$$C_f^2 = C_{X'}^2 = \frac{N-n}{N-1}\frac{C^2}{n} \tag{7}$$

provided we define C^2 properly for each estimate, as follows:

$$C^2 = \frac{\sigma^2}{a^2} = \frac{1}{N}\sum_1^N \left(\frac{a_i - a}{a}\right)^2 \qquad \text{in Eq. 6}$$

$$C^2 = \frac{1}{N}\sum_1^N \left(\frac{a_i - \phi b_i}{a}\right)^2 \qquad \text{in Eq. 7}$$

The similarity between the 2 sets of rel-variance is also striking and easy to remember if one makes geometrical interpretations of the 2 definitions of C^2.   For $\bar{x}$ and X, C^2 is simply the average square of the N line-segments dropped from a_i to a, each segment being measured in units of a, the average of the N values of a_i.   For f and X', C^2 is the average square of the N line-segments dropped from the N points in Fig. 25 to the radial line of slope $\phi = a/b$, each segment being again measured in units of a or Ex_i.   Table 9 displays the comparison of formulas.

TABLE 9

Comparison of formulas for $\bar{x}$ and for $f = \bar{x}/\bar{y}$.
The sampling plan begins on page 419

Denominator fixed	Denominator random
n = number of sampling units, fixed in advance	n = number of sampling units, fixed in advance

$$\bar{x} = \frac{\sum_{1}^{n} x_i}{n} \quad \text{[The } x\text{-population per sampling unit in the sample]}$$

$$f = \frac{\sum_{1}^{n} x_i}{\sum_{1}^{n} y_i} \quad \text{[The } x\text{-population per unit of } y\text{-population in the sample]}$$

$$E\bar{x} = \frac{A}{N} = a$$

$$Ef = \frac{A}{B} - \frac{1}{n}(\rho C_x C_y + C_y{}^2)$$

[See Table 2, page 395]

N = number of sampling units in the frame B = y-population in the frame

$X = N\bar{x}$, an estimate of A, the total x-population in the frame $X' = Bf$, an estimate of A, the total x-population in the frame

$$= N\frac{\sum_{1}^{n} x_i}{\sum_{1}^{n} 1} \quad \text{[Note that } N \text{ and the denominator have the same physical dimensions]}$$

$$= B\frac{\sum_{1}^{n} x_i}{\sum_{1}^{n} y_i} \quad \text{[Note that } B \text{ and the denominator have the same physical dimensions]}$$

For the rel-variances   For the rel-variances

$$C_X{}^2 = C_{\bar{x}}{}^2 = \frac{N-n}{N-1}\frac{1}{n}\frac{1}{N}\sum_{1}^{N}\left[\frac{a_i - a}{a}\right]^2$$

$$C_{X'}{}^2 = C_f{}^2 = \frac{N-n}{N-1}\frac{1}{n}\frac{1}{N}\sum_{1}^{N}\left[\frac{a_i - \phi b_i}{a}\right]^2$$

(This is an alternative form of Eq. 20, page 383.)

For an estimate of the rel-variances   For an estimate of the rel-variances

$$\hat{C}_X{}^2 = \hat{C}_{\bar{x}}{}^2 = \frac{N-n}{N}\frac{1}{n(n-1)}$$

$$\hat{C}_{X'}{}^2 = \hat{C}_f{}^2 = \frac{N-n}{N}\frac{1}{n(n-1)}$$

$$\times \frac{\sum_{1}^{n}(x_i - \bar{x})^2}{\bar{x}^2} \quad \text{[Derivation on page 436]}$$

$$\times \frac{\sum_{1}^{n}(x_i - fy_i)^2}{\bar{x}^2} \quad \text{[Derivation on page 439]}$$

The derivations of the estimates of these variances that we may make from the sample itself are in Section E of this chapter.

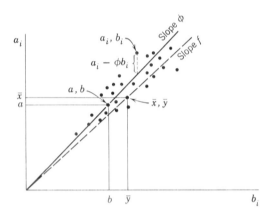

Fig. 25. Illustrating the use of a ratio in which the denominator is a random variable. In the frame there are N sampling units. When we draw a sample of n sampling units, we are in effect drawing n of these points at random. Each sampling unit has an x-population equal to a_i and a y-population equal to b_i. The center of the frame is the point a, b, where a and b are the average x- and y-populations per sampling unit in the frame. The center of the sample is the point $\bar{x}, \bar{y}$, where $\bar{x}$ and $\bar{y}$ are the average x- and y-populations per sampling unit in the sample. The 2 centers determine respectively the slopes of the 2 lines ϕ and f. The slope f is a random variable. If the N points lie fairly close to the line of slope ϕ (i.e., if the correlation between the populations a_i and b_i is high), then the coefficients of variation of f and of X' will be small.

Exercise. Show that

$$C_x^2 - 2C_{xy} + C_y^2 = \frac{1}{N} \sum_1^N \left\{ \frac{a_i - \phi b_i}{a} \right\}^2$$

where C_x^2 is the rel-variance between the x-populations of the N sampling units of the frame, and C_{xy} is rel-covariance between the x- and y-populations. a_i is the x-population of sampling unit i. $\phi = a/b$. C_{xy} may be written as $\rho C_x C_y$, where ρ is the correlation between the x- and y-populations in the sampling units of the frame.

SOLUTION

The proof is algebra, not statistical theory, once we start with the definitions of C_x^2, C_{xy}, and C_y^2. Term by term the definitions give

$$C_x^2 - 2C_{xy} + C_y^2 = \frac{1}{n} \sum_1^N \left\{ \left(\frac{a_i - a}{a} \right)^2 - 2 \frac{a_i - a}{a} \frac{b_i - b}{b} + \left(\frac{b_i - b}{b} \right)^2 \right\}$$

$$= \frac{1}{N} \sum_1^N \left\{ \frac{a_i - \phi b_i}{a} \right\}^2 \qquad \text{[By algebraic rearrangement]}$$

424 PART III. SOME THEORY USEFUL IN SAMPLING

Which estimate? We now have 2 estimates of A, viz.,

$$X = N\bar{x} \quad \text{[Formula X]}$$

and

$$X' = Bf \quad \text{[Formula X']}$$

Which estimate is better, X or X'? The answer is that

$$C_{X'} \gtrless C_X$$

if respectively

$$\rho \lesseqgtr \frac{C_y}{2C_x} \tag{8}$$

If $C_x = C_y$, as is sometimes nearly true, then the ratio-estimate X' is preferable if $\rho > \frac{1}{2}$. That is, Formula X will be less precise than Formula X' if $\rho > \frac{1}{2}$, more precise if $\rho < \frac{1}{2}$, and about the same if ρ is near $\frac{1}{2}$.

The reader may wish to show that Var $X' >$ Var X whenever $C_y > 2C_x$, even if $\rho = 1$. Moreover, Var $X' >$ Var X whenever $C_y > 1.5C_x$ and $\rho < .75$. Fortunately, one may usually defer the choice of estimate until he has a chance to study the variances and correlation in the first chunk of returns. A chart and illustration on pages 169 and 171 of Hansen, Hurwitz, and Madow's *Sample Survey Methods and Theory*, Vol. I (Wiley, 1953) have been very helpful to the author in this connexion. I may mention in passing that replication facilitates study of the returns, as the subsamples furnish points for a scatter-diagram that displays graphically the variances, the correlation, and the bias if any.

There is sometimes more than one choice of ratio to use to estimate a total, as when the total y-population B is known, and also the total z-population, C. The sample will provide x/y and also x/z, and it may be that the estimate $X'' = C(x/z)$ has a lower variance than $X' = B(x/y)$. It is fairly simple but important to test the 2 formulas against X, and to choose the better one.* As in Chapter 14 (p. 271) it is important not to be fooled on the variance.

PROOF OF EQ. 8

$$C_{X'}^2 = C_f^2 = \frac{N-n}{N-1}\frac{1}{n}[C_x^2 + C_y^2 - 2\rho C_x C_y] \quad \text{[P. 393]}$$

and

$$C_X^2 = C_{\bar{x}}^2 = \frac{N-n}{N-1}\frac{C_x^2}{n}$$

The comparison between $C_{X'}$ and C_X depends therefore on whether $C_y^2 - 2\rho C_x C_y$ is 0, less than 0, or greater than 0. Thus, $C_{X'}$ will be equal to C_X if $C_y^2 - 2\rho C_x C_y = 0$; i.e., if

$$\rho = \frac{C_y}{2C_x}$$

* Kinichiro Saito, "Maximum likelihood estimate of proportion using supplementary information," *Bull. Math. Statist.*, vol. 7, 1956: pp. 11–17 (Research Association of Statistical Sciences, Fukuoka, Japan).

If ρ has a bigger value, $C_{X'}$ will be less than C_X.

$$Q. E. D.$$

The occasional bias in the ratio f.* The ratio f may occasionally have some slight bias in it, especially when there are only a few sampling units in the sample. In practice, if there are many sampling units in the sample, and if care is taken to estimate f by the use of the whole sample, this bias is of negligible magnitude. In fact, if the "line of best fit" passes through the origin, f will be unbiased even if there is only 1 sampling unit in the sample, which is to say, if f and y are uncorrelated.

It is not necessary to specify exactly what one means by "the line of best fit" on the scatter-diagram that represents the x- and y-coordinates of a sample of 10 or more points. Ordinarily, a line drawn by eye will suffice. If the line passes through or near the origin, or more specifically, if f and y are uncorrelated, there is little or no bias in f to neglect, even if n is small. Fig. 26 exhibits the various possible combinations.

PROOF†

$$f_i = \frac{x_i}{y_i} \qquad [i = 1, 2, \ldots, n]$$

where x_i and y_i are the coordinates of a random point i.

$$x_i = f_i y_i$$

$$Ex_i = Ef_i Ey_i + \rho\sigma_f\sigma_y \qquad [\text{P. 364}]$$

where ρ is the correlation between f_i and y_i. Now if $\rho = 0$, then the last equation reduces to

$$Ef_i = \frac{Ex_i}{Ey_i} = \phi$$

which says that if f_i and y_i are uncorrelated, f_i is an unbiased estimator of ϕ. The same proof holds for a sample of n points. Fig. 26 will assist the practicing statistician to recognize the conditions under which this bias exists. It is fortunately easy, when one uses the replicated method, to correct the bias, if any (next section).

Detection and correction of bias by the use of replicated samples.‡ One of the most important features of the replicated method is that it furnishes

---

* The following references will assist the reader who wishes further reading on the bias and variance of the ratio-estimate: (1) Hansen, Hurwitz, and Madow, *Sample Survey Methods and Theory*, Vol. II (Wiley, 1953): p. 112; (2) P. V. Sukhatme, *Sampling Theory of Surveys with Applications* (Iowa State College Press, Ames; and the Indian Society of Agricultural Statistics, New Delhi, 1953): pp. 139–146 (excellent); (3) William G. Cochran, *Sampling Techniques* (Wiley, 1953): pp. 130 ff.

† From my colleague William N. Hurwitz, about 1942.

‡ This correction is explained by Howard L. Jones in his paper "Investigating the properties of a sample-mean by employing random subsample means," *J. Amer. Statist. Ass.*, vol. 51, 1956: pp. 54–83; p. 77 in particular. The suggestion came originally verbally from Professor John Tukey, though Quenouille has priority (p. 427).

a correction for the bias in an estimate.   This bias, when there is any, contains the factor $1/n$ and therefore decreases rapidly as n increases (Eq. 32, page 369).   We see at once that the replicated method with 2 samples per zone gives us estimates of f with 2 sample-sizes, one estimate made from the average of the 2 individual subsamples, and another estimate

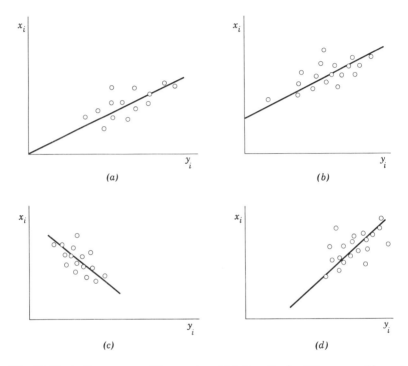

(a)

(b)

(c)

(d)

Fig. 26.   Illustrating some conditions under which the ratio $f = \bar{x}/\bar{y}$ possesses bias as an estimate of Ex/Ey.   The bias, if it exists at all, will diminish rapidly as the number n of sampling units in the sample increases.   I am indebted to my colleague William N. Hurwitz, who first exhibited to me about 1945 the principles displayed by these charts. (a) Unbiased; f_i and y_i uncorrelated.   (b) Biased; f_i and y_i correlated negatively; $Ef > \phi$.   (c) Biased; f_i and y_i correlated negatively; $Ef > \phi$.   (d) Biased; f_i and y_i correlated positively; $Ef < \phi$.

made by combining the 2 subsamples.   The replicated method with 10 samples per zone offers us the possibility of estimating f from a number of sample-sizes, such as the average of the 10 individual subsamples, the average of 2 groups of 5 subsamples each, the average of 5 groups of 2 subsamples each, and finally, the entire 10 combined into one big sample. One may plot the estimates so obtained against the reciprocal of the number of subsamples; then draw the line of best fit.   Extrapolation to 0 gives a quick estimate of the ratio f that one would obtain by increasing indefinitely the size of the sample.

A suggestion made in 1949 by Quenouille* applies directly to 2 subsamples, and leads to a simple calculation.   Let f denote the result of the 2 subsamples combined, and let f_1 and f_2 be the result of the 2 subsamples individually.   Then (as the student may wish to take as an exercise) the estimate

$$f' = 2f - \tfrac{1}{2}(f_1 + f_2) \qquad (9)$$

is equivalent to the extrapolation mentioned in the text above.

For the first example we may take the average rent in the small urban area surveyed in Chapter 10.   The results of the 2 subsamples are in Table 10.

TABLE 10

RESULTS FROM PAGES 176 AND 178

Characteristic	Sample 1	Sample 2	Both samples
Total rent	$2720	$2350	$5070
Total number of dwelling units	33	31	64
Average rent	$82.42	$75.81	$79.22

Quenouille's scheme gives $f' = 2 \times \$79.22 - \tfrac{1}{2}(\$82.42 + \$75.81) = \$79.33$, which is so near $f = \$79.22$ that we make no correction.

Another example is in Table 11, which shows the results of some psychological tests in 146 households, laid out in 10 subsamples.   We could plot a chart as follows:

1. At 1, $f = .339$, this being the average of the 10 individual f-values (.250, .091, .538, etc.).

2. At .5, $f = .333$, this being the average of 5 subsamples formed by combining 1 and 2, 3 and 4, 5 and 6, 7 and 8, 9 and 10.

3. At .2, $f = .337$, this being the average of 2 subsamples, formed by combining 1-5, and 6-10.

4. At .1, $f = .327$, this being the result of combining all 10 subsamples.

There is a faint indication of bias here if one has imagination.   Whatever be the bias, it is certainly small in comparison with the sampling

* M. H. Quenouille, *J. Royal Statist. Soc.*, Series B, vol. 11, 1949: pp. 68–84; p. 70 in particular.   *Biometrika*, vol. 43, 1956: pp. 353–360.   My friend Dr. Churchill Eisenhart kindly called my attention to Quenouille's papers.

error in f. I would adopt the estimate $f = .327$ (or $f = .33$), without adjustment.

Remark 2.   The standard error of f is about .05, as one may see by use of the range.   The maximum f is .600; the minimum f is .091, whence $\hat{\sigma}_f = (.600 - .091)/10 = .05$.

TABLE 11

THE OCCURRENCE OF MENTAL RETARDATION IN A SAMPLE
OF 110 HOUSEHOLDS: 10 SUBSAMPLES

Characteristic	1	2	3	4	5	6	7	8	9	10	All 10
x retarded	3	1	7	2	2	4	3	6	5	3	36
y retarded plus not retarded	12	11	13	15	10	9	12	10	9	9	110
f	.250	.091	.538	.133	.200	.444	.250	.600	.555	.333	.327

Table 6 on page 121 furnishes another example in 10 subsamples.   We saw there the average of the estimates obtained from the 10 individual subsamples, alongside the estimate obtained from all 10 subsamples combined, and we concluded that the bias, if any, was negligible.   The reader may make similar calculations with other tables of results here and there in the book.

Remark 3.   When we say that the bias, if any, decreases with the size of the sample (as we learned from Eq. 32 on page 369), we must be careful to recall Remark 8 on page 370 and measure n as the number of sampling units in the sample, and not as the number of dwelling units, people, or items within these sampling units.   Whatever be the sampling unit, it is entirely valid to compute sampling biases with relative values of n, which is exactly what we did in the tests above, wherein we took n as 1 for 1 subsample, 2 for 2 subsamples combined, 5 for 5 subsamples combined, and 10 for all 10 subsamples combined.

Correction of the bias in an estimate of the area of a circle.   We return now to the n measurements $x_1, x_2, \ldots, x_n$ on the radius of a circle that we dealt with on page 368.   Let us assume that the measurements behave like random variables with mean a and with standard deviation σ.   We saw that the estimate $\pi \bar{x}^2$ of the area of the circle is on the average too high by the amount $\pi \sigma^2/n$.   Suppose that we adopt the following procedure to correct for the bias: (i) estimate σ^2 from the n measurements; (ii) subtract $\pi \hat{\sigma}^2/n$ from $\pi \bar{x}^2$.   Show that, if $\hat{\sigma}^2$ is unbiased, then:

   a. The new estimate $\pi \bar{x}^2 - \pi \hat{\sigma}^2/n$ is unbiased.

   b. This estimate may, however, unless n is large, have a greater mean square error than the biased estimate $\pi \bar{x}^2$.

This circumstance arises in the fact that an estimate of σ^2 is itself a random variable with approximate variance $\sigma^4(\beta_2 - 1)/n$ (see page 439); hence unless n is large enough, a well-intended attempt to eliminate the bias in the manner just described may be a delusion and may leave us with a worse estimate than if we had let $\pi\bar{r}^2$ stand without correction.*

Note, however, that the procedure that we have already learned in the preceding section for correcting the bias in an estimate is effective here. We regard the n measurements (r_i) as n replications, and proceed as follows: (i) plot $F_n = \pi\bar{r}^2$ on the vertical axis at abscissa $1/n$; (ii) plot F_1, the average of the n values of πr_i^2, at abscissa 1; (iii) draw a line through the 2 points and extrapolate to $1/n = 0$. The vertical intercept $F_n - (F_n - F_1)/(n - 1)$ is an unbiased estimate of the area πa^2.

The student may wish to show that the variance of the correction term $(F_n - F_1)/(n - 1)$ is of order $1/n^3$. Thus, the procedure of extrapolation is safe and effective.

Exercise. Show that if (a) y_i is either 0 or 1, and that if (b) $x_i = 0$ when $y_i = 0$, and that if (c) we draw a sample of size n by reading out n random numbers between 1 and N, then the ratio $f = \Sigma x_i/\Sigma y_i$ will be an unbiased estimate of A/B, where A is the x-population in the entire frame and B is the number of units of y in the entire frame.

This result is important when the sample may contain empty sampling units (sampling units that contain no member of the universe). An example occurred in Chapter 8 where a sampling unit was a card in the file. A card might or might not lead to a contract motel. A contract motel might answer "Frequently" or something else. The sample gave an unbiased estimate of the number of contract motels per sampling unit (card), and an unbiased estimate of the proportion of motels that would have answered "Frequently" had the sample been 100%, even though the number of motels in the sample was a random variable.

Lahiri's scheme for an unbiased ratio.† If we carry out the following steps in the selection of the sample, the fraction f will be unbiased under any set of relations between x_i and y_i.

1. Draw n sampling units by reading out n unduplicated random numbers between 1 and N. This step selects 1 sample out of the $\binom{N}{n}$ possible samples, and gives every sample the same probability as another.

2. Compute $\sum_1^n y_i$ for this sample. Whether you will retain this sample

---
* Called to my attention by my friend Dr. Churchill Eisenhart of the National Bureau of Standards.

† D. B. Lahiri, "A method of sample selection providing unbiased ratio estimates," *Bull. Int'l Stat. Inst.*, vol. xxxiii, Part ii, 1951: pp. 133–140.

or draw another one, depends on the outcome of the side-play in the next step.

3. Draw a random number w between 1 and G, where G represents now a number as big or bigger than the largest possible denominator of the form $\sum_1^n y_i$. (There will be no trouble about determining this maximum: as a matter of fact, one may use for G the sum of the n biggest y-populations in the frame.)

4. Retain the sample of n farms selected in Step 1 if w falls between 1 and the denominator $\sum_1^n y_i$. Reject it otherwise, and repeat Step 1, then Step 2 in cycles until the random number w falls between 1 and Σy_i.

Now let f be the ratio of the x-population to the y-population in the first sample that you retain by this scheme. Then f will be an unbiased estimate of ϕ, the quantity that we are trying to measure. In symbols,

$$Ef = \phi \qquad \text{[Under the Lahiri scheme]}$$

Remark 4. We see again that one must know not only the formula for an estimate, *but how the sample is to be drawn.* Thus, the ratio f may under certain conditions have a slight bias in it (p. 425). Yet the same formula, when the drawings are made from the same frame by the Lahiri scheme, is unbiased.

Remark 5. The Lahiri scheme requires advance knowledge of the y-populations b_i in the frame, which go into the summation Σy_i in Step 2. We may unfortunately not be in possession of this information for some specified ratio for which we desire an unbiased estimate.

Remark 6. If we apply the Lahiri scheme to obtain an unbiased estimate of one specified ratio, such as the increase or decrease in the proportion of all arable acres cultivated, we do not necessarily obtain an unbiased estimate for some other ratio, such as the ratio of the acres in wheat to the acres in oats.

General description of some usual experience in the use of ratios in fine classes. We learned in Section D some of the theory for the ratio $f = x/y$, and how to choose between $X' = Bf$ and $X = N\bar{x}$. It is sometimes possible to reap further gains in precision by estimating the overall ratio f by building it up from finer classes. An example is the estimates of employment and unemployment put forth in the Monthly Report of the Labor Force from the Bureau of the Census in Washington, and by the Census in Ottawa, and elsewhere in the world. The sample contains a random number of males 20–29 employed; also a random number of all males 20–29, employed or not. The sample then estimates the ratio f between these 2 populations in every stratum; hence, gives an estimate of this ratio over all strata—that is, over the whole country.

Now the total B number of males 20–29 is known accurately from the last Census and from death rates and immigration.* Multiplication of the ratio f by the known total B, exactly as we saw it in Eq. 4 on page 420 then gives an estimate of the number of males 20–29 employed. The improvement in precision is in this instance considerable, the variance being about half what it would be without the use of the ratio in fine classes. This example is only an illustration; it should not convey the impression that a ratio-estimate in fine classes will always be an improvement; it may be, or it may not be.

We may make a simple formulation of the procedure as follows. Form from a sample in Stratum 1 an estimate of the number of males 20–29 in a certain occupation; also the number of all males 20–29. Then do the same for Stratum 2, and for Stratum 3, etc. We shall reuse our symbols.

$X = X_1 + X_2 +$ etc. where X_1, X_2, etc. are estimates of the x-population in Strata 1, 2, etc. X is an estimate of the x-population in the whole frame, all strata combined. For example, the x-population might be males 20–29, engaged in transport (a specific occupation).

$Y = Y_1 + Y_2 +$ etc. where Y_1, Y_2, etc. are estimates of the y-population in Strata 1, 2, etc. Y is an estimate of the y-population in the whole frame, all strata combined; e.g., all males 20–29.

With the estimates X and Y we may form the ratio

$$f = \frac{X}{Y} \tag{10}$$

Interest will sometimes lie in the ratio f itself. The sample will give us this ratio. At other times interest will lie in a total figure, such as the number of males 20–29 in a specific occupation, in the whole frame. Then, IF we know by external information (such as a recent census) what is the total number B of males of age 20–29 in the whole frame, then we may again form the *ratio-estimate*

$$X' = Bf \qquad \text{[Eq. 4, page 420]}$$

for the total number of males 20–29 in the specific occupation in the whole frame. This estimate will very often have a smaller standard error than the estimate X formed without use of the ratio.

The standard errors of f and of X' are related as in Eq. 5 on page 421 because, again, the constant B is not a random variable.

* Hansen, Hurwitz, and Madow, *Sample Survey Methods and Theory*, Vol. 1 (Wiley 1953): p. 570. Frank Yates, *Sampling Methods for Censuses and Surveys* (Griffin, 1949): pp. 215–217.

E. ESTIMATION OF THE STANDARD ERROR

Need for postevaluation of the precision. A sample-design is a plan worked out in advance, by the aid of the theory that we have been studying. The plan aims at some desired degree of precision. After the results have come in, and have been tabulated, comes the question: what precision was actually attained? This question can only be answered by a post-evaluation of the precision from the results of the survey. The sample-design comes in advance: the postevaluation of the precision comes afterward. Every survey thus possesses:

The precision aimed at. This is the precision built into the sample-design.

The precision actually attained. This is the indisputable label of precision estimated from the results of the survey itself. The theory therefor is the purpose of this section of the chapter.

Why do we not always get exactly the precision that we aim at? Because we do not often possess in advance the exact values of the variances between the sampling units, in which case we must fill in the gaps with some inexact values drawn from experience in order to proceed with the sample-design. For other surveys there may be a considerable background of experience, by which we may tailor the sample very exactly to the requirements, and hit close to the precision aimed at.

There are 2 very good reasons for appraising the precision that has actually been attained:

1. A statement of the result obtained (X, $\bar{x}$, f, X', etc.) in any survey is incomplete and may even be misleading unless it is accompanied by an index of its precision.

2. An appraisal of the precision attained provides information on variances and other statistical characteristics of the material that can be used to good advantage in the planning of future surveys.

Theory for estimating the variance between the sampling units of the frame. Read out n random numbers between 1 and N to draw n sampling units with equal probabilities from the frame. Denote the populations of these n sampling units by

$$x_1, x_2, \ldots, x_n$$

Their mean is $\bar{x}$, and their variance is by definition

$$s^2 = \frac{1}{n} \sum_1^n (x_i - \bar{x})^2 \tag{1}$$

Both $\bar{x}$ and s^2 are random variables, because if we return the n sampling units to the frame (either one at a time or after all n sampling units are drawn), and then redraw the sample, we get new values of $\bar{x}$ and s^2. Repetitions of the sampling procedure will generate distributions of $\bar{x}$ and of s^2. We already know that the mean of the theoretical sampling distribution of $\bar{x}$ is $E\bar{x} = a$, and that its variance is σ^2/n, modified by the finite multiplier if we draw without replacement. σ^2 is here, as before, the variance between the sampling units in the frame. In a few minutes, however, when we adapt our equations to replicated sampling, σ^2 will be the variance between the results of repeated subsamples.

The mean Es^2 of the theoretical distribution of s^2 is given by the following equations:

$$\left.\begin{array}{ll} \dfrac{n}{n-1}\, Es^2 = \sigma^2 & \text{with replacement} \\[3mm] \dfrac{n}{n-1}\, Es^2 = \dfrac{N}{N-1}\, \sigma^2 & \text{without replacement} \end{array}\right\} \quad (2)$$

A proof is on the next page.

To get an unbiased estimate of σ^2, we merely remove E and replace σ^2 by $\hat{\sigma}^2$. The above equations then give

$$\left.\begin{array}{ll} \hat{\sigma}^2 = \dfrac{n}{n-1}\, s^2 & \text{with replacement} \\[3mm] = \dfrac{N-1}{N}\, \dfrac{n}{n-1}\, s^2 & \text{without replacement} \end{array}\right\} \quad (3)$$

We may usually in practice omit the factor $(N-1)/N$ and rewrite the last equations as

$$\hat{\sigma}^2 = \frac{1}{n-1} \sum_1^n (x_i - \bar{x})^2 \qquad (4)$$

The factor $n-1$ in the denominator in all the above estimates of variance and in other estimates to follow is known as the *number of degrees of freedom* in the estimate.* It is merely the number of x_i that are free, once we fix $\bar{x}$. The greater the number of degrees of freedom, the better the estimate of the variance, as we shall see when we come to Eq. 17 and the accompanying text.

* The use of $n-1$ in place of n in the denominator, in order to get an unbiased estimate of the variance, was introduced first by Gauss, *Theoria Combinationis Observationum Erroribus Minimis Obnoxiae*, Pars posterior (Göttingen, 1823; vol. 3 of his Werke): Art. 38. He made also the more general correction $n-2$ in curve-fitting in the evaluation of 2 parameters; $n-3$ for 3; etc.

Remark 1. All these formulas give an unbiased estimate of σ^2, regardless of the shape of the distribution of the populations of the sampling units in the frame. It is interesting to note, though, that $\sqrt{\hat{\sigma}^2}$ is a biased estimate of σ.

Remark 2. We may sometimes to advantage simplify the calculation of the sum of squares by adopting some convenient whole number d near $\bar{x}$ and computing

$$\Sigma(x_i - d)^2 - (\bar{x} - d)^2$$

for $\Sigma(x_i - \bar{x})^2$. If we set $d = 0$, this reduces to $\Sigma x_i^2 - \bar{x}^2$, but we must then be careful to use enough decimals in $\bar{x}$, as any approximation thereto is greatly magnified (a point often overlooked).

Remark 3. The reader should bear in mind that all the above estimates of σ are valid only when the denominator n in $\bar{x} = x/n$ is fixed in advance. They require modification for the variance of the ratio $f = x/y$ if x and y are both random (p. 437).

Remark 4. Suppose that n is some large number, and that we need only n' sampling units to form an estimate of σ^2. We may (a) draw without replacement a subordinate sample of size n' from the main sample of size n, by reading out n' random numbers between 1 and n; then (b) calculate

$$\hat{\sigma}^2 = \frac{1}{n' - 1} \sum_1^{n'} (x_i - \bar{x}')^2 \tag{5}$$

where $\bar{x}'$ is the mean of the n' sampling units. The number of degrees of freedom in this estimate is $n' - 1$.

PROOF OF EQ. 2 (DUE TO GAUSS, 1813)

Write for any one sample the identities

$$\left.\begin{aligned}
x_1 - a &= (x_1 - \bar{x}) + (\bar{x} - a) \\
x_2 - a &= (x_2 - \bar{x}) + (\bar{x} - a) \\
&\cdot \\
&\cdot \\
&\cdot \\
x_n - a &= (x_n - \bar{x}) + (\bar{x} - a)
\end{aligned}\right\} \tag{6}$$

Square each member, add the n equations, divide by n: the result is

$$\frac{1}{n} \sum_1^n (x_i - a)^2 = s^2 - (\bar{x} - a)^2 \tag{7}$$

Take the expected value now of each term for repeated samples. The result is

$$\begin{aligned}
\sigma^2 &= Es^2 - \sigma_{\bar{x}}^2 \\
&= Es^2 - \frac{N - n}{N - 1} \frac{\sigma^2}{n} \tag{8}
\end{aligned}$$

The rest is mere algebraic rearrangement.

EXERCISES

1. Read out n random numbers between 1 and N to draw a sample of n chips from a bowl that contains initially Np red chips and Nq white chips. Suppose that the sample shows r red and $n - r$ white. Define the sample as

$$x_1, x_2, \ldots, x_n$$

as heretofore, where now $x_i = 1$ if the chip is red, and $x_i = 0$ if it is white.

a. Show that $\sum_1^n x_i = r$, and that the mean ($\bar{x}$) of the sample is merely $\hat{p}$, where $\hat{p} = r/n$.

b. Define $s^2 = \frac{1}{n}\sum_1^n (x_i - \bar{x})^2$. Show that $s^2 = \hat{p}\hat{q}$ where $\hat{p} = r/n$ and $\hat{p} + \hat{q} = 1$.

1ST SOLUTION

We know that $\sigma^2 = pq$ for the frame of 2 cells, 0 and 1 (p. 404). Then $s^2 = \hat{p}\hat{q}$, as these 3 symbols have the same definitions for the sample that σ^2, p, q do for the frame.

2D SOLUTION

$$s^2 = \frac{1}{n}\sum_1^n (x_i - \bar{x})^2 = \frac{1}{n}\sum_1^n x_i^2 - \bar{x}^2$$

$$= \frac{1}{n}\sum_1^n x_i - \hat{p}^2 \quad [x_i^2 = x_i \text{ because } x_i \text{ is either 0 or 1}]$$

$$= \frac{r}{n} - \hat{p}^2 = \hat{p} - \hat{p}^2 = \hat{p}(1 - \hat{p}) = \hat{p}\hat{q}$$

c. Show that $\dfrac{N-1}{N} E\hat{p}\hat{q} = \dfrac{n-1}{n} pq$ (without replacement).

1ST SOLUTION

Turn to Eq. 3 on page 433 and replace σ^2 by pq, and s^2 by $\hat{p}\hat{q}$.

2D SOLUTION

$$E\hat{p}\hat{q} = E\hat{p} - E\hat{p}^2 \quad [\hat{q} = 1 - \hat{p}]$$
$$= p - [\sigma_{\hat{p}}^2 + (E\hat{p})^2] \quad [\text{P. 380}]$$
$$= p - \sigma_{\hat{p}}^2 - p^2$$
$$= p(1 - p) - \frac{N-n}{N-1}\frac{pq}{n}$$

The rest is algebra.

d. Show that

$$\sigma_{\hat{p}}^2 = \frac{N-1}{N}\frac{\hat{p}\hat{q}}{n-1}$$

$$\doteq \frac{\hat{p}\hat{q}}{n-1} \qquad \text{[As } N \text{ is large in practice]}$$

is an unbiased estimate of $\sigma_{\hat{p}}^2 = pq/n$.

Hint: Solve for pq/n in terms of $E\hat{p}\hat{q}$, from part *c*. Then remove E for an estimate.

2. Read out random numbers between 1 and n to draw with replacement a sample of size n from a frame of N sampling units, mean a, variance σ^2. Restore the n sampling units of the sample to the frame; draw a new sample; and repeat. Let $\bar{x}$ be the mean of a sample, and s^2 its variance. Let $\bar{x}'$ be the mean of the $N - n$ sampling units that remain in the frame, and s'^2 their variance.

a. Show that

$$E\bar{x} = E\bar{x}' = a$$

b. Show that

$$Es^2 = \frac{N}{N-1}\frac{n-1}{n}\sigma^2 = \frac{N}{N-1}\frac{N-n-1}{N-n}Es'^2$$

c. If n is $\frac{1}{2}N$,

$$Es^2 = Es'^2 \qquad \text{exactly}$$

d. If N is very big compared with n,

$$Es^2 \doteq \frac{n-1}{n}Es'^2 \qquad \text{very nearly}$$

e. If n is big, and N much bigger,

$$Es^2 = Es'^2 \qquad \text{very nearly}$$

What these equations say is that the mean of a sample and the mean of the remainder of the frame both provide unbiased estimates of a. The variance of the sample and the variance of the remainder also both provide estimates of σ^2. In other words, what remains in the frame must be as good a sample of the frame as the sample that we remove. (First stated verbally to the author by J. Stevens Stock and Lester R. Frankel, about 1937.)

Theory for estimating the variance of a mean. We learned on page 383 that if we draw a sample of size n by reading out n random numbers between 1 and N, and calculate $\bar{x}$ as usual, then the variance of $\bar{x}$ will be

$$\sigma_{\bar{x}}^2 = \frac{\sigma^2}{n} \qquad \text{with replacement}$$

$$= \frac{N-n}{N-1}\frac{\sigma^2}{n} \qquad \text{without replacement} \qquad (9)$$

If we substitute herein the estimate of σ^2 derived in Eq. 4, we find the estimate

$$
\left.
\begin{aligned}
\hat{\sigma}_{\bar{x}}^{2} &= \frac{1}{n(n-1)} \sum_1^n (x_i - \bar{x})^2 & \text{with replacement} \\
&= \frac{N-n}{N} \frac{1}{n(n-1)} \sum_1^n (x_i - \bar{x})^2 & \text{without replacement}
\end{aligned}
\right\} \quad (10)
$$

This is a very important formula. It enables us to estimate from a single sample of size n the variance of the distribution of $\bar{x}$ in repeated samples drawn and processed by the stated sampling plan. We have already used this formula a number of times in Part II.

If n is large, we may decrease the labor of computation by calculating the sum of squares for only a random portion of appropriate size n': see Remark 4.

When we adapt the above estimate of $\sigma_{\bar{x}}^2$ to replicated designs, x_i will stand for an estimate from Subsample i, and $\bar{x}_1$ will stand for the estimate made from all subsamples combined; n is the number of subsamples. The symbol $\hat{\sigma}^2$, which we estimated in Eq. 3, will be the variance between all the possible estimates x_i that could be formed by the sampling procedure prescribed. The reader should satisfy himself that Eqs. 2 and 14 in Chapter 11 (p. 197) are applications of Eq. 10 above.

Remark 5.   We have already used also, in earlier chapters, the estimate[*],[†]

$$
\hat{\sigma}_{\bar{x}} = \frac{w}{n} \tag{11}
$$

where w is the range between n individual estimates. This formula gives good results for n between 3 and 10. It holds for a rectangular distribution about as well as for a normal distribution.

Theory for estimating the variance of a ratio.   Each subsample in a replicated sampling design furnishes an estimate of a ratio. Thus, as in Chapter 6 and elsewhere, all 10 subsamples combined gave the estimate $f = x/y$, and each subsample gave its own estimate f_i. Use of Eq. 25 (p. 200), namely, the range between the highest and the lowest f_i divided by 10, gave an estimate of the standard error of f. This is a useful estimate, but there are circumstances when we need to estimate Var f by the sum of squares, in order to add variances or to compare variances.

We deal with the same sampling plan that we had in the preceding pages; viz., draw the sample of size n by reading out n random numbers between 1 and N; form $\bar{x}$ and $\bar{y}$ and $f = \bar{x}/\bar{y}$.

[*] L. H. C. Tippett, "On the extreme individuals and range of samples taken from a normal population," *Biometrika*, vol. 117, 1925: p. 364.

[†] Nathan Mantel, "On a rapid estimation of standard errors for the means of small samples," *Amer. Statistician*, vol. 5, Oct. 1951: pp. 26–27.   M. H. Quenouille, *Rapid Statistical Calculations* (Hafner, 1959): pp. 5, 6, 7.

We recall from Eq. 13 on page 393 the approximation

$$C_f^2 \doteq C_{\bar{x}}^2 + C_{\bar{y}}^2 - 2C_{\bar{x}\bar{y}}$$

$$= \frac{N-n}{N-1}\frac{1}{n}\{C_x^2 + C_y^2 - 2C_{xy}\} \tag{12}$$

where $C_x^2 = E(x_i - a)^2/a^2$ is the rel-variance of the x-populations of the N sampling units of the frame, and C_y^2 is defined likewise with the y-populations in the frame. $C_{xy} = E(x_i - a)(y_i - b)/\sigma_x\sigma_y$ is the rel-covariance of the x- and y-populations in the frame. The following estimates

$$\hat{C}_x^2 = \frac{\hat{\sigma}_x^2}{a^2} = \frac{N-1}{N}\frac{1}{n-1}\sum_1^n \left\{\frac{x_i - \bar{x}}{\bar{x}}\right\}^2 \tag{13}$$

$$\hat{C}_y^2 = \frac{\hat{\sigma}_y^2}{b^2} = \frac{N-1}{N}\frac{1}{n-1}\sum_1^n \left\{\frac{y_i - \bar{y}}{\bar{y}}\right\}^2 \tag{14}$$

come from Eq. 2 on page 433 and by the use of $\hat{\sigma}^2/\bar{x}^2$ as an estimate of $C_x^2 = \sigma_x^2/a^2$, which commits no more error than is already in Eq. 12 through neglect of terms that contain $1/n^2$ as a multiplier. In any event, the approximation improves rapidly as more sampling units come into the sample.

We need next an estimate of C_{xy}, and we can get it by following out the same idea that we used on page 434 to derive Eq. 2. Write

$$(x_i - a)(y_i - b) = \{(x_i - \bar{x}) - (\bar{x} - a)\}\{(y_i - \bar{y}) - (\bar{y} - b)\}$$

for the ith member of the sample. Then sum over all n members of the sample to find that

$$\frac{1}{n}\sum_1^n (x_i - a)(y_i - b) = \frac{1}{n}\sum_1^n (x_i - \bar{x})(y_i - \bar{y}) + (\bar{x} - a)(\bar{y} - b)$$

Now take the expected value of each term, and recall Eq. 26 on page 386 of this chapter for the last term on the right. Thence comes

$$E(x_i - a)(y_i - b) = E\frac{1}{n}\sum_1^n (x_i - \bar{x})(y_i - \bar{y})$$

$$+ \frac{N-n}{N-1}\frac{1}{n}E(x_i - a)(y_i - b)$$

whence by algebra

$$\frac{N-1}{N}E(x_i - a)(y_i - b) = \frac{n-1}{n}E\sum_1^n (x_i - \bar{x})(y_i - \bar{y}) \tag{15}$$

which bears a remarkable resemblance to Eq. 2 on page 434. We now insert all our estimates into Eq. 12 to get

$$
\hat{C}_f^2 \doteq \frac{N-n}{N} \frac{1}{n(n-1)} \sum_1^n \left\{ \left(\frac{x_i - \bar{x}}{\bar{x}} \right)^2 + \left(\frac{y_i - \bar{y}}{\bar{y}} \right)^2 - 2 \frac{x_i - \bar{x}}{\bar{x}} \frac{y_i - \bar{y}}{\bar{y}} \right\}
$$

$$
= \frac{N-n}{N} \frac{1}{n(n-1)\bar{x}^2} \sum_1^n \{ (x_i - \bar{x}) - f(y_i - \bar{y}) \}^2 \qquad (16a)
$$

$$
= \frac{N-n}{N} \frac{1}{n(n-1)\bar{x}^2} \sum_1^n (x_i - fy_i)^2 \qquad [\text{If } f = \bar{x}/\bar{y}] \qquad (16b)
$$

This is the estimate that we were seeking. We have used it in Chapters 10 and 11 and elsewhere.

Remark 6. The reader will recognize $\dfrac{1}{n-1} \displaystyle\sum_1^n \left(\dfrac{x_i - fy_i}{\bar{x}} \right)$ as the average square of the distances of the n sample-points measured vertically to the line $x = fy$ (Fig. 25, page 423), each distance being measured in units of $\bar{x}$.

Exercise 1. Make use of the estimate just written for $\hat{C}_f^2$ to derive Eq. 18 in Chapter 11 on page 199 for k subsamples.

Exercise 2. Use the same estimate to derive Eq. 9 in Chapter 11 on page 198 for 2 subsamples.

Exercise 3. Show that $C_{y/x} \doteq C_{x/y}$; and that $C_{1/x} \doteq C_x$.

The standard error of an estimate of a standard error. We have learned how to use a sample to estimate the variance of the theoretical distribution of the means of all possible samples that we could draw and process by the sampling plan that we were using. The estimate of a variance made from a sample is itself a random variable. What is the standard deviation of the distribution of this estimate? The following formulas apply to a sample of size n drawn without replacement from a frame of N sampling units by reading out n random numbers between 1 and N:

$$
C_{\hat{\sigma}_{\bar{x}}} = C_{\hat{\sigma}} = \sqrt{ \frac{1}{2(n-1)} \left\{ 1 + \frac{n-1}{2n} (\beta_2 - 3) \right\} } \qquad (17)
$$

$$
\left.
\begin{aligned}
&= \frac{1}{2} \sqrt{ \frac{\beta_2 - 1}{n} } \qquad &&[\text{For large } n] \\[2mm]
&= \sqrt{ \frac{1}{2(n-1)} } \qquad &&[\text{For any } n \text{ if } \beta_2 = 3]
\end{aligned}
\right\} \qquad (18)
$$

The derivation is Exercise 8 ahead (p. 445).   The symbol

$$\beta_2 = \frac{1}{\sigma^4}\frac{1}{N}\sum_1^N (a_i - a)^4$$

$$= E\left\{\frac{x_i - a}{\sigma}\right\}^4 \qquad (19)$$

is the standardized 4th moment coefficient of the distribution of the x-populations a_i in the sampling units in the frame. $\beta_2 = 3$ for a normal distribution, but it may attain high values in extreme skewness (see Exercise 6 on page 445, where $\beta_2 = 1/pq - 3$ for the binomial: obviously β_2 will be large if p is small).   Just as the Var $\bar{x}$ depends on σ, so the variance of an estimate of Var $\bar{x}$ depends on β_2.

As $\beta_2 - 3 = 0$ for a normal curve, the quantity inside the braces in Eq. 17 shows how much the nonnormality of the sampling units in the frame affects the standard error of the estimate of $\sigma_{\bar{x}}$ [quoted from Cochran's *Sampling Techniques* (Wiley, 1953): p. 28].

What we need next is the variance of an estimate of Var $\bar{x}$ in a replicated design.   Let c_j be the number of thin zones in Thick Zone j, m the number of thick zones, k the number of subsamples, and $n = Mk$ the number of sampling units in the sample; σ_i^2 and β_{2i} refer to Thin Zone i.   The formula is*

$$C_{\hat{\sigma}_X} = C_{\hat{\sigma}_{\bar{x}}} = \frac{1}{2}\sqrt{\frac{2}{k-1}\frac{\sum_{j=1}^m\left[\sum_{i=1}^{c_j}\sigma_i^2\right]^2}{\left[\sum_1^M\sigma_i^2\right]^2} + \frac{\sum_{i=1}^M \kappa_{4i}}{k\left[\sum_1^M\sigma_i^2\right]^2}} \qquad \text{[Zindler]} \quad (20)$$

$$= \frac{1}{2}\sqrt{\frac{2}{m(k-1)} + \frac{\beta_2 - 3}{n}} \qquad \begin{array}{l}\text{[If } c_j = M/m \text{ for all} \\ j, \text{ and if } \beta_{2i} = \beta_2 \\ \text{and } \sigma_i = \sigma \text{ for all } i]\end{array} \quad (21)$$

$$= \sqrt{\frac{1}{2m(k-1)}} \qquad \text{[If } n \text{ is large, or if } \beta_2 = 3] \quad (22)$$

The derivation is in the last exercise ahead.

In a design where there are 10 subsamples and 1 thick zone (as in Chapter 6 and in other examples), and where β_2 is constant and equal to 3 throughout (the normal value of β_2), the coefficient of variation of the standard error of $\bar{x}$ is $\sqrt{1/2(10-1)} = 23\%$.

A calculation with a higher value of β_2 may be helpful.   Suppose that

* Hans-Joachim Zindler, "Über die Genauigkeit von Streungsschätzungen durch Gruppensummen," *Mitteilungsblatt für Mathematische Statistik*, Jahrgang 8, 1956: pp. 192–201; "Über einige Aspekte des Demingplanes," *ibid.*, Jahrgang 9, 1957: pp. 55–72.   I thank Herr Zindler for showing me this equation and assisting me with the derivation thereof.

there were 50 thin zones, 10 thick zones, 2 subsamples, and that $\beta_2 = 20$ throughout.   Then $n = 50 \times 2 = 100$ and Eq. 20 gives

$$C_{\hat{\sigma}_X} = C_{\hat{\sigma}_{\bar{x}}} = \frac{1}{2} \sqrt{\frac{2}{10(2-1)} + \frac{20-3}{100}}$$

$$= \frac{1}{2} \sqrt{.20 + .17} = 30\%$$

We may ask what would happen if we should reduce the number of thick zones to 4, and take 5 subsamples instead of 2.   Then would

$$C_{\hat{\sigma}_X} = C_{\hat{\sigma}_{\bar{x}}} = \frac{1}{2} \sqrt{\frac{2}{4(5-1)} + \frac{20-3}{100}}$$

$$= \frac{1}{2} \sqrt{.13 + .17} = 26\%$$

which with the same number of tabulations, is a bit better than before, with the risk, however, of some possible slight increase in Var $\bar{x}$, owing to the wider zones necessary for 5 subsamples.

A high value of β_2, or great variation in β_2, may indicate need of stratification.   The main purpose of stratification is usually to reduce Var $\bar{x}$, but stratification has also fringe benefits such as reducing β_2, thus normalizing the distribution of $\bar{x}$ and producing a better estimate of Var $\bar{x}$. However, if we are sampling for a rare population that is scattered widely and not concentrated in a few sampling units, stratification will not reduce β_2, except at great expense.   Exercise 5 shows that if the characteristic is scattered like a binomial variate, β_2 will be $1/pq - 3$, and there is not much that we can do about it.   We are in any case fortunate to have some theory so that we may understand better how to use an estimate of a standard error, and how to improve it if there be any possibility.

EXERCISES

1. Prove that the coefficients of variation of s, $\hat{\sigma}$, and $\hat{\sigma}_{\bar{x}}$ are all equal; i.e.

$$C_s = C_{\hat{\sigma}} = C_{\hat{\sigma}_{\bar{x}}} \tag{23}$$

if $\hat{\sigma}$ and $\hat{\sigma}_{\bar{x}}$ are both multiples of s.

2. Show that if x_1 and x_2 are the populations of 2 random sampling units drawn with replacement, then

$$\text{Var}(x_1 + x_2) = E(x_1 - x_2)^2 \tag{24}$$

It follows by removal of E that $(x_1 - x_2)^2$ furnishes an estimate of Var $(x_1 - x_2)$ that is unbiased regardless of the distribution of the x_i.   This form, due to Keyfitz,* is adaptable immediately to replications in 2

---

* The theorems for this exercise and for the next were published by Nathan Keyfitz, "Estimates of sampling variance where two units are selected from each stratum," *J. Amer. Statist. Ass.*, vol. 52, 1957: pp. 503–510.

subsamples for estimates of Var $\bar{x}$ and Var X.   Although equivalent to Eq. 4 on page 433, Keyfitz's result is the easier to remember and to apply.

SOLUTION (KEYFITZ)

$$\text{Var}\,(x_1 + x_2) = E\{(x_1 + x_2) - E(x_1 + x_2)\}^2$$
$$= E\{(x_1 - Ex_1) + (x_2 - Ex_2)\}^2$$
$$= E(x_1 - Ex_1)^2 + E(x_2 - Ex_2)^2$$
$$+ 2E(x_1 - Ex_1)(x_2 - Ex_2)$$

But as x_1 and x_2 are drawn with replacement, there is no correlation between them; hence the last term is 0.   We may in fact reverse its sign, as it is 0 anyway, and write

$$\text{Var}\,(x_1 + x_2) = E(x_1 - Ex_1)^2 + E(x_2 - Ex_2)^2$$
$$- 2E(x_1 - Ex_1)(x_2 - Ex_2)$$
$$= E\{(x_1 - Ex_1) - (x_2 - Ex_2)\}^2$$
$$= E(x_1 - x_2)^2$$

as $Ex_1 = Ex_2$. $Q.E.D.$

The student will observe that, with 2 subsamples, every thick zone gives a numerical value of $(x_1 - x_2)^2$.   A succession of m thick zones gives

$$\hat{C}_{\bar{x}}^2 = \frac{\Sigma(x_1 - x_2)^2}{\{\Sigma(x_1 + x_2)\}^2} \tag{25}$$

for an estimate of the rel-variance of $\bar{x}$.   This equation is equivalent to Eq. 5 on page 197 in Chapter 11 which gave $\hat{C}_{\bar{x}} = R/x$, R^2 being the numerator in the above equation, and x being $\Sigma(x_1 + x_2)$.   Introduce the finite multiplier $(Z - 2)/Z$ before $(x_1 - x_2)^2$ if required.

A fundamental equation for the variance of duplicate samples with equal or unequal sampling fractions λ and μ is due to Yates* in the form

$$\text{Var}\,\Sigma(x_1 + x_2) = \lambda\mu\left(1 - \frac{n}{N}\right)\Sigma(x_1 - x_2)^2 \qquad [\lambda + \mu = 1] \tag{26}$$

which the careful student will wish to establish for himself, by a proof similar to Keyfitz's above, or otherwise.   The sums (Σ) run, as before, over all thick zones or other areas of tabulation.   Eq. 26 reduces to Eq. 24 if $\lambda = \mu = \frac{1}{2}$.

3. Show that

$$\text{Var}\,\{\widehat{\text{Var}}\,(x_1 + x_2 + x_3 + x_4)\}$$
$$= E\{(x_1 - x_2)^2 - (x_3 - x_4)^2\}^2 \tag{27}$$

It follows by removal of E, that $\frac{1}{16}\{(x_1 - x_2)^2 - (x_3 - x_4)^2\}^2$ furnishes an

* Frank Yates, *Sampling Methods for Censuses and Surveys* (Griffin, 1949): p. 242.

estimate of the variance of $\hat{\mathrm{V}}\mathrm{ar}\ \bar{x}$, which is unbiased regardless of the shape of the distribution of the x_i, where $\bar{x} = \frac{1}{4}(x_1 + x_2 + x_3 + x_4)$.

SOLUTION (KEYFITZ)

$$\mathrm{Var}\ \{\hat{\mathrm{V}}\mathrm{ar}\ (x_1 + x_2 + x_3 + x_4)\}$$
$$= \mathrm{Var}\ \{\hat{\mathrm{V}}\mathrm{ar}\ (x_1 + x_2) + \hat{\mathrm{V}}\mathrm{ar}\ (x_3 + x_4)\}$$

$$\text{[As } x_1 + x_2 \text{ is independent}$$
$$\text{of } x_3 + x_4]$$

$$= \mathrm{Var}\ \{(x_1 - x_2)^2 + (x_3 - x_4)^2\}$$
$$= E\ \{(x_1 - x_2)^2 - (x_3 - x_4)^2\}$$

by the result of the preceding exercise. $\quad Q.E.D.$

This elegant theorem also due to Keyfitz has great utility. If x_1 and x_2 come from one thick zone, and x_3 and x_4 from the next, then together they furnish one value of $\{(x_1 - x_2)^2 \pm (x_3 - x_4)^2\}$. A succession of pairs of thick zones would furnish

$$\hat{C}^2_{\hat{\mathrm{V}}\mathrm{ar}\ \bar{x}} = \frac{\Sigma\{(x_1 - x_2)^2 - (x_3 - x_4)^2\}^2}{[\Sigma\{(x_1 - x_2)^2 + (x_3 - x_4)^2\}]^2} \tag{28}$$

for an estimate of the rel-variance of $\hat{\mathrm{V}}\mathrm{ar}\ \bar{x}$ derived from the same sample. An estimate of the coefficient of variation of $\hat{\sigma}_{\bar{x}}$ is then

$$\hat{C}_{\hat{\sigma}_{\bar{x}}} \doteq \frac{1}{2} \sqrt{\hat{C}^2_{\hat{\mathrm{V}}\mathrm{ar}\ \bar{x}}}$$

which comes from Exercise 10, page 446. An application appears on page 449.

THE FOLLOWING EXERCISES ARE FOR REFERENCE
AND FOR ADVANCED STUDENTS*

4. Read out n random numbers between 1 and N with replacement to draw a sample of size n from a frame in which a is the mean population per sampling unit, and σ^2 is the variance between the N sampling units. Compute the mean $\bar{x}$ of the sample so drawn. Define

$$\Delta\bar{x} = \bar{x} - E\bar{x}$$

Denote by μ_3 and μ_4 the 3d and 4th moment coefficients of the populations of the sampling units in the frame. Show that

$$E\ \Delta\bar{x} = 0, \quad \text{by definition of } \Delta\bar{x}$$

$$E\ (\Delta\bar{x})^2 = \frac{\sigma^2}{n}, \quad \text{the 2d moment coefficient of } \bar{x} \quad \text{[From Eq. 20,}$$
$$\text{page 383]}$$

$$E\ (\Delta\bar{x})^3 = \frac{1}{n^2}\mu_3, \quad \text{the 3d moment coefficient of } \bar{x} \tag{29}$$

* An excellent reference to the calculation of the expected values required here is Oskar Anderson's *Probleme der Statistischen Methodenlehre* (Physica Verlag, Würzburg, 1957): Ch. VI. Also Harald Cramér, *Mathematical Methods o f Statistics* (Princeton, 1946): pp. 345–347.

$$E\,(\Delta\bar{x})^4 = \frac{3\sigma^4}{n^2} + \frac{1}{n^3}\,(\mu_4 - 3\sigma^4), \qquad \text{the 4th moment coefficient of } \bar{x} \;(30)$$

These are results that we referred to in Chapter 16.   The factor $1/n^2$ in the expected values of the 3d and 4th moments of $\bar{x}$ justifies neglect of the terms 3d and 4th moments of $\Delta\bar{x}$ in Eq. 31 in Chapter 16 on page 369.   In general, $E\,(\Delta\bar{x})^{2k-1}$ and $E\,(\Delta\bar{x})^{2k}$ both commence with $1/n^k$ (Cramér, p. 346).

If the skewness of the distribution of $\bar{x}$ be defined as proportional to $E\,(\Delta\bar{x})^3$ or to $E\,(\bar{x} - E\bar{x})^3$, then the factor $1/n^2$ indicates a rapid decrease in the skewness of $\bar{x}$ as the size n of the sample increases.   (See page 443 in the author's *Some Theory of Sampling* (Wiley, 1950) for a pictorial illustration.)

As a corollary,

$$\beta_2(\bar{x}) = \frac{E(\bar{x} - E\bar{x})^4}{\sigma^4/n^2} = \frac{E\,(\Delta\bar{x})^4}{\sigma^4/n^2}$$

$$= 3 + \frac{1}{n}\,(\beta_2 - 3) \tag{31}$$

where β_2 has the definition in Eq. 18.

This corollary shows that $\beta_2(\bar{x})$ approaches rapidly the normal value 3 as n increases, no matter what be the original β_2 of the frame.   High values of β_2 in the frame do not therefore produce high values of $\beta_2(\bar{x})$ if n is large enough.

5.   Let $F = x_1 + x_2 + \cdots + x_n$, where x_i is a random variable with mean a_i and variance σ_i^2.   Assume that the x_i are independent.   Let M_j denote the jth moment coefficient of F, and let μ_{ij} denote the jth moment coefficient of x_i.   All sums (Σ) run from 1 to n.   Show by simple extension of the results of the preceding exercise, or otherwise, that

$$M_1 = \Sigma\mu_{i1}$$
$$M_2 = \Sigma\mu_{i2}$$
$$M_3 = \Sigma\mu_{i3}$$
$$M_4 - 3M_2{}^2 = \Sigma(\mu_{i4} - 3\mu_{i2}{}^2)$$

The left-hand sides of the above equations are known as the *cumulants* of F.   The terms under the summations on the right are the cumulants of the x_i.   If K_j denotes the jth cumulant of F, and if κ_{ji} denotes the jth cumulant of x_i, then

$$K_j = \Sigma\,\kappa_{ji} \tag{32}$$

The importance of cumulants lies in this simple additive property.   Cumulants were discovered by Thiele,* and called by him *half-invariants*.   We shall use the cumulants in Exercise 11.

---

* T. N. Thiele, *The Theory of Observations* (Copenhagen, 1889); reprinted in *Ann. Math. Statist.*, vol. ii, May 1931.
See also M. G. Kendall, *The Advanced Theory of Statistics* (Griffin, 1945): pp. 61–64.

6. Show that for the binomial

$$\beta_2 = \frac{1}{pq} - 3$$

The mean of the binomial is at p, and the proportions of 0 and 1 are q and p, hence

$$\beta_2 = \{q(0 - p)^4 + p(1 - p)^4\} \div p^2 q^2$$

The rest is algebraic reduction.   *Q.E.D.*

Note that β_2 may attain very high values when p is small, in which case the binomial is badly skewed.   However, as we know from Eq. 31, $\beta_2(\hat{p})$ nevertheless approaches the normal value of 3 as n increases.

7. Show that for the sampling plan described in Exercise 4,

$$\text{Var } s^2 = \sigma^4 \left\{ \frac{\beta_2 - 1}{n} - 2\frac{\beta_2 - 2}{n^2} + \frac{\beta_2 - 3}{n^3} \right\} \tag{33}$$

$$\doteq \sigma^4 \frac{\beta_2 - 1}{n} \qquad \text{[For large } n\text{]}$$

$$= \frac{2\sigma^4}{n} \qquad \text{[For any } n \text{ if } \beta_2 = 3\text{]} \tag{34}$$

where s^2 and β^2 are defined as heretofore.   This result applies to samples drawn with replacement.   For a proof, the reader may wish to consult the reference in Remark 7, or the author's *Some Theory of Sampling* (Wiley, 1950): p. 339.

8. The variance of an estimate of a variance.   Define $\hat{\sigma}^2 = [n/(n - 1)]s^2$ as in Eq. 3, page 433.   Show that

$$\text{Var } \hat{\sigma}^2 = \left(\frac{n}{n-1}\right)^2 \text{Var } s^2 = \frac{2\sigma^4}{n-1}\left\{1 + \frac{n-1}{2n}(\beta_2 - 3)\right\} \tag{35}$$

$$\doteq \sigma^4 \frac{\beta_2 - 1}{n} \qquad \text{[For large } n\text{]}$$

$$= \frac{2\sigma^4}{n-1} \qquad \text{[For any } n \text{ if } \beta_2 = 3\text{]} \tag{36}$$

Note: The last form is exact for a normal frame, for which $\beta_2 = 3$.

Remark 7.   The arrangement of terms in Eq. 35 is Cochran's, as I mentioned on page 440.   The student may wish to transform Var $\hat{\sigma}^2$ into the alternate form

$$\text{Var } \hat{\sigma}^2 = \frac{\sigma^4}{n}\left\{\beta_2 - \frac{n-3}{n-1}\right\} \tag{37}$$

as given by Hansen, Hurwitz, and Madow, *Sample Survey Methods and Theory* (Wiley, 1953): Vol. I, pp. 134 and 427; Vol. II, p. 101; also by

Oskar Anderson, *Probleme der Statistischen Methodenlehre* (Physica Verlag, Würzburg, 1957): p. 285. Both books extend the formula to sampling without replacement.

9. The rel-variance of an estimate of a variance. Show that for the sampling plan described in Exercise 4,

$$C_{\hat{\sigma}^2}^2 = C_{s^2}^2 = \frac{2}{n-1}\left\{1 + \frac{n-1}{2n}(\beta_2 - 3)\right\} \tag{38}$$

$$
\left.
\begin{aligned}
&\doteq \frac{\beta_2 - 1}{n} && \text{[For large } n\text{]} \\[2ex]
&= \frac{2}{n-1} && \text{[For any } n \text{ if } \beta_2 = 3\text{]}
\end{aligned}
\right\} \tag{39}
$$

(*Hint:* divide Var $\hat{\sigma}^2$ in the last exercise by $(E\hat{\sigma}^2)^2$ or σ^4.)

10. Show that for any sampling plan,

$$C_{\hat{\sigma}_X} = C_{\hat{\sigma}_{\bar{x}}} = C_s \doteq \tfrac{1}{2}C_s^2 = \tfrac{1}{2}C_{\hat{\sigma}^2} \tag{40}$$

This relation comes from Table 1 on page 394, which shows that $C_{\sqrt{x}} \doteq \tfrac{1}{2}C_x$.

11. Derive Eq. 20, page 440.

SOLUTION (ZINDLER)

The population x_{jr} in Subsample r in Thick Zone j is built up by a random drawing from each of the c_j thin zones that constitute Thick Zone j. By the theorem of cumulants in Exercise 5, Var x_{jr} is the sum of the c_j individual variances. That is,

$$\sigma_j^2 = \text{Var } x_{jr} = \kappa_2(x_{jr}) = \sum_{i=1}^{c_j} \kappa_{2i} = \sum_{i=1}^{c_j} \sigma_i^2 \tag{41}$$

as $\kappa_{2i} = \sigma_i^2$, being the variance in Thin Zone i in Thick Zone j. Also, for the 4th cumulant of x_{jr},

$$\kappa_4(x_{jr}) = \sum_{i=1}^{c_j} \kappa_{4i} \tag{42}$$

Now set $x_j = \sum_{r=1}^{k} x_{jr}$ for the total population in all k subsamples combined in Thick Zone j, and set $\bar{x}_j = x_j/k$ for the average population per subsample. An estimate of Var x_j is

$$\hat{\sigma}_j^2 = \frac{1}{k-1} \sum_{r=1}^{k} (x_{jr} - \bar{x}_j)^2 \tag{43}$$

which comes from Eq. 4 on page 433.   (I am omitting for convenience the finite multipliers.)   Now for the variance of $\hat{\sigma}_j{}^2$ we turn to Eq. 35 and remember that σ^4 there will now be $(\sigma_j{}^2)^2 = \left(\sum_{i=1}^{c_j} \sigma_i{}^2 \right)^2$, and that $\sigma^4(\beta_2 - 3)$ will be $\kappa_4 (x_{jr})$.   Then

$$\text{Var } \hat{\sigma}_j{}^2 = \frac{2}{k-1} \left[\sum_{i=1}^{c_j} \sigma_i{}^2 \right]^2 + \frac{1}{k} \sum_{j=1}^{c_j} \kappa_{4i} \tag{44}$$

Now let $x = \sum_{j=1}^{m} x_j$ be the total population in all subsamples through all thick zones.   Again by the addition of variances,

$$\sigma_x{}^2 = \sum_{j=1}^{m} \sigma_j{}^2 \tag{45}$$

Moreover, our estimate of $\sigma_x{}^2$ is the sum of the m individual estimates $\hat{\sigma}_j{}^2$, one for each thick zone.   That is,

$$\hat{\sigma}_x{}^2 = \sum_{j=1}^{m} \hat{\sigma}_j{}^2 \tag{46}$$

hence again by the addition of variances

$$\text{Var } \hat{\sigma}_x{}^2 = \text{Var} \sum_{j=1}^{m} \hat{\sigma}_j{}^2 = \sum_{j=1}^{m} \text{Var } \hat{\sigma}_j{}^2$$

$$= \frac{2}{k-1} \sum_{j=1}^{m} \left[\sum_{i=1}^{c_j} \sigma_i{}^2 \right]^2 + \frac{1}{k} \sum_{j=1}^{m} \sum_{i=1}^{c} \kappa_{4i}$$

$$= \frac{2}{k-1} \sum_{j=1}^{m} \left[\sum_{j=1}^{c_j} \sigma_i{}^2 \right]^2 + \frac{1}{k} \sum_{i=1}^{M} \kappa_{4i} \tag{47}$$

Now for the rel-variance of $\hat{\sigma}_x{}^2$, we divide Var $\hat{\sigma}_x{}^2$ by $(E\hat{\sigma}_x{}^2)^2 = (\sigma_x{}^2)^2 = \left[\sum_{i=1}^{M} \sigma_i{}^2 \right]^2$.   Moreover, the rel-variance of $\hat{\sigma}_x{}^2$ is the same as the rel-variance of $\hat{\sigma}_X{}^2$, where $X = (1/k)Zx$ as in Chapter 11; whereupon we have derived Eq. 20, page 440.

                                                                        Q.E.D.

Standard error of a ratio in fine classes.   A further theorem published by Keyfitz provides a simple calculation to estimate the variance of a ratio obtained from fine classes.

Suppose that we break up the data from a survey on the characteristics of the labor force into 7 age groups and 2 sexes, 14 classes in all, for which we may use the index $j = 1, 2, \ldots, 14$.   Let the index i designate the stratum: $i = 1, 2, \ldots, 8$ perhaps, as it might be for 4 census regions, in and out of metropolitan areas within a region.   Let w_j be the weight of the *j*th age-sex class as a fraction of the total adult population for the whole of Canada, as known from Census data, and from figures on births, deaths, and immigration.   Put

$$f_j = \frac{\sum\limits_i (x_{ij1} + x_{ij2})}{\sum\limits_i (y_{ij1} + y_{ij2})} \tag{48}$$

for the estimated ratio of the unemployed (the *x*-population) in Age-Sex Class *j* to the total number of people (the *y*-population) in that class, employed and unemployed.   The figures 1 and 2 in the subscripts refer to the 2 subsamples.   As Keyfitz says, we may think of 14 different surveys, each one to determine one of the f_j.   The weighted average f_j will be

$$f = \sum_i \sum_j w_j f_j \tag{49}$$

which is the estimated average ratio of unemployment in the whole country.   Introduce now the symbol

$$h_{ij} = (x_{ij1} - x_{ij2}) - f_j(y_{ij1} - y_{ij2}) \tag{50}$$

Then the simple formula

$$\mathrm{Var}\, f \doteq \sum_i \{\sum_j w_j h_{ij}\}^2 \qquad \text{[Compare with Eq. 12, p. 198.]} \tag{51}$$

estimates the variance of the ratio *f*.   This beautiful theorem evaluates automatically the correlated terms that arise between the random sample-sizes of the age-sex classes.   (The correlation exists because if one age-sex class is too big in the sample, another class will usually also be too big.) The proof of Eq. 51 depends on another elegant theorem due to Keyfitz. If x_1 and x_2 are random variables as a result of drawings with replacement and with equal probabilities, and y_1 and y_2 likewise, then

$$\mathrm{Var}\,(x_1 + x_2) = E(x_1 - x_2)^2 \tag{52}$$

$$\mathrm{Cov}\,(x_1 + x_2)(y_1 + y_2) = E(x_1 - x_2)(y_1 - y_2) \tag{53}$$

whereupon the approximation

$$C_{x/y}^2 \doteq C_{\bar{x}}^2 + C_{\bar{y}}^2 - 2C_{\bar{x}\bar{y}} \qquad \text{[P. 393]}$$

reduces to

$$C_{x/y}^2 \doteq E\left\{\frac{x_1 - x_2}{E(x_1 + x_2)} - \frac{y_1 - y_2}{E(y_1 + y_2)}\right\}^2 \tag{54}$$

x/*y* is here $(x_1 + x_2)/(y_1 + y_2)$.   Extension to the overall weighted *f* is fairly obvious.

Numerical example to estimate the variance of a variance. Exercises 8 and 9 on pages 445 and 446 gave formulas that will enable us to understand the effect of β_2 on the estimate of a variance. Eq. 28 in Exercise 4 gave a simple formula, due to Keyfitz, which estimates the variance of the estimate of a variance, based on the results of the sample itself, without the intermediate task of estimating β_2. We now apply Eq. 28 to the table of results on page 184, for the number of dwelling units in 2 subsamples. We consolidate successive pairs of zones, and assume that thick zone 1 furnishes the random variables x_1 and x_2, that thick zone 2 furnishes x_3 and x_4, that thick zones 3 and 4 furnish another set of x_1, x_2, x_3, x_4, etc.; and thus we construct Table 12 (next page). The numerical values in the table give

$$\hat{C}_{\bar{x}} = \frac{\sqrt{3525 + 3706}}{651 + 624 + 595 + 591} = .034 \qquad \text{[As on page 184]}$$

$$\hat{C}^2_{\hat{\text{Var}}\,\bar{x}} = \frac{\Sigma\{(x_1 - x_2)^2 - (x_3 - x_4)^2\}^2}{[\Sigma\{(x_1 - x_2)^2 + (x_3 - x_4)^2\}]^2} \qquad \text{[Eq. 28, Keyfitz]}$$

$$= \frac{9,489,253}{(3525 + 3706)^2} = .251 \qquad (60)$$

whence

$$\hat{C}_{\hat{\sigma}_{\bar{x}}} = \tfrac{1}{2}\sqrt{.251} = .25 \qquad (61)$$

We note that $C_{\hat{\sigma}_{\bar{x}}}$ would be $1/\sqrt{2 \times 16} = 17.8\%$ if the quantities x_1, x_2, x_3, x_4 in the thick zones were distributed normally. We may attribute the difference to non-normality, although some of the difference could arise from sampling error in the estimate of $C_{\hat{\sigma}_{\bar{x}}}$.

Eq. 28 is not restricted to 2 subsamples. Thus, for 10 subsamples, one may draw, from any thick zone, 4 of the 10 sample-values for a set of x_1, x_2, x_3, x_4, with a further drawing for another set from the same zone, if desired.

For the effect of skewness on estimates of the coefficient of variation, see Hansen, Hurwitz, and Madow, *Sample Survey Methods and Theory*, Vol. I (Wiley, 1953), p. 429.

The reader may wish to show from Eq. 21, page 440, that the average value of β_2 for the distribution of the random variables x_1, x_2, x_3, x_4 in the above example is 7.

Calculation of an error band, or of a margin of error.* Any sample that we draw is 1 sample from the equal complete coverage; and if we had more time, we could in principle at least draw many more samples—

* The term margin of error for a prescribed probability originated with Mahalenobis, about 1936.

usually hundreds or thousands from the same complete coverage.*   Let a be the result of the complete coverage for some specified characteristic, and let $\bar{x}$ denote the random estimate of a.   In repeated samples (assuming normality), about 33 % of the results of the samples will lie beyond $a \pm \sigma_{\bar{x}}$; only about 4 % of them will lie beyond $a \pm 2\sigma_{\bar{x}}$; and practically none will fall beyond $a \pm 3\sigma_{\bar{x}}$.   Other percentages for a normal distribution appear in Table 13 (next page).   Fig. 27 defines P_u.

The interval $a \pm 3\sigma_{\bar{x}}$ is called the 3-sigma error band, or the 3-sigma margin of error.   As we learned in Chapter 4, this band represents the

TABLE 12

(All figures but those in the last column come from page 184)

Thick zones	x_1	x_2	x_3	x_4	$(x_1 - x_2)^2$	$(x_3 - x_4)^2$	$\{(x_1 - x_2)^2 - (x_3 - x_4)^2\}^2$
1 and 2	65	93	81	81	784	0	614 656
3 and 4	84	80	119	74	16	2025	4 036 081
5 and 6	83	62	78	87	441	81	129 600
7 and 8	69	82	75	65	169	100	4 761
9 and 10	88	71	69	75	289	36	64 009
11 and 12	74	89	75	77	225	4	48 841
13 and 14	89	88	74	112	1	1444	2 082 249
15 and 16	99	59	24	20	1600	16	2 509 056
Sum	651	624	595	591	3525	3706	9 489 253

practical limit of variation from repeated samples.   We may also form other error bands or margins of error, for any probability desired, and on one side only, if we wish (see Remark 9, page 453), such as 2-sigma above, or 3-sigma below, or 2.33-sigma above (for 1 % probability).

For testing, as when one wishes to be sure not to overlook any source of trouble (e.g., in comparing interviewers or inspectors, or in comparing the results of the audit with the main sample), it may pay to use 2-sigma limits.   But for estimation, the aim of this book, management and the expert in the subject-matter need a figure to use, and they need to know how far it may differ from the equal complete coverage.   The 3-sigma limits answer this question (p. 55).

In practice, we often know not σ, but the exciting thing about modern statistical theory is that any one sample with enough degrees of freedom (p. 433) will provide a good enough estimate of the width of an error

* A plot of 100 results appears on page 563 in the author's, *Some Theory of Sampling* (Wiley, 1950).

band.   When we estimate σ from a sample, we thereby estimate the
approximate width of an error band, or margin of error.

An estimate of σ is a random variable, as we have learned, and we must
often be satisfied with a fairly small number of degrees of freedom (such
as 9) in this estimate.   It so happens, however, that in practice, one hardly
ever need know a probability exactly.   Few conclusions will change if a

TABLE 13

THE NORMAL INTEGRAL

u (in units of the standard deviation)	P_u	$2P_u$
0	0.5	1.00
0.6745	.25	0.50
1	.16	.32
1.28	.10	.20
1.64	.05	.10
1.96	.025	.05
2	.02	.04
2.33	.01	.02
3	Practically 0	

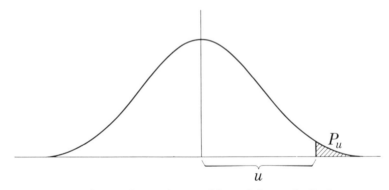

Fig. 27.   Showing P_u, the area in one tail beyond the standardized error u.

probability is incorrectly calculated as .04 when it should be .03 or .05;
or if a probability is incorrectly calculated as .001 when it should be .0005
or .003.

Another important point is that in any case the probabilities fall off
with extreme rapidity as we increase the multiple of σ.   Table 13
illustrates this point for a normal distribution, but the same principle
holds for other sampling distributions.   Thus, an adjustment from

3 to 3.2 or 3.3 will take care of a low number of degrees of freedom and a considerable amount of skewness as well.*

Confidence intervals. There is another way to relate the result of a sampling procedure to the result of the equal complete coverage, which is especially useful in testing hypotheses. Suppose that we enquire how to compute from a sample an upper limit $\bar{x}_U$ such that the inequality

$$\bar{x}_U > a \tag{55}$$

will fail in only 1 % of a long succession of samples from an equal complete count whose mean is a. If we knew σ, and hence $\sigma_{\bar{x}}$, and if the distribution of $\bar{x}$ were normal, the calculation would be simple: we could merely set

$$\bar{x}_U = \bar{x} + 2.33\sigma_{\bar{x}} \tag{56}$$

and then write the inequality $\bar{x}_U > a$. We may speak of the interval from $\bar{x}$ to $\bar{x} + 2.33\sigma_{\bar{x}}$ as a confidence interval for the probability 1 % in the lower tail. The figure 2.33 comes from the simple table of the normal integral on page 451.

The student should satisfy himself that if $\bar{x}_U$ be calculated for each of a long succession of samples all drawn from the same equal complete count, and all processed by the same sampling procedure, then (a) the distribution of the random variable $\bar{x}_U$ will have exactly the same shape as the distribution of $\bar{x}$; (b) its standard deviation will be $\sigma_{\bar{x}}$; (c) if the distribution of $\bar{x}$ be normal, the mean of the distribution of $\bar{x}_U$ will be $a + 2.33\sigma_{\bar{x}}$; and (d) only 1 % of the lower tail of the distribution of $\bar{x}_U$ will fall below a.

We could in like manner calculate the random variable

$$\bar{x}_L = \bar{x} - 2.33\sigma_{\bar{x}} \tag{57}$$

for each sample and assert that if the distribution of $\bar{x}$ be normal, the inequality

$$\bar{x}_L > a \tag{58}$$

* An excellent and concise treatment of the effect of departures from the normal on probabilities occurs in Owen L. Davies, *Design and Analysis of Industrial Experiments* (Oliver and Boyd, 1956): pp. 51–56.

The original reference is to A. K. Gayen, "The distribution of Student's *t* from non-normal universes," *Biometrika*, vol. 36, 1949: p. 353; "The distribution of the variance-ratio from non-normal universes," *Biometrika*, vol. 37, 1950: p. 236; "Significance of difference between the means of two non-normal samples," *Biometrika*, vol. 37, 1950: p. 399.

See also G. E. P. Box, "Non-normality and tests on variances," *Biometrika*, vol. 40, 1953: p. 318.

will fail in only 1% of the samples, and that the inequalities

$$\bar{x}_L < a < \bar{x}_U \qquad (59)$$

will fail in only 2% of the samples.*

Remark 8. The probability of failure of the inequalities that we have been dealing with remains the same even if a and n vary in any manner from sample to sample. We may thus use these inequalities in one experience after another, being sure that in all this experience the expected proportion of failures of the inequalities are precisely as stated. We may even select a random sample from our experience, with no change in the probabilities associated with the inequalities. The probability would of course be vitiated if we were to select only those experiences in which the inequalities look suspiciously like a failure, or are almost certain to be fulfilled. This would be like looking to see if we have heads or tails, and then calling it; the game would be unfair (i.e., predictable).

Remark 9. Confidence limits have some utility in problems of estimation, if we interpret them carefully. Thus, we may say that a "99% safe" upper limit for the mean a of the equal complete coverage is $\bar{x} + 2.33\,\sigma_{\bar{x}}$. If the mean a of the equal complete coverage were above $\bar{x} + 2.33\sigma_{\bar{x}}$, then fewer than 1% of the means of samples would be as low as the observed $\bar{x}$. Similarly, the "99% safe" lower limit for the mean a of the equal complete coverage is $\bar{x} - 2.33\sigma_{\bar{x}}$.

Remark 10. The 3-sigma interval $\bar{x} \pm 3\sigma_{\bar{x}}$ contains the range of possible admissible alternatives for the mean a of the equal complete coverage. See calculations on page 466.

As we know well, we must often use an estimate of $\sigma_{\bar{x}}$ derived from the sample. Not only $\bar{x}$, but $\hat{\sigma}_{\bar{x}}$ also, will then be a random variable. When there are only a few degrees of freedom in the estimate of $\sigma_{\bar{x}}$, we may wish to enlarge the factor in our confidence interval to preserve a specified area (such as .01) in the tail of the distribution of $\bar{x}_U$ or of $\bar{x}_L$. This is easy to do with the factors in Table 14, which take account of the skewness and of the additional variance contributed by $\hat{\sigma}_{\bar{x}}$.

The difference between the factors in the table and those on the bottom line (which correspond to $\sigma_{\bar{x}}$ known) is obviously not of great importance unless the number of degrees of freedom is very small.

* This formulation, commonly called confidence intervals (after Neyman and Pearson), was invented by Stanislos Millot, "Sur la probabilité à posteriori," *Comptes rendus*, Paris, vol. 176, 1923: p. 30; *Théorie nouvelle de la probabilité des causes* (Gauthier-Villars, Paris, 1925). An excellent treatment appears in Oskar Anderson's *Einführung in die mathematische Statistik* (Julius Springer, Wien, 1935): pp. 114–116. I am indebted to Professor A. Hald of Copenhagen for the references to Millot and to Anderson.

TABLE 14

VALUES OF t BEYOND WHICH LIE 2.5%, 1%, AND PRACTICALLY NONE OF
THE DISTRIBUTION OF t, FOR A FEW SELECTED DEGREES OF FREEDOM,
AND FOR SAMPLES DRAWN FROM A NORMALLY DISTRIBUTED FRAME

Degrees of freedom	Probability		
	2.5%	1%	Practically 0
5	2.57	3.36	5.5
9	2.26	2.82	4.1
18	2.10	2.55	3.5
30	2.04	2.46	3.3
∞	1.96	2.33	3 (normal values)

Remark 11. The factors in the table are valid for samples drawn from a normally distributed frame. They came from Pearson and Hartley's *Biometrika Table for Statisticians* (Cambridge University Press, 1954), vol. I, Table 12, with interpolation for the factors in the last column. The distribution of t was first published by Student, with short tables.*

* The distribution of t was worked out first by Student, "On the probable error of a mean," *Biometrika*, vol. vi, 1908: pp. 1–25; *ibid.*, vol. xi, 1915–17: pp. 414–417. Birge once called my attention to the fact that Student nowhere in these papers mentioned the probable error of the mean. Student's papers are now available in *Student's Collected Papers*, edited by E. S. Pearson and John Wishart (Office of Biometrika, University College, London, W.C.1; 1942).

CHAPTER 18

The Poisson Series
and the Square-Root Transformation

The greatest achievements are those not accomplished.—Ralph Roeder, *Juarez and His Mexico* (Viking, 1947); page 179.

Examples of the Poisson series. The Poisson series applies to problems in sampling when a sampling unit may contain any number of people or other subunits of a specified characteristic, and in which the occurrence of these characteristics is independent within a sampling unit. Rare characteristics frequently follow closely the Poisson series. For example, Lester Frankel and J. Stevens Stock* noted in 1937 that the number of vacant dwelling units per block or for any large area of nearly constant size distributed themselves pretty closely on the Poisson series. The number of people per segment of area, in some specific age group (20–29) that follow some specific occupation will usually be closely Poisson within strata. The number of flaws in a test-piece of a square meter of finish or of cloth follows the same distribution: departures signify a special cause which can perhaps be identified and removed. Feller† mentions the number of misprints on a page, and the number of raisins in a loaf of bread, which will also be Poisson (provided that there are not too many raisins, he should add). The number of genes destroyed per 100 seeds in the irradiation of seeds; the number of twins or of triplets born per year in an area that contains (e.g.) 100,000 people; the number of accidents over a period of time; the number of errors per set of punched cards or of ledgers or of entries of any kind are further examples. Arrivals for service, or the demand per day or per week for a product, often follow the Poisson series, although the random demand may be superimposed on

---

* Private communication, about 1938.

† William Feller, *An Introduction to Probability Theory and Its Applications*, Vol. 1, 2d ed., (Wiley, 1957): pp. 111–114.

455

a constant or seasonal demand.   Radioactive disintegration is Poisson. Demand for service on a line or other piece of apparatus in a telephone office over a test-period of time is often Poisson superimposed on a diurnal cycle.

If one event increases or diminishes the chance of another, the distribution will depart more or less from Poisson.   Thus, the distribution of the number of accidents per week from some set of causes will be Poisson if the accidents are independent, but the distribution will not be Poisson if several accidents come from 1 explosion, or if people are extra-careful for the week immediately following an accident.

Departures from the Poisson distribution (recognized with the aid of statistical tests such as the Shewhart charts, or tests of significance) signify the existence of special or additional causes.   Extra-difficult passages in printing may cause misprints to occur in bunches, and the distribution will depart from Poisson (Feller).

Besides its utility in sampling for rare and independent characteristics, the Poisson series solves expeditiously much work in comparisons of medical treatments, in the incidence of disease, in comparisons of machines, methods, and product, and in comparisons of sales (see the exercises further on).   Numerical work in the calculation of the most economical inventory is largely use of the Poisson series.*

One of the important uses of the Poisson series is to approximate the binomial distribution.†   The Poisson series also approximates the hypergeometric series (p. 406) under certain conditions.   Many of the operating characteristics of acceptance sampling plans have been calculated by the approximation afforded by the Poisson series.

A probability mechanism to generate the Poisson series.   A square metre of painted surface may have 0 or 1 or more flaws.   A square metre of cloth may have 0 or 1 or more flaws.   A lot of 10,000 items may have 0 or 1 or more defective items.   A lot of 100,000 seeds, when irradiated by X-rays, may show 0 or 1 or more seeds affected (flaws).   Repetitions of the process of painting, manufacturing, or irradiating will produce varying numbers of flaws—0, 1, 2, 3, or more.   The number of flaws may in fact reach a very high value.   Why is the number of flaws ever 0? Why is it not always 1000?   We suppose that there is a mechanism whereby each flaw that happened might not have happened, and whereby each flaw that did not happen might have happened.   Look at 4 flaws: X X X X.   If we delete any one of them by 0, we have 3 flaws, and there

---

* W. Edwards Deming, *Some Theory of Sampling* (Wiley, 1950): p. 421.

† See the reference to Morton S. Raff on page 407; see also Churchill Eisenhart, "Inverse sine transformations of proportions," being Ch. 16 in the book by Eisenhart, Hastay, Wallis, et al., *Techniques of Statistical Analysis* (McGraw-Hill, 1947).

are 4 ways to delete 1 of the 4:

$$
\begin{array}{cccc}
0 & X & X & X \\
X & 0 & X & X \\
X & X & 0 & X \\
X & X & X & 0
\end{array}
$$

A test-panel with r flaws and 0 more is a test-panel with $r - 1$ flaws plus 1. There are r ways in which the 1 additional flaw may take its place amongst the $r - 1$ flaws. Let $P(r)$ be the probability that there will be r flaws. Then if there is complete independence from one flaw to another, what we have observed leads us to say that

$$P(r - 1)P(1) = rP(r) P(0) \tag{1}$$

If

$$m = \frac{P(1)}{P(0)} \tag{2}$$

then Eq. 1 will take the form

$$P(r) = \frac{m}{r} P(r - 1) \tag{3}$$

This is a recursion formula that will generate the Poisson series of probabilities.

$P(0) = P(0)$ The probability of 0 flaws

$r = 1, \quad P(1) = \dfrac{m}{1} P(0)$,,   ,,   ,, 1 ,,

$r = 2, \quad P(2) = \dfrac{m}{2} P(1) = \dfrac{m^2}{1.2} P(0)$,,   ,,   ,, 2 ,,

$r = 3, \quad P(3) = \dfrac{m}{3} P(2) = \dfrac{m^3}{1.2.3.} P(0)$,,   ,,   ,, 3 ,,

.

.

.

$r = r, \quad P(r) = \dfrac{m^r}{r!} P(0)$,,   ,,   ,, r ,,

.

.

.

The limit of the sum of these terms to infinity must be 1; that is,

$$P(0)\left\{1 + m + \frac{m^2}{2!} + \frac{m^3}{3!} + \cdots\right\} = 1 \qquad (4)$$

because r must take some one of the values 0, 1, 2, 3, ...

Now the series in the braces has a definite limit for any value of m. This limit is known by the symbol e^m. Hence

$$P(0) = e^{-m} \qquad (5)$$

and the Poisson series of probabilities may be written in the following form

$$r = 0, \quad 1, \quad 2, \quad 3, \quad \ldots, \quad r, \quad \ldots$$

$$e^{-m}\left\{1, \quad m, \quad \frac{m^2}{2!}, \quad \frac{m^3}{3!}, \ldots, \quad \frac{m^r}{r!}, \cdots\right\}$$

or shortly

$$P(r) = \frac{m^r}{r!} e^{-m} \qquad (6)$$

first invented and used by Poisson.*

The symbol m, as the student will prove in Exercise 1 ahead, is the weighted average value of r, the weight of r being $P(r)$. In other words, m is the expected value of the series.

Remark 1. Table 1 shows 4 comparisons between Poisson terms for $m = np$, and binomial terms for $p = \frac{1}{20}$, $n = 10, 20, 50, 100$. The comparison is remarkably good, though what appears to be good term by term is often not so good when one requires the probability in a tail of the distribution. One should remember, though, that in tests of significance, if one series leads to recognition of significance, the other series may also, even though the absolute probabilities reckoned by the 2 series may differ by 2- or 3-fold (e.g., .01 vs. .005).

Remark 2. There are tables of the Poisson series that show the terms for $r = 0, 1, 2, 3$, etc., for various values of m; also the sum of all the terms from some specified value of r on out to infinity. Some useful tables are these:

E. C. Molina, *Poisson's Exponential Binomial Limit* (Van Nostrand, 1942).

E. S. Pearson and H. O. Hartley, Biometrika *Tables for Statisticians*, Vol. 1 (Cambridge University Press, 1954), Tables 39 and 40.

Burington and May, *Handbook of Probability and Statistics with Tables* (Handbook Publishers, Sandusky, Ohio, 1953): Table VII (p. 259), the Poisson function $P(r)$; Table VIII (p. 263), sums of the Poisson function.

* Poisson, *Récherches sur la Probabilité des Jugements* (Paris, 1837): p. 206.

TABLE 1

COMPARISON OF THE POISSON WITH THE BINOMIAL

$$p = .05, q = .95 \text{ in the binomial } B = \binom{n}{r} q^{n-r} p^r$$

$$m = np \text{ in the Poisson term } P = e^{-m} m^r / r!$$

r	B $n = 10$	P $m = .5$	B $n = 20$	P $m = 1$	B $n = 50$	P $m = 2.5$	B $n = 100$	P $m = 5$
0	.598 737	.606 531	.358 486	.367 879	.076 945	.082 085	.005 921	.006 738
1	.315 125	.303 265	.377 353	.367 879	.202 487	.205 212	.031 161	.033 690
2	.074 635	.075 816	.188 677	.183 940	.261 101	.256 516	.081 182	.084 224
3	.010 475	.012 636	.059 582	.061 313	.219 875	.213 763	.139 576	.140 374
4	965	.001 580	.013 328	.015 328	.135 975	.133 602	.178 142	.175 467
5	61	158	.002 245	.003 066	.065 841	.066 801	.180 018	.175 467
6	2	13	295	511	.025 990	.027 834	.150 015	.146 223
7		1	031	73	.008 598	9 941	.106 025	.104 445
8			003	9	2 432	3 106	.064 871	.065 278
9				1	597	863	.034 901	.036 266
10					129	216	.016 716	.018 133
11					25	49	7 198	8 242
12					4	10	2 810	3 434
13					1	2	1 001	1 321
14							327	472
15							99	157
16							28	49
17							7	14
18							2	4
19								1
20								

The calculations of the binomial were made by my friend Mr. B. R. Stauber in 1934, in connexion with a problem in the sampling of land. The Poisson terms came from Molina's tables.

EXERCISES

1. Show that m in Eq. 4 is the mean of the Poisson series.

SOLUTION

By definition, the mean of the series is its expected value

$$Er = \Sigma r P(r)$$
$$= m \Sigma P(r - 1) = m \qquad (7)$$

This last step follows from the fact that the sum of all probabilities is 1. Interpret $P(-1)$ as 0.

2. Show that (*a*) the variance of the Poisson series is *m*; (*b*) its standard deviation is $\sqrt{m}$; (*c*) its coefficient of variation is $1/\sqrt{m}$.

SOLUTION

From Eq. 3,

$$r(r - 1)P(r) = m(r - 1)P(r - 1) \tag{8}$$

whence

$$Er(r - 1) = \Sigma r(r - 1)P(r) = m\Sigma(r - 1)P(r - 1) = m^2 \tag{9}$$

because $\Sigma(r - 1)P(r - 1) = \Sigma rP(r)$, either of which is *Er* and is *m*. By definition, the variances of the Poisson series will be

$$\sigma^2 = E(r - Er)^2 = Er^2 - (Er)^2$$
$$= E[r(r - 1) + r] - (Er)^2$$
$$= m^2 + m - m^2 = m \tag{10}$$

Q.E.D.

3. Sum of 2 Poisson variables.   Let *r* and *s* be 2 independent Poisson variables, having means *m* and *n* respectively.   Then their sum is a Poisson variable with mean *m* + *n* and variance *m* + *n*.

SOLUTION

The simultaneous distribution of *r* and *s* is

$$P(r, s) = e^{-r}\frac{m^r}{r!} e^{-s}\frac{n^s}{s!} \quad \text{[From Eq. 6]} \tag{11}$$

Let *t* = *r* + *s*.   The distribution of *t* for any *r* and *s* whose sum is *t* will be

$$P(t) = e^{-t}\sum_{r,s}\frac{m^r n^s}{r!s!} \quad [r \text{ and } s \text{ run from 0 to } \infty]$$
$$= \frac{e^{-t}}{t!}\sum_{r=0}^{t}\binom{t}{r} m^r n^{t-r}$$
$$= e^{-t}\frac{(m + n)^t}{t!} \tag{12}$$

Q.E.D.

The above theorem extends itself easily to 3 or more Poisson variables.

4. Let *r* and *s* be 2 independent Poisson variates, having means *m* and *n*.   Then the linear function *ar* + *bs* is a random variable with expected value *am* + *bn* and with variance $a^2m + b^2n$.   *a* and *b* are any constants. In particular, the mean of *r* − *s* is *m* − *n* and its variance is *m* + *n*.

As the mean and the variance of *ar* + *bs* are equal only if *a* = *b* = 1, *ar* + *bs* is not a Poisson variable unless *a* = *b* = 1, when it degenerates to *r* + *s*,

The distribution of $ar + bs$, hence of $r - s$ and of $r + s$, will be nearly normal if m and n are separately big enough, which for most purposes means 10 or more.

Quick calculation of the size of sample required to estimate a Poisson variate.   Suppose that a sample is to be drawn to estimate some characteristic (x) whose distribution within sampling units is known to follow pretty closely a Poisson distribution.   The coefficient of variation desired is $\frac{1}{6}$.   Consequently we set $1/\sqrt{m} = \frac{1}{6}$, whence m must be 36.   The sample must be big enough to produce an expected x-value of 36.   If (e.g.) the best advance estimate were 5 defects in 100 test-pieces, the size of sample should be $36 \times 100/5 = 720$ test-pieces.

Conversely, suppose that the number of occurrences in a sample is x. The estimate of the standard error is $\sqrt{x}$, and the estimate of the coefficient of variation is $1/\sqrt{x}$.   Thus, suppose that a sample of dwelling units yields 100 vacant dwelling units, and that the number of vacant dwelling units per segment follows pretty well a Poisson distribution.   Then the standard error of the estimate is $\sqrt{100} = 10$ d. us.

The square-root transformation.   An easy way to use the Poisson series is to transform it to the square root.   It is a remarkable fact that if x be a random variable with a Poisson distribution, then $\sqrt{x}$ is nearly normal about $\sqrt{Ex}$ with standard deviation equal very nearly to $\frac{1}{2}$, even when Ex is small (*vide* Exercise 1 on page 464).   The chief use of the transformation is for quick calculations of the probability beyond a given value of r.   One only need ask himself whether $\sqrt{r} - \sqrt{m} > 1$ for 2-sigma significance, or $> 1\frac{1}{2}$ for 3-sigma.   Such tests will be reliable indicators of significance even when m is as low as 3 or 4, especially if one introduces a correction for continuity (see the reference to Hald on page 462).

Examples in the use of the square-root transformation. *

EXAMPLE 1.   There are 2 sources of supply, Manufacturers A and B, for the same piece-part.   The inspection-records show that 1000 pieces from A contained 121 defectives, and that 1000 pieces from B contained 144 defectives.

a. Is there any reason to assert that the production-processes A and B are different?

b. Is there any reason to give the next order to A?

---

* The reader will profit at this point by turning to A. Hald, *Statistical Theory with Engineering Applications* (Wiley, 1952): Sec. 22.7, p. 725.

Also, Allan Birnbaum, "Statistical methods for Poisson processes and exponential functions," *J. Amer. Statist. Ass.*, vol. 49, 1954: pp. 254–266.

SOLUTION

Treat each number defective as a Poisson variable whose square root has variance $\frac{1}{4}$.   The variance of the difference is then $\frac{1}{2}$ (Exercise 1, page 391).

$$\sqrt{144} = 12$$
$$\sqrt{121} = 11$$

$$\text{Difference} = 1$$

whence

$$t = 1 \times \sqrt{2} = 1.41$$

The value of t so computed will be distributed very nearly on a normal curve.   Use of the Mosteller-Tukey paper gives the same t in a few seconds, without square roots (Fig. 28).

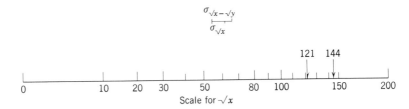

Fig. 28. Showing the solution to Example 1 by plotting the 2 observed numbers (144 and 121) on a scale for $\sqrt{x}$.   If the scale is so chosen that $x = 1$ lies 10 mm. from $x = 0$, then $\sigma_{\sqrt{x}}$ will be 5 mm., and $\sigma_{\sqrt{x}-\sqrt{y}}$ will be 7.1 mm.   The distance between the 2 points 121 and 144 is 10 mm., so $t = 10/7.1 = 1.4$.

The reader may also obtain t by recalling from Exercise 1 on page 391 that the variance of $r - s$ is $m + n$.   If we take 144 and 121 as estimates of m and n, then*

$$t = \frac{r - s}{\sqrt{r + s}}$$
$$= \frac{144 - 121}{\sqrt{144 + 121}} = \frac{23}{\sqrt{265}} = 1.41$$

which agrees with the t obtained above.   The numbers 144 and 121 are so big that we may safely use the normal curve.   However we compute t, it is not significant.   That is, there is no reason to assert on the basis of these figures alone that one manufacturer is doing a better job than another. The figures could easily reverse themselves next month, as they could both have come from the same system of chance causes.

---

* A. Hald, *op. cit.*, p. 725, makes a correction for continuity by taking off $\frac{1}{2}$ unit at each end of the interval between r and s.   This correction gives $r - s - 1$ for the numerator where I have $r - s$.   Hald's correction is helpful at small values of r and s.

However, if I had to make a decision right now, and if both manufacturers could supply me, and at the same price, I should choose A.   This is a case of a *decision without definite evidence* that A will be the better of the 2 manufacturers, or even that one manufacturer is better than another.

EXAMPLE 2.   Accidental deaths from motor vehicles were 73 in Connecticut over the 1st quarter of 1955, and 64 over the same period of time in 1956.   There was much publicity about the Governor's pressure on speeding, which he began to apply in 1955.   Do the figures give grounds for attributing the decrease to the Governor's action?   The answer is no; the same figures could come from the same cause-system.

SOLUTION

$$\sqrt{73} = 8.5$$
$$\sqrt{64} = 8.0$$

Difference $= .5$

$t = .5\sqrt{2} = .71$ (not significant)

The same result comes instantly from the Mosteller-Tukey paper. Obviously the figures cited give no grounds whatever for attributing the decrease to the Governor's action.   Remember that the figure had to go up or down, with a small chance that it could be the same.   If the figure had been higher for 1956, there would have been no noise.

EXAMPLE 3.   The sales-manager in Philadelphia of a fine computing machine sold 5 machines one year, laid in a supply for parts to build 5 machines next year; actually could have sold 8.   He therefore laid in a supply of parts for 8 the 3d year; actually sold 4.   He complained dolefully about the vagaries of business, despondent of his own inability to forecast better.   Was he justified?   What could you tell him?

Comment: The average of 5, 8, 4 is 6.   The estimated 3-sigma variation in $\sqrt{x}$ is $(\sqrt{6} \pm 3/2)^2$, which gives 16 for the upper limit and 1 for the lower limit.   The same result comes instantly from the Mosteller-Tukey paper. In a business of this kind, where demand is dictated by the vagaries of chance, there will be a terrific fluctuation, as the limits indicate.   The solution is (*a*) to balance the possible losses of a sale against the cost of the inventory required to meet it, and thus to arrive at an economical and rational inventory; or (*b*) to offer a discount to a prospective customer if he will wait 6 months or a year.

Use of the Mosteller-Tukey paper to compare 2 Poisson variables.   We have just learned that the square-root transformation $\sqrt{x}$ of a Poisson variable x is very nearly a normal curve of standard deviation $\frac{1}{2}$ centered close to $E\sqrt{x}$.   Hence, to see if 2 results known to be Poisson variables came from significantly different Poisson distributions, we need only to:

1. Plot the 2 results x_1 and x_2 on a square-root scale, so marked off that the distance from 0 to x is $\sqrt{x}$.

The Mosteller-Tukey paper is ready-made for this purpose, even though we need only the horizontal scale. The point for $x = 1$ lies 10 mm. from $x = 0$, and $x = 100$ lies 10 cm. from $x = 0$. Then $\sigma_{\sqrt{x}} = 5$ mm. (the radius of the sigma-circle), and $\sigma_{\sqrt{x} - \sqrt{y}} = 5\sqrt{2} = 7.1$ mm. (See Exercise 2 on page 465). There is also a ruler made for this purpose, with a sigma-scale, so that one may use any blank paper (obtainable from The Free Press, Glencoe, Illinois).

2. Measure in millimeters the distance between the 2 points.

3. Divide this distance by 7.1 to get t.

To use the sigma-circle on the Mosteller-Tukey paper, mark off about 1.4 radii = 7 mm on the edge of a card, for the unit of measurement. (This unit is 1 standard error of a random difference.) Now transfer the card to the scale where you plotted the 2 points, and measure roughly, in the unit marked on the edge of the card, the distance between the 2 points. This distance is t. We need only know whether t is under 2 or under 3, depending on what level of significance suits the purpose.

Fig. 28 carries out this solution for Example 1, which we saw on page 461.

EXERCISES

1. If x is a Poisson variable with expected value $Er = m$, then

a. $\sigma_{\sqrt{x}} = \frac{1}{2}$ very nearly   (13)

b. $E\sqrt{x} = \sqrt{m - \text{Var } \sqrt{x}} \doteq \sqrt{m - \frac{1}{4}} \doteq \sqrt{m}\left(1 - \frac{1}{8m} + \cdots\right)$ (14)

SOLUTION

By definition,

$$\text{Var } \sqrt{x} = E(\sqrt{x} - E\sqrt{x})^2$$
$$= Ex - (E\sqrt{x})^2$$
$$= m - (E\sqrt{x})^2$$

whence

$$E\sqrt{x} = \sqrt{m - \text{Var } \sqrt{x}}$$

We learned in Chapter 17 (p. 391) that

$$\sigma_y^2 = \left(\frac{dy}{dx}\right)^2 \sigma_x^2$$

Put

$$y = \sqrt{x}$$

Then

$$\frac{dy}{dx} = \frac{1}{2y} = \frac{1}{2\sqrt{m}} \text{ at } y = \sqrt{Ex}$$

whence

$$\sigma_y^2 = \left(\frac{1}{2\sqrt{m}}\right)^2 m = \frac{1}{4}$$

$$\sigma_y = \sigma_{\sqrt{x}} = \frac{1}{2}$$

Q.E.D.

This is an approximation, but a good one.

2. Show that if x and y are 2 independent Poisson variates, then

$$\text{Var}\,(\sqrt{x} - \sqrt{y}) = \tfrac{1}{2}$$

$$\sigma_{\sqrt{x} - \sqrt{y}} = \frac{1}{\sqrt{2}}$$

SOLUTION

We know from page 391 that if x and y are random and uncorrelated,

$$\text{Var}\,(x - y) = \text{Var}\,x + \text{Var}\,y$$

Hence

$$\text{Var}\,(\sqrt{x} - \sqrt{y}) = \text{Var}\,\sqrt{x} + \text{Var}\,y$$

$$= \tfrac{1}{4} + \tfrac{1}{4} = \tfrac{1}{2}$$

Q.E.D.

The standard error of $\sqrt{x} - \sqrt{y}$ is thus $\sqrt{\tfrac{1}{2}} = .71$, which is about 7 mm. on the Mosteller-Tukey paper.

3. If 1 birth in 80 on the average is twins (1 person in 40 having a twin brother or sister), what is the expected number of people who have a twin brother or sister somewhere, also the upper and lower 3-sigma limits, in a district that contains 20,000 inhabitants? Assume that the life expectancy at all ages is no different for twins than for other people.

4. Turn to Table 2 on page 105 in Chapter 7. Test whether the variation of the results in the 10 subsamples for each characteristic could reasonably be ascribed to chance alone.

Illustration: Take the column under 1–11 rooms. The average number of motels per subsample is 23.8. $\sqrt{23.8} = 4.9$. $23.8 \pm 2 \times 4.9 = 34.6$ and 14.0. The 10 numbers are all well within these limits, and there is thus no reason to suspect, on the basis of these figures, any departure from the sampling procedure prescribed.

An easier solution comes by use of the Mosteller-Tukey paper. Mark off 2-sigma, and measure off left and right from 23.8: read 35 and 15.

The student should use a table of the Poisson distribution to test the "attractions" in the extreme right-hand column of Table 2, where the numbers are small, and should compare his conclusion with the use of the Mosteller-Tukey paper, and with Hald's correction (p. 462). One would lead to the same conclusion as the other: the variations could well come from chance alone.

5. Would you use 2-sigma limits or 3-sigma limits in the last exercise for the test of the sampling procedure?

This is a problem in testing a hypothesis. It is not a problem in estimation (for which I would use 3-sigma limits). In testing a hypothesis, one should fix his action-limits in relation to the cost of failing to discover something wrong. In my own practice I probe the procedure if a result falls outside the 2-sigma limits, in order to fail only rarely to discover a flaw in the procedure, yet not to probe too often when there is no flaw.

6. Show that the results by subsample in Table 2 show an uncomfortable amount of variation (beyond 2-sigma). Examination of the sampling procedure disclosed some misunderstandings, which when corrected, brought the figures within the limits permissible in Poisson variation. (This exercise illustrates an advantage of replication. See remark 4, page 72.)

TABLE 2

Subsample	Result
1	2
2	8
3	14
4	15
5	14
6	3
7	7
8	12
9	11
10	9
Average	9.5

7. A careful investigation of 435 sampling units drawn by random numbers for an audit of a main sample of accounts showed 8 nonsampling errors of a certain type (similar to Table 5 on page 161).

a. What is the upper 3-sigma limit for the number of errors of this type in repeated samples of 435?

b. What is the upper 3-sigma limit for the proportion of errors of this type in the main sample?

a. Use the square-root-transformation, with 8 as the lower 3-sigma limit.

$$(\sqrt{8} + 1.5)^2 = 18.5$$

The Mosteller-Tukey paper also gives 18.5, as the student should verify.

b. The corresponding expected proportion is $18.5/435 = .043$.

The interpretation of the upper 3-sigma limit for the proportion of errors of the type under consideration is this: if the actual proportion in the main sample were greater than .043, then only a very unusual audit-sample of 435 items would show as few as 8 errors.

Or, if we were to calculate repeatedly $p_U = (\sqrt{r} + 1.5)^2/435$, where r is the observed number of defects in an audit-sample of 435 items, only an insignificant number of the values of p_U so calculated would fall above the proportion in the main sample (refer back to page 451 on confidence intervals).

Either way, we may accept .043 as a practical maximum for the proportion in the main sample.

8. An inventory of telephone apparatus carried out by sampling methods by a telephone company showed in the sample a total of 979 telephones of various types, 22 of which were so-called "wall sets."   What are the maximum and minimum 3-sigma proportions of wall sets in the total frame that was used for the sample? (Assume that the appearance of wall sets in samples follows a Poisson distribution.)

SOLUTION

Use the square-root-transformation.   Let p be the proportion sought. Then the upper and lower 3-sigma limits of p will be calculable from the equations

$$\sqrt{979 p_U} - 1.5 = \sqrt{22}$$

$$\sqrt{979 p_L} + 1.5 = \sqrt{22}$$

the solutions being by simple algebra

$$p_U = \frac{(\sqrt{22} + 1.5)^2}{979} = 5.3\%$$

$$p_L = \frac{(\sqrt{22} - 1.5)^2}{979} = 1.8\%$$

Note that these limits are not symmetrical about $22/979 = 2.3\%$.   The limits are symmetrical on the scale for $\sqrt{r}$, but not for r.

The interpretation of the upper and lower 3-sigma limits is this: if the actual proportion of wall sets in the frame were greater than 5.3%, the observed number of wall sets in a sample of this size would very rarely be as low as 22; the observed result could hardly have arisen by chance.   If the actual proportion of wall sets in the frame were smaller than 1.8%, the number of wall sets in a sample of this size would very rarely be as high as 22.

CHAPTER 19

Optimum Number of Segments per Block

Ye suffer fools gladly, seeing that ye yourselves are wise.—ii
Corinthians 11: 19.

Statement of the problem. We have learned how to draw replicated
samples of employees, of records, and of segments of area, and how to
compute the estimates and their standard errors. We recall that a sampling
unit is by definition what 1 random number draws from the frame. We
have learned that there may be by design an average of 1, 2, 3, or more
employees, lines, or segments in a sampling unit. We have not yet
learned, however, what is the best number of employees, records, or
segments of area to include in a sampling unit, nor what is the best size
of segment, nor how to distribute them. This chapter will make a start
on some of these questions, and will give references on the others, for
further study. Fortunately, failure to achieve the optimum design does
not introduce bias.

Theory. To begin, we need some notation. The word block will
denote a bounded and definite amount of material, divisible into segments.
Every block in the frame contains a certain number of sampling units.
A sampling unit may be a segment, or it may be 2 or more segments
drawn at random from the block. The frame gives a serial number to
every sampling unit. When a random number draws a sampling unit,
we can see from the frame which block contains this sampling unit.

Let us continue to use $\sigma_w{}^2$ for the average variance between the segments
within the blocks of the city that is to be sampled. Suppose now that
we form a sampling unit which shall consist of 2 segments drawn at random
from each block. Then the average variance between the populations
per segment in these new sampling units within blocks will be $\frac{1}{2}\sigma_w{}^2$. If
the new sampling unit consisted of 3 segments drawn at random from
each block, the average variance between these new sampling units
within blocks would be $\frac{1}{3}\sigma_w{}^2$.

468

Let $\bar{N}$ be the average number of segments per block, and suppose for the moment that all blocks contain $\bar{N}$ segments. Group now the $\bar{N}$ segments in each block into sampling units, $\bar{n}$ segments to the sampling unit: do this by a random ordering of the serial numbers $1, 2, 3, \ldots$, $\bar{N}$ in each block. Then we have:

M blocks in the whole city

$\bar{N}$ segments in a block

$N = M\bar{N}$ segments in the whole city

$\bar{n}$ segments in a sampling unit

$\bar{N}/\bar{n}$ sampling units in a block

$M\bar{N}/\bar{n}$ or $N/\bar{n}$ sampling units in the whole city

In practice, we form a sampling unit by drawing 1 segment from every successive $\bar{N}/\bar{n}$ segments in a block, as in Chapter 10 and elsewhere, but the random ordering assumed above will simplify the calculation of variance.

The average variance between the sampling units within blocks, reckoned on the basis of the mean x-population per segment, is

$$\sigma^2_{w(\bar{n})} = \frac{\bar{N} - \bar{n}}{\bar{N} - 1} \frac{\sigma_w^2}{\bar{n}} \qquad \text{[An application of Eq. 20, page 383]} \qquad (1)$$

The variance σ_b^2 between blocks, reckoned likewise on the basis of the mean x-population per segment, is unaffected by our choice of sampling unit; it remains σ_b^2. It follows that the total variance between all $N/\bar{n}$ sampling units in the city, on the same basis, must be

$$\sigma^2_{(\bar{n})} = \sigma_b^2 + \sigma^2_{w(\bar{n})} = \sigma_b^2 + \frac{\bar{N} - \bar{n}}{\bar{N} - 1} \frac{\sigma_w^2}{\bar{n}} \qquad (2)$$

this being the sum of the variance between blocks and the variance within blocks (an application of Eq. 32 on page 486).

Suppose now that we read out m unduplicated random numbers between 1 and $N/\bar{n}$ to draw a sample of m distinct sampling units, and that we form the estimate

$$\bar{x} = \frac{x}{m\bar{n}} \qquad (3)$$

where x is the sum of the m average x-populations, derived from the m sampling units in the sample. Then $\bar{x}$ is an unbiased estimate of the mean population per segment in the city. Its variance will be just $1/m$ times the variance between the segments. Or, with adjustment for the finite number of segments,

$$\text{Var } \bar{x} \doteq \frac{N/\bar{n} - m}{N/\bar{n} - 1} \left\{ \frac{\sigma_b^2}{m} + \frac{\bar{N} - \bar{n}}{\bar{N} - 1} \frac{\sigma_w^2}{m\bar{n}} \right\} \qquad (4)$$

The overall sampling fraction is $m\bar{n}/M\bar{N}$. The zoning interval for segments, in a replicated design, will be $2M\bar{N}/m\bar{n}$ for 2 subsamples, or $10M\bar{N}/m\bar{n}$ for 10 subsamples. σ_b^2 and σ_w^2, in a replicated design, refer to variances between blocks and within blocks, within zones.

We now relax the requirement that the blocks all contain the same number of segments.   We define σ_b^2 as in Eq. 27 on page 485, and define $\sigma_i^2 = (1/N_i)\sum_{j=1}^{N_i}(a_{ij} - a_i)^2$ for the variance between the N_i segments in Block i.   Then in place of Eq. 4, and by the same reasoning, we find that*

$$\text{Var } \bar{x} = \frac{N/\bar{n} - m}{N/\bar{n} - 1}\left\{\frac{\sigma_b^2}{m} + \frac{1}{m\bar{n}}\frac{1}{M\bar{N}}\sum_1^M \frac{N_i - \bar{n}}{N_i - 1}N_i\sigma_i^2\right\} \qquad (4a)$$

This equation reduces to Eq. 4 if $N_i = \bar{N}$ in every block.

Every sampling unit lies in some block.   Two or more random numbers may fall in the same block.   The number of blocks in the sample is thus a random variable, usually close to m (p. 473).

When there is no information on the sizes of the blocks, we may treat all blocks as equal, as we did in Part F of Chapter 11 (p. 219).   To each block we ascribe $\bar{N}$ segments, and $\bar{N}/\bar{n}$ sampling units, where $\bar{n}$ is a figure agreed upon as about the optimum number of segments to draw out of every $\bar{N}$ consecutive segments.   The sampling units take serial numbers. A random number falls in a block.   We thereupon create segments in this block and draw without replacement a proportionate sample of these segments, the sampling fraction for segments being $\bar{n}$ in $\bar{N}$ in every block.

Or, we may draw a sample of blocks, create segments in these blocks, and draw without replacement a proportionate sample ($\bar{n}$ in $\bar{N}$) of these segments.   The zoning interval for blocks will be $2M/m$ for 2 subsamples, $10M/m$ for 10 subsamples.   The finite multiplier outside the braces in Eqs. 4 and 4a no longer applies; the variance between subsamples will overestimate the variance of the sampling procedure more than this multiplier indicates if m is appreciable in comparison with M.

Derivation of the optimum number of segments per block.   Let c_1 be the cost of preparing an average block for sampling.   This will ordinarily be the average marginal cost of preparing the extra map and the boundaries of a block, the cost of sending an interviewer to the block, plus her time to create segments.   If the m random numbers draw m blocks into the sample, this cost will be mc_1.

Let c_2 be the cost of interviewing in 1 segment, including the recalls, and the average cost of travel between segments within the block (often negligible, but not strictly independent of $\bar{n}$ in large areas).   The total number of segments in the sample will be $m\bar{n}$, so we write

$$K = c_1 m + c_2 m\bar{n} \qquad (5)$$

for the total variable cost.

We seek to find m and $\bar{n}$ such that Var $\bar{x}$ will be a minimum for a given cost K.   Let

$$\bar{n} = \frac{\sigma_w}{\sigma_b}\sqrt{\frac{c_1}{c_2}}\ (1 + R) \qquad (6)$$

* I thank my colleagues Morris H. Hansen and William N. Hurwitz for the derivation of this equation.

where R is a small quantity to be determined to make Var $\bar{x}$ a minimum. Eq. 6 gives

$$m = \frac{K}{c_1 + c_2\bar{n}} \tag{7}$$

which when substituted into Eq. 4 gives

$$\text{Var } \bar{x} = \frac{c_1 + c_2\bar{n}}{K}\left\{\sigma_b^2 + \frac{\sigma_w^2}{\bar{n}}\right\} \qquad \begin{array}{l}\text{[Finite multipliers}\\\text{omitted]}\end{array}$$

$$K \text{ Var } \bar{x} = c_2\sigma_w^2 + c_1\sigma_b^2 + c_2\sigma_b^2 \frac{\sigma_w}{\sigma_b}\sqrt{\frac{c_1}{c_2}}(1 + R)$$

$$+ c_1\sigma_w^2\left(\frac{\sigma_b}{\sigma_w}\sqrt{\frac{c_2}{c_1}}\right)(1 - R + R^2 + \text{higher powers})$$

$$\doteq c_2\sigma_w^2 + c_1\sigma_b^2 + \sigma_b\sigma_w\sqrt{c_1c_2}(2 + R - R + R^2)$$

$$= (\sigma_w\sqrt{c_2} + \sigma_b\sqrt{c_1})^2 + \sigma_w\sigma_b\sqrt{c_1c_2}R^2 \tag{8}$$

This variance is obviously a minimum if $R = 0$, whereupon*

$$\bar{n} = \frac{\sigma_w}{\sigma_b}\sqrt{\frac{c_1}{c_2}} \qquad \text{[Optimum]} \tag{9}$$

must be the optimum $\bar{n}$. To calculate the Var $\bar{x}$, we adopt a value of $\bar{n}$ at or near the optimum, and use Eq. 4 or 4a with m from Eq. 7.

Applications

EXAMPLE 1. Suppose that the population characteristic of interest is the number of readers of a certain journal. The intended size of the segment is to be 5 d. us., for various reasons depending on statistical efficiency and smoothness of operation. Suppose that for this size of segment, $\sigma_w^2 : \sigma_b^2$ is about 1:1, σ_w^2 being as before the average variance between segments within the blocks of the city. The costs c_1 and c_2, we suppose, stand in the ratio $c_1 : c_2 = 4:1$. Then Eq. 9 gives

$$\bar{n} = \sqrt{\frac{1}{1}\frac{4}{1}} = 2$$

for the optimum number of segments per block.

Remark 1. Calculations like this, even with rough values of the ratios, indicate the neighborhood of the optimum, which is all that we require as the increase in variance in the neighborhood of the optimum is very slow.

* First published independently and in the same year by L. H. C. Tippett, *The Methods of Statistics* (Williams and Norgate, 1931): p. 177; and by Walter A. Shewhart, *The Economic Control of Quality of Manufactured Product* (Van Nostrand, 1931): p. 389.

EXAMPLE 2.   Spools of cotton yarn are tested for breaking strength by subjecting the yarn to increasing tension until it breaks. Suppose that the aim is to determine the average breaking strength of a shipment of 20,000 spools of cotton. How many breaks are optimum per spool? c_1 is the cost of a spool of yarn, plus the cost to put it into place to test it. c_2 is the cost of recording a break. Take $\sigma_w:\sigma_b = 1:2$, $c_1:c_2 = 20:1$.

SOLUTION

By Eq. 9,

$$\bar{n} = \tfrac{1}{2}\sqrt{\frac{20}{1}} = 2.2$$

which we round off to 2.   In the actual example, the company that brought up the problem in a course of lectures that I gave in Frankfurt had been making 50 breaks per spool—an illustration of what practice may be without the use of theory.   The use of $\bar{n} = 50$ rendered the testing program expensive and inefficient.

EXAMPLE 3.   Wool in bales is tested for clean content, and settlement is made and duty paid on these tests. The test-specimen is a core, cut out by a machine. How many cores per bale is optimum? Assume that for the wool under consideration (Australian wool, highly homogeneous) $\sigma_w:\sigma_b$ is known to be about $1:1$, and that $c_1:c_2$ is about $4:1$.

SOLUTION

$$\bar{n} = \sqrt{4} = 2$$

In some testing laboratories, with similar wool, I have seen men take 20 cores per bale, again an expensive and ineffective testing program.

How many blocks in the sample?   We note in the first place from Eq. 9 that $\bar{n}$ is independent of m, and of the cost K, and of Var $\bar{x}$.   We may therefore compute $\bar{n}$ before we settle on m, or on the cost K, or on Var $\bar{x}$. What decides the number m of blocks is either (a) the total allowable cost (how many blocks will the funds permit us to go into?, answered by Eq. 8, once K is fixed), or (b) the prescribed value of Var $\bar{x}$, which gives m through Eq. 4 or 4a.

Remark 2.   Strictly, these equations give m as the number of sampling units, but the number of blocks and the number of sampling units will be nearly the same (see Exercise 1 on the next page).

Remark 3.   If we prescribe in advance the standard error of $\bar{x}$, the procedure for the computation of m and of K will be first to compute $\bar{n}$ from Eq. 9; then to compute K from Eq. 8 with $R = 0$; then m from Eq. 7. A still better way is to put $R = 0$ in Eq. 8 and to compute a few points on the curve that connects K and $\sigma_{\bar{x}}$, to see what the cost K would be for any prescribed value of $\sigma_{\bar{x}}$.   If the cost K is too big for the standard error desired, it will be necessary to relax on the precision.   The only alternative would be to abandon the survey, although one should re-examine the variances and costs in Eq. 8 before making such a drastic decision.

Remark 4.   I may mention here an error that is easy to fall into without guidance from theory.   I refer to the error of drawing too few blocks in a sample of a city, too few counties or too few cities in a national sample, and then drawing heavy samples from within the block, county, or city.   In symbols, such a procedure makes m too small.   A sample that consists of 12 blocks is still but a sample of 12 blocks, even though they be covered 100%.   A national sample that consists of 6 cities and 6 counties is only a sample of 12 units at the best.   If the sample within 1 of these units is less than 100%, the sample is in effect even weaker than it was to start with.

Theory tells us the correct procedure: compute the optimum $\bar{n}$, and use it.   The sample will extend into as many blocks or other primary units as the budget will permit, or as the prescribed Var $\bar{x}$ requires.

If the primary unit (block, city, county) is very big, a complete theory does point to a very slight reduction in m, as there is under such conditions a weak connexion between m and $\bar{n}$.*

A similar principle emerges from a simpler cost function.   Suppose that the cost c of investigating 1 sampling unit is independent of where the sampling unit is located.   Then the total cost will be simply $K = m\bar{n}c$. Let us fix K; then the total sample $m\bar{n}$ is also fixed.   If m decreases, $\bar{n}$ must decrease to keep K constant.   Substitution into Eq. 4 gives

$$\text{Var } \bar{x} = \frac{\sigma_b{}^2}{m} + \frac{c\sigma_w}{K}$$

What value of m will make Var $\bar{x}$ a minimum?   The 2d term on the right is independent of m, being fixed by K.   Only the 1st term on the right may vary.   This term (and hence Var $\bar{x}$ also) is obviously a minimum if m is a maximum; and thus emerges the principle that if the total cost is exactly or nearly $m\bar{n}c$, then the optimum plan is to make $\bar{n} = 1$ (the lowest value possible) and to make m as big as funds permit.

The cost function $K = m\bar{n}c$ that we have just used is a limiting case of Eq. 6 with $c_1 = 0$.   What comes out of it is a similar principle, viz., make $\bar{n}$ small, m big.†

EXERCISES

1.   A city contains M blocks, and in each block are $\bar{N}$ segments.   There are thus $M\bar{N}$ sampling units altogether.   You draw m sampling units without replacement by reading out m unduplicated numbers between 1 and $M\bar{N}$.   Two numbers, or even 3, could strike the same block.   Show that the expected number of blocks in the sample is very nearly $m - m(m - 1)/2M$, which may be taken as m unless $m > \sqrt{M}$.   Thus, it is only in large samples ($m > \sqrt{M}$) that the number of blocks may be $m - 1$.

* Hansen, Hurwitz, and Madow, *Sample Survey Methods and Theory*, Vol. I, (Wiley, 1953): pp. 290 ff.

† My colleague Morris H. Hansen showed me this derivation and principle in 1939.

2. A company has offices in 100 counties. In each county are 80 employees—8000 employees total. A random sample of employees is to be drawn for a study of the time required to perform certain types of work. The sampling unit will have an intended size of 8 employees. Suppose that we draw 10 sampling units without replacement by reading out 10 random numbers between 1 and 1000, duplicates not permitted. Show that the expected number of counties in the sample is 9.60. In other words, in repeated samples of 10 groups of 8 employees each we shall find 10 counties represented in about 60% of the samples, and 9 counties in about 40%. (The probability of 8 counties is small.)

SOLUTION

There are 1000 sampling units, and 10 of them belong to any one county. So, the probability that any one county will not be drawn in 10 draws running is

$$\frac{990}{1000} \frac{989}{999} \frac{988}{998} \cdots \frac{981}{991} = .9040$$

The probability that any county will go into the sample is therefore .0960, and this is the same for all counties. The expected number of counties is therefore just $100 \times .0960 = 9.60$.

Q.E.D.

3. One sometimes hears the term unrestricted random sample, in which one draws the n sampling units from the frame of N sampling units by reading out n random numbers between 1 and N. One sometimes also hears the statement that to draw an unrestricted random sample of 100 sampling units, one must make a list of all N units; then draw 100 random numbers between 1 and N. This procedure will indeed fulfill the requirements, but it may be very expensive if N is large. Moreover, it would be, for many purposes, one of the poorest samples that one could construct, as it would reap none of the possible benefits of stratification in zones, which costs nothing extra (Ch. 15).

Show that the following procedure is equivalent to an unrestricted random sample, and that it will in many circumstances cost much less than to list all N units.

1. Divide the entire frame into 1000 parts, all parts having the same number of sampling units. Call these primary units.

2. Number the primary units 1 to 1000.

3. Read out 100 random numbers between 001 and 000. Accept duplicates.

4. Now make a list of the sampling units in each primary unit that the random numbers hit.

5. Draw 1 sampling unit from each list.   If a primary unit was struck twice, draw 2 sampling units without replacement from the list.

SOLUTION (HANSEN)

We had in Eq. 4 (p. 469)

$$\sigma_{\bar{x}}^2 = \frac{\sigma_b^2}{m} + \frac{\sigma_w^2}{m\bar{n}}$$

Put $\bar{n} = 1$ and this reduces to

$$\sigma_{\bar{x}}^2 = \frac{1}{m}(\sigma_b^2 + \sigma_w^2) = \frac{\sigma^2}{m}$$

which is precisely the variance of a plan that draws m sampling units from the entire list of N sampling units.   The statement will still be true when we insert the finite multipliers.   In other words, the plan described, and an unrestricted random sample, belong to exactly the same probability system.

4. Refine Eq. 9 by retaining the 2d multiplier $(\bar{N} - \bar{n})/(\bar{N} - 1)$ in Eq. 4*a*.   The result is

$$\bar{n} = \frac{\sigma_w}{\sigma_b} \sqrt{\frac{c_1}{c_2}} \, \frac{1}{\sqrt{1 - \sigma_w^2/\bar{N}\sigma_b^2}} \tag{10}$$

Example of advance calculation of variance of a ratio.   Each block in the frame receives a measure of size, early in the preparation of the sample (e.g., Step 2 of Chapter 10, p. 169).   This measure of size is an estimate of the number of sampling units in the block, based on advance information concerning the number of dwelling units in the block.   For example, if the intention is to interview in 1 segment per block, the number of sampling units in a block will be equal to the estimate of the number of segments in the block.   If the intention is to interview in 2 segments per block, the number of sampling units in a block will be equal to half the estimate of the number of segments in the block.   If there is no advance information, each block receives the average size $\bar{N}$, *vide infra*.

A random number strikes a block, whereupon the next step is to create segments in that block, then to draw (by random numbers in a sealed envelope, page 175) 1 segment in every segment-interval.   Repetitions of the sampling procedure will draw different blocks, and varying numbers of segments.   The number of segments in a block is thus a random variable, usually concentrated with small variance about the intended number, which might be 1, or might be 2. Likewise the number of dwelling units and the number of people of any characteristic in a sampling unit are also random variables.

These variations will be accentuated if our rough counts of the number of dwelling units, block by block, fluctuate widely above and below the number found at the time of the survey. The sampling procedure nevertheless gives equal probabilities of selection to every dwelling unit and to every person of every characteristic.

Suppose that we conduct a survey to estimate (e.g.) the number of people unemployed in some area. Call the unemployed the x-population, and call the number of dwelling units the y-population. Then the sample will give us

$$f = \frac{x}{y} = \frac{\text{The unemployed in the sample}}{\text{The number of dwelling units in the sample}} \tag{11}$$

An estimate of the total unemployed is then

$$X' = Nf \tag{12}$$

where N is the number of dwelling units in the area covered by the frame, known from outside sources. (I am saying nothing about the complexities of the definitions of employed and unemployed, which are the same whether we use a sample or a complete count.)

We shall have no trouble to estimate Var X' if we use a replicated design. The variance of X' comes wholly from the variance of f, as N is not a random variable for the purposes of this survey. Hence, $C_{X'} = C_f$.

In the design of the sample, we need to know in advance whence comes the variance of f. In the first place, f has a numerator x, and a denominator y, and we know from Chapter 17 (p. 393) that

$$C_{X'}{}^2 = C_f{}^2 = C_{x/y}^2 \doteq \frac{1}{m} \{ C_x{}^2 + C_y{}^2 - 2C_{xy} \} \tag{13}$$

where m is the number of sampling units in the sample. Here, $C_x{}^2$ is the rel-variance of the number of unemployed in repeated samples of size $m = 1$. $C_y{}^2$ is the rel-variance of the number of dwelling units for $m = 1$. C_{xy} is the rel-covariance between x and y for $m = 1$.

Let us try to see where the sampling variation in x and y may come from. First, the number of people unemployed varies from one dwelling unit to another. This will cause variation in x. Second, the number of dwelling units in a segment varies from one segment to another, which causes both x and y to vary. Third, more variation comes from the fact that the number of segments that a random number will draw is a random variable, not only (a) because the number of segments in a block will sometimes not be an exact multiple of the segment-interval ascribed to that block in Step 2, page 169, even if the rough count was correct, but also (b) because the rough count is usually rough, owing to the growth in one block, and removals from another, or merely because the rough count was intended only as a quick approximation.

If the distribution of the number of people unemployed per segment were a Poisson variable, then the rel-variance of this distribution would be $1/a$ (Ch. 18), where a is the average number of people unemployed per segment.   But there is usually a clustering effect, because of door-to-door correlation, which raises the variance between segments within the block, of the number of people unemployed in a segment.   This clustering effect in a segment of 6 d. us. may increase the rel-variance per segment from 1 to $1.3/a$ or even $1.5/a$.   The clustering effect in a segment of 20 d. us. may increase the rel-variance to $1.7/a$.

To proceed with our calculations, we shall deal with segments that have an intended size of 15 d. us., with rel-variance between segments equal to $1.6/a$.   If the magnitude of unemployment were 1 person in 15 d. us., on the average, then a would be 1, and the rel-variance between our segments would be 1.6.   This would be the value of $C_x{}^2$ in Eq. 13, and the other 2 terms therein would be 0, were there no other variations.

To investigate the effect of the variations in the number of dwelling units per segment, and the variation in the number of segments per sampling unit, we write, for sampling unit i,

$$y_i = K(1 + r_i)(1 + s_i)(1 + t_i) \qquad (14)$$

K is the average number of dwelling units per sampling unit; r_i is the relative excess number of dwelling units above the overall average number of dwelling units per segment; s_i is the relative excess above the average number of segments that there would be in Sampling Unit i if the rough count were exact (by which 1 means the same as the interviewer obtains today); t_i is the relative excess number of segments that arises because the rough count is not exact.   We have chosen r_i, s_i, and t_i so that the means of their distributions are 0; i.e.,

$$Er_i = Es_i = Et_i = 0 \qquad (15)$$

We assume that r_i, s_i, and t_i are uncorrelated, wherefore, by Eq. 4 on page 391,

$$C_y{}^2 = C_{1+r}^2 + C_{1+s}^2 + C_{1+t}^2 \qquad (16)$$

We now try to predict numerical values of these terms, under plausible possible conditions.   The sizes of the segments might vary 50% above and below the average size, on a rectangular distribution like Panel B on page 260.   The rel-variance of this distribution would be $(1.5 - .5)^2/12 = \frac{1}{12}$, and this is the contribution C_{1+r}^2.   We record it as .10 for good measure.

There is yet to evaluate the effect of the variation in the number of segments that 1 sampling unit will bring into the sample.   (*a*) Suppose first that the rough count is correct.   Even then, when the intended number of segments is 1 per sampling unit, the actual number that get into the sample may be 0, 1, 2, owing to variation in the sizes of the segments (as we saw in Chapters 10 and 11).   To calculate the contribution from this source, we suppose that s_i takes on the values 0, 1, 2 with frequencies .15, .70, .15.   The rel-variance of such a distribution is .30, and this is the contribution C_{1+s}^2.   (*b*) Wrong information in the rough counts will add the rel-variance

$$C_{1+t}^2 = \frac{1}{M} \sum_1^M \frac{(N_i - N_i')^2}{\bar{N}^2} \tag{17}$$

where N_i is the number of dwelling units in Block i, and N_i' is the rough count. M is the number of blocks in the frame.   If N_i varied on a rectangular distribution from $.5N_i$ to $1.5N_i$ (Panel B on page 260), then $C_{1+t}^2 = \frac{1}{12}$.   The variance that arises from wrong information in the rough counts is thus in practice likely to be one of the smallest of all the contributions to the total variance.   To be sure that we have not understated its possible numerical value in this example, I raise it to .15.

We may now construct the accompanying table, to see the effects of the various sources of variation.   We ignore, safely I believe, the possibility of correlations between these effects, wherefore the contributions to C_x^2 will be the same as the contributions to C_y^2.   The total rel-variances are then the simple sums at the foot of the table.

Now comes the term $-2C_{xy}$ in Eq. 13.   If the correlation between x and y were .4, then would

$$2C_{xy} = 2 \times .4C_x C_y = .8\sqrt{2.25 \times .65} = .97 \tag{18}$$

Eq. 13 gives

$$C_{X'}^2 = \frac{1}{m} \{2.25 + .65 - .97\}$$

$$= \frac{1.93}{m} \tag{19}$$

If we desire $C_{X'}$ to be about .05, then would

$$m = \frac{1.93}{.05^2} = 772 \tag{20}$$

SOURCES OF VARIATION AND OF POSSIBLE MAGNITUDES

The x-population is the number unemployed. The y-population is the number of dwelling units.

Source	C_x^2	C_y^2
1. Variation within the segment, number of dwelling units fixed	1.60	xxx
2. Variation in the number of dwelling units in the segment	.10	$.10 = C_{1+r}^2$
3. Variation in the number of segments per sampling unit a. When the rough count of dwelling units is correct	.40	$.40 = C_{1+s}^2$
b. Addition from error in the rough count	.15	$.15 = C_{1+t}^2$
All sources	2.25	.65

which means that about 775 random numbers would suffice. If the average size of segment were 15 d. us., and if each random drawing brought in 1 segment, on the average, there would be 11,600 d. us. in the sample, in 775 segments.

Other conditions and other characteristics would of course yield different variances. For example, the student may wish to change the size of the segment to 6 d. us., with the factor 1.4 for the increase in variance owing to clustering. If a is 1 for a segment of 15 d. us., it will be $\frac{6}{15} = .4$ for a segment of 6 d. us., wherefore the variance between segments of 6 d. us. would be $1.4/a = 3.5$. Then if the other variances remain unchanged, a recalculation similar to Eq. 20 would give $C_{x'}^2 = 3.5/m$. The number m required to produce a standard error of .05 would then be 1400 segments, in which would be located $6 \times 1400 = 8400$ d. us. The cost of this sample might be less than the cost of the other one.

Remark 5. The above type of analysis is very helpful in the planning of surveys of many kinds. One must of course adjust the numerical values of the rel-variances to each problem.

The treatment above dealt tacitly with a small fairly homogeneous region. It applies, however, to any stratum of a national sample, and hence to a whole national sample. The Censuses in Washington and Ottawa now use for national estimates of changes month-to-month in the employed, unemployed, and their characteristics estimates in fine classes, a word on which occurred near the end of Chapter 7 (p. 447).

Variance of the direct estimate. Suppose that we have no rough counts, and that we can not be sure of the figure N for the total number of dwelling units; that consequently we can not use the ratio-estimate Eq. 12. We therefore decide to use the following plan: (1) draw m blocks at random, with equal probabilities m/M, which might be (e.g.) 1 in 20; (2) create segments in each of these blocks; (3) draw in each of these blocks a sample of segments, with a uniform probability in all blocks, fixed at (e.g.) 1 segment in 10; (4) interview in the segments so drawn; (5) compile the results; (6) form the estimate

$$X = \tfrac{20}{1} \tfrac{10}{1} x = 200x \tag{21}$$

We should of course alter the above design to replicate it, to be able to estimate Var X with ease.

Note that the direct estimate X has to stand on its own feet; it receives no benefit from the correlation between (a) the number of people unemployed in a sampling unit, and (b) the number of dwelling units in the segment.

The variation in the number of dwelling units per segment will be the same as before. The only change will come in the addition from error in the rough count, which causes more variation in the number of segments per sampling unit. This will come from C_{1+t}^2, in which we must substitute $\bar{N}$ for every N_i', as we have, by hypothesis, no rough counts at all here, and must give every block equal probability with another. To evaluate C_{1+t}^2, we could assume that the actual counts vary from near 0 to some maximum, possibly in the shape of Panel B on page 260, the rel-variance of which is $\tfrac{1}{3}$. This figure replaces .15 for C_{1+t}^2 in the table. Our new value of C_x^2 would then be $1.60 + .10 + .40 + .33 = 2.43$. Designate this number by $\tilde{C}_x^2$. The comparison between the 2 estimates X' and X will then be

$$\frac{C_{X'}^2}{C_X^2} = \frac{C_x^2 + C_y^2 - 2C_{xy}}{\tilde{C}_x^2}$$

$$= \frac{1.93}{2.43} = 100:126 \tag{22}$$

which means that Var X would be about 126 times as big as Var X', or that σ_X would be 12% bigger than $\sigma_{X'}$. The saving of 26% comes from the use of the ratio-estimate, which was possible because of the rough counts of dwelling units.

The reader may note that if the rough counts were exact, the rel-variance of the ratio-estimate X' would drop only .15, and the ratio $C_{X'}:C_X$ would be $1.82:2.43 = 100:134$ in place of $100:126$.

Remark 6. The student may be astonished at the small gain that arose from use of a ratio-estimate in this example. The reason for the small gain is that the main contribution (1.7) to $C_x{}^2$ was inherent in the variability of the unemployed from segment to segment. Variations in the size of segment, and in the number of segments per sampling unit, added relatively little more variance. The correlation between x and y was not much help, because C_x was so much bigger than C_y.

The reader may see from Eq. 13 that in the extreme case of perfect correlation between x and y, $C_{X'} = C_x - C_y$. The correlation is thus most effective in the reduction of the variance if C_x and C_y are of nearly the same magnitude.

One may perceive that if we wished to estimate a characteristic whose variance between segments of equal size was considerably less than the one that we just worked with, there would be some point in trying to decrease the other variances C_{1+r}^2, C_{1+s}^2, C_{1+t}^2. In particular, one might be tempted to cut C_{1+s}^2 by creating a designated number of segments in a block. The only trouble is that this procedure in the field might lead to relaxation on the requirement of definite boundaries for the segments, with bias that might well outweigh the sampling error that arises from the variation in the number of segments per sampling unit.

Exercise. Show that if the rough counts of the blocks vary on a rectangular distribution within a range of $17\frac{1}{2}\%$ above and below the actual counts, then the contribution to $C_{\bar{x}}^2$ from the inaccuracy in the rough counts does not exceed 1%.

This exercise emphasizes a point already in the text, to the effect that obsolete rough counts, if they are still approximately correct, may be almost as useful as more exact information. However, as we just saw on page 480, if the roughness $N_i - N_i'$ runs from 0 to double the average, the contribution may rise to $\frac{1}{3}$.

Optimum size of segment. We have found the optimum number of segments in a block or other primary unit, but we have so far done nothing on the optimum size of segment—how many dwelling units, how many lines in a ledger, how many successive transactions, how many successive items of product per segment. We leave a full treatment of this question to Hansen, Hurwitz, and Madow, *Sample Survey Methods and Theory*, Vol. I (Wiley, 1953); pp. 259 ff. The optimum size of segment is tied up with the door-to-door correlation of the statistical characteristics that the survey is to measure. There are extreme cases for which one may see the answer without a formal calculation. For example, rent is highly correlated door-to-door; likewise income, savings, and total expenditures (but not brands of household products bought with income, as people usually make up their own minds on such points, without help from the neighbors). In surveys on income and savings, the segment should be small, such as 2 or 3 d. us.

At the other extreme there are cases of rare characteristics. For example, the universe for the survey might be the readers of a certain magazine, or the users of a certain product, and this proportion might be roughly 1 household out of 10, or 1 person out of 10. In such a case, one would be foolish to use a segment smaller than 20 or 30 d. us. UNLESS there is a strong possibility that the readers or users are clumped solidly so that there would be danger of running into segments of solid readers or solid users, while other segments contain none at all. This could be described in symbols as a case where $p = .1$, but where the serial correlation door-to-door within segments is unity.

I am at this moment designing a survey of female readers of a certain newspaper where sales of the paper point to 1 female in 10 as a reader. I prescribed therefore 35 d. us. for the size of segment. It would be wasteful to use a smaller segment. The size 35 is, however, about as economical as a bigger one, because in most cases the segment will be an entire block or 2 blocks tied together. In heavily populated blocks there may occasionally be 2, 3, or 4 segments per block, but big segments in a heavily populated block are cheap to create on the spot. It is unlikely that a bigger segment would be cheaper; it might even be more expensive, requiring the tying of more blocks which may be adjacent on the list of blocks, but not adjacent on the ground.

Mixtures of rare and solid and medium density are common. I encountered a survey where the main aims were to discover (1) how many dwelling units in the country are connected to a gas main, and (2) how many of these dwelling units cook with electricity. Dwelling units connected to the gas main occur in solid segments—they all do or they all don't. However, the dwelling units that have gas may or may not cook with electricity, with little door-to-door correlation. Thus, there was for this survey a conflict of interest—a small segment would be best to estimate the number of dwelling units that have gas, and a bigger one would be best to estimate the number of these dwelling units that cook with electricity. As the second of these characteristics was more important to the client than the first one, the best choice seemed to be a medium-sized segment of 6 to 10 d. us. Perhaps even 20 d. us. would not have been too many.

The statistician should point out such conflicts to the client, so that he may decide whether he wishes to spend his money for a good estimate of one characteristic, and which one it shall be, or to get more information on a number of characteristics. He is paying for the survey, and the decision should be his.

Exercise 1. There are $\bar{N}$ segments in an area, there being g consecutive dwelling units per segment. The sampling plan is as follows: (a) draw 1 segment at random from the $\bar{N}$ segments; (b) enumerate the x-population therein $(x_1, x_2, \ldots, x_g$ dwelling unit by dwelling unit); (c) form the mean

$\bar{x} = \dfrac{1}{g}\sum_1^g x_j$, for the average population per dwelling unit in the sample-segment.   Show that for this sampling plan

$$E\bar{x} = a \qquad \text{[The sampling plan is therefore unbiased]}$$

and $\qquad \sigma_{\bar{x}}^2 = \dfrac{\sigma^2}{g}\{1 + (g-1)\rho\} \qquad\qquad (23)$

where σ^2 is the total variance between all Ng d. us. in the frame, and

$$\rho = \frac{1}{Ng(g-1)\sigma^2}\sum_{i=1}^{N}\sum_{\substack{j'\neq j=1}}^{g}(a_{ij}-a)(a_{ij'}-a) = \frac{\sigma_b^2 - \sigma_w^2/(g-1)}{\sigma^2} \qquad (24)$$

is the average intraclass correlation between the x-populations of the g d. us. in a segment. a_{ij} is the x-population in Dwelling Unit j in Segment i.

$$a = \frac{1}{Ng}\sum_i^N\sum_j^g a_{ij} \qquad \text{[The average x-population per d. u.]}$$

$$\sigma^2 = \frac{1}{Ng}\sum_i^N\sum_j^g (a_{ij}-a)^2 \qquad \text{[The total variance between the N d. us.]}$$

This result for $E\bar{x}$ and $\sigma_{\bar{x}}^2$ is a formal statement of what we know already. We see now that a sampling plan in which we draw segments of g d. us. each is bigger by the relative amount $(g-1)\rho$ than if we were to draw the same number of dwelling units singly at random and hence scattered over the entire frame of Ng d. us.   In both plans, $\bar{x}$ is an unbiased estimate of a.

ρ varies with the characteristic measured, and with the size g of the segment.   As g increases, ρ decreases.   Hansen, Hurwitz, and Madow* give some numerical values of ρ (their δ) for a number of characteristics of human populations, for several sizes of segment. ρ is confined mathematically between -1 and $+1$.   We saw on page 187, and again at the bottom of page 481, cases where ρ may be high, and the term $(g-1)\rho$ would add considerably to the variance if g were 3 or 4 or more.   In most of my experience, ρ has been so small that a segment of 4 or 6 d. us., on up to 35, has been administratively smooth and economical in view of the ease and speed in the demarcation of big segments, and the reduced cost of travel. Negative values of ρ are, in practice, rare.   Examples occur, nevertheless, as when revolution of a roller in a factory turns out a unit of product: variation of thickness within the unit, owing to defects in the roller, may be greater than variation of average thickness between units.

Eq. 19 on page 383 shows that if a characteristic be scattered amongst the $\bar{N}g$ d. us. independently at random, then the expected value of ρ is $-1/(\bar{N}g - 1)$, which for practical purposes is 0.

---

* Hansen, Hurwitz, and Madow, *Sample Survey Methods and Theory*, Vol. I (Wiley, 1953): Ch. 6, Part G.

SOLUTION

$$E\bar{x} = E\frac{1}{g}\sum_{j=1}^{g} x_{ij}$$

$$= \frac{1}{N}\sum_{1}^{N}\frac{1}{g}\sum_{j=1}^{g} a_{ij} \quad \text{[Because all segments have the probability } 1/N\text{]}$$

$$= a$$

The sampling plan is therefore unbiased.

$$\sigma_b^2 = \sigma_{\bar{x}}^2 = E(\bar{x} - E\bar{x})^2 \quad \text{[For a sample of 1 segment]}$$

$$= \frac{1}{N}\sum_{i=1}^{N}\left(\frac{1}{g}\sum_{j=1}^{g} a_{ij} - a\right)^2$$

$$= \frac{1}{Ng^2}\sum_i \begin{bmatrix} (a_{i1} - a)^2 & (a_{i1} - a)(a_{i2} - a) & (a_{i1} - a)(a_{i3} - a) \\ (a_{i2} - a)(a_{i1} - a) & (a_{i2} - a)^2 & (a_{i2} - a)(a_{i3} - a) \\ (a_{i3} - a)(a_{i1} - a) & (a_{i3} - a)(a_{i2} - a) & (a_{i3} - a)^2 \end{bmatrix}$$

[Connect all terms with + signs. There will be a matrix like this for every segment. Extension to more than 3 d. us. per segment is obvious. Note that there are g terms on the diagonal; $g(g - 1)$ off the diagonal.]

$$= \frac{1}{g}\sigma^2 \quad \text{[average term on the diagonal]}$$

$$+ \frac{1}{g}(g - 1)\rho\sigma^2 \quad \text{[average term off the diagonal]}$$

$$= \frac{\sigma^2}{g}\{1 + (g - 1)\rho\}$$

$$Q.E.D.$$

Remark 7. A sample of n segments, containing ng d. us., drawn by reading out n random numbers between 1 and N (where N is the total number of segments in the frame) would lead to the variance

$$\sigma_{\bar{x}}^2 = \left(1 - \frac{n}{N}\right)\frac{\sigma^2}{ng}\{1 + (g - 1)\rho\} \tag{25}$$

where σ^2 is, as above, the variance between dwelling units.

Exercise 2. Show that the average variance between dwelling units within the N segments of the preceding exercise is

$$\sigma_{w(g)}^2 = \sigma^2 - \sigma_b^2 = \sigma^2\left(1 - \frac{1}{g}\right)(1 - \rho) \tag{26}$$

This is the quantity σ_w^2 in the first part of this chapter. If g were 10, and if ρ for this size of segment and for the characteristic being counted were .02, then σ_w^2 would be $.882\sigma^2$, while $\sigma_{\bar{x}}^2$ for a sample of 1 segment would be, by Eq. 23, $\dfrac{\sigma^2}{10}(1 + 9 \times .02) = \dfrac{\sigma^2}{10}(1 + .18) = .118\sigma^2$, which is in fact also $\sigma_b{}^2$. The term .18 is the relative increase in $\sigma_{\bar{x}}{}^2$ that arises from use of a sample of segments instead of a sample of the same number of dwelling units drawn singly at random from the entire frame of Ng d. us. We saw another application on page 122.

Remark 8. There are important extensions of Eqs. 4 and 9 to more terms, along with the theory in the last exercise for the effect of serial correlation of adjacent items. For example, one may include time as 1 dimension, with the month as the sampling unit in the case of monthly surveys like the *Monthly Report on the Labor Force* published by the Bureau of the Census. Theory, along with empirical data on variances, shows how much of the sample should turn over each month, in view of the cost and of the gain in moving into a new segment compared with the cost and the gain of staying 1 more month in an old one. One may view this problem as one in the optimum allocation of the sample to space and to time. The same theory applies in the measurement of the number of television or radio sets tuned to a certain program day after day, or week after week. There is an optimum length of time to leave a diary or a recording instrument with the set. In the sampling of coal and of other commercial materials, a primary sampling unit may be a random cross-section on a conveyor-belt; the above theory then provides a guide to the optimum number of pounds in 1 scoop. The reader will find extensions in Hansen, Hurwitz, and Madow's *Sample Survey Methods and Theory*, Vol. I, (Wiley, 1953): Ch. 12, especially Section 9, contributed by Max Bershad, Bureau of the Census.

Allowance for varying size of block.* If the number of segments N_i varies from block to block, we may define

$$\sigma_b{}^2 = \frac{1}{M\bar{N}} \sum_1^M N_i(a_i - a)^2 \qquad (27)$$

$$\sigma_w{}^2 = \frac{1}{M\bar{N}} \sum_{i=1}^{M} \sum_{j=1}^{N_i} (a_{ij} - a_i)^2 \qquad (28)$$

These definitions give weight to Block i in proportion to N_i, the number of sampling units therein. Here a_{ij} is the population of Segment j in Block i; a_i is the average population per segment in Block i; a is the average population per segment in all blocks. Varying size of block frequently makes little difference in the variances $\sigma_b{}^2$ and $\sigma_w{}^2$. A condition therefor is no correlation between N_i and $(a_i - a)^2$, nor between N_i and $\sigma_{wi}{}^2$, the variance between the sampling units within Block i. The reader may wish to turn his attention to the proofs, as the following exercises invite him to do.

The fact remains, though, that varying size of block can introduce either an increase or a decrease in $\sigma_b{}^2$ and $\sigma_w{}^2$, and one should be on guard.

* The student may omit this section unless he has special interest in it.

Exercise 1.   Show that $\sigma_b{}^2$ as defined above reduces to

$$\sigma_b{}^2 = \frac{1}{M} \sum_1^M (a_i - a)^2 \tag{29}$$

exactly as if the sizes of the blocks were uniform, provided N_i is uncorrelated with $(a_i - a)^2$. This is a condition that varying size of block create no increase nor decrease in $\sigma_b{}^2$. (Compare with pages 377 ff. in Chapter 16, and with pages 397 ff. in Chapter 17.)

Hint: The correlation between N_i and $(a_i - a)^2$ is proportional to $\Sigma(N_i - \bar{N})(a_i - a)^2$. If the correlation is 0, then we may replace $\Sigma N_i(a_i - a)^2$ in Eq. 27 by $\bar{N}\Sigma(a_i - a)^2$, and the proof is complete.

Exercise 2.   Show that $\sigma_w{}^2$ as defined above reduces to

$$\sigma_w{}^2 = \frac{1}{M} \sum_1^M \sigma_{wi}{}^2 \tag{30}$$

exactly as if the sizes of the blocks were uniform, provided N_i is uncorrelated with $\sigma_{wi}{}^2$. This is a condition that varying size of block create no increase nor decrease in $\sigma_w{}^2$. Here

$$\sigma_{wi}{}^2 = \frac{1}{N_i} \sum_1^{N_i} (a_{ij} - a_i)^2 \tag{31}$$

is the variance between the N_i sampling units in Block i.

Exercise 3.   Show that

$$\sigma_b{}^2 + \sigma_w{}^2 = \sigma^2 \tag{32}$$

where $\sigma_b{}^2$ and $\sigma_w{}^2$ have the definitions in Eqs. 27 and 28, and

$$\sigma^2 = \frac{1}{M\bar{N}} \sum_{i=1}^M \sum_{j=1}^{N_i} (a_{ij} - a)^2 \tag{33}$$

is the total variance between all $M\bar{N} = N$ segments in the city.

Exercise 4.   Rewrite Eq. 9 in the form

$$\frac{\bar{N}}{\bar{n}} = \frac{\sigma_b\sqrt{\bar{N}}}{\sigma_w} \sqrt{\frac{c_2\bar{N}}{c_1}} \tag{34}$$

with a similar change in Eq. 10. Interpret (a) $\bar{N}/\bar{n}$; (b) the factor $\sigma_w/\sigma_b\sqrt{\bar{N}}$; (c) the factor $c_2\bar{N}$. (*Hint:* $\bar{N}/\bar{n}$ will be the segment-interval within blocks, whether the blocks are of equal or unequal size. $\bar{N}$ will be the average size of block. $\sigma_w/\sigma_b\sqrt{\bar{N}}$ is a quantity that would be 1 if the segments were scattered at random throughout the entire city, but which is usually greater than 1 because of serial correlation. $c_2\bar{N}$ is the average cost of interviewing an entire block.)

CHAPTER 20

Theory for the Formation of Strata*

I held my tongue and spake nothing: I kept silence, yea, even
from good words; but it was pain and grief to me.—Psalm 39,
5:3.

Discretion in the formation of strata. We assumed when we studied
stratification in Chapter 15 that the strata had been defined before we
ever decided which type of stratified sampling to use, or whether to use
any. The serious student will wonder how the strata came into existence
in the first place. The fact is that (*a*) one may form strata in many
different ways, and that (*b*) it is just as important to form strata to the
best advantage as it is to sample them effectively after they are formed.
One will find the 2 questions to be interlocked—the best way to form the
strata will depend to some extent on which plan of stratified sampling he
will use. In Plans D and H the statistician need not define the strata
until after he has drawn the sample—in fact, he may defer the decision
until the results come in.

Some principles in the formation of strata. It is possible to state some
principles that one may use as a guide. These principles arise from the
theory to appear further on in this chapter. The 1st principle is that most
frames possess a certain amount of natural stratification, free of charge,
through the order in which the units appear in the frame. This order
often provides excellent geographic stratification, and often also stratification by type and by size. Material never arrives ready-mixed.

A good example is almost any table of census data, wherein the areas
are shown by type and by size, by urban and rural, and (as in the case of
statistics by block or by tract) in geographic order as well. The subscribers of a magazine are often arranged in geographic order, alphabetic
within areas. The customers of a department store or of a dairy may be
alphabetic by route.

* I express here my thanks and admiration to my colleague Tore Dalenius of Stockholm for writing the sections on theory in this chapter, and for many helpful suggestions.

In the sampling of human populations (including agriculture and business), geographic stratification is usually excellent, as it automatically forms strata by characteristics that are correlated with geography—e.g., rent, income, urban and rural, color, race, and even broad occupation. Some of these characteristics are in turn correlated with purchases, opinions, occupation, education, even political party, and to some extent, religion.

If the frame is not already prepared, one may take advantage of the possibility of introducing some effective stratification, without effort and without cost.

The method of replicated sampling described in Part II of this book exploits the natural stratification of the frame.   Each zone is a stratum of Z sampling units.   The zones provide high geographic stratification, which is often all that one requires.   Any census area or factory or other unit in the frame that is big enough to cover a full zone can hardly escape contributing 2 sampling units to the sample.   Actually, one need concern himself only with the order of listing the units that are smaller than a zone. The order of listing will of course often be dictated by convenience in tabulation rather than by consideration of statistical efficiency.   We saw all this on page 188.

The method of replicated sampling reaps the benefit of the natural stratification in the frame whether we use random starts in the 1st zone and form each subsample as a patterned sampling unit, with skip-interval Z, or whether we draw fresh random numbers in every zone to form the subsamples.

One will certainly exploit the stratification that exists already free of charge in the frame, or which can readily be built into it, and the only question will then be whether some additional rearrangement of the sampling units (stratification or classification) will be worth its cost, before or after the interviewing or testing.   Some further principles and the theory further on will help to answer this question.

As a 2d principle, the number of strata additional to the natural stratification of the frame should be only enough to do the job efficiently in view of the cost of preparation.   In case of doubt about whether to use stratified sampling at all, or whether to form another stratum, my advice is to proceed with caution, and to introduce further stratification than already exists only if we are sure that the extra cost and trouble will yield a substantial net saving.

In my own work I find very often that 2 strata give a good gain over no stratification, but that 3 strata would be but little better than 2.   Moreover, especially when the work will be done by a client whose office is not statistical in nature, I am careful to keep the plan simple and foolproof:

the use of 3 strata with different ratios as in Plan F might cause administrative collapse unless I am willing to spend a lot of time on the job to teach and to supervise.

As a 3d principle, if the distribution is highly skewed, so that a large part of the characteristic to be measured lies in a few sampling units, it will nearly always be a good idea to cut off the tail of the distribution to form a stratum of big units, and to sample them heavily if not indeed 100%. It may in some cases be wise to break the distribution in several places so as to form 3 or more strata. We saw examples in Chapter 8, Section B, and in Chapter 9 where we set off a stratum of big accounts for 100% treatment.* Another example occurred in Chapter 15, page 310. Samples for estimating industrial employment, or capital investment or profit, will rely on similar principles of stratification. If there are several modes in the distribution of size, the best breaking points are usually obvious (next section, followed by theory later on).

We sometimes encounter administrative restrictions that will not permit application of this 3d principle. For example, in Chapter 8, Section A (p. 110), there was not time to search the 220,000 accounts in the ledgers for the big accounts, which were mingled indiscriminately with the small ones. Fortunately the purpose of the study was only to estimate roughly how many accounts there were over $2000, and roughly the amount of money in them. A preliminary sample, thinned by the methods of the preceding chapter, would have been possible only with more time and talent than the administrative restrictions permitted.

As a 4th principle, it will usually not pay to form a small stratum (small P_i) unless this stratum is greatly different from all the other strata in respect to mean or standard deviation, or both. We may see this from Eqs. 39–42 on page 298. Thus, in a sample for the labor force in America (wherein each person counts 1 or 0), it would not pay to identify a stratum for the people born in China. The reason is that the proportionate contribution of this segment is small, and that the means and the variances of their characteristics for the labor force are not greatly different from equal characteristics of the rest of the population.

Formation of strata by inspection. The decision on the formation of strata is sometimes almost obvious, in view of the foregoing principles. An example is the sampling of farms for the acreage in wheat or in peaches, or for the population of pigs, when one knows that there is a host of medium and small farms, and that there are some extremely big farms that contain 6, 8, 10, or 20 times the number of acres or the number of pigs in the average farm. It will usually then be highly desirable, if there

* The reader may wish to refer back at this point to the illustration on page 131, which makes use of a helpful rule given by Hansen, Hurwitz, and Madow.

are enough big farms, and if they are big enough, to set them off into a class for separate treatment; and the breaking point will be obvious. There may be a list of the big farms, or a partial list that can be made more complete by the expenditure of effort. We may sample this list heavily, perhaps 100%, without regard to location. We may then sample the remaining farms by drawing a sample of areas, and by canvassing the farms in these areas.

There is very often a distinct bunching of big units, like the skyscrapers that stand out on the horizon when one views New York from a distance. An extreme example is the rubber industry in the U. S., where 4 firms account for 90% of the volume of goods manufactured. There is no doubt about what to do in a survey of the production of rubber: these 4 firms will form a stratum for a 100% sample; the lesser ones will form a stratum for a sample of much less than 100%, perhaps 1 in 10.

Need for a theory to help define the strata. The above rules do not answer the questions, where do we break the distribution? what is a big farm? a big account? They give no help in a case where there is no abrupt change in size, nor when a few firms or a few farms do not dominate the industry. Fortunately, this problem has received consideration in several papers by Dalenius and Margaret Gurney.* This chapter has the benefit of their results. The student will derive some help in the exercises that commence on page 493.

The following theory treats only the case where the population y, which we use for stratification, is related linearly to x, which is the population whose total we are trying to estimate. The frame may have any distribution in x.

Formation of strata for proportionate allocation. For Plan B, the point α of division for 2 strata should be chosen so that

$$\alpha = \tfrac{1}{2}(a_1 + a_2) \tag{1}$$

where a_1 is the mean in Stratum 1 and a_2 is the mean in Stratum 2, as they were in Chapter 15. Of course, as we move α to the right or to the left, a_1 and a_2 change, but there is one position that satisfies Eq. 1. At this point, Var X will be a minimum. Figs. 29 and 30 ahead depict what happens to Var X for various positions of α, and it shows that the shape of the curve is broad near the minimum (the optimum breaking point).

* Tore Dalenius, "The problem of optimum stratification," *Skandinavisk Aktuarietid-skript*, 1950: pp. 203–213.

Tore Dalenius and Margaret Gurney, "The problem of optimum stratification, ii," *ibid.*, 1951: pp. 133–148.

Tore Dalenius, "The problem of optimum stratification in a special type of design," *ibid.*, 1952: pp. 61–70.

One need therefore not be in possession of exact information on where to draw the line between strata; it suffices to know the neighborhood of the minimum.

We may generalize this result to any number of strata. Thus, the point α_i between the ith and $(i + 1)$th stratum should fulfill the condition

$$\alpha_i = \tfrac{1}{2}(a_i + a_{i+1}) \tag{2}$$

Formation of strata for Neyman sampling. For Neyman allocation, the optimum point α_i of division between the ith and the $(i + 1)$th stratum should fulfill the condition

$$\frac{\sigma_i^2 + (\alpha_i - a_i)^2}{\sigma_i} = \frac{\sigma_{i+1}^2 + (\alpha_i - a_{i+1})^2}{\sigma_{i+1}} \tag{3}$$

Suppose that the sampling plan uses only 2 strata, that Class 1 contains the "big" units and is sampled 100%, then the point α of stratification should fulfill the condition

$$(\alpha - a_1)^2 = \frac{P_1 \sigma_1^2}{n/N - 1 + P_1} > 0 \tag{4}$$

where P_1 is as heretofore the relative size of the 1st stratum. The above formulas apply strictly to sampling with replacement, but they serve well for samples drawn without replacement when $n/N < 20\%$.

Theory for the number of strata. * We have now discovered theory by which to deal with 2 of the 3 problems in the theory for the formation of strata. We have assumed heretofore that the number of strata has somehow been fixed. We shall now look into the problem of determining the best number M of strata. We have already learned some principles at the beginning of this chapter. We shall now see the theory that justifies them and permits us to make decisions that would not be obvious without calculation by formula.

If one has opportunity to stratify the frame by a characteristic that is correlated perfectly with the x-population, which we are trying to estimate, the variance of Plan C with M strata optimally defined is related to the variance of Plan A in the following way:

$$\mathrm{Var}_{M,C}\, X \doteq \frac{1}{M^2}\, \mathrm{Var}_A\, X \tag{5}$$

M being the number of strata (see Exercise 1, page 493). When the

* Tore Dalenius, "The economics of one-stage stratified sampling," *Sankhyā*, 1953: pp. 351–356.

population by which we stratify is only moderately correlated with the x-population, the relation is

$$\text{Var}_{M,C}\, X \doteq \frac{1}{M^2}\, \text{Var}_A\, X + \text{constant} \tag{6}$$

If the cost of carrying out the survey is

$$K = Mc_M + nc \tag{7}$$

where c_M is the cost of increasing the number of strata by 1, and c is the cost of increasing the total sample-size by 1 (this cost being assumed the same regardless of M), then

$$M_{\text{opt}} = \frac{2K}{3c_M} \tag{8}$$

$$n_{\text{opt}} = \frac{K}{3c} \tag{9}$$

Eq. 6 shows clearly the futility of using many strata in the usual situation: there is a limit, given by the constant, below which it is impossible to decrease the variance of X by increasing the number M of strata. The lower the correlation between the population of stratification and the x-population that we are trying to estimate, the higher be this limiting variance.

Further note on the futility of too many strata. It might seem that a large number of strata would always be excellent, and that one might well carry the idea to the extreme, down to 2 sampling units per stratum (Exercise 1, next page). This idea will in fact be good under conditions where the characteristics of the sampling units do not change readily, and where strata formed on the basis of data derived from a census or from tests conducted some time in the past will still have means and standard deviations in approximately the same relations today as they had at the earlier date. The best examples probably lie in the characteristics of areas. The bigger the area, the slower it changes, relatively, although areas as small as blocks still retain for years considerable similarity to the figures shown therefor by the last census; hence the utility of the so-called "block statistics" published by the Bureau of the Census.

The characteristics of individual farms and of business firms usually change much more rapidly than the characteristics of areas. A farmer may shift from a huge acreage of wheat to 0, as he changes his main crop from wheat to alfalfa or to soybeans; or he may shift back again. A farmer may have 6 pigs one month, none the next. A business firm may change size from big to little, or the converse, especially if size is measured by the output or by the sales of one specific product, and not by the combined output of sales of several products.   Areas that over one period of time showed intense activity in the construction of new homes will be stable a few years thereafter and show but little such activity, for the simple reason that there is no more land left vacant to build on.

For such reasons, stratification of such materials in fine classes will usually be a disappointment.   Eq. 6 foretold this result: the constant on the right-hand side put a floor under the variance and we can not go below it.

EXERCISES ON THE FORMATION OF STRATA

1. Divide a rectangular frame (Fig. 29) into M equal parts or strata. *a.* Show that the variance within strata is

$$\sigma_w^2 = \frac{\sigma^2}{M^2}$$

and that the variance between strata is

$$\sigma_b^2 = \left(1 - \frac{1}{M^2}\right)\sigma^2$$

where $\sigma^2 = h^2/12$ (Fig. 16, page 260).

b. Hence if a sample of size n be divided equally amongst the M strata (Plan B or C), then

$$\text{Var } \bar{x} = \frac{\sigma^2}{M^2 n} = \frac{A}{M^2}$$

where A is the variance of Plan A (no stratification).

One might be tempted to conclude from this result that stratification in fine classes will always show huge gains; and that the finer the classes, the bigger the gain.   Unfortunately, the characteristics of the sampling units are in practice in a state of flux, some increasing in size, some decreasing, and the formation of fine classes on the basis of past data (and we have nothing else) does not form fine classes in the present sizes, nor for next month or next year when the survey will be made.   We saw this in Eq. 6.

2. *a.* Show that if you are going to split a rectangular frame (p. 260) into 2 strata, then the best place to split it for either Plan B or Plan C is in the middle.

b. When split exactly in the middle, Plans B and C are identical, and both plans produce a gain of 75%, and the size of sample required to reach a specified precision by Plans B and C is only 25% of the size required by Plan A.

c. When split elsewhere Plans B and C are not identical; Plan C has always the smaller variance.

d. Neglect the finite multiplier and prove that for any split α the ratio of the variances of Plans B and A is

$$\frac{B}{A} = \left(\frac{\alpha}{h}\right)^3 + \left(1 - \frac{\alpha}{h}\right)^3$$

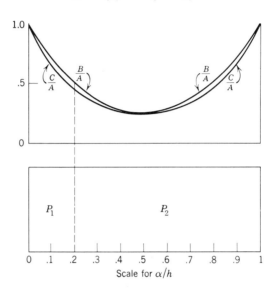

Fig. 29. The abscissa α divides the rectangle into 2 parts, Stratum 1 and Stratum 2, the number of sampling units in the 2 strata being in the proportions P_1 and P_2. Samples are drawn from these strata by proportionate sampling (Plan B) and by Neyman sampling (Plan C). The curves show the ratios of the variances of the plans to the variance of unstratified sampling (Plan A). Thus, at $\alpha/h = .2$, $B/A = .52$ and $C/A = .46$. Note that neither Plan B nor Plan C gives much gain when P_1 or P_2 is small, as we observed on page 302 of Chapter 15. (A, B, C denote the variances obtained with Plans A, B, C respectively.)

e. For Neyman sampling the ratio is

$$\frac{C}{A} = \left[\left(\frac{\alpha}{h}\right)^2 + \left(1 - \frac{\alpha}{h}\right)^2 \right]^2$$

f. The variance of Plan B stays within 10% of its minimum value so long as the split (α/h) stays within the interval $.50 \pm .09$, and that the variance of Plan C stays within 10% of its minimum value so long as the split stays within $.50 \pm .11$.

Remark 1. In both variances you may interchange $1 - \alpha/h$ with α/h. Both variances are thus symmetrical about the midpoint $z = \frac{1}{2}h$.

Remark 2. If you were to plot 2 curves, to show B/A vs. α/h, and C/A vs. α/h, the curve for C/A would be the flatter of the 2 at the minimum, which occurs at $\alpha/h = .5$. Why? What is the interpretation?

g. Show that your answers to part *a* satisfy Eqs. 1 and 3.

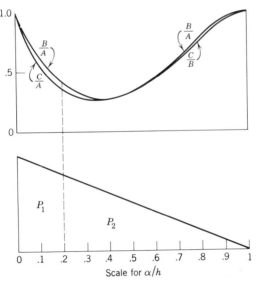

Fig. 30. The abscissa α divides the triangle into 2 parts, Stratum 1 and Stratum 2, the number of sampling units in the 2 strata being in the proportions P_1 and P_2. Samples are drawn from these strata by proportionate sampling (Plan B) and by Neyman sampling (Plan C). The curves show the ratios of the variances of these plans to the variance of unstratified sampling (Plan A). Thus, at $\alpha/h = .2$, $B/A = .43$ and $C/A = .36$. The 2 plans B and C are equal at $\alpha = .45$. Note that neither Plan B nor Plan C gives much gain when P_1 or P_2 is small, as we observed on page 302 of Chapter 15 (A, B, C denote the variances obtained with Plans A, B, C respectively.)

3. Calculate the variances of Plans B and C for sampling the rectangular distribution in Fig. 29, which is split into 2 portions by the distance α measured from the left-hand end. Plot the curves in Fig. 29 as functions of α.

Hint: By the equations in the preceding exercise, $B:A = P_1^3 + P_2^3$ and $C:A = (P_1^2 + P_2^2)^2$.

4. Show that as the size of Stratum 1 decreases, or as the size of Stratum 2 decreases, the difference between Plans B and C decreases, and that both Plans B and C approach Plan A.

5. Show that the results just derived for Plans B and C will hold also for Plans D and E.

6. The populations a_i $(i = 1, 2, \ldots, N)$ in a frame have a distribution that is triangular, as in Fig. 30.   Divide the frame into 2 strata, with proportions P_1 and P_2, where the subscript 1 will denote the portion near the large end of the triangle; the subscript 2 the portion that ends in the vertex.

a. Show that the minimum variance by Plan B occurs if the triangle is split at a distance $.38h$ from the large end, where h is the base.   Ignore the finite multiplier.

b. At this point $P_1 = .62$ and $P_2 = .38$, and the size of the sample required to reach a prescribed precision by Plan B is only 29 % of the size required by Plan A.

c. Show that the minimum variance by Plan C occurs if the triangle is split at a distance of $.34h$ from the large end.

d. At this point, $P_1 = .57$, $P_2 = .43$, and the size of the sample required to reach a prescribed precision by Plan C is only 26.5 % of the size required by Plan A.

e. Show that your results for parts *a* and *c* satisfy Eqs. 1 and 3.

7. Show that the results just derived for Plans B and C will hold also very close for Plans D and E.

8. Calculate the variances of Plans B and C for sampling the triangular distribution in Fig. 30, which is split into 2 portions at the distance α measured from the left-hand end.   Plot the curves in Fig. 30 as functions of α.

CHAPTER 21

Choice of Zoning Interval
and Number of Subsamples

The advantage of obtaining an explicit mathematical statement
of the optimum is that it aids in thinking about good sample design,
since it points out which factors are effective in determining the
amount of information obtained per dollar.—Hansen, Hurwitz,
and Madow, *Sample Survey Methods and Theory*, Vol. I (John
Wiley and Sons, 1953): page 287.

Reasons for choosing the proper number of subsamples. We have already
learned to regard the zones in the frame as strata. The natural stratification in the frame is often pretty effective, although, as we learned in
Chapter 15, it may pay to rearrange the sampling units in the frame, or
in the sample, to gain further precision from additional stratification.
The more subsamples one uses, the wider must be the zone, and the less
effective will be the stratification in the frame. Zones wide enough to
accommodate·5 or 10 subsamples may cause noticeable loss in precision.
One has a choice of 2 or of 10 subsamples or of any number intermediate.
  Where the material is likely to vary much from zone to zone, as it often
does in a survey of human populations over a large region of the country,
one might well hesitate to use 10 subsamples, although a lot depends on
the characteristic to be estimated. One might use 10 subsamples, however,
over a small area where the zones will be narrower. The survey of the
Cincinnati area in Chapter 11, for example, where we used 2 subsamples
and 9 subtotals (9 thick zones), might very well have been accomplished
with but little loss had we used 10 subsamples and no subtotals. In the
sampling of accounts, after one has stratified by size of account, one may
usually use 10 subsamples, I believe, with little loss. In the inventory of
materials in process in Chapter 9, we used 2 subsamples. Our survey of
the small urban areas in Chapter 10 had 2 subsamples, the main reason
for the choice being to gain experience with the subtotals by thick zone.

497

One must remember that there is always a compromise of opposing forces to settle.   If one uses 2 subsamples, he must divide the frame and the tabulation into thick zones for subtotals, in order to gain sufficient degrees of freedom in the estimate of the variance.   If one uses 10 subsamples, with no subtotals, he has 9 degrees of freedom, a very convenient plan, but he runs the risk of some loss in efficiency.   One break in the frame in the middle, when one uses 10 subsamples, will produce 18 degrees of freedom, which is a point to remember.

One might be tempted to settle the argument by prescribing always 2 subsamples per zone; never any more.   There would then never be a loss from zones wider than necessary.   This would be a good rule were it not for the competing convenience of 10 subsamples, even at the risk of some loss in precision.   In my own practice, I hesitate to depart from 10 subsamples unless the loss is likely to be 10% or more; yet 10% or even more might not be too high a price to pay in some instances for convenience and smoothness of operation.

In any event, it is always a comfort to know that whether we made a wise choice or not in the number of subsamples, there is never any bias from the choice; and we always have, from our replication, an estimate of the precision actually attained.

It is possible, by theory that we know now, to discover from the results of a replicated sample whether and how rapidly the precision decreases with the size of the zone—that is, with the number of subsamples.

The best choice for one kind of material will not be the best for another. It is thus desirable to build up experience through occasional calculations in order to have some actual figures as a guide, so that one may balance rationally losses against convenience.

This chapter will give an actual example.   The purpose of this example is to demonstrate a procedure by which one may study data from past experience to guide him in the future.   The purpose is not to make any general recommendation of 10 subsamples, 2 subsamples, or any other number, but to show how one may use data from previous surveys to make a numerical evaluation that will assist one to make a rational choice.

The example here came from a problem in accounting, in which the size of the accounts varied from small negative amounts up to several hundred dollars (Table 3).   These dollar-values came from freight bills. The aim of the original sample was to estimate revenue.   Each entry refers to a carload.*   Freight bills show some homogeneity from car to car, from train to train, and from one week to another.   One would thus expect some loss from the use of 10 subsamples rather than 5, or 2.   This

---

* In railway parlance, these were not freight bills, but items on the "consist" of a train as it leaves a terminal.   The consist is usually transmitted by telegraph.

expectation turns out to be fulfilled, though the loss is too small, in my judgment, to warrant the use of fewer than 10 subsamples in this particular problem.

Exercise. *a.* Show that if the material in the frame be thoroughly mixed (as it never is), one may use any number of subsamples, with no loss and no gain.

b. This is no longer true if one rearranges the sampling units to form strata.

Formulas for the estimation of the variances within and between zones. We first rewrite some of the formulas of Chapter 17 in a form that is adapted to the calculations proposed here.   Let there be M thin zones in all, $i = 1, 2, \ldots, M$.   There are Z sampling units in each zone, Z being the original zoning interval.   Let

$$c_{ij} = a_i + a_{ij} \qquad (1)$$

be the population of sampling unit ij, the jth unit in Zone i. a_i is the mean of Zone i.   The a_i may all be different.   The a_{ij} in Zone i are distributed about 0 with variance σ_{wi}^2.   (In what follows, the population will be dollars of revenue.)   Define now

$$a = \frac{1}{M} \sum_1^M a_i, \qquad \text{the overall average number of dollars per} \qquad (2)$$
$$\text{sampling unit}$$

$$\sigma_b^2 = \frac{1}{M} \sum_1^M (a_i - a)^2, \qquad \text{the variance between the means}$$
$$\text{of the } M \text{ zones} \qquad (3)$$

$$\sigma_{wi}^2 = \frac{1}{Z} \sum_1^Z a_{ij}^2, \qquad \text{the variance within Zone } i \qquad (4)$$

$$\sigma_w^2 = \frac{1}{M} \sum_1^M \sigma_{wi}^2, \qquad \text{the average variance within all zones} \qquad (5)$$

Draw a zone by reading out a random number between 1 and M. There are 10 random sampling units in this zone.   Record the dollars for these 10 sampling units, and lay out the results in order.   Table 1 shows the form; Table 4 shows some actual figures.   Let z_{ij} be the entry in the random Zone i, Subsample j.   Then put

$$z_{ij} = z_{i.} + x_{ij} \qquad (6)$$

where $z_{i.}$ is the mean of the sample in Zone i, and x_{ij} is a random deviation from the mean.   Draw m zones altogether, and have m equations like this, for $i = 1, 2, \ldots, m$.

Each zone furnishes the estimate

$$\hat{\sigma}_{wi}^2 = \frac{1}{9} \sum_{j=1}^{10} (z_{ij} - z_{i.})^2 \qquad \text{[Zone } i\text{]} \tag{7}$$

of the variance within Zone i. This estimate comes by application of Eq. 4 on page 433, with the finite multiplier $(N - n)/N$ replaced by 1. All m zones give the final estimate

$$\hat{\sigma}_w^2 = \frac{1}{9m} \sum_{i=1}^{m} \sum_{j=1}^{10} (z_{ij} - z_{i.})^2 \tag{8}$$

Similarly, Subsample j gives an estimate of the total variance, $\sigma_b^2 + \sigma_w^2$, wherefore we may write

$$\hat{\sigma}_{bj}^2 = \frac{1}{m-1} \sum_{i=1}^{m} (z_{ij} - z_{.j})^2 - \hat{\sigma}_w^2 \qquad \text{[Subsample } j\text{]} \tag{9}$$

All 10 subsamples give the final estimate

$$\hat{\sigma}_b^2 = \frac{1}{10(m-1)} \sum_{i=1}^{m} \sum_{j=1}^{10} (z_{ij} - z_{.j})^2 - \hat{\sigma}_w^2 \tag{10}$$

We may now lay out a procedure of calculation as follows. Put $m = 10$.

1. Draw 10 zones at random.   Record them (Table 4).
2. Form the 10 estimates $\hat{\sigma}_{wi}^2$, $i = 1, 2, \ldots, 10$, by use of Eq. 7 (Table 5).
3. Find the average $\hat{\sigma}_w^2$ of these 10 estimates.   Record it at the foot of the table.
4. Form the 10 estimates $\hat{\sigma}_{bj}^2$, $j = 1, 2, \ldots, 10$, by use of Eq. 9, inserting the estimate $\hat{\sigma}_w^2$ just found.
5. Find the average σ_b^2 of these 10 estimates (Eq. 15 *infra*).

The user of the method may make suitable modifications if the original data are from fewer than 10 subsamples, as we do here later on for 5 subsamples.

We have now our final estimates $\hat{\sigma}_b^2$ and $\hat{\sigma}_w^2$.   Our estimate of the gain of the proportionate stratified sampling (Plan B in Ch. 15) which the zones furnish automatically, over no stratification at all (Plan A), is then

$$\frac{A - B}{A} = \frac{\hat{\sigma}_b^2}{\hat{\sigma}_b^2 + \hat{\sigma}_w^2} \qquad \text{[From p. 298; applied on p. 506]} \tag{11}$$

Numerical illustration.   It is possible to include here an illustration of some calculations that show (*a*) a comparison between wide zones and narrow zones, and (*b*) the gain of the stratification by zones.*   The

---

* I am indebted to Mr. Leon Kilbert for the calculations in this section, and for important suggestions in the exposition throughout the chapter.

calculations follow the procedure suggested in the preceding section.   The application here is to a sample of accounts.   Each sampling unit was a railway car.   Table 3 shows a portion of the results of the sample.   The zoning interval was 150 cars.

TABLE 1

Random zone	Subsample (j)				Mean
	1	2	$\cdots$	10	
$i = 1$	z_{11}	z_{12}	$\cdots$	$z_{1,10}$	$z_{1.}$
2	z_{21}	z_{22}	$\cdots$	$z_{2,10}$	$z_{2.}$
.	.				
.	.				
.	.				
m	z_{m1}	z_{m2}	$\cdots$	$z_{m,10}$	$z_{10.}$
Mean	$z_{.1}$	$z_{.2}$	$\cdots$	$z_{.10}$	$z_{..}$

The procedure was to draw at random 10 zones from all the work-sheets for January 1956.   Table 4 shows the 10 zones.   F is for forwarded traffic, R for received, I for intermediate.   The order of the zones in the table is the order in which the random numbers made the selections.

TABLE 2

SCHEMATIC ARRAY OF THE ESTIMATES $\sigma_{wi}{}^2$

Random zone	Estimate of $\hat{\sigma}_{wi}{}^2$
1	$\hat{\sigma}_{w1}{}^2$
2	$\hat{\sigma}_{w2}{}^2$
.	
.	
m	$\hat{\sigma}_{wm}{}^2$
Mean	$\hat{\sigma}_w{}^2$

Column 2 in Table 5 shows the 10 estimates of the variance within the full zoning-interval of 150 cars.   These estimates are labeled $\sigma_{w(10)}^2$. Columns 3 and 4 show estimates of variances within zones about half as wide.   The 5 cars with the 5 lowest random numbers constitute half a zone, and the other 5 constitute the other half.

TABLE 3

EXAMPLE OF ORIGINAL DATA

CLASS OF TRAFFIC _FORWARDED_   DATE _FEB. 1956_

ZONE	SUBSAMPLE									
	1	2	3	4	5	6	7	8	9	10
1–	123	105	121	050	061	098	147	072	087	031
150	121.36	60.94	58.62	34.62	69.39	199.09	72.00	152.16	92.88	87.22
151–	261	256	215	294	271	187	248	233	217	278
300	84.04	57.60	32.33	127.09	40.00	99.38	40.24	27.50	39.91	74.40
30/–	338	371	394	399	417	379	333	440	309	429
450	18.00	70.68	35.04	42.88	234.36	58.57	122.12	29.68	89.95	33.72
451–	519	548	457	577	520	580	459	468	455	587
600	74.42	33.65	49.68	22.70	24.24	104.68	49.68	160.16	41.76	71.27
601–	699	607	652	661	619	739	609	646	632	745
750	–15.89	77.11	233.33	44.22	29.42	104.49	97.84	–33.22	92.98	68.43
751–	823	762	807	849	885	783	814	758	774	756
900	79.92	65.67	103.04	139.03	–80.06	FREE	89.85	106.26	58.63	131.20
901–	1046	1030	967	901	919	958	1008	974	906	964
1050	256.12	96.40	154.70	83.55	89.02	144.36	103.34	34.31	70.95	65.94
105/–	1148	1145	1099	1199	1159	1195	1071	1091	1140	1183
1200	54.24	32.81	94.48	49.68	33.46	97.04	81.70	58.63	28.06	27.21
120/–	1241	1230	1246	1346	1337	1338	1345	1273	1258	1330
1350	93.78	132.04	106.95	24.28	94.72	83.58	94.09	111.02	98.66	72.06
135/–	1362	1355	1492	1359	1472	1488	1462	1389	1388	1474
1500	–58.52	34.77	71.18	32.24	48.26	111.62	138.54	29.86	32.38	89.91
150/–	1535	1647	1616	1626	1537	1512	1642	1561	1537	1541
1650	22.58	197.65	90.42	425.88	27.62	93.06	294.72	64.38	285.00	74.38
165/–	1685	1792	1687	1691	1733	1700	1662	1673	1741	1796
1800	85.12	308.70	23.74	133.78	126.64	26.39	161.82	–9.46	202.85	332.14
	815.17	1168.02	1053.51	1159.95	737.07	1122.26	1345.94	731.28	1134.01	1127.88

A detailed calculation of the figure 5415 in Column 2 may be helpful. It came from Zone 1 in Table 4 by the use of Eq. 7.   Let the subscript $w(10)$ refer to the full zone i of 10 subsamples.   Then

$$\hat{\sigma}^2_{w(10)} = \frac{1}{9} \sum_1^{10} (z_{ij} - z_{i.})^2$$

$$= \frac{1}{9} [(62 - 124.9)^2 + (98 - 124.9)^2 + \cdots + (140 - 124.9)^2]$$

$$= \frac{1}{9} \left[(62^2 + 98^2 + \cdots + 140^2 - \frac{1249^2}{10} \right]$$

$$= 5415 \tag{12}$$

TABLE 4

DISPLAY OF THE RESULTS OF 10 ZONES DRAWN AT RANDOM FROM THE ESTIMATING
WORK-SHEETS FOR JANUARY 1956

The results under the 10 subsamples are in dollars. A negative sign indicates that the company gave up some money received in an earlier movement of the material in the car

Zone number (in order drawn)	Zone and class in the main sample	Subsamples										Total
		1	2	3	4	5	6	7	8	9	10	
1	R–24 3451–3600	62	98	114	257	249	72	86	44	127	140	1249
2	F–91 3151–3300	71	79	128	125	165	59	45	21	214	107	1014
3	F–83 1951–2100	74	32	–19	44	49	29	116	31	66	52	474
4	R–14 1951–2100	30	157	125	119	89	144	93	66	84	105	1012
5	F–71 151– 300	62	104	290	49	113	46	50	292	44	29	1079
6	F–77 1051–1200	11	–4	45	129	37	25	64	41	138	42	528
7	I–65 5701–5850	115	82	35	112	154	43	35	87	68	49	780
8	I–54 4051–4200	120	48	104	72	68	110	82	68	88	47	807
9	R–26 3751–3900	99	237	76	50	25	56	166	85	34	63	891
10	R–04 451– 600	99	79	76	93	91	267	51	31	114	96	997
	Total	743	912	974	1050	1040	851	788	766	977	730	8831

TABLE 5

THE ESTIMATES OF $\sigma_w{}^2$ CALCULATED FOR JANUARY 1956 BY SPLITTING
A FULL ZONE OF 10 SUBSAMPLES INTO 2 PARTS, DETERMINED
BY THE LOWEST 5 OF THE 10 RANDOM NUMBERS,
AND BY THE HIGHEST 5

The results of this table enable us to decide whether a smaller
zone with 5 subsamples would be more efficient
than a full zone of 10 subsamples

Zone number from the preceding table	$\hat{\sigma}^2_{w(10)}$	$\hat{\sigma}^2_{w(5)}$	
		Lower 5	Upper 5
1	5 415	998	9 269
2	3 432	528	2 440
3	1 228	2 504	226
4	1 407	630	2 368
5	10 018	10 694	11 816
6	2 167	193	4 462
7	1 585	1 155	2 385
8	623	927	249
9	4 281	6 601	2 840
10	4 046	1 152	6 325
Average	3 420	2 538	4 238

Average $\hat{\sigma}^2_{w(5)} = \tfrac{1}{2}(2538 + 4238) = 3388$

We transfer this figure to Zone 1 of Table 5. The 9 other estimates of $\sigma^2_{w(10)}$ came likewise. The average of these 10 estimates is 3420, as shown at the foot of Table 5.

We need also the average variance within zones half as wide as the full zones of 10 subsamples. We therefore divide the full zone into 2 parts, the 5 cars drawn by the lowest 5 random numbers, and the other 5 cars. We then go to the sampling table (not shown) for this zone and observe that the lowest 5 random numbers drew the cars that produced the figures 62, 98, 114, 72, 140, whose total is 486. Hence, with the subscript $w(5)$ for a zone of 5 subsamples, we calculate for this subzone,

$$\hat{\sigma}^2_{w(5)} = \frac{1}{5-1}\left[62^2 + 98^2 + 114^2 + 72^2 + 140^2 - \frac{486^2}{5}\right]$$
$$= 998 \tag{13}$$

The other 19 estimates of $\sigma^2_{w(5)}$ in Table 5 follow in the same way. The average of these 20 estimates is 3388, this being $\tfrac{1}{2}(2538 + 4238)$.

The 5 lowest random numbers do not divide a zone into exact halves in every instance, but the simplicity of the random division is appealing, and the results suffice.

Choice of the width of zone and of the number of subsamples.   Table 6 shows the averages of the estimates $\sigma^2_{w(10)}$ and $\sigma^2_{w(5)}$ for January, February, May, June, and July, 1956, derived in the manner described.   The smaller zone shows persistent gain, as all 5 ratios are greater than 1, even under the handicap of the random division; and one might be tempted to jump at once to recommend the narrower zone with 5 subsamples instead of 10.   However, on further thought, the evidence is not compelling.   The average ratio, 1.06, could be an overestimate of the gain: it could also be an underestimate.   The question is whether the gain, whatever it be, is worth the cost to capture it.

The use of 5 subsamples would require tabulations in 2 thick zones for the month, in order to produce 8 degrees of freedom.   The use of 10 subsamples with 1 thick zone gives 9 degrees of freedom.   Division by 5 is not so simple as division by 10.   These may seem like small points, but simplicity is often pretty important.   In summary, in my opinion the small gain from 5 subsamples in a narrower zone is not sufficient to repay the cost of the additional trouble; and my advice would be, on this job, to continue the use of 10 subsamples.   The reader must not assume, though, that 10 subsamples is *ipso facto* going to be the correct number in all accounting problems.

TABLE 6

SOME COMPARISONS OF THE ESTIMATED VARIANCES WITHIN
ZONES OF FULL WIDTH FOR 10 SUBSAMPLES, AND WITHIN
ZONES OF HALF WIDTH FOR 5 SUBSAMPLES

Derived from 10 zones of full width drawn at random from
the main sample for the months specified in the table

Month, 1956	$\hat{\sigma}^2_{w(10)}$	$\hat{\sigma}^2_{w(5)}$	$\hat{\sigma}^2_{w(10)}/\hat{\sigma}^2_{w(5)}$
January	3420	3388	1.01
February	3032	2929	1.03
May	7275	6650	1.09
June	3168	2803	1.13
July	3502	3374	1.04
Average	4079	3829	1.06

The variance between zones and the gain of the natural stratification.
Table 7 shows the 10 estimates of the total variance, derived from the 10
zones drawn from the main sample for January 1956.   For Subsample 1
the estimate of total variance is 1250, derived from Table 4 as follows:

$$\hat{\sigma}^2 = \frac{1}{10 - 1} \left[62^2 + 71^2 + 74^2 + \cdots + 99^2 + 99^2 - \frac{743^2}{10} \right]$$

$$= 1250 \quad [j = 1] \tag{14}$$

Subsamples 2, 3, . . . , 10 give the remaining 9 estimates in Table 7.
The average of the 10 estimates of the total variance is 3855.   We had
the figure 3420 for our final estimate of the variance within zones, as the
reader may recall from Table 5.   Hence

$$\hat{\sigma}_b{}^2 = \hat{\sigma}^2 - \hat{\sigma}_w{}^2 = 3855 - 3420 = 435 \tag{15}$$

is our final estimate of the variance between zones of full width, 150 cars.
The estimated gain that arises from the natural stratification in zones
of 150 cars is

$$\frac{A - B}{A} = \frac{\hat{\sigma}_b{}^2}{\hat{\sigma}^2} = \frac{435}{3855} = 11\% \quad [\text{P. 298}] \tag{16}$$

TABLE 7

TEN ESTIMATES OF THE TOTAL VARIANCE FOR JANUARY 1956

Each estimate is the 1st term on the right of Eq. 12.
The original data came from Table 4

Subsample	$\hat{\sigma}^2$
1	1250
2	4512
3	6690
4	3937
5	4752
6	5437
7	1570
8	6260
9	2820
10	1320
Average	3855

This was for the month of January 1956. February, May, June, and July give further estimates as shown in Table 8. Zones of full width (150 cars) thus show an average gain of 13% over no stratification at all. Zones half as wide, we learned in the preceding section, might show an additional gain of 6%, or a total gain of 19%. The stratification that we speak of here is the natural stratification that exists in the frame, free of charge.

TABLE 8

ESTIMATION OF THE GAIN FROM STRATIFICATION

The estimated gain is the estimated variance between
zones divided by the estimated total variance

Month 1956	Total variance	Variance within zones	Variance between zones	Gain from stratification
January	3855	3420	435	11%
February	3368	3032	336	10
May	8629	7275	1354	16
June	3584	3168	416	12
July	3976	3502	474	12
Average	4682	4079	603	13

Index

509

Senator Pepper 78
Periodicities and patterned (systematic) selection 79, 94, 95, 125, 126
John Perrin 331
Persistent error 62, 66, 159, 160, 162, 317
Physical condition of property 344
Pigs in Norway, count of, 34
Poisson series of probabilities 457, 458; compared with binomial terms 459
Poisson variable 146, 455 ff.; sum of 2 Poisson variables 460, 465
Alfred Politz 67, 68, 213, 241, 301, 326
Population, definition of, 52, 95, 96, 179, 196
Portions, splitting big block, 204
George C. Pratt 263
Preferred technique 62, 63
Primary frame 187, 216
Leon Pritzker 17, 251
Probability, knowledge of probabilities not sufficient, 39
Probability proportionate to size 196
Probability sampling, definition 23, 24; misconceptions 28; standards of 55
Probable error 55
Probes in administration, use of samples, 74
A. A. Procassini 76
Procedure for creation of segments 171, 195, 202 ff., 231
Product, variance of, 393
Propagation of variance 390 ff.

M. H. Quenouille 425, 427, 437
Quick calculation of margin of sampling error 116
Quota sample 31
Quotient, variance of, 393; variance of a ratio 181, 198, 437–439

Radio, measurement of number of FM owners 256, 301; measurement of number of listeners or viewers 485
Radioactive disintegration 456
Morton S. Raff 407, 456
Railway, sample of interline abstracts, 130
Des Raj 384
Clarence B. Randall 87
Random numbers 25, 28; fresh random

numbers in every zone 95; starting point in use of table 94; *see also* Examples of sampling tables
Randomness 54, 55
Range, use of range for standard error, 95, 108, 116, 179, 181, 199, 437
Rare characteristics, sampling for, 190, 409
Ratio-estimates, example 95, 98, 117, 424, 480; example of instructions 143; theory for variance of a ratio 181, 198, 437–439, 449; correction for bias 425
Record of interviews 239
Recorded deeds, sample of, 261
Rel-covariance, definition, 371
Rel-variance, definition, 124, 224; of an estimate of a variance 446; example 449
Relative gains in stratified sampling 298; example 317
Replicated sampling, advantages of, 87
Replication, general statement, 88, 186, 187; *see also* Applications of sampling
Report, statistician's, 12, 14, 16–22, 109, 117, 162
Report to management 12, 14, 16–22, 109, 117, 162
Responsibilities, of the client 15; of the statistician 10–15, 492
Retail stores, sample of, 324
Reverse internal variance 323, 324
Harry V. Roberts 387
Ralph Roeder 455
Rough counts 92, 169, 203; theory of 478–481
Allan L. Rudell 151
David D. Rutstein 13

Kinichiro Saito 267, 424
Sampling, advantages of, 26, 163; disadvantages of 28
Sampling error, definition, 52
Sampling new material 255; sequentially 270, 271; decision on type of stratified design 301, 341, 367
Sampling plan, elements of, 39
Sampling plan not helpful to the user of the data 15
Sampling, probability, definition, 23, 24; misconceptions 28, 55, 280